BIOLOGICAL AND MEDICAL PHYSICS
PHYSICS
BIOMEDICAL ENGINEERING

BIOLOGICAL AND MEDICAL PHYSICS BIOMEDICAL ENGINEERING

The fields of biological and medical physics and biomedical engineering are broad, multidisciplinary, and dynamic. They lie at the crossroads of frontier research in physics, biology, chemistry, and medicine. The Biological and Medical Physics/Biomedical Engineering series is intended to be comprehensive, covering a broad range of topics important to the study of the physical, chemical, and biological sciences. Its goal is to provide scientists and engineers with textbooks, monographs, and reference works to address the growing need for information.

Continued After Index

Martin Beckerman

Molecular and Cellular Signaling

With 227 Figures

Martin Beckerman
Y12 National Security Complex
Oak Ridge, TN 37831-7615
USA
beckermanm@y12.doe.gov

Library of Congress Cataloging-in-Publication Data
Beckerman, Martin.
    Molecular and cellular signaling/Martin Beckerman.
      p.   cm.—(Biological and medical physics, biomedical engineering, ISSN 1618-7210)
    Includes bibliographical references and index.
    ISBN 0-387-22130-1 (alk. paper)
    1. Cellular signal transduction.   I. Title.   II. Series.
  QP517.C45B43 2005
  571.7′4—dc22   2004052556

ISBN-10: 0-387-22130-1   Printed on acid-free paper.
ISBN-13: 978-0-387-22130-4

Printed in the United States of America.   (BS/EB)

9 8 7 6 5 4 3 2 1   SPIN 10948309

springeronline.com

Series Preface

The fields of biological and medical physics and biomedical engineering are broad, multidisciplinary, and dynamic. They lie at the crossroads of frontier research in physics, biology, chemistry, and medicine. The Biological and Medical Physics/Biomedical Engineering series is intended to be comprehensive, covering a broad range of topics important to the study of the physical, chemical, and biological sciences. Its goal is to provide scientists and engineers with textbooks, monographs, and reference works to address the growing need for information.

Books in the series emphasize established and emergent areas of science including molecular, membrane, and mathematical biophysics; photosynthetic energy harvesting and conversion; information processing; physical principles of genetics; sensory communications; and automata networks, neural networks, and cellular automata. Equally important will be coverage of applied aspects of biological and medical physics and biomedical engineering, such as molecular electronic components and devices, biosensors, medicine, imaging, physical principles of renewable energy production, advanced prostheses, and environmental control and engineering.

Oak Ridge, Tennessee

ELIAS GREENBAUM
Series Editor-in-Chief

Preface

This text provides an introduction to molecular and cellular signaling in biological systems. Cells partition their core cellular processes into a fixed infrastructure and a control layer. Proteins in the control layer, the subject of this textbook, function as signals, as receptors of the signals, as transcription factors that turn genes on and off, and as signaling transducers and intermediaries. The signaling and regulatory proteins and associated small molecules make contact with the fixed infrastructure responsible for metabolism, growth, replication, and reproduction at well-defined control points, where the signals are converted into cellular responses.

The text is aimed at a broad audience of students and other individuals interested in furthering their understanding of how cells regulate and coordinate their core activities. Malfunction in the control layer is responsible for a host of human disorders ranging from neurological disorders to cancers. Most drugs target components in the control layer, and difficulties in drug design are intimately related to the architecture of the control layer. The text will assist students and individuals in medicine and pharmacology interested in broadening their understanding of how the control layer works. To further that goal, there are chapters on cancers and apoptosis, and on bacteria and viruses. In those chapters not specifically devoted to pathogens, connections between diseases, drugs, and signaling are made.

The target audience for this text includes students in chemistry, physics, and computer science who intend to work in biological and medical physics, and bioinformatics and systems biology. To assist them, the textbook includes a fair amount of background information on the main points of these areas. The first five chapters of the book are mainly background and review chapters. Signaling in the immune, endocrine (hormonal), and nervous systems is covered, along with cancer, apoptosis, and gene regulation.

Biological systems are stunningly well engineered. Proof of this is all around us. It can be seen in the sheer variety of life on Earth, all built pretty much from the same building blocks and according to the same assembly rules, but arranged in myriad different ways. It can be seen in the relatively modest sizes of the genomes of even the most complex organisms, such as ourselves. The genomes of worms, flies, mice, and humans are roughly comparable, and only a factor of two or three larger than those of some bacteria. The good engineering of biological systems is exemplified by the above-mentioned partition of cellular processes into the fixed infrastructure and the control layer. This makes possible machinery that always works the same way in any cell at any time, and whose interactions can be exactly known, while allowing for the machinery's regulation by the variable control layer at well-defined control points.

Another example of good engineering design is that of modularity of design. Proteins, especially signaling proteins, are modular in design and their components can be transferred, arranged, and rearranged to make many different proteins. The protein components interact with one another through their interfaces. There are interfaces for interactions with other proteins and with lipids DNA and RNA. Modularity is encountered not only in the largely independent components, but also in the DNA regulatory sequences. These sequences serve as control points for the networks that regulate gene expression. The DNA regulatory sequences can also be rearranged in a multitude of ways along the chromosomes, and these rearrangements, rather than the genes themselves, are largely responsible for the richness of life on Earth. Two of the key objectives of the text are to examine how modularity in design is used and how interfaces are exploited. X-ray crystal structures and nuclear magnetic resonance (NMR) solution structures provide insights at the atomic level of how the interfaces between modules operate, and these will be looked at throughout the text.

One of the great conceptual breakthroughs in explorations of the control layer was the idea that signaling proteins involved in cell-to-cell communication are organized into signaling pathways. In a signaling pathway, there is a starting point, usually a receptor at the plasma membrane, and an endpoint (control point), more often than not a transcription regulatory site in the nucleus, and there is a linear route leading from one to the other. In spite of the enormous complexity of metazoans, there are only about a dozen or so such pathways. These will be explored in the context of where they are most strongly associated. For example, some pathways are prominent during development and are best understood in that context. Other pathways are associated with stress responses and are best understood within that framework, and still others are associated with immune responses.

Signaling and the cellular responses to signals are complex. The responses are controlled by a plethora of positive and negative feedback loops. The presence of feedback complicates the simple picture of a linear pathway, but this aspect is an essential part of the signaling process. Positive feedback ensures that once the appropriate thresholds are passed there will be a firm commitment to a specific action and the system will not jump back and forth between alternative responses. Negative feedback generates the thresholds that ensure random excursions and perturbations do not unnecessarily commit the cell to some irreversible response when it ought not to, and it permits the cells to turn off the signaling once it has served its purpose. These feedback loops will be examined along with the discussions of the linear signaling pathways.

The goal of this textbook is to provide an introduction to the molecular and cellular signaling processing comprising the control layer. The topic is a vast one, and it is not possible to cover every possible aspect and still keep the text concise and readable. To achieve the stated goal, material of a historical nature has been omitted, as have lengthy descriptions of all proteins identified as being involved in the particular aspect of signaling being considered. In place of such an encyclopedic approach, selected processes are presented step-by-step from start to end. These examples serve as simple models of how the control process is carried out.

Oak Ridge, Tennessee   MARTIN BECKERMAN

Contents

Guide to Acronyms

This Guide to Acronyms contains a list arranged alphabetically of commonly encountered acronyms all of which are discussed in the text. There are a number of instances where the same acronym has more than one usage. In some cases, the correct meaning can be discerned from the way the acronym is denoted, but in other cases, the correct usage must be deduced from the context. In the text, proteins are written starting with a capital letter, while the genes encoding the proteins are written all in lowercase letters. Protein names are, for the most part, not included in the list of acronyms. Proteins appearing in the list with names ending in numerals such as Ste2 are entered once; names of proteins of the same spelling with different numerals (e.g., Ste7, Ste11 in the case of Ste2) can be readily deduced.

5-HT	5-hydroxytryptamine (serotonin)
AA	arachidonic acid
AC	adenylyl cyclase
ACE	angiotensin-converting enzyme
ACF	ATP-dependent chromatin assembly and remodeling factor
ACh	acetylcholine
ACTH	adrenocorticotropic hormone
ADAM	a disintegrin and metalloprotease
ADHD	attention-deficit hyperactivity disorder
ADP	adenosine diphosphate
AFM	atomic force microscopy
AGC	PKA, PKG, PKC family
AHL	acetyl homoserine lactase
AIDS	acquired immunodeficiency syndrome
AIF	apoptosis inducing factor
AIP	autoinducing peptides
AKAP	A-kinase anchoring protein
ALK	activin receptor-like kinase
ALS	amytrophic lateral sclerosis

AMP	adenosine monophosphate
AMPA	α-amino-3-hydroxyl-5-methyl-4-isoxazole propionate acid
AMPK	AMP-dependent protein kinase
ANT	adenosine nucleotide translocator
APC	adenomatous polyposis coli
APC	antigen-presenting cell
APP	amyloid β protein precursor
ARC-L	activation-recruited coactivator-large
Arf	ADP-ribosylation factor
ARF	alternative reading frame (of exon 2)
ARR	Arabidopsis response regulator
ATM	ataxia-telangeictasia mutated
ATP	adenosine triphosphate
ATR	ATM and Rad3-related
AVN	atrioventricular node
AVP	vasopressin
Bcl-2	B cell leukemia 2
BCR	B-cell receptor
BDNF	brain-derived neurotrophic factor
BER	base excision repair
BFGF	basic fibroblast growth factor
BIR	baculoviral IAP repeat
BLV	bovine leukemia virus
BMP	bone morphogenetic protein
BRCA1	breast cancer 1
BRCT	BRCA1 C-terminal
BRE	TFIIB recognition element
bZIP	basic region leucine zipper
C1	protein kinase C homology-1
CAD	caspase-activated deoxyribonuclease
CaM	calmodulin
CaMKII	calcium/calmodulin-dependent protein kinase II
cAMP	cyclic AMP
CAP	catabolite activator protein
CAPRI	calcium-promoted Ras inactivator
CaR	extracellular calcium receptor
CARD	caspase recruitment domain
CASK	CaMK/SH3/guanylate kinase domain protein
CB	Cajal body
CBP	complement binding protein
CBP	CREB binding protein
CD	cluster of differentiation
Cdc25	cell division cycle (protein) 25
Cdk	cyclin-dependent kinase

cDNA	complementary DNA
CFP	cyan fluorescent protein
CFTR	cystic fibrosis transmembrane conductance regulator
cGMP	cyclic guanosine monophosphate
CHRAC	chromatin accessibility complex
Chromo	chromatin organization modifier
Ci	cubitus interruptus
Ck2	casein kinase-2
ClC	chloride channel of the CLC family
Clk	cyclin-dependent kinase-like kinase
CMGC	CDK, MAPK, GSK-3 CLK, CK2
CNG	cyclic nucleotide-gated
CNS	central nervous system
CNTF	ciliary neurotrophic factor
CoA	acetyl coenzyme A
COX	cyclo-oxygenase
CPG	central pattern generator
CR	consensus repeat
CRD	cysteine-rich domain
CRE	cAMP response element
CREB	cAMP response element-binding protein
CRF	corticotropin-releasing factor
CRH	corticotropin-releasing hormone
CRSP	coactivator required for Sp1 activation
CSF	cerebrospinal fluid
cSMAC	central supramolecular activation cluster
CST	cortistatin
DA	dopamine
DAG	diacylglycerol
DAT	dopamine transporter
dATP	deoxyadenosine triphosphate
DC	dendritic cell
DCC	deleted in colorectal cancer
DD	death domain
DED	death effector domain
DEP	disheveled, egl-10, and pleckstrin
DFF	DNA fragmentation factor
Dhh	desert hedgehog
DIABLO	direct IAP binding protein with low pI
DISC	death-inducing signaling complex
DIX	disheveled and axin
DLG	discs large
DNA	deoxyribonucleic acid
DNA-PK	DNA-dependent protein kinase
DPE	downstream promoter element

DR	death receptor
DSB	double-strand break
DSL	delta/serrate/lin
dsRNA	double-stranded RNA

E	epinephrine (adrenaline)
ECF	extracytoplasmic function
ECM	extracellular matrix
EEG	electroencephalographic
EGF	epidermal growth factor
EGFR	epidermal growth factor receptor
eIF	eukaryotic initiation factors
EPEC	enteropathogenic *E. coli*
ER	endoplasmic reticulum
ERK	extracellular signal-regulated kinase
ESCRT	endosomal-sorting complexes required for transport
ESE	exonic splice enhancer
ESI	electrospray ionization
ESS	exonic splice silencer
EVH1	enabled/vasodilator-stimulated phosphoprotein homology-1

FA	focal adhesion
FADD	Fas-associated death domain
FAK	focal adhesion kinase
FAT	focal adhesion targeting
FH	forkhead
FHA	forkhead associated
FNIII	fibronectin type III
FRAP	fluorescence recovery following photobleaching
FSH	follicle-stimulating hormone
FYVE	Fab1p, YOTB, Vac1p, Eea1

GABA	γ-aminobutyric acid
GAP	GTPase-activating protein
GAS	group A streptococcus
GAS	interferon-gamma activated site
GDI	GDP dissociation inhibitors
GDNF	glial-derived neurotrophic factor
GDP	guanosine diphosphate
GEF	guanine nucleotide exchange factor
GFP	green fluorescent protein
GFR	growth factor receptor
GH	growth hormone
GHIH	growth hormone-inhibiting hormone
GHRH	growth hormone-releasing hormone

GIRK	G protein-linked inward rectified K$^+$ channels
GKAP	guanylate kinase-associated protein
GPCR	G protein-coupled receptor
GPI	glycosyl phosphatidyl inositol
GRH	gonadotropin-releasing hormone
GRIP	glutamate receptor interacting protein
GRK	G protein-coupled receptor kinase
GSK-3	glycogen synthase kinase-3
GTP	guanosine triphosphate
HA	histamine
HAT	histone acetyltransferase
HDAC	histone deacetylase
hGH	human growth hormone
HGT	horizontal gene transfer
Hh	hedgehog
HHV	human herpesvirus
HIV	human immunodeficiency virus
HK	histidine kinase
HLH	helix-loop-helix
HNC	hyperpolarization-activated cyclic nucleotide gated
hnRNP	heterogeneous nuclear RNP
HOG	high osmolarity glycerol
HPt	histidine phosphotransfer
HR	homologous recombination
Hsp	heat shock protein
HSV-1	herpes simplex virus type 1
hTERT	human telomerase reverse transcriptase
HTH	helix-turn-helix
HTLV-1	human T lymphotropic virus type 1
hTR	human telomerase RNA
HtrA2	high temperature requirement factor A2
IAP	inhibitor of apoptosis
ICAD	inhibitor of CAD
ICAM	intercellular cell adhesion molecule
ICE	interleukin-1β converting enzyme
IEG	immediate early gene
IFN	interferon
Ig	immunoglobulin
IGC	interchromatin granule clusters
IgCAM	immunoglobulin cell adhesion molecule
IGluR	inhibitory glutamate receptor ion channel
IGluR	ionotropic glutamate receptor
Ihh	Indian hedgehog

IL	interleukin
ILP	IAP-like protein
IN	integrase
Inr	initiator
InsP$_3$R	inositol (1,4,5) triphosphate receptor
IP	Ischemic preconditioning
IPSP	inhibitory postsynaptic potential
IRAK	IL-1R-associated kinase
IRES	internal ribosomal entry site
IRF	interferon regulatory factor
IS	immunological synapse
IS	intracellular stores
ISE	intronic splice enhancer
ISRE	interferon stimulated response element
ISS	intronic splice silencer
ISWI	imitation SWI
ITAM	immunoreceptor tyrosine-based activation motif
Jak	Janus kinase
JNK	c-Jun N-terminal kinase
KSHV	Kaposi's sarcoma-associated herpesvirus
L	late (domain)
LAMP	latency-associated membrane protein
LANA-1	latency-associated nuclear antigen type 1
LH	luteinizing hormone
LNR	lin/notch repeat
LNS	laminin, neurexin, sex hormone-binding globulin
LPS	lipopolysaccharides
LRR	leucine-rich repeat
LTD	long-term depression
LTP	long-term potentiation
LTR	long terminal repeat
LZ	leucine zipper
MA	matrix
MAGE	melanoma-associated antigen
MALDI	matrix-assisted laser desorption ionization
MAOI	monoamine oxidase inhibitor
MAP	mitogen-activated protein
MAPK	mitogen-activated protein kinase
MCP	methyl-accepting chemotaxis protein
MD	molecular dynamics
MH1	mad homology-1
MHC	major histocompatibility complex

MIP	macrophage inflammatory protein
MM	molecular mechanics
MMP	matrix metalloproteinase
MMR	mismatch repair
MRI	magnetic resonance imaging
mRNA	messenger RNA
MSH	melanocyte-stimulating hormone
MVB	multivesicular body
NAc	nucleus accumbens
nAChR	nicotinic acetylcholine receptor
NADE	p75-associated cell death executioner
NAIP	neuronal inhibitory apoptosis protein
NBS	Nijmegem breakage syndrome
NC	nucleocapsid
NCAM	neural cell adhesion molecule
NE	norepinephrine (noradrenaline)
NER	nucleotide excision repair
NES	nuclear export signal (sequence)
NFAT	nuclear factor of activated T cells
NF-κB	nuclear factor kappa B
NGF	nerve growth factor
NH	amide (molecule)
NHEJ	nonhomologous end joining
NICD	notch intracellular domain
NKA	neurokinin A
NKB	neurokinin B
NLS	nuclear localization signal (sequence)
NMDA	N-methyl-D-aspartate
NMR	nuclear magnetic resonance
NPC	nuclear pore complex
NRAGE	neurotrophin receptor-interacting MAGE homolog
NRIF	neurotrophin receptor-interacting factor
NSAID	nonsteroidal anti-inflammatory drug
NSF	N-ethylmaleimide-sensitive fusion protein
NURF	nucleosome remodeling factor
OCT	octopamine
OPR	octicopeptide repeat
OT	oxytocin
PACAP	pituitary adenylate cyclase-activating polypeptide
PAGE	polyacrylamide gel electrophoresis
PBP	periplasmic binding protein
PCP	planar cell polarity
PCR	polymerase chain reaction

PDB	protein data bank
PDE	phosphodiesterase
PDGF	platelet-derived growth factor
PDK	phosphoinositide-dependent protein kinase
PDZ	PSD-95, DLG, ZO-1
PGHS	endoperoxide H synthase
PH	pleckstrin homology
PIC	pre-initiation complex
PIH	prolactin-inhibiting hormone
PIKK	phosphoinositide 3-kinase related kinase
PIP	phosphatidylinositol phosphatase
PKA	protein kinase A
PKB	protein kinase B
PKC	protein kinase C
PKG	protein kinase G
PKR	protein kinase R
PLA$_2$	phospholipase A$_2$
PLC	phospholipase C
PMCA	plasma membrane calcium ATPase
PNS	peripheral nervous system
POMC	pro-opiomelanocortin
PP-II	polyproline (helix)
PRH	prolactin-releasing hormone
PRL	prolactin
PS	pseudosubstrate
PSD	postsynaptic density
PSD-95	postsynaptic density protein of 95 kDa
pSMAC	peripheral supramolecular activation cluster
PTB	phosphotyrosine binding
PTH	parathyroid hormone
PTHrH	parathyroid hormone related protein
PTPC	permeability transition pore complex
PYD	pyrin domain
QM	quantum mechanics
RACK	receptor for activated C-kinase
RAIP	Arg-Ala-Ile-Pro (motif)
RE	responsive (response) element
REM	rapid eye movement
RF	radiofrequency
RGS	regulator-of-G-protein signaling
RH	RGS homology
RHD	rel homology domain
RIP	receptor-interacting protein

RNA	ribonucleic acid
RNP	ribonucleoprotein
ROS	reactive oxygen species
RPA	replication protein A
RR	response regulator
RRE	rev response region
RRM	RNA recognition motif
rRNA	ribosomal RNA
RSC	remodels the structure of chromatin
RT	reverse transcriptase
RTK	receptor tyrosine kinase
RyR	ryanodine receptor
S/T	serine/threonine
S6K	ribosomal S6 kinase
SAGA	Spt-Ada-Gen5 acetyltransferase
SAM	S-adenosyl-L-methionine
SAM	sterile α motif
SAN	sinoatrial node
SARA	smad anchor for receptor activation
SC1	Schwann cell factor-1
SCR	short consensus repeat
SDS	sodium dodecyl sulfate
SE	spongiform encephalopathies
SERCA	sarco-endoplasmic reticulum calcium ATPase
SH2	Src homology-2
Shh	sonic hedgehog
SIV	simian immunodeficiency virus
Ski	Sloan–Kettering Institute proto-oncogene
Smac	second mitochondrial activator of caspases
SMCC	SRD- and MED-containing cofactor complex
SN	sunstantia nigra
SNAP	soluble NSF-attachment protein
SNARE	soluble NSF-attachment protein receptor
SNF	sucrose nonfermenting
SnoN	ski-related novel gene N
snRNA	small nuclear RNA
snRNP	small nuclear ribonucleoprotein particle
SODI	superoxide dismutase
Sos	Son-of-sevenless
SP	substance P
SSRI	selective serotonin reuptake inhibitor
SST	somatostatin
STAT	signal transducer and activator of transcription
Ste2	sterile 2

STG	stomatogastric ganglion
STRE	stress responsive element
STTK	serine/threonine and tyrosine kinase
SUMO	small ubiquitin-related modifier
SWI	(mating type) switch
TACE	tumor necrosis factor-α converting enzyme
TAF	TBP-associated factor
TAR	transactivating response (region)
TBP	TATA box binding protein
TCA	tricyclic antidepressants
TCR	T-cell receptor
TF	transcription factor
TGF-β	transforming growth factor-β
TGIF	TG3-interacting factor
TM	transmembrane
TNF	tumor necrosis factor
TOF	time-of-flight
TOP	terminal oligopyrimidine
TOR	target of rapamycin
TOS	Phe-Glu-Met-Asp-Ile (motif)
TRADD	TNF-R-associated death domain
TRAF	TNF receptor-associated factor
TRAIL	TNF-related apoptosis-inducing ligand
TRAP	thyroid hormone receptor-associated protein
TRF1	telomeric repeat binding factor 1
tRNA	transfer RNA
TSH	thyroid-stimulating hormone
UP	upstream (sequence)
UPEC	uropathogenic E. coli
UTR	untranslated region
VAMP	vesicle-associated membrane protein
VDAC	voltage-dependent anion channels
VEGF	vascular endothelial growth factor
VIP	vasoactive intestinal peptide
Vps	vascular protein sorting
VTA	ventral tegmental area
Wg	wingless
XIAP	X-chromosome-linked inhibitor of apoptosis
YFP	yellow fluorescent protein
ZO-1	zona occludens 1

1
Introduction

Life on Earth is remarkably diverse and robust. There are organisms that live in the deep sea and far underground, around hot midocean volcanic vents and in cold arctic seas, and in salt brines and hot acidic springs. Some of these creatures are methanogens that synthesize all their essential biomolecules out of H_2, CO_2 and salts; others are hyperthermophiles that use H_2S as a source of hydrogen and electrons, and still others are halophiles that carry out a form of photosynthesis without chlorophyll. Some of these extremeophiles are animallike, while others are plantlike or funguslike or like none of these.

What is significant about this diversity is that although the details vary from organism to organism, all carry out the same core functions of metabolism, cell division and signaling in roughly the same manner. The underlying unity extends from tiny parasitic bacteria containing minimal complements of genes to large differentiated multicellular plants and animals. Each organism has a similar set of basic building blocks and utilizes similar assembly principles. The myriad forms of life arise mostly through rearrangements and expansions of a basic set of units rather than different biochemistries or vastly different parts or assembly rules.

1.1 Prokaryotes and Eukaryotes

There are two basic forms of cellular organization, prokaryotic and eukaryotic. *Prokaryotes*—bacteria and archaeons—are highly streamlined unicellular organisms. Prokaryotes such as bacteria are small, typically 1 to 10 microns in length and about 1 micron in diameter. They may be spherical (coccus), or rod shaped (bacillus), or corkscrew shaped (spirochette). Regardless of their shape, prokaryotic cells consist of a single compartment surrounded by a plasma membrane that encloses the cytoplasm and separates outside from inside. The genetic material is contained in a small number, usually one, of double-stranded, *deoxyribonucleic acid* (DNA) molecules, the chromosomes that reside in the intracellular fluid medium (cytosol). Bacterial chromosomes are typically circular and are compacted into a nucleoid region of the cytosol. Many bacterial species contain additional (extra-chromosomal) shorter, circular pieces of DNA called *plasmids*.

The bacterial plasma membrane contains the molecular machinery responsible for metabolism and the sensory apparatus needed to locate nutrients. When nutrients are plentiful the bacterial cell organization is ideally suited for rapid growth and proliferation. There are two kinds of bacterial cell envelopes. The envelopes of *gram-positive bacteria* consist of a thick outer cell wall and an inner plasma membrane. Those of *gram-negative bacteria* consist of an outer membrane and an inner plasma membrane. A thin cell wall and a periplasmic space are situated between the two membranes. The plasma membrane is an important locus of activity. In addition to being sites for metabolism and signaling, the plasma membrane and cell wall are sites of morphological structures extending out from the cell surface of the bacteria. These include flagellar motors and several different kinds of secretion systems.

Eukaryotic cells are an order of magnitude larger in their linear dimensions than prokaryotic cells. Cells of *eukaryotes*—protists, plants, fungi, and animals—differ from prokaryotes in two important ways. First, eukaryotic cells have a *cytoskeleton*, a highly dynamic meshwork of protein girders that crisscross these larger cells and lend them mechanical support. Second, eukaryotic cells contain up to ten or more *organelles*, internal compartments, each surrounded by a distinct membrane and each containing their own complement of enzymes. In contrast to prokaryotes, core cellular functions such as metabolism are sequestered in these compartments.

1.2 The Cytoskeleton and Extracellular Matrix

The cytoskeleton and extracellular matrix perform multiple functions. The *cytoskeleton* provides structural support, and serves as a transportation highway and communications backbone. Chromosomes, organelles, and vacuoles are transported along actin filaments and microtubules of the cytoskeleton. Actin filaments are used for short distance transport, while microtubules serve as a rail system for delivering cargo over long distances. Signal molecules are anchored at sites along the cytoskeleton, and the cytoskeleton functions as a communications backbone linking signaling molecules in the plasma membrane and extracellular matrix (ECM) to signaling units in the cell nucleus.

The extracellular matrix consists of an extended network of polysaccharides and proteins secreted by cells. The ECM provides structural support for cells forming organs and tissues in multicellular eukaryotes. In plants, the ECM is referred to as the *cell wall* and serves a protective role. Cells of animals secrete a variety of signaling molecules onto the extracellular matrix, and these molecules guide cellular migration and adhesion during development. The ECM is not a simple passive medium. Instead, signaling between ECM and the cytoskeleton is maintained throughout development and into adult life.

The existence of a transport system in which large numbers of molecules can be moved along the cytoskeleton to and from the plasma membrane is important for signaling between cells in the body. In the immune system, transport vacuoles move signal molecules called *cytokines* (anti-inflammatory agents such as histamines, and antimicrobial agents that attack pathogens) to the cell surface where they are secreted from the cell. In the nervous system, neurotransmitters are moved over long distances down the axon and into the axon terminal via transport vesicles. In addition to outbound trafficking, there is inbound trafficking. Surface components are continually being recycled back to the internal organelles, where they are then either reused or degraded.

1.3 Core Cellular Functions in Organelles

In prokaryotes, a single outer membrane is sufficient for membrane-dependent processes such as photosynthesis and oxidative phosphorylation (respiration), and protein and lipid synthesis. However, a single membrane is not adequate in eukaryotes because of the large, cubic increase in cell volume. Nature's solution to this design problem is a system of *organelles* surrounded by membranes that perform membrane-specific cell functions and sequester specific sets of enzymes. There are more than a half dozen different kinds of organelles in a typical multicellular eukaryote. Organelles present in typical multicellular eukaryotic cells are listed in Table 1.1, along with their cellular functions.

TABLE 1.1. Organelles of the eukaryotic cell: The principal functions of the proteins sequestered in these organelles are listed in the second column.

Organelle	Function
Mitochondria	Respiration
Chloroplasts	Photosynthesis (plants)
Nucleus	Stores DNA; transcription and splicing
Endoplasmic reticulum	Protein synthesis-translation
Golgi apparatus	Processing, packaging, and shipping
Lysosomes	Degradation and recycling
Peroxisomes	Degradation
Endosomes	Internalization of material

Organelles are characterized by the mix of enzymes they contain and by the assortment of proteins embedded in their membranes. Three kinds of proteins—pores, channels, and pumps—embedded in plasma and organelle membranes allow material to enter and leave a cell or organelle.

- *Pores*: Pore-forming proteins, or *porins*, are membrane-spanning proteins found in the outer membrane of gram-negative bacteria, mitochondria and chloroplasts. They form water-filled channels that enable hydrophilic molecules smaller than about 600 Da to pass through the membrane in and out of the cell or organelle. For example, bacterial porins allow nutrients to enter and waste products to exit the cell while inhibiting the passage of toxins and other dangerous materials.
- *Ion channels*: These are membrane-spanning proteins forming narrow pores that enable specific inorganic ions, typically Na^+, K^+, Ca^{2+} or Cl^-, to pass through cell membranes. Ion channels are an essential component of the plasma membranes of nerve cells, where they are responsible for all electrical signaling. Ion channels regulate muscle contractions and processes associated with them, such as respiration and heartbeat, and regulate osmobalance and hormone release.
- *Pumps*: Pumps are membrane-spanning proteins that transport ions and molecules across cellular and intracellular membranes. While ion channels allow ions to passively diffuse in or out of cells along electrochemical gradients, pumps actively transport ions and molecules. Thus, they are able to act against electrochemical gradients, whereas ion channels cannot, and maintain homeostatic balances within the cell. The transport involves the performance of work and must be coupled to an energy source. A variety of energy sources are utilized by pumps, including adenosine triphosphate (ATP) hydrolysis, electron transfer, and light absorption.

1.4 Metabolic Processes in Mitochondria and Chloroplasts

In all cells, energy is stored in the chemical bonds of *adenosine triphosphate* (ATP) molecules. In metabolism, enzymes break down large biomolecules into small basic components, synthesize new biomolecules out of those basic components, and produce ATP. In catabolic processes such as glycolysis and oxidative phosphorylation, large polymeric molecules are disassembled into smaller monomeric units. The intermediates are then further broken down into cellular building blocks such as CO_2, ammonia, and citric acid. Key goals of the catabolic processes are the production of ATP and reducing power needed for the converse, anabolic processes—the assembly of cellular building blocks into small biomolecules, the synthesis of components, and their subsequent assembly into organelles, cytoskeleton, and other cellular structures.

Glycolysis takes place in the cytoplasm while the citric acid cycle occurs in the mitochondrial matrix. Five complexes embedded in the inner mito-chondrial membrane carry out oxidative phosphorylation (respiration). The constituents of the five respiratory complexes—enzymes of the electron transport chain—pump protons from the matrix to the cytosol of the mito-chondria, then use the free energy released by these actions to produce ATP from adenosine diphosphate (ADP). Their photosynthetic counterparts, photosystems I and II, function in chroloplasts.

Chloroplasts and mitochondria are enclosed in double membranes. The inner membrane of a mitochondrion is highly convoluted, forming struc-tures called *cristae*. A similar design strategy is used in chroloplasts. The inner membrane of a chloroplast encloses a series of folded and stacked thylakoid structures. These designs give rise to organelles possessing large surface areas for metabolic processes. The ATP molecules are used not only for anabolism, but also in other core cellular processes, including signaling, where work is done and ATP is needed.

The relocation of the machinery for metabolism from the plasma mem-brane to internal organelles is a momentous event from the viewpoint of signaling. It not only provides for a far greater energy supply but also frees up a large portion of the plasma membrane for signaling. In eukaryotic cells, the plasma membrane is studded with large numbers of signaling proteins that are either embedded in the plasma membrane, running from the outside to the inside, or attached to one side or the other by means of a tether.

1.5 Cellular DNA to Chromatin

Cellular DNA is sequestered in the nucleus where it is packaged into chro-matin. As shown in Figure 1.1, all organisms on Earth today use DNA to encode instructions for making proteins and use RNA as an intermediate stage. This fundamental aspect of all of biology was firmly established by Crick and Watson in their pioneering study in the mid-twentieth century. In the first step—transcription—protein machines copy selected portions of a DNA molecule onto mRNA templates. In prokaryotes the riboso-mal machinery operating concurrently with the transcription apparatus translates the mRNA molecules into proteins. In eukaryotes there is an

$$\text{DNA} \xrightarrow{\text{transcription}} \text{RNA} \xrightarrow{\text{translation}} \text{Proteins}$$

FIGURE 1.1. Genes and proteins: Depicted is the two-step process in which DNA nucleotide sequences, or genes, are first transcribed onto messenger RNA (mRNA) nucleotide sequences, and then these templates are used to translate the nucleotide sequences into amino acid sequences.

intermediate step: Protein machines known as *spliceosomes* edit the initial RNA transcripts called pre-mRNA molecules, and produce as their output mature mRNA molecules. The ribosomal machines then translate the mature mRNAs into proteins.

In eukaryotes, cellular DNA is sequestered within the nucleus, and this organelle is the site of transcription and splicing. The nucleus is enclosed in a concentric double membrane studded with large numbers of aqueous pores. The pores enable the two-way selective movement of material between the nucleus and cytoplasm. Since proteins are synthesized in the cytoplasm, nuclear proteins—proteins that carry out their tasks inside the nucleus—are imported from the cytoplasm to the nucleus, while messenger RNAs and ribosomal subunits are exported. A variety of structural and regulatory proteins regularly shuttle back and forth between nucleus and cytoplasm. The pores, referred to as *nuclear pore complexes*, are composed of about 100 proteins and are approximately 125 MDa in mass. All particles entering or exiting the nucleus pass through these large pores. Small particles passively diffuse through the pores while large macromolecules are actively transported in a regulated fashion.

The sequestering of the DNA within a nucleus is advantageous for several reasons. It insulates the DNA against oxidative byproducts of normal cellular processes taking place in the cytoplasm and from mechanical forces and stresses generated by the cytoskeleton. It separates the transcription apparatus from the translation machinery, thereby allowing independent control of both, and it makes possible the intermediate ribonucleic RNA editing (splicing) stage.

Eukaryotic DNA is wrapped in proteins called *histones* and tightly packaged into a number of chromosomes in the nucleus. As a result of sequestering and packaging, far more information can be stored in eukaryotic DNA than in prokaryotic DNA. The wrapping up of the DNA to form chromatin enables the cells to regulate transcription of its genes in a particularly simple way that is not possible in prokaryotes. When the DNA is wrapped tightly about the histones the DNA cannot be transcribed since the sites that need to be accessible to the transcription machinery are blocked. When the wrapping is loosened, these sites become available and transcription can be carried out. A large number of eukaryotic regulators of transcription operate on a chromatin-level of organization, making transcription easier or harder by manipulating chromatin.

1.6 Protein Activities in the Endoplasmic Reticulum and Golgi Apparatus

The *endoplasmic reticulum* (ER) encompasses more than half the membrane surface of a eukaryotic cell and about 10% of its volume. It is the primary site of protein synthesis (translation), fatty acid and lipid synthesis, and bilayer assembly. It is divided into a rough ER and a smooth ER. The rough ER gets its name from the presence of numerous ribosomes bound to its cytosolic side. The rough ER is the site where membrane-bound proteins, secreted proteins, and proteins destined for the interior (lumen) of organelles are synthesized. The smooth ER lacks ribosomes. It is the site where lipids are synthesized and assembled and where fatty acids such as steroids are synthesized. It stores intracellular Ca^{2+} and assists in carbohydrate metabolism and in drug and poison detoxification.

Not all ribosomes are bound to the endoplasmic reticulum. Instead, there are two populations of ribosomes, bound and free. Bound ribosomes are attached to the rough ER, but free ribosomes are distributed in the cytosol. The free ribosomes are otherwise identical to their membrane-bound counterparts, and they synthesize cytosolic proteins.

In order for a protein to carry out its physiological function it must fold into and maintain its correct three-dimensional shape. Proteins are subject to several different kinds of stresses. Abnormal conditions, such as elevated or reduced temperatures and abnormal pH conditions, can result in the denaturization (unfolding) or misfolding of proteins so that they no longer have the correct shape and cannot function. Another type of condition that can affect the shape of the protein is molecular crowding. A group of small protein-folding machines called *heat shock proteins* or *stress proteins* or *molecular chaperones* guide nascent polypeptide chains to the correct location and maintain the proteins in folded states that permit rapid activation and assembly. They also refold partially unfolded proteins. In those cases where the proteins cannot be returned to a proper state the misfolded proteins are tagged for destruction by another set of small protein machines called *proteases*. These proteolytic machines enable a cell to degrade and recycle proteins that are no longer needed, as well as those that are damaged and cannot be refolded properly by the stress proteins.

Newly synthesized proteins are processed, subjected to quality control with respect to their folding, and then shipped to their cellular destinations. Prosthetic groups—sugars and lipids—are added to proteins destined for insertion in the membrane to enable them to attach to the membranes. These modifications are made subsequent to translation in several stages, as the proteins are passed through the ER and Golgi apparatus. The overall process resembles an assembly line that builds up the proteins, folds them, inserts them into membranes, sorts them, labels them with targeting sequences, and ships them out to their cellular destinations (Figure 1.2).

The Golgi apparatus consists of a stacked system of membrane-enclosed sacs called *cisternae*. Some of the polysaccharide modifications needed to make glycoproteins are either made or started in the rough ER. Proteins, especially signaling proteins destined for export (secretion) from the cell or for insertion into the plasma membrane, are sent from the rough ER to the smooth ER where they are encapsulated into transport vesicles pinched off from the smooth ER. The transport vesicles are then sent to the Golgi

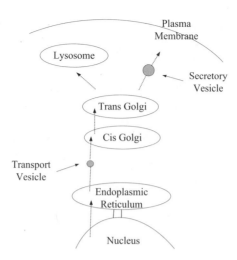

FIGURE 1.2. Movement of proteins through the endoplasmic reticulum and Golgi apparatus: Proteins synthesis and processing start with the export of mRNAs from the nucleus to the ribosome-studded rough endoplasmic reticulum. Nascent proteins synthesized in ribosomes are processed and then shipped in transport vesicles to the Golgi. They pass through the cis (nearest the ER) and trans (furthest from the ER) Golgi, and the finished products are then shipped out to their lysosomal and the plasma membrane destinations.

for further processing and eventual shipping to their cellular destinations. The Golgi apparatus takes the carbohydrates and attaches then as oligosaccharide side chains to some of these proteins to form glycoproteins and to complete modifications started in the rough ER. Both proteins and lipids are modified in the Golgi. Other proteins, synthesized as inactive precursor molecules, are processed to produce activated forms in the Golgi. Modified proteins are enclosed in transport vesicles, pinched off from the Golgi, and shipped to destinations such as the plasma membrane and the extracellular matrix (Figure 1.2).

1.7 Digestion and Recycling of Macromolecules

Digestion and the recycling of macromolecules take place in a network of transport and digestive organelles. The last three organelles listed in Table 1.1 are involved in digestion. *Peroxisomes* and *lysosomes* contain sets of enzymes used for digestion of macromolecules. In these highly acidic environments, macromolecules are broken down into smaller molecules. By sequestering enzymes in these compartments the rest of the cell is protected from the digestive properties of the enzymes. Lysosomes are small organelles that degrade ingested bacteria and nonfunctional organelles.

Perixosomes are utilized for the sequestering of oxidative enzymes. Their digestive enzymes degrade fatty acids to small biomolecules. Peroxisomes are a diverse collection of organelles, each with its own mix of enzymes. Some peroxisomes detoxify harmful substances. Others, in plants, convert fatty acids to sugars and carry out photorespiration.

Endosomes, the final set of eukaryotic organelles listed in Table 1.1, facilitate the transport of extracellular material and membrane proteins from the plasma membrane to lysosomes for degradation. Several kinds of organelles—early endosomes, carrier vesicles, and late endosomes—form a transport and sorting system that moves ingested foodstuffs, captured pathogens, dead material, and ligand-bound receptors and lipid plasma membrane components to the lysosomes and other cellular compartments for either degradation or reuse.

In summary, cells are highly dynamic entities; materials are continually being brought in and out of the cell, and moved back and forth to the surface. There is a continual flow of outbound traffic of cargo from the ER and Golgi to organelles and the cell surface, and there is a continual flow of inbound traffic from the cell surface to lysosomal and peroxisomal compartments. Signal proteins destined for the plasma membrane and for secretion are packaged into vacuoles. These transient structures are formed by pinching off portions of membrane. The vacuoles are moved over the rail system and fused with membranes at the destination (*exocytosis*). Similarly, materials from the cell surface are captured, packaged into vesicles, and shipped to digestive compartments for processing (*endocytosis*).

1.8 Genomes of Bacteria Reveal Importance of Signaling

Insights into the importance of signaling can be obtained from analyses of the composition of the genomes of bacteria. Prokaryotes are tiny organisms that tell us a lot about signaling. Prokaryotic genomes range in size from 0.5 MBp to more than 12 MBp. The *Mycoplasmas* sit at the bottom of this range; they are minimal organisms. They occupy very limited ecological niches, are restricted in their metabolic capabilities, and have the smallest genomes of any organisms. Genes encoding signal proteins are largely nonexistent, taking up no more than about 1% of their genomes. Prokaryotes with somewhat larger genomes include the archaeal extremophiles mentioned at the beginning of the chapter, and many obligate bacterial parasites that are the causal agents of diseases in humans. These organisms live in fairly constant and unvarying environments, and as a result their requirements on signaling and control are modest. Their signaling proteins account for no more than a few percent of their genomes.

Prokaryotes that can alter their metabolic and reproductive strategies to match their changing environmental conditions have larger genomes than those that live under constant conditions. Because of their ability to adapt their physiology to their environment, these bacteria may be referred to as *environmentalists*. *Escherichia coli* and *Psuedomonas aeruginosa* are typical environmentalists. Their genomes are five to ten times larger than the *Mycoplasmas* and encode ensembles of signaling and regulatory proteins that are 50 to 100 times larger than those of the *Mycoplasmas*. As a result, they are able to thrive in a variety of environments—soil, water, air—and they deal with many different stresses. Thus, not only are the genomes becoming larger as the bacteria become more versatile and adaptive, but the fractions of the genomes devoted to regulatory functions are increasing as well.

Colony-forming bacteria have even larger genomes. These bacteria can not only cope with environmental changes and stresses, but also they can assemble into colonies and exhibit a limited form of differentiation. Their genomes are the largest of all the prokaryotes, and the fraction of their genomes devoted to signaling and control approaches or exceeds 10%. An example of a prokaryote that exhibits environmental diversity and colonial behavior is *Streptomyces coelicolor*. This versatile soil bacterium is used for the production of antibiotics such as tetracycline and erythromycin. Its signaling component accounts for 12% of its genome.

As might be expected, the genomes for multicellular plants and animals are larger than those for even the most sophisticated prokaryotes, but not by as much as one might expect. The first estimates from the complete sequencings of the human genome are in the range of 26,000 genes, of which roughly one quarter is devoted to signaling. There are still some genes that remain to be identified, and when further analyses are completed there may be as many as 32,000 to 40,000 genes. This number appears to be astonishingly low. It is scarcely a factor of three larger than the genomes for some of the bacteria. Furthermore, the genome for the bacterium *S. coelicolor* is nearly as large as that for the highly differentiated, multicellular fruit fly *Drosophila melanogaster*.

1.9 Organization and Signaling of Eukaryotic Cell

Eukaryotic cell organization and expanded signaling capabilities make multicellularity possible. The significance of these observations is that something more than a simple increase in genome size produced the greatly increased complexity associated with multicellular plants and animals. The answer to the question of what is happening has several parts. It involves the way eukaryotic cells are organized and the way the expanded repertoire of signaling proteins is organized and used. There are four broad categories of environmental and regulatory signals. These are as follows:

- Physical and chemical sensations indicative of external conditions.
- Contact signals indicative of ECM-to-cell and cell-to-cell adhesion.

- Signals sent from one cell to another that allow the sender to regulate gene expression and other cellular responses in the recipient.
- Signals and sensations indicative of internal stresses and balances.

The first category includes a diverse set of physical and chemical signals. Most organisms can neither alter their environments nor move over large distances. Instead, they must continually adapt their metabolism and growth strategies to match the environmental conditions in which they find themselves. In this grouping of signals are environmental cues important to unicellular organisms, such as light, temperature, osmolarity, pH, and nutrients. Also included in this category are signals such as odorants and tastants detected by sensory organs in multicellular animals, or *metazoans*.

The next category encompasses contact signals between surfaces and is specific to multicellular organisms. Proteins embedded in the plasma membrane and in the ECM convey contact signals. These signals allow cells to establish and maintain physical contact with supporting structures within the body, and mediate two-way communication between cells in physical contact. This category is greatly expanded in vertebrates, and includes elements of the immune system such as antibodies that have evolved from adhesion molecules.

The third category of signals consists of the cell-to-cell messages. This category includes *pheromones*, chemical signals that promote mating and colonial behavior in unicellular organisms, and it includes the signals in multicellular plants and animals that allow cells in tissues and organs to work together. In humans, there are several systems of tissues, organs, and glands that continually send and receive chemical messages. Cells of the immune systems send and receive cytokines; cells residing in glands of the endocrine system secrete hormones; and neurons in the nervous system communicate using neurotransmitters and neuromodulators. Embryonic development is controlled from cell division to cell division by programs of gene expression. The category of cell-to-cell signals encompasses the signals that help establish cell fate and polarity during development, and the growth factor and hormonal signals that shape and guide the programs of cell growth and differentiation.

The last category consists of signals generated within the cell that help maintain proper internal balances, or *homeostasis*. This category includes signals indicative of internal stresses such as improper pH conditions, excessive temperatures, and water imbalances. Macromolecules such as DNA and proteins are marginally stable under physiological conditions. Cellular DNA can be damaged by ultraviolet radiation, by ionizing radiation, and by oxidative byproducts of normal cellular processes. In addition, DNA strand breaks can occur during the DNA replication stage that precedes mitosis and meiosis. All cells, prokaryotic and eukaryotic, possess DNA repair systems that continually sense and repair single and double strand breaks. In multicellular organisms, whenever DNA damage is detected and found to be irreparable, the cell is targeted for destruction. The process of eliminating a cell that is damaged, or infected, or deemed to be no longer needed is called *apoptosis*. The apoptosis, or cell suicide, machinery cuts up (cleaves) and disassembles large cellular components—DNA and membranes—and packages the cellular contents in such a way that they do no harm to neighboring cells. The last category of signals includes those that integrate growth and survival signals with repair and apoptosis signals, determining whether a cell grows and proliferates, repairs itself, or dies.

1.10 Fixed Infrastructure and the Control Layer

The proteins responsible for signaling and control form a *control layer*. This layer sits on top of a lower layer, the *fixed infrastructure*, which is responsible for core cellular functions such as metabolism and replication. Proteins belonging to two layers carry out their cellular roles synergistically. Proteins belonging to the control layer make contact with elements of the fixed infrastructure at well-defined loci, or control points, where they exert their regulatory functions, but don't otherwise interfere with the machine-like operations of those proteins. In turn, eukaryotic architecture, with its organelles and cytoskeleton, is especially well suited for signaling and is extensively exploited for that purpose.

Unlike the proteins of the fixed infrastructure, the proteins belonging to the control layer are not sequestered within a single compartment or organelle. Instead, they form a meshwork of signaling pathways that extend throughout the cytoplasm and into organelles, most notably the nucleus, but other as well. Each signaling pathway has a start point and an end point. The start points are typically proteins that function as sensors and as receivers of signals from other cells. These proteins are often associated with the plasma membrane, where outside meets inside, and are referred to as *receptors*. The receptors not only detect the signals but also convert them into forms that can be understood and processed further within the cell. This conversion process is called *signal transduction*. The signaling pathways terminate at sites where the elements of the control layer come into contact with the components of the fixed infrastructure. These are the control points where the environmental and regulatory signals are converted into cellular responses.

The cell nucleus contains large numbers of control points, and when these sites are the end points the signaling process is termed *gene regulation*. Alternatively, and more generally, the control processes carried out by signaling proteins is referred to as *cell regulation*. One of the key factors making possible the emergence of complex multicellular organisms is the formation of gene regulatory networks composed of transcription regulating proteins, or transcription factors, and the DNA sequences they bind.

TABLE 1.2. Comparison of proteins in the fixed infrastructure and control layer.

Property	Fixed infrastructure	Control layer
Location	Machines/factories in organelles	Complexes in subcompartments
Mobility	Little	Considerable
Lifetime	Longer	Shorter
Structure	Fixed	Variable
Function	Unifunctional	Multifunctional

Changes in how these networks are built rather than in the genes they control underlie the increased complexity of multicellular life.

Signal proteins are not only the messengers but also the messages. Since the messages must be conveyed from one place to another, mobility is a key property of the proteins. Mobility is less important for functions such as metabolism carried out as part of large molecular machines in the organelles. Another way signal proteins differ from the other proteins is in the presence of post-translational modifications. Signal proteins are subjected to a host of post-translational modifications. Some are made in the ER and Golgi as part of the finishing process, but others, to be discussed in the next chapter, are part of the signaling process itself. These alterations endow the proteins with switchlike response properties, turning them on and off. This property is absolutely essential for a signaling element, keeping it available for conveying a message, but in an off-configuration until an activating signal arrives.

There are several other ways that proteins belonging to the control layer and fixed infrastructure differ from one another (Table 1.2). The function of a protein, that is, what it does, is determined by its associations, and by where and when it establishes them. Proteins that function as part of the transcription machinery in the nucleus or as part of the electron transport chain in the mitochondria, usually have a single purpose, and they carry out this task over and over. The signaling proteins form complexes too. However, the protein complexes are smaller than the large machines used for metabolism and replication, and the associations and interactions are more variable. They can be different in different cell types and will even vary somewhat over time in the same cell type. Because of this flexibility, signaling proteins are multifunctional, or *pleiotropic*, in their actions.

1.11 Eukaryotic Gene and Protein Regulation

Eukaryotic genes and proteins can be regulated in several ways. One of the most important consequences of switch from prokaryotic to eukaryotic cell organization is the creation of a large number of ways of controlling the mix of proteins being expressed at any given time in a cell. In place of DNA regulation of prokaryotic gene and protein expression there is now a multiplicity of eukaryotic control mechanisms. Gene and protein expression can be controlled through

- DNA regulation
- histone modification
- splicing regulation
- translation regulation
- nuclear import/export

Histone modification has already been discussed. Nuclear import and export refers to a strikingly simple and widely used form of regulation. Many transcription factors are parked in convenient locations in the cytoplasm awaiting activating signals. When activated they diffuse to the nucleus, where they carry out their transcriptional activities. Exporting the proteins back out of the nucleus into the cytoplasm is an equally simple way of terminating their activities.

Translational regulation allows for the placement of messenger RNAs (mRNAs) in sites where the proteins they encode might be needed at a later time. Asymmetric distributions of mRNAs and proteins in cells early in development lead to offspring that are dissimilar. Localized populations of mRNAs are utilized as part of adult physiology. They are used, for example, in nerve cells to avoid long time delays arising when signals must be sent over long distances to the nucleus and the resulting proteins shipped back out to the distant extrema.

The observation that for every one unique gene there is one unique protein is correct as far as it goes, but in eukaryotes one must append the equally true statement that the unique proteins come in several "*flavors*". Alternative splicing permits adjustments to be made to the proteins expressed in different cell types. Eukaryotic genes are larger than their prokaryotic counterparts and contain greater numbers of units called *domains* (these are discussed in the next chapter). In vertebrates there are, on the average, about three alternative spliced forms, or flavors, for each protein. Signals sent to the control points in the splicing machinery help determine which spliced variant, or *isoform*, gets made at that particular time in that cell type.

A piece of the mystery of the relatively small size of the eukaryotic genome is resolved by the observation that alternative splicing creates many variants from a single gene, alleviating the need to store the instructions for making each variant in the DNA. Alternative splicing and post-translational modifications are extensively utilized in the control layer. If all of these flavors are counted as distinct items, the resulting signaling protein numbers would comprise most of the genome, and the eukaryotic totals would be far more impressive. One of the consequences of this multiplicity is that study of the control layer is more difficult than it would be otherwise. Because of the large numbers of structural variations, and also because of the multifunctionality, mobility, low copy numbers, and short lifetimes of the signaling proteins (Table 1.2), understanding of the control layer is not as advanced as that of the proteins in the fixed infrastructure involved in, for example, metabolism.

Lying at the heart of biology is the following observation, succinctly stated by Francois Jacob in a 1977 article in *Science*: "What distinguishes a butterfly from a lion, a hen from a fly, and a worm from a whale is much less about differences in chemical constituents than in the organization and distribution of their components." In other words, all creatures, great and small, carry out metabolism, replication, multiplication, and division in much the same way. They differ from one another mainly in the way their parts are arranged. Multicellular eukaryotes such as flies and worms are clearly far more complex than bacteria. Yet it only takes a factor of two or three more genes to get from one to the other. The answer to how this can be possible is not in the numbers of genes or in a new type of biochemistry, but rather in the way that the genes and their products, the proteins, are used, and in the way eukaryotic cells are organized.

1.12 Signaling Malfunction Central to Human Disease

Malfunctions in molecular and cellular signaling lie at the heart of human diseases. Proteins belonging to the control layer are involved in a host of human disorders. They are key elements in cancers and in neurodegenerative disorders in the elderly and mood disorders in the young. Signaling processes make complex organisms like humans possible, but when there are malfunctions, the signaling processes give rise to diseases in those very same organisms.

Improper expression levels and malfunctions of signaling proteins are responsible for a host of human cancers. The underlying causes of cancers are mutations and other alterations in DNA. These aberrations produce malfunctions and inappropriate expression levels of genes encoding proteins that either promote growth or restrain it, or direct the apoptosis machinery, or are responsible for DNA damage repair and signaling, and chromatin remodeling. Erroneous signaling conveys inappropriate growth signals, fails to turn on the body's cell suicide program when it is needed, and fails to repair DNA damage when it occurs.

The brain is the most complex organ in the body. A substantial portion of the human genome is taken up with encoding brain-specific signaling proteins. Some of these, such as the ion channels, endow the neurons with the ability to generate action potentials, which are used to signal other neurons and control muscle cells. Improper and excessive rhythms resulting from imbalances between the excitation and inhibition of neurons are responsible for epileptic seizures and a host of attention, learning, and mood disorders. Proteolytic processing is a prominent part of the signaling routes activated during embryonic development. But some of the same processing elements that are crucial for embryonic development early in life contribute to neurological disorders such as Alzheimer's disease late in life.

Receptors, the proteins that reside in the plasma membranes of cells and receive signals from other cells are key targets of therapeutic drugs. These drugs act as *ligands* for the receptors, and are intended to elicit one of two kinds of actions: (1) By binding the receptor, the drug may activate the signaling pathway into the cell, stimulating processes that are otherwise not properly working; (2) or alternatively, the drug may serve as a *null ligand*—one that can bind the receptor but not stimulate signaling when doing so. This second type of action is that of a blocker, since it ties up the receptor, preventing other ligands from binding it and activating the signaling pathway.

The preeminent family of receptors in humans is the G protein-coupled receptor family. They are responsive to hormones, neuromodulators, and neurotransmitters. Some 40 to 60% of all drugs target G protein-coupled receptors. Some of the best known of these are the serotonin and adrenergic receptor-targeted drugs that treat depression, and the dopamine receptor-directed drugs that treat schizophenia. Other examples are the vasopressin receptor-mediated drugs that act as antidiuretics and the angiotensin receptor-targeted drugs that treat hypertension. Three more examples are: the histamine receptor-targeted drugs that alleviate allergic symptoms, the opioid receptor-targeted drugs that alter mood, and the neurokinin receptor-targeted drugs that alleviate pain.

1.13 Organization of Text

The textbook is organized into several parts. Chapters 2 through 5 serve as an introduction, providing background information helpful for an understanding of signaling. Chapter 2 gives an overview of the control layer and its relationship to cellular, nuclear, DNA, and protein organization. Chapter 3 examines the principal experimental methods used to probe the structure of signaling proteins and their interactions with one another. One of the most interesting and intensively studied processes in all of science is the folding of newly synthesized proteins into their physiologically functional three-dimensional shape. Chapter 4 has as its focus energy considerations. It covers how energy considerations drive protein folding and binding, and how proteins fold in the cell with the assistance of molecular chaperones. Chapter 5 deals with the macromolecular forces that underlie not only how proteins assume their functional forms but also how they interact with each other.

The next three chapters serve as an introduction to signaling. A good starting point for any discussion of signaling is the plasma membrane; it is the place where environmental conditions are sensed and most cell-to-cell signals are received. The yeast stress and pheromone signaling systems, the focus of Chapter 6, and the bacterial chemotaxis system, the main subject of Chapter 7, are archetypical signaling systems. Yeasts must constantly adapt to changing conditions in their environment, and, similarly, bacteria must locate nutrients and sense and respond to changes in their external environments. These systems exhibit many of the properties and principles that characterize signaling in more complex organisms. For many years the plasma membrane was regarded as a fairly homogeneous and passively fluid substrate into which signaling proteins were embedded. That view has undergone considerable change over the past few years with the realization that the plasma membrane is organized into distinct signaling domains, and that several kinds of lipids serve as signaling intermediaries. This new picture of the plasma membrane, along with the role of small molecules— lipid and nonlipid—in signaling is explored in Chapter 8.

The immune system provides several layers of defense against viral, bacterial, and eukaryotic pathogens. These defenses are coordinated and regulated by networks of signaling molecules. These signaling molecules are the subjects of Chapter 9. In response to certain signals, leukocytes, or white blood cells, converge upon sites of an infection and kill pathogens. Chapter 10 explores the role of signaling events taking place at the cell surface that make possible the directed movement of leukocytes into an infection site. Leukocytes are not the only cells requiring motility. During development cells must move about and aggregate into tissues, and nascent nerve cells must send out growth messages to connect with other nerve cells. These signaling processes are examined, too, in Chapter 10.

Cells secrete growth factors and hormones in order to coordinate cellular growth and proliferation. The mechanisms whereby polypeptide growth factors are received and transduced into cellular responses are discussed in Chapter 11. G protein-coupled receptors transduce a remarkably diverse spectrum of messages into the cell. Among these are light, gustatory, odorant, pheremone, pain, immunological, endocrine, and neural signals. Signaling through these receptors is covered in Chapter 12.

During embryonic development cell-to-cell signals coordinate the genetic programs of growth, differentiation, and proliferation. These signaling events, along with those that regulate motility, guide the formation of organs and tissues with well-defined boundaries composed of functionally specialized cells derived from less specialized progenitors. The determination of which tissue or organ a particular cell becomes part of, or *cell fate*, is the focus of Chapter 13.

As noted in the last section, cancer and signaling are intimately connected. Cancer can be regarded as a collection of diseases associated with malfunctions of key elements of the control layer and DNA repair machinery. The relationships between cancer and aberrant signaling are explored in Chapter 14. During the past few years the subject of programmed cell death, or apoptosis, has moved to the forefront of cancer research. The goal is to create drugs that kill cancers by forcing the cancerous cells to undergo apoptosis. This topic is explored in Chapter 15.

The main way that cells respond to environmental and cellular signals is to alter gene expression, turning some genes on and others off. In eukaryotes, the primary terminus of the signaling pathways leading from the cell surface into the cell interior is the transcription and the splicing machinery located in the cell nucleus. The mechanisms involved in turning on and off specific genes are the focus of Chapter 16.

The next two chapters, 17 and 18, deal with gene regulation in bacteria and by viruses. Bacterial cell-to-cell communication and gene regulation are discussed in Chapter 17. Viruses suborn host defenses in order to gain entry into a cell and once inside the cell create environments conducive to their replication. Several examples of how elements of the host and viral control layers interact are presented in Chapter 18.

The last three chapters of the textbook are devoted to signaling in the nervous system. Nerve cells, or *neurons*, send signals to other neurons and to muscle cells. The main goal of Chapter 19 is to explore how ion channels open and close and work together to generate action potentials, the hallmark of nerve cell signaling. The next chapter (Chapter 20) describes how nerve cells working together generate rhythmic activities, some associated with sleep and other with awakening, some with rhythmic motor activities such as walking and chewing, and others with heartbeat and breathing. The last chapter (Chapter 21) deals with how complex organisms ranging from worms to flies to humans learn and remember and forget and learn again.

General References

Alberts B, Johnson A, Lewis J, Raff M, Roberts K, and Walker P [2002]. *Molecular Biology of the Cell* (4th ed). New York: Garland Publishers.

Jacob F [1977]. Evolution and tinkering. *Science*, 196: 1161–1166.

Matthews CK, van Holde KE, and Ahem KG [2000]. *Biochemistry*, 3rd ed. San Francisco: Pearson Benjamin Cummings.

Stryer L [1995]. *Biochemistry*, 4th edition. New York: W.H. Freeman and Company.

References and Further Reading

Molecular Machines

Cramer P, et al. [2000]. Architecture of RNA polymerase II and implications for the transcription mechanism. *Science*, 288: 640–649.

Hirokawa N [1998]. Kinesin and dynein superfamily proteins and the mechanism of organelle transport. *Science*, 279: 519–526.

Rout, MP, et al. [2000]. The yeast nuclear pore complex: Composition, architecture, and transport mechanism. *J. Cell Biol.*, 148: 635–651.

Saraste M [1999]. Oxidative phosphorylation at the fin de siecle. *Science*, 283: 1488–1493.

Vale RD, and Milligan RA [2000]. The way things move: Looking under the hood of molecular motor proteins. *Science*, 288: 88–95.

Ubiquitin-Preotosome Complex

Ciechanover A [1998]. The ubiquitin-proteosome pathway: On protein death and cell life. *EMBO J.*, 17: 7151–7160.

Organelle Function and Trafficking

Mellman I, and Warren G [2000]. The road taken: Past and future foundations of membrane traffic. *Cell*, 100: 99–112.

Presley JF, et al. [1997]. ER-to-Golgi transport visualized in living cells. *Nature*, 389: 81–85.

Quality Control

Ellgaard L, Molinari M, and Helenius A [1999]. Setting the standards: Quality control in the secretory pathway. *Science*, 286: 1882–1888.

Lindahl T, and Wood RD [1999]. Quality control by DNA repair. *Science*, 286: 1897–1905.

Wickner S, Maurizi MR, and Gottesman S [1999]. Posttranslational quality control: Folding, refolding and degrading proteins. *Science*, 286: 1888–1893.

2
The Control Layer

Eukaryotic cell organization impacts signaling by the control layer in several ways. One of the most important of these is to provide flexibility by increasing the number of control points. The expression of a gene can be regulated through interactions with the DNA and the transcription machinery, with the histones, and with the splicing machinery. This property of the eukaryotic control layer is explored further in the first part of this chapter. The nucleosome, the basic repeating unit of chromatin, will be discussed first and then nuclear architecture will be examined. Chromosomes are not randomly distributed through the nucleus. Rather, there is a considerable amount of nuclear order in support of transcription and splicing. The arrangement of chromosomes into distinct structures and compartments will be looked at.

Signaling proteins are highly modular, and most signaling proteins appear to have been built by the forming of different arrangements of a set of building block domains. Modularity is a good example of efficient coding—if proteins are constructed from almost independent pieces, their parts can be reused to make other proteins. This method of design facilitates genetic rearrangements in which new proteins can be created without the body having to first design new building blocks. There is a hierarchy of protein structures from primary to quaternary. This organization will be the presented in the middle part of the chapter.

Proteins belonging to the control layer are themselves controlled through the addition and removal of small groups of atoms. These changes are reversible. They are referred to as *post-translational modifications* since they occur subsequent to translation. They serve several signaling purposes. The modifications influence the location of the signaling proteins within the cell; they regulate their signaling activities by turning catalytic activities on and off, and exposing and hiding interfaces, and they alter the messages they convey. The main types of posttranslational modifications will be presented in the last part of the chapter.

2.1 Eukaryotic Chromosomes Are Built from Nucleosomes

The double-stranded DNA molecules of the eukaryotic cell nucleus form associations with proteins called *histones*. The resulting DNA-histone material is known as *chromatin*. It was initially thought that DNA molecules were uniformly wrapped in histones. The main function of the largely undifferentiated histones was in support of supercoiling, which allowed the chromatin to pack tightly into the nucleus. The current picture differs radically from the earlier one. In the new picture, histones are not distributed uniformly. Instead, there is a fundamental repeating unit of DNA plus histone called the *nucleosome*, and these nucleosomes are strung together much like individual beads on the string. The histones have a quite specific stoichiometry and act in a dynamic manner to regulate gene transcription. They are not merely a passive set of girders for the DNA.

Nucleosomes are composed of 146 base pairs of DNA wrapped about a histone octamer. The duplex DNA is wrapped about a histone core composed of H2A, H2B, H3, and H4 histone pairs. The core nucleosome unit (shown in Figure 2.4) is connected by a linker segment (called H1) of chromatin to the next nucleosome core unit. Each nucleosome has the DNA wrapped about 1.75 times about the histone core forming a bead-like structure roughly 11 nm in diameter. The nucleosomes are themselves organized into 30-nm diameter solenoidal structures, and these next larger structural units are further wound into loops, and then into minibands, and finally into tightly coiled chromosomes. By this means a meter or so of DNA is organized into a compact unit that fits into a region a few microns in each linear dimension.

Growing crystals suitable for use in X-ray crystallography is often the most difficult step in applying the technique to biomolecules, especially large ones. In the case of the nucleosome, the challenge was in getting the phases aligned. The exact positioning of the nucleosome along a length of DNA is known as the *phase*. In order to create a crystal suitable for high-resolution X-ray crystallography the DNA and histone core of each nucleosome must have the same phase. This problem was solved through the development of a bacterially derived palindrome ("reading" the same forwards and backwards) DNA sequence that is able to attach to the histone core in a repeatable (same phase) way. Multiple copies of these sequences were inserted into bare histone cores taken from chicken red blood cell nuclei. The resulting nucleosome and histone core structures are displayed in Figures 2.1 and 2.2.

The transcription machinery and regulatory elements must have access to DNA in order for transcription to occur. In its fully wound form, most of the DNA binding sites are blocked, and are inaccessible. The DNA material that is to be transcribed or replicated must be unwound and separated

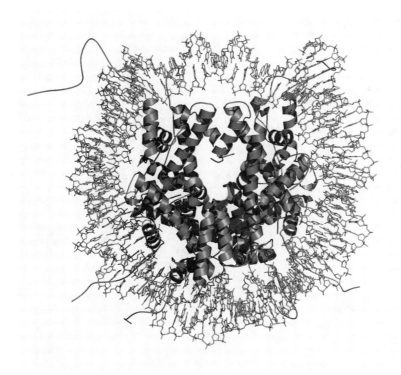

FIGURE 2.1. Structure of the nucleosome revealed by X-ray crystallography at 2.5 Å resolution: Shown in the figure is the double-stranded DNA wrapped around a histone core. The DNA molecule is depicted in a stick model highlighting the base pairings between complementary strands, and the histones are shown as ribbons and strings. [From Harp et al. [2000]. *Acta Cryst. D*, 56: 1513–1534. Reprinted with permission from the authors.]

sufficiently to expose the binding sites. Steric blockages are relieved, or alternatively enhanced, by proteins that interact with and modify DNA structure, and also by proteins that alter histone structure. In eukaryotes, the ability to influence gene expression by modifying histone structure adds a further layer of control to that available through the protein-DNA binding. Multiple control points are situated on histone tails that extend out from the histone core beyond the DNA. As can be seen in Figure 2.2, these sites are well exposed, thereby providing regulatory elements with ease of access.

2.2 The Highly Organized Interphase Nucleus

The interphase nucleus is a highly organized structure. The cycle of cell growth and division passes through four stages. The first of these stages (G_1) is a cellular growth stage that prepares the way either for entry into a cell

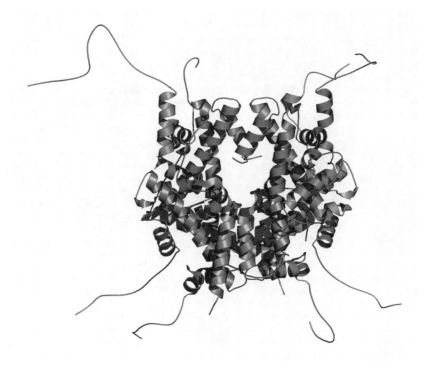

FIGURE 2.2. Structure of the histone octamer revealed by X-ray crystallography at 2.5 Å resolution: Depicted are the H2A, H2B, H3, and H4 core histones without the surrounding DNA. Tails from the H2A, H2B, H3, and H4 histones are represented by strings. These structures extend out from the core and interact with DNA and with neighboring nucleosomes. [From Harp et al. [2000]. *Acta Cryst. D* 56: 1513–1534. Reprinted with permission from the authors.]

division series of stages (S, G_2, and M), to growth arrest (G_0), or to apoptosis (cell death). Cells that do not undergo growth arrest or apoptosis enter a synthesis (S) stage where the DNA is replicated in preparation for mitosis. This stage is followed by a further mitosis preparatory stage (G_2) where RNAs and proteins required for mitosis are synthesized. Finally, the cell enters into mitosis (M) where the cell divides to produce two offspring. The three stages preceding mitosis are collectively referred to as *interphase*.

Chromosomes are not randomly distributed in the nucleus during interphase. Instead, they occupy distinct chromatin territories and are positioned in a way that reflects their replication status. The chromatin territories are partitioned into ~1Mbp-chromatin domains. Interchromatin spaces separate the chromosome territories from one another. Gene-rich early replicating chromatin is sequestered from gene-poor chromatin, and from mid-to-late replicating chromatin. Early replicating chromosomes are located in the nuclear interior, while inactive and late-replicating chromosomes are situated at the periphery of the nucleus near or in contact with the nuclear envelope.

The nuclear architecture facilitates transcription and splicing. Highly condensed chromatin serves to repress transcription while an open formation allows access of transcription and splicing factors to the sites of transcription. Chromosomes containing active regions of transcription are more open in their organization than those not actively undergoing transcription. The active regions typically lie on the outer portion of the chromosome territories. They extend into the interchromatin spaces to permit a maximal exposure of the surface of active genes to transcription and splicing factors; the two activities, transcription and splicing, are synchronized with one another.

The nucleus has several other kinds of structures in addition to chromatin territories. The additional structures—nucleoli, coiled (Cajal) bodies, and interchromatin granule clusters—are nonmembrane-associated organelles. They are used for sequestering and storage, and for preparation and assembly of materials used for gene transcription, pre-mRNA splicing, and protein synthesis. They are highly dynamic structures that form and reform about local concentrations of materials that process RNA molecules. In more detail:

- The nucleolus is the site of ribosome assembly, and forms in response to transcription of ribosomal DNA (rDNA), the genes that encode ribosomal RNA. Preribosomal RNA (pre-rRNA) units associate with the small nucleolar RNAs (snoRNAs), and are processed to form mature rRNAs. Pretransfer RNAs (pre-tRNAs) are imported into the nucleolus and processed there. The mature rRNAs associate with the tRNAs and a large ensemble of ribosomal proteins to form ribosomes. Once assembled the ribosomes are exported to the cytoplasm.
- Cajal bodies (CBs) are the sites of assembly of transcription complexes, and contain high concentrations of RNA Pol I, Pol II, and Pol III. Recall that there are three kinds of eukaryotic polymerases. RNAP Pol II transcribes pre-mRNAs and most splicing RNAs. The other RNA polymerases transcribe rRNA subunits, pre-tRNAs, and the U6 splicing RNAs. A cell may contain up to ten Cajal bodies. Once assembled within a CB, pol I complexes translocate to the nucleoli, where they transcribe rRNA genes. Pol III assemblages diffuse to the U6 small nuclear RNA (snRNA) involved in splicing, 5S rRNA and tRNA transcription sites, and Pol II complexes diffuse out of the CBs and over to mRNA and snRNA transcription sites.
- Interchromatin granule clusters (IGCs), or speckles, are distributed throughout the inter-chromatin spaces. These organelles are enriched in U1, U2, U4/U6, and U5 small nucleolar ribonuclear particles (snRNPs) and other components of the splicing machinery. (These will be discussed in detail in Chapter 15.) The localization of splicing factors within the

IGCs is not static, but instead changes in time in a way that reflects the transcriptional activity underway. Splicing factors diffuse in and out of the granules to and from sites of transcription activity.

2.3 Covalent Bonds Define the Primary Structure of a Protein

Proteins have a primary structure, a secondary structure, and a tertiary structure. Proteins constructed from more than one polypeptide chain have a quaternary structure, as well. A protein is a linear chain of amino acids, connected to one another by means of peptide bonds. Twenty different amino acids contribute to the formation of proteins. All have the same basic architecture shown in Figure 2.3. There is a central carbon atom, called an *alpha carbon*. Tied to it are an amino group, a carboxyl group, and a hydrogen atom. The fourth bond is with atoms belonging to the side chain (R). Two forms, uncharged and charged, are depicted in Figure 2.3.

The starting point in determining a protein's structure is the linear sequence of amino acids. These sequences are unique for each type of protein. In a protein, each amino acid residue in a sequence is linked to the next amino acid residue by means of a covalent peptide bond, and for that reason proteins are commonly referred to as *polypeptides*. The primary structure of a protein macromolecule is its covalent structure, including all covalent disulfide bonds that form during folding. The core of the amino acid is the repeating $NC_{\alpha}C$ unit. These units, covalently linked to one another by peptide bonds, form the main chain, or backbone, of the protein in which a carboxyl group of one amino acid and an amino group of the next amino acid in the sequence are linked together by a peptide bond. The result of joining amino acids by means of peptide bonds is depicted in Figure 2.4. By comparing Figures 2.3 and 2.4 one can see that two hydrogen atoms and an oxygen atom are removed during peptide bond formation. These are restored during hydrolysis of the peptide bond.

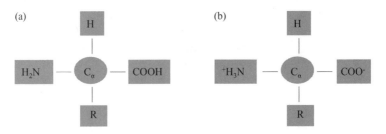

FIGURE 2.3. Amino acid structure: In amino acids, the bound organic groups, denoted by the "R" symbols, are termed side chains. (a) Form of an uncharged amino acid. (b) Form of a dipolar, or *zwitterion*, amino acid.

FIGURE 2.4. Two amino acids joined together by means of a peptide bond.

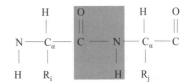

TABLE 2.1. Properties of common secondary structure: The second column lists the psi (ψ) values and the third column the phi (φ) values for the ideal structural elements. The last column list the number of residues per helical or strand turn.

Secondary structure	psi	phi	n
Alpha helix	−57.8	−47.0	3.6
3.10 helix	−74.0	−4.0	3.0
Pi helix	−57.1	−69.7	4.4
Beta strand	−139.0	+135.0	2.0

The peptide CN link is partially double bond in character. Consequently, the atoms cannot rotate freely about the CN bond axis. Instead, the four atoms highlighted in Figure 2.4 form a rigid planar peptide unit. The single bonds on each side of the peptide bond, that is, the $C_\alpha C$ and the NC_α bonds, are quite flexible with respect to rotations (torsions) about the bond axes. The angle of rotation about the $C_\alpha C$ axis is usually denoted as *psi* (Ψ) and that about the NC_α axis as *phi* (φ). The identification of (Ψ, φ) for each amino acid residue completes the specification of the main chain conformation. Not all values of Ψ and φ are allowed, due to steric constraints associated with the van der Waals radii. Instead, a Ramachandran plot of the (Ψ, φ) values for a given protein will have areas of high density of points indicative of that protein's β sheet and α helix secondary structural content. Large regions of the Ramachandran plot will be empty corresponding to (Ψ, φ) values that are sterically forbidden. Characteristic psi and phi values for several kinds of helical structures, and for the beta sheet, are presented in Table 2.1. The tabulated values differ considerably from one another, and data points falling about these values will be well separated in a Ramachandran plot.

2.4 Hydrogen Bonds Shape the Secondary Structure

Portions of the polypeptide chain that occur near one another tend to form geometrically regular, repeating structures during the process of folding into a three-dimensional functional form. The most commonly encountered regular structures are alpha helices (α *helices*), beta sheets (β *sheets*) and *turns*. Sequences that do not form one of these kinds of regular structures are grouped together in a general category called *random coils*. The arrangement of these features within the protein constitutes the protein's secondary structure.

Alpha helices and beta sheets are stabilized by hydrogen bonds between amide (NH) and carbonyl (CO) groups. In alpha helices, the hydrogen bonds form between a carbonyl group on the *i*th residue and the amide group on the (i + 4)th residue lying below it. In beta sheets, the hydrogen bonds form between the carbonyl group lying on one strand and the amide group situated immediately adjacent to it on the other strand. The beta strands can be oriented in either a parallel or an antiparallel manner. Some proteins are mostly alpha helix; other mostly beta sheet, and some are a mixture of the two. These are referred to as α/β if the two types of structural element are mixed and $\alpha + \beta$ if they remain distinct.

Alpha helices are about 10 residues in length and these structures account for about a third of the amino acid residues in a typical protein. Beta sheets are typically 6 amino acid residues in length and they account for a quarter of the residues. Turns allow the chain to reverse direction. They along with loops are usually located on the protein surface and thus contain polar and charged residues. Protein structures tend to be compact with little space left open in the interior. Hydrogen bonds formed in the protein interior neutralize buried polar groups. Structural compactness and the use of hydrogen bonds to neutralize interior polarity drive the formation of the alpha helices and beta sheets.

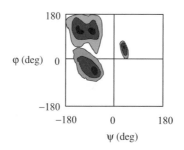

FIGURE 2.5. Ramachandran plot: Shown is a stereotypic contour plot of the distribution of main chain torsion angles φ and Ψ determined from high-resolution X-ray crystallographic data. Darker shading denotes more highly favored torsion angle combinations. Blank (white) regions represent conformations that are sterically forbidden. The most favorable conformations for alpha helices and beta sheets are concentrated into the three main regions. Beta sheet torsion angles appear as a double-peaked distribution in the upper left quadrant of the plot. The distribution of torsion angles for right handed alpha helices peaks in the lower left quadrant, and that for left handed alpha helices peaks in the upper right quadrant. Each of the 20 amino acids populates similar, but slightly different, portions of the Ramachandran plot.

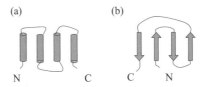

FIGURE 2.6. Topology of two common structural motifs: (a) The four-helix bundle, and (b) the Greek key. In the plots, cylinders represent alpha helices, and arrows oriented in the amino-to carboxyl-terminal direction denote beta strands. The helices and strands are connected by short turns and longer loops. The four-helix bundle can be regarded as a pair of helix-turn-helix structures. The Greek key is assembled from two pairs of antiparallel beta strands.

2.5 Structural Motifs and Domain Folds: Semi-Independent Protein Modules

In the alpha helix, the backbone forms the inner portion of structure while the side chains rotate outward. Several different arrangements of helices can be formed. One arrangement is as a stand-alone helix. Another arrangement is as coiled coils in which two or more helices are twisted about one another to make a particular stable structure. This type of structural motif is common in muscle and hair. Yet another kind of structure is formed when two helices are connected by a flexible loop. This structure, a helix-loop-helix, is commonly encountered in DNA-binding proteins. More generally, when secondary structure elements associate with other secondary structures they form supersecondary structures. In this process, strands, loops, and turns connect the alpha helices and beta sheets, and form different kinds of stereotypic compact structures. When these structures involve sequential arrangements of two or three elements they are referred to as *structural motifs*.

Domains are structural units formed by sequential combinations of more than three secondary structures, and by stable associations of two or more structural motifs. Typical examples of a domain is the four-helix bundle formed by two pairs of alpha helices connected by a loop; another typical domain is the five-element, alternating arrangement of beta sheets and alpha helices known as a *Rossman fold*, and still another is a two pairs of beta sheets known as a *Greek key*. Domains are semi-independent folding units, and for that reason they are often referred to as *domain folds*.

2.6 Arrangement of Protein Secondary Structure Elements and Chain Topology

A number of empirical rules can be formulated that summarize the packing of amino acid sequences into three-dimensional folding domains. These rules are quite general and summarize the observation that secondary structures pack into one of a small number of geometries and the chains assume one of an equally small number of topologies. For example, beta sheets form layered structures with either alpha helices or other beta sheets on their faces. Alpha helices either distribute themselves about a core or form layered structures. The packing of helices about a core can be described by regular polyhedra. Three helices pack into an octahedron, four helices describe a dodecahedron, five helices form a hexadecahedron, and six helices pack into a icosahedron. The packing of sheets is more variable. Most beta sheets pack into two-layered structures, with the sheets either aligned or orthogonal. Some beta sheets, notably in α/β proteins, form barrels. In barrel patterns the β sheets coil around to form a cylinder and the α helices are arranged on the cylinder surface.

The rules for chain topology state that knots in the polypeptide chain do not form nor do secondary structure elements cross through one another. Instead, the chain topology is minimally convoluted. Secondary structure elements that are adjacent in the polypeptide sequence prefer to pack in an antiparallel manner, and groupings of the form β-X-β, where X is either an α-helix or a β-strand in an adjacent sheet, are right handed. The helices pass through the folding domain and so the connections between helices form on the outside along the ribs of the polyhedra. Beta sheet topology forms hairpin, Greek key, or jellyrolls, depending on the number of strands (respectively, two, four, six).

2.7 Tertiary Structure of a Protein: Motifs and Domains

Proteins involved in signaling tend to be fairly large. They are constructed from a number of structural motifs and domains connected by loops that serve as flexible linkers. The Src protein shown in Figure 2.7 serves as an example of domain organization. Domains operate as functional units, and when proteins are formed with a specific set of domains they inherit those functions. Some signaling proteins contain just a few domains, while others contain large numbers of domains and because of that are termed *mosaic proteins*. Domain folds typically contain from 100 to 250 amino acids, but can be as small as 25 to 30 amino acids. In mosaic proteins, small domains along with larger ones often appear as tandem repeated sequences. Prokaryotic genes average 850 to 1000 bp in length while eukaryotic genes average 1400 to 1450 base pairs (bp) in size, and thus contain a greater number of these semiautonomous folding units.

The tertiary structure of a protein refers to the way the constituent domains and supersecondary structure elements come together to form the folded and physiologically functional protein. For most proteins this is the final layer of protein organization, and proteins are classified into structural families based on their domain composition. Listed in Table 2.2 are some of the more common families of proteins encoded in the human genome as

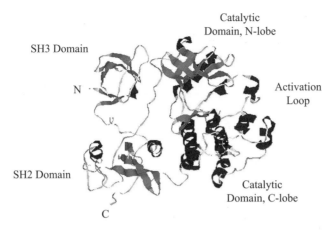

FIGURE 2.7. Domain organization of the Src protein, a nonreceptor tyrosine kinase: It consists of four domains—an N-terminal SH3 domain, a C-terminal SH2 domain, and two catalytic domains, connected by linker segments. Amino acid residues that form well-defined secondary structure elements are drawn as (dark) coiled ribbons and as (grey) planar ribbons or arrows. The coiled ribbons, like the cylinders used in the previous figure, denote alpha helices. The planar ribbons or arrows form layers that denote beta sheet secondary structures, either parallel or antiparallel. The SH2 domain is of the form $\alpha + \beta$; its secondary structure resembles a two-layered sandwich with two short alpha helices plus a central beta sheet. The SH3 domain contains a pair of antiparallel beta sheets. Tyrosine kinases catalyze the transfer of phosphoryl groups to selected tyrosine residues. As can be seen in the figure the catalytic domains are mostly alpha helical, except for the N-lobe, where a prominent beta sheet component is present. Tyrosine kinases will be discussed in Chapter 11. The figure was generated using Protein Explorer with the Brookhaven Protein Data Bank (PDB) entry (accession number) 2Src containing the atomic coordinates of Src determined by x-ray crystallography (to be discussed in Chapter 3).

defined by the presence of one or more of the domains listed in the first column. The sizes vary from as few as 25 to 33 amino acid residues for the minidomains to 120 amino acid residues for the PH domain. The domains are not mutually exclusive, and a single protein will usually contain one or more of several different kinds of domains. Minidomains are often arranged in the proteins as tandemly repeated elements. Including the term "repeat" as part of the name reflects this propensity.

Several kinds of changes have taken place in the human genome. New families of vertebrate genes appear that encode proteins belonging to the immune and nervous systems, and there is a new family (KRAB) of zinc finger transcription factors. A more widespread kind of change is the creation of new architectures, that is, of new arrangements of domains that can subsume new functions. A third type of change is the large expansion of existing families. Listed in Table 2.2 are representative examples of each of

TABLE 2.2. Commonly encountered domains in the human: Sizes of the domains are given in terms of the numbers of amino acid residues. The fourth column lists the number of genes encoding the domains. All of these domains have signaling roles. The last column indicates the most prominent activities associated with proteins containing these domains. Im: immune system function; Extra: Extracellular adhesion; Trans: transcription regulation.

Domain	Symbol	Size	Genes*	Function
Immunoglobulin	Ig	100	765	Im, Extra
EGF-like	EGF	35	222	Im, Extra
Fibronectin Type III	Fn3	90	165	Im, Extra
EF hand	EF	40	242	Im, Extra
Ankyrin repeat	Ankyrin	33	276	Im, Extra, Signal
Leucine-rich repeat	LRR	25	188	Im, Extra, Signal
Cadherin	Cadherin	110	114	Im, Extra, Signal
WD-40 repeat	WD40	50	277	Signal
Pleckstrin homology	PH	120	193	Signal
Src homology 3	SH3	60	143	Signal
Src homology 2	SH2	100	119	Signal
PDZ	PDZ	80	162	Signal
C2H2 zinc finger	C2H2	30	706	Trans
Homeobox	Homeobox	60	267	Trans
RING finger	RING	50	210	Trans
Krueppel-associated box	KRAB	75	204	Trans

* Data from International Human Genome Sequencing Consortium [2001]. *Nature* 409: 860–921; Venter JC et al. [2001]. *Science*, 291: 1304–1351.

these categories. The table includes domain families prominently associated with immune and extracellular functions such as adhesion to the extracurricular matrix, the ECM, and also large numbers of domains that mediate interactions between control layer proteins and between proteins and DNA. These domains and the signaling proteins that contain them will be examined in detail in later chapters.

2.8 Quaternary Structure: The Arrangement of Subunits

Signaling proteins are frequently assembled from distinct polypeptide chains. In these situations, the protein is said to have a *quaternary structure*. "Quaternary structure" describes how the individual chains, or subunits, are arranged in the protein. Several kinds of multichain proteins play important roles in the control layer. One type of multichain protein is the ion channel, where several subunits form a membrane-spanning pore that permits passage of ions and small molecules across the membrane. In these proteins, the subunits are tightly bound to one another. Other classes of proteins embedded in the plasma membrane, and serving as receivers of cell-to-cell signals, are assembled from multiple subunits that are much more loosely bound to one another. This kind of protein organization is widely encountered in the immune system, and in these cases, unlike the ion channels, the different chains may perform distinct functions.

Some of the large signaling proteins that reside in the cytosol and serve as central organizers of the signal pathways are constructed from multiple subunits. Specific functions are sequestered in the different subunits of the proteins, and the ability of the protein's subunits to associate and dissociate from one another in a key part of how the signaling system works. Alternative splicing is often used to generate different forms of specific subunits. This makes possible the creation of many different combinations of subunits; each one specialized for a specific cell or tissue type. By this means many proteins of a similar kind, but each performing a slightly different task, can be constructed from a small instruction set.

2.9 Many Signaling Proteins Undergo Covalent Modifications

Proteins can associate with the plasma membrane in several ways. If they pass completely through the plasma membrane they are able to convey a signal from one side to the other. Most receptors work this way relaying signals from outside the cell to the inside. Alternatively, the proteins may attach to either the extracellular side or the intracellular side by means of the *tether*. The tether, a type of post-translational modification, passes into one of the leaflets but does not pass all the way through to the other side of the plasma membrane. The proteins so anchored tend to form clusters of signaling elements, which may relay signals to nearby proteins that pass through the plasma membrane or not, depending on the composition of the cluster.

One of the most striking features of the control layer is its widespread use of post-translational modifications. Some of the modifications are made during protein processing and finishing as discussed in the last chapter. This type of modification is common in proteins that function in the plasma membrane. But many others are made later and are part of the signaling process itself. There are several different kinds of modifications all involving either the covalent addition of a group or structure, or the cleavage of the protein or a part thereof.

The main modifications are:

- Covalent attachment of anchors that tether signal proteins to one side or the other of the plasma membrane.
- Addition of sugar groups to the extracellular region of transmembrane signaling proteins.
- Proteolytic cleavage of anchors and proteins to free up the proteins for movement and conveyance of a message, or, alternatively, to keep them inactive and degrade them.
- Covalent addition and removal of phosphoryl, methyl, and acyl groups to cytosolic proteins.

2.10 Anchors Enable Proteins to Attach to Membranes

The region at and just below the plasma membrane is an important locus of signaling molecules. In response to the onset of signaling, many cytosolic proteins translocate to the plasma membrane where they are anchored to the cytosolic face and become activated. There are four types of modifications to cytosolic proteins that enable them to anchor to the plasma membrane and to the membranes of organelles: myristoylation, palmitoylation, farnesylation, and geranylgeranylation (Table 2.3, and Figures 2.8 and 2.9).

A crucial feature of acyl and prenyl anchors is that they are weak. Electrostatic interactions between N-terminal basic (+) residues and acidic (–) lipids contribute along with the anchor to the attachment. Because of the weak attachment the proteins can be detached easily by altering the electrostatic environment. This is usually done by phosphorylation, the covalent attachment of a phosphoryl group (with two negative charges). The modification to the basic residues weakens the electrostatic forces sufficiently to enable the protein to detach from the membrane and translocate to the cytoplasm to carry out its signaling function. This allows for reversible attachment and operation as an electrostatic switch.

The addition of the aforementioned fatty groups to proteins makes possible the proteins' anchoring to the inner, or cytoplasmic, leaflet of the plasma membrane. A different kind lipid modification is made to proteins to allow them to be anchored to the outer, or exoplasmic, leaflet. These

TABLE 2.3. Membrane anchors: Abbreviations—cysteine (C); aliphatic residue (residues with long hydrocarbon side chains such as leucine and isoleucine) (a); X = serine (S), methionine (M), alanine (A), or glutamine (Q); leucine (L).

Location and type of anchor	Attachment
Inner Leaflet	
Acyl type	
Myristoyl	Preferentially attaches to glycine residues at the N-terminus
Palmitoyl	Preferentially attaches to cysteine residues variably located through a thioester link
Prenyl type	Covalently attaches through a thioester link to a cysteine residue located four residues from the C-terminus. The last three residues are removed and the new C-terminus group is methylated.
Farnesyl	C-terminal sequence CaaX
Geranylgeranyl	C-terminal sequence CaaL
Outer Leaflet	
Glycosylphosphatidylinositol (GPI)	Attaches to the C-terminal amino acid; directed by a signal peptide that is removed and replaced by the anchor

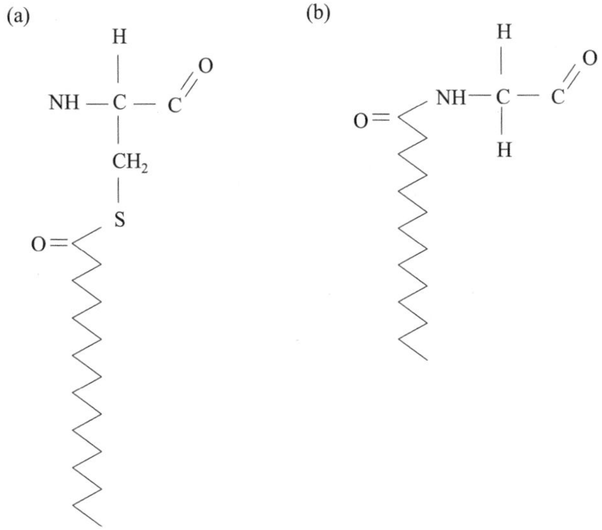

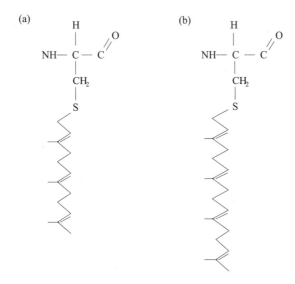

FIGURE 2.8. Acyl anchors covalently attached to membrane proteins: (a) A 16-C palmitoyl anchor is attached to a cysteine residue. (b) A 14-C myristoyl anchor is covalently attached to a glycine residue.

FIGURE 2.9. Prenyl anchors covalently attached to membrane proteins: (a) A 15-C farnesyl anchor is attached to a cysteine residue. (b) A 20-C geranylgeranyl anchor is covalently attached to a cysteine residue.

anchors are made from a complex sugar plus a phosphatidylinositol grouping, and are called GPI (glycosyl phosphatidyl inositol) anchors. The GPI anchors are preformed in the endoplasmic reticulum (ER) and attached to the newly synthesized proteins. Proteins bearing these anchors are localized on the cell exterior, where they can be easily freed up by proteolytic proteins called *sheddases*. These proteolytic proteins are given that particular name because they cleave, or shed, the exoplasmic domains, or ectodomains of their targets. The proteins, once freed of their anchors, function as soluble proteins. Dual form proteins, membrane-associated and soluble, are frequently encountered in immune and endocrine (hormonal) signaling.

2.11 Glycosylation Produces Mature Glycoproteins

Most proteins destined for insertion in the plasma membrane contain covalently linked oligosaccharides that extend out from their extracellular side. These proteins are referred to as *glycoproteins*. Their post-translational modifications are started in the ER and finished in the Golgi apparatus. There are two forms of modification, *N-linked* and *O-linked*. In N-linked glycoproteins, a carbohydrate is added to a side chain NH_2 group of an asparagine amino acid residue. In O-linked glycoproteins, the oligosaccharide chain is appended to a side chain hydroxyl group of a serine or threonine amino acid residue.

These modifications alter the binding properties of the signaling proteins. By adding or removing the carbohydrate groups, the affinity for a particular ligand can be either increased or decreased, and can even be shifted to favor one ligand over another. These properties have been explored for a prominent signaling protein called *Notch* that is involved in embryonic development. Notch signaling and other developmentally important processes will be explored in Chapter 13. What is of interest here is that rather than genetically encoding a variety of Notchlike proteins, each with slightly different ligand-binding properties, a small number of Notch proteins are expressed and their properties are then altered post-translationally as the need arises.

2.12 Proteolytic Processing Is Widely Used in Signaling

Notch undergoes additional modifications subsequent to ligand binding. It is proteolytically processed to form a diffusible messenger protein that translocates to the nucleus where it functions as a transcription factor. Proteolytic processing of membrane proteins to create mobile messengers is not limited to Notch, but instead is used in several signaling pathways. In this way, a single protein performs multiple tasks within a single signaling pathway. Because several signaling intermediates are eliminated, the signaling process is a rapid one requiring few steps to go from the initiating sensing stage to a terminating control point.

Proteolytic processing is not restricted to membrane proteins. It is utilized heavily to change the properties of cytosolic proteins from immobilized forms to mobile ones. Proteolytic processing is a key component of signaling pathways involved in regulating the cell cycle, embryonic development, and immune function. The mechanism is similar in all of the pathways. A crucial signaling element is parked in a specific location in the cell through its binding to an inhibitory protein. In response to activating signals, proteolytic enzymes (the 26S proteosome) chop up the inhibitory proteins, thereby freeing the signaling proteins for movement into the nucleus where they promote gene transcription. These mechanisms will be discussed in more detail when the specific pathways in which they appear are examined.

2.13 Reversible Addition and Removal of Phosphoryl Groups

Reversible protein phosphorylation is the preeminent mechanism used by all cells to convey a signal. It is used to direct cellular responses to the binding of hormones, growth factors, and neurotransmitters to receptors and ion channels at the cell surface. It is used to regulate metabolism, growth and differentiation, and learning and memory. In this process, a phosphoryl group is covalently attached to a specific residue on a target protein thereby modifying that protein's activity. ATP serves as the donor of the phosphoryl group. Protein kinases are enzymes that catalyze the transfer of a phosphoryl groups from the ATP molecules to protein targets. Another class of signaling molecules, protein phosphatases, does the opposite. Protein phosphatases catalyze the removal of phosphoryl groups, that is, they catalyze their hydrolysis.

Different amino acid residues serve as the primary recipients of phosphoryl groups in bacteria and eukaryotes. In bacteria, phosphoryl groups are transferred to aspartate and histidine residues forming His-Asp phosphorelays. In eukaryotes, there are two classes of protein kinases. One group catalyzes the covalent attachment of phosphoryl groups to serine and threonine residues and the other promotes the attachment to tyrosine residues (Figure 2.10). Protein kinases are central elements in most signal pathways and are present in large numbers of metazoan and plant genomes. The human genome encodes close to 900 eukaryotic protein kinases. These signaling elements will be introduced again in Chapters 6 (bacteria) and 7 (yeasts and other eukaryotes).

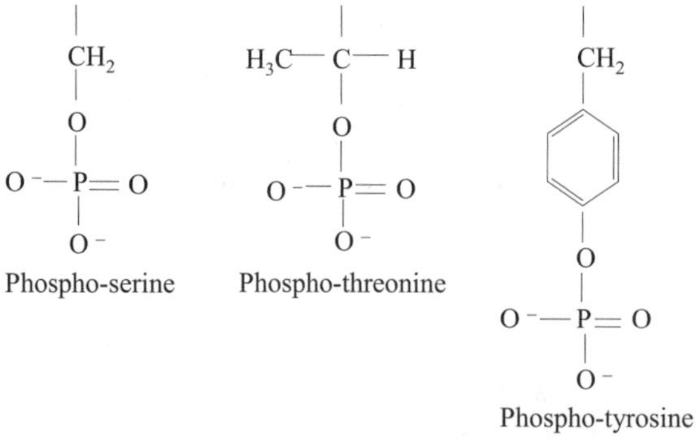

FIGURE 2.10. Phosphorylation of serine, threonine, and tyrosine side chains: Hydroxyl (OH) groups located at the ends of the side chains provide sites for covalent attachment of phosphoryl groups.

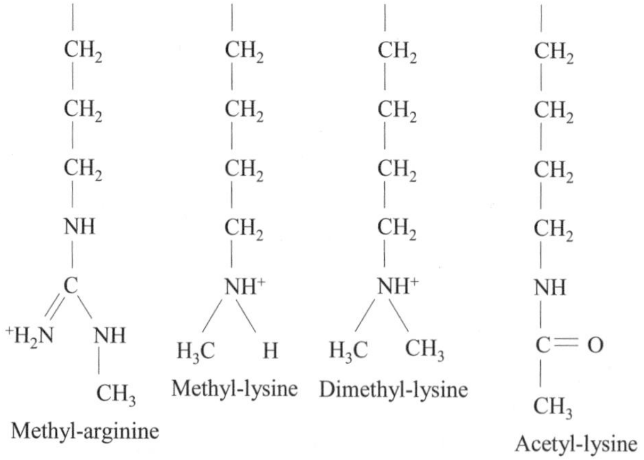

FIGURE 2.11. Covalent attachment of methyl and acetyl groups to arginine and lysine side chains: Amino groups located at the ends of arginine and lysine side chains provide sites for attachment of one or more methyl or acetyl groups.

2.14 Reversible Addition and Removal of Methyl and Acetyl Groups

Phosphoryl groups are not the only groups that are added and removed from signaling proteins as part of their cellular function. Methyl and acetyl groups are added and removed, as well. Whereas side chain hydroxyls provide binding sites for the phosphoryl groups, side chain nitrogens do the same for the methyl and acetyl groups. The transfer targets are the amino groups lying at the ends of the side chains of lysine and arginine residues. The amino groups provide multiple attachment sites for methyl groups. Several examples of methylation are presented in Figure 2.11. Two nitrogens are present at the ends of arginine side chains, allowing for a symmetric distribution of methyl groups between the two nitrogens, or, alternatively, for one or the other of the nitrogens to have most if not all of the methyl groups.

The enzymes that catalyze the transfer of methyl groups to arginine and lysine residues are referred to as *arginine methyltransferases* and as *lysine methyltransferases*, respectively. These enzymes use an endogenous cellular molecule called S-adenosyl-L-methionine, or SAM, as the methyl group donor. SAM is synthesized in cells of the body from methionine and ATP, and readily donates its methyl group to the guanidine groups on the arginine side chains and to the amino groups on the lysine side chains.

The amino groups at the end of lysine side chains are the main attachment sites for acetyl groups. The enzymes responsible for catalyzing the transfer of acetyl groups to lysines are known as *acetyltransferases*. These enzymes use acetyl coenzyme A (acetyl CoA) as the acetyl group donor. The acetyl group is attached by a sulfur atom to CoA forming a high-energy thioster bond, making it easy to transfer to acceptors such as lysine.

2.15 Reversible Addition and Removal of SUMO Groups

Small ubiquitin-related modifier (SUMO) is a member of the ubiquitin family of regulatory proteins. Like ubiquitin, it is covalently attached to a variety of proteins through the sequential actions of three sets of enzymes. Recall that ubiquitin is first activated in an ATP-dependent way by E1 enzymes. A thioester bond is formed between the C-terminal of the ubiquitin protein and the E1 ubiquitin-activating enzyme. The ubiquitin protein is then transferred to an E2 ubiquitin-conjugating enzyme, and then to the E3 ubiquitin protein ligase, which then transfers it to either the substrate or to multiubiquitin chains formed there. The substrate so tagged by one or more ubiquitin molecules is then degraded by the 26S proteosome.

The SUMO system of enzymes operates in a somewhat similar fashion. There is an E1 SUMO-activating enzyme, which consists of a heterodimer in place of the single polypeptide chain found for the ubiquitin E1. There is an E2 SUMO-conjugating enzyme, and there is an E3 SUMO protein ligase, which accelerates the direct transfer of SUMO from the E2 to the substrate. These attachments are illustrated in Figure 2.12.

Both ubiquitination and sumoylation have regulatory roles. Whereas ubiquitination of a protein generally tags that protein for destruction by the 26S proteosome, attachment of a SUMO group has a different role. It is a reversible process. It helps stabilize proteins and their interactions with

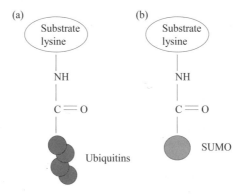

FIGURE 2.12. Ubiquitination and sumoylation: Conjugates are formed between lysine NH group and carbonyl groups at the C-terminal of ubiquitins and SUMO.

other proteins, and assists in localizing proteins to specific compartments within the cell.

2.16 Post-Translational Modifications to Histones

Post-translational modifications to histones are an important regulatory mechanism. Chromatin structure plays an important role in determining which genes are to be transcribed at any given moment in time. There are two forms of chromatin. Transcriptionally inactive chromatin, called *hete-rochromatin*, is tightly compacted. In heterochromatin, sites where transcription initiation and regulation take place are largely inaccessible to the proteins responsible for these actions. In contrast, transcriptionally active chromatin, or *euchromatin*, has a far more open shape. In euchromatin, sites where transcription factors and the transcription machinery bind are accessible to the responsible proteins.

As discussed earlier in this chapter, nucleosomes contain two each of several core histones. The amino terminals of these core histones extend out from the nucleosomes to form histone tails. These tails provide a means for other proteins to influence transcription. The histone tails provide sites for attachment of methyl groups, acetyl groups, and SUMO groups. The predominant targets of the post-translational modifications are the amino groups lying at the end of lysine side chains. The lysine residues, along with arginine residues, can be methylated, acetylated, sumoylated, or ubiquitinated. As shown in Figure 2.13, the tails of the histones are enriched in these residues, and thus supply multiple sites for attachment of these groups. Attachment of acetyl and other groups neutralizes the net positive charge on the tail regions, and this reduction weakens the attraction between the tails and the DNA. As a consequence the histone-DNA interactions are

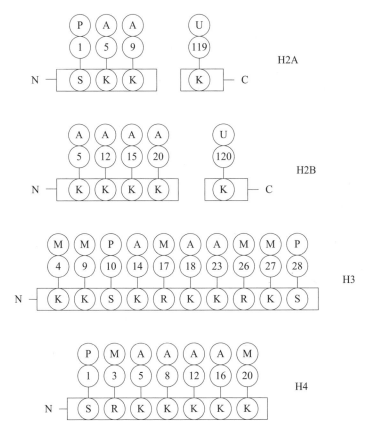

FIGURE 2.13. Covalent modifications of histone tails: Shown are regulatory sites modified by acetylation (A), methylation (M), phosphorylation, or ubiquitination (U). As is customary the sites are numbered in ascending order starting from the N-terminal. One-letter codes for the amino acids are arginine (R), lysine (K), and serine (S). (Three- and one-letter abbreviations for the amino acids are listed in Table 4.1.)

lessened allowing for a greater access of the DNA to transcription regulators. Similarly, addition of net negative charge through phosphorylation on serine residues also serves to decondense chromatin.

Many of the proteins that act in a supporting role to help activate or suppress transcription interact directly with the histones rather than with the DNA. Some of these enzymes function as histone acetyltransferases (HATs) or alternatively as histone deacetylases (HDACs), adding or removing acetyl groups from histone tails, using acetyl coenzyme A (acetyl-CoA) as the donor. Other coregulatory enzymes operate in a similar manner, adding or removing methyl groups or phosphoryl groups or SUMO groups or ubiquitin groups.

General References

Alberts B, Johnson A, Lewis J, Raff M, Roberts K, and Walker P [2002]. *Molecular Biology of the Cell*, 4th edition. New York: Garland Publishers.

Matthews CK, van Holde KE, and Ahem KG [2000]. *Biochemistry*, 3rd edition. San Francisco: Pearson Benjamin Cummings.

Stryer L [1995]. *Biochemistry*, 4th edition. New York: W.H. Freeman and Company.

References and Further Reading

Chromatin and the Nucleosome

Harp JM, et al. [2000]. Asymmetries in the nucleosome core particle at 2.5 Å resolution. *Acta Cryst.*, D56: 1513–1534.

Kornberg RD, and Lorch YL [1999]. Twenty-five years of the nucleosome, fundamental particle of the eukaryote chromosome. *Cell*, 98: 285–294.

Nuclear Organization

Lamond AI, and Earnshaw WC [1998]. Structure and function in the nucleus. *Science*, 280: 547–553.

Lewis JD, and Tollervey D [2000]. Like attracts like: Getting RNA processing together in the nucleus. *Science*, 288: 1385–1389.

Chromosome Organization

Cremer T, and Cremer C [2001]. Chromosome territories, nuclear architecture and gene regulation in mammalian cells. *Nat. Rev. Genet.*, 2: 292–301.

Manuelidis L [1990]. A view of interphase chromosomes. *Science*, 250: 1533–1540.

Protein Organization

Hovmöller S, Zhou T, and Ohlson T [2002]. Conformations of amino acids in proteins. *Acta Cryst.*, D58: 768–776.

Kleywegt GJ, and Jones TA [1996]. Phi/Psi-chology: Ramachandran revisited. *Structure*, 4: 1395–1400.

Post-Translational Modifications

Fortini ME [2000]. Fringe benefits to carbohydrates. *Nature*, 406: 357–358, and references cited therein.

McLaughlin S, and Aderem A [1995]. The myristoyl-electrostatic switch: A modulator of reversible protein-membrane interactions. *Trends Biochem. Sci.*, 20: 272–276.

Milligan G, Parenti M, and Magee AI [1995]. The dynamical role of palmitoylation in signal transduction. *Trends Biochem. Sci.*, 20: 181–186.

Peschon JJ, et al. [1998]. An essential role for ectodomain shedding in mammalian development. *Science*, 282: 1281–1284.

Resh MD [1999]. Fatty acylation of proteins: New insights into membrane targeting of myristoylated and palmitoylated proteins. *Biochim. Biophys. Acta*, 1451: 1–16.

Regulated Proteolysis

Brown MS, et al. [2000]. Regulated intramembrane proteolysis: A control mechanism conserved from bacteria to humans. *Cell*, 100: 391–398.

Maniatis T [1999]. A ubiquitin ligase complex essential for the NF-B, Wnt/Wingless, and Hedgehog signaling pathways. *Genes Dev.*, 13: 505–510.

Townsley FM, and Ruderman JV [1998]. Proteolytic ratchets that control progression through mitosis. *Trends Cell Biol.*, 8: 238–244.

Sumoylation

Müller S, et al. [2001]. SUMO, ubiquitin's mysterious cousin. *Nature Rev. Mol. Cell Biol.*, 2: 202–210.

Histone Modifications

Grunstein M [1997]. Histone acetylation in chromatin structure and transcription. *Nature*, 389: 349–352.

Jenuwein T, and Allis CD [2001]. Translating the histone code. *Science*, 293: 1074–1080.

Problems

These problems require use of a PC to run the *Protein Explorer*, the free software used to model the Src protein as shown in Figure 2.7. The first step is to install "Chime" to enable your browser to talk to the *Protein Explorer*. Then bring up the *Protein Explorer*. (You may want to save the URL under your "Favorites" pull-down.)

2.1 Bring up the atomic coordinates for the Src protein displayed in Figure 2.7. Its Protein Data Bank (PDB) accession number is 2src. Next remove the water molecules and ligands leaving just Src. Then display the image in the form shown in Figure 2.7. Hint: Go to "Quick Views" and use the "Display" feature to bring up the "cartoon" depiction of the secondary structure. Then drag on the image to rotate into the orientation presented in Figure 2.7.

2.2 Click on "Mol Info" and then the "Header" button to bring up additional information about the protein Src. Note the list of amino acids comprising the primary structure.

2.3 Returning to the image of the Src protein. Bring up a "backbone" display of the protein. What does this image show?

2.4 Now, bring up a "ball-and-stick" model of Src. What do the balls represent and what do the sticks denote? Click on some of the balls shown in the image. What happens when this is done?

2.5 Change the display to a "space-fill" model. What do the spheres now mean; that is, what do their radii represent? How is hydrogen treated?

3
Exploring Protein Structure and Function

A variety of methods have been developed that enable researchers to study the structure and function of proteins at the atomic, molecular, intermolecular and cellular levels. Some methods exploit differences in charge and mass of the proteins in order to distinguish one kind of protein from another. Other methods exploit the many kinds of interactions occurring between electromagnetic radiation and biomolecules. Depending on the wavelength and the properties of the target material, electromagnetic radiation will be scattered and diffracted, absorbed and emitted. In all, an ensemble of methods is used in the laboratory not only to explore protein structure, but also to investigate posttranslational modifications, intermolecular interactions, cellular localization, and pleiotropy.

The preeminent methods for exploring the shape and internal structure of proteins at atomic level detail are X-ray crystallography and nuclear magnetic resonance (NMR). These techniques provide detailed three-dimensional information on how the proteins are organized into their functionally distinct domains and motifs, and what happens when one protein binds to another. They identify which amino acid residues are critical for protein-protein and protein-DNA binding, and they reveal the functional consequences of mutations of specific amino acid residues. As will be seen in later chapters, proteins involved in intracellular signaling often possess catalytic domains that stimulate the transfer of phosphoryl groups from one protein to another. X-ray crystallography and NMR show how these catalysts work, and how their activities are regulated.

The goal of techniques such as gel electrophoresis and DNA microarrays is to examine which proteins are expressed at higher levels and which ones at lower levels in response to specific kinds of signaling events and conditions. That is, they provide intermolecular and cellular level details. The mass spectrograph, often used in conjunction with these methods, permits the researcher to determine with high resolution the masses of proteins that have been isolated by gel electrophoresis and the microarrays. Another method, the yeast two-hybrid method, has become the leading method for determining which sets of proteins interact with one another to form the

TABLE 3.1. Methods using electromagnetic interactions to explore protein structure and interactions.

Experimental method	Level of detail	Process
X-ray crystallography	Atomic	X-ray diffraction
Circular dichroism	Molecular	Absorption of polarized UV light
Fluorescence resonance energy transfer	Intermolecular	Visible light absorption and emission
IR and Raman spectroscopy	Molecular	Absorption (IR) and scattering (Raman) of IR light
NMR spectroscopy	Atomic	Nuclear spin flips

TABLE 3.2. Physical methods used to explore protein structure and interactions.

Experimental method	Level of detail	Process
Yeast two-hybrid	Intermolecular	Protein-protein interactions
Gel electrophoresis	Molecular, Intermolecular	Mass/charge separation
Mass spectrograph	Molecular	Mass/charge separation
DNA microarrays	Cellular	Complementary base-pairing

signaling pathways in the cell. And another method, fluorescence resonance energy transfer, is used to explore protein interactions and view the movements of the signaling proteins in the cell. All of the methods listed in Tables 3.1 and 3.2 will be discussed in this chapter, some to a greater extent than others. The discussion will start with a review of the electromagnetic spectrum and how electromagnetic waves interact with matter. X-ray crystallography, NMR, and FRET will be explored next, followed by discussions of the physical methods.

3.1 Interaction of Electromagnetic Radiation with Matter

Electromagnetic radiation interacts with matter in a variety of ways. Recall that the wavelength and frequency of an electromagnetic wave are inversely proportional to one another, and the speed of light is the constant of proportionality. That is,

$$v = \frac{c}{\lambda}, \tag{3.1}$$

where λ (cm) is the wavelength, v (cycles/s) denotes the frequency, and c represents the speed of light, equal to 2.99×10^{10} cm/s. Electromagnetic radiation is quantized into discrete packets called *photons*. The energy E of each packet is given by *Planck's formula*:

$$E = h\nu, \tag{3.2}$$

where $h = 6.626 \times 10^{-27}$ erg-sec is Planck's constant. According to Eqs. (3.1) and (3.2), as the frequency increases, or equivalently the wavelength decreases, the photon energy goes up.

Electromagnetic radiation is not limited to a narrow range of wavelengths, but rather spans a broad range of wavelengths from less than a nanometer to more than a meter. The shortest waves are the gamma rays (γ-rays) emitted by atomic nuclei, and the longest waves are radio waves emitted by charged particles as they move back and forth in, for example, interstellar gases. The continuum of different kinds of radiation, each characterized by a unique wavelength, is known as the *electromagnetic spectrum*. The middle portion of the electromagnetic spectrum contains the infrared and ultraviolet regions with the visible range sandwiched in between. These three regimes are the most important portion of the spectrum with respect to biological systems.

According to the formulas just presented, as wavelengths decrease, photon energies increase. The energies corresponding to the different wavelength regimes are presented in Figure 3.1 in units of kcal/mol to facilitate comparison to familiar bonding energies, which are usually given in these units—the energies of covalent bonding are on the order of 100 kcal/mol. Noncovalent interactions such as hydrogen bonds have energies in the range 1 to 10 kcal/mol, and thermal energies are roughly 0.6 kcal/mol.

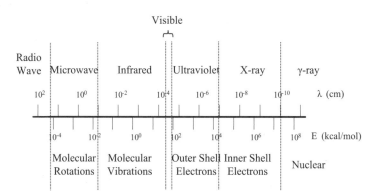

FIGURE 3.1. Electromagnetic spectrum: Shown in the upper portion of the figure are the types of radiation emitted. The kinds of transitions, or motions, that absorb and emit the radiation are shown in the lower part of the figure. Vertical dashed lines delineate the boundaries between the different regimes. The actual boundaries between radio waves and microwaves, and between ultraviolet and X-rays, are not sharp but instead the regimes merge into one another.

Molecules are not static entities but instead are in constant thermal motion, gaining and losing energy through random collisions with other molecules. The kinetic energy absorbed in the collisions is converted into vibrations and rotations about bond angles. These so-called *collective modes* can also be excited when the atoms and molecules absorb radiation in the appropriate wavelength range. The correspondence between wavelength regimes of electromagnetic radiation and protein motions (transitions) that produce the radiation is presented in the lower portion of Figure 3.1.

Recall that electrons move in specific atomic and molecular orbits, each with a well-defined energy. The lowest energy state of the electrons in their various orbits around a nucleus is called the *ground state*, and all others are referred to as *excited states*. Besides inducing vibrational and rotational activity, electrons can transition into excited states when radiation of the correct wavelength is absorbed. Electromagnetic radiation of well-defined energies is involved whenever electrons, atoms, and molecules undergo transitions from one state to another, either higher or lower. The energy of the absorbed and emitted radiation is equal to the difference in energies between the two energy levels (states), and its frequency (wavelength) is given by the following form of Planck's formula:

$$E_2 - E_1 = \Delta E = h\nu. \tag{3.3}$$

The absorption and emission of photons is depicted in Figure 3.2.

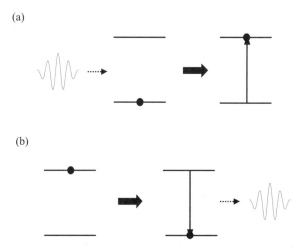

FIGURE 3.2. Absorption and emission of electromagnetic radiation: Horizontal lines represent energy levels of a simplified molecule possessing just a single ground state and a single excited state. From left to right: (a) Absorption—The molecule, initially in its ground state, absorbs a photon and undergoes a transition to an excited state. (b) Emission—The molecule, initially in its excited state, undergoes a transition to the ground state and emits a photon in the process.

3.2 Biomolecule Absorption and Emission Spectra

Biomolecules such as amino acids, peptides, and proteins absorb and scatter light in the ultraviolet and visible portions of the electromagnetic spectrum. Many of these biomoleucles selectively absorb light of particular wavelengths while scattering light at other wavelengths. These molecules act as pigments, imparting color to the materials in which they reside. The specific groups of atoms responsible for these absorption properties are known as *chromophores*, or "color bringers." The chlorophylls are prominent examples of chromophore-bearing molecules. They selectively absorb light in the blue and red portions of the spectrum, and scatter light in the green range. The result is the pronounced green color of plants.

Another common example of a biomolecule with striking chromatic properties is hemoglobin. Its absorption/scattering properties impart a red color to fully oxygenated blood while conveying a bluish tint to deoxygenated blood. Those electrons in a chromophore responsible for the light absorption are sensitive to their local environments, especially the presence of nearby charged groups. In the case of hemoglobin, the heme group forming the chromophore is sensitive to the presence or absence of bound oxygen atoms.

Photoreceptors form another class of chromophore-bearing proteins. The class includes *ospins* found in the mammalian retina along with *phytochromes* and *cryptochromes*, which enable bacteria, plants, and animals to adapt their cellular responses to local lighting conditions and undergo circadian rhythms. Opsins will be discussed in Chapter 12. Phytochromes and cryptochromes will be explored in Chapter 7. Protein chromophores can be built about the peptide bond, or about side chains, or upon prosthetic groups covalently attached chains that are nominally not part of the protein. Examples of prosthetic groups operating as chromophores are the heme groups and opsins.

When atomic groups absorb light, electrons are promoted into excited states (orbits). One of three things can then happen. The electrons may deexcite to the ground state without emitting radiation in a process called *internal conversion*. Alternatively, the excess energy may be lost through emission of photons having energies less than those of the absorbed light, with the remaining energy lost in a radiationless manner. If this happens rapidly, in a time scale of nanoseconds, the process is referred to as *fluorescence* (Figure 3.3), and the chromophores are called *fluorophores*. If, on the other hand, the lifetime of the excited state is long-lived, in the millisecond to second range, the process of absorbing and then emitting light is called *phosphorescence*.

3.3 Protein Structure via X-Ray Crystallography

Because of their short wavelength, X-rays can be used to explore the arrangements of individual atoms in proteins and other biomolecules. X-rays are produced whenever swiftly moving electrons strike a solid target.

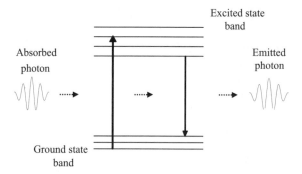

FIGURE 3.3. Schematic depiction of fluorescence: Shown are sets of related states, or bands, along with vertical arrows denoting transitions between states. From left to right, a packet of electromagnetic radiation is absorbed by a fluorophore, producing a transition from the ground state to an excited state. Shortly thereafter, the fluorophore deexcites back to the ground state, emitting a photon with a slightly lower energy and longer wavelength. Energy not carried off by the emitted photon is lost through radiationless transitions between vibrational states within the excited state band and within the ground state band.

In an X-ray tube, a beam of electrons is generated that strikes a metallic anode (typically copper) to produce an X-ray beam. Two types of X-rays are produced—characteristic X-rays that generate a line spectrum, containing a set of narrow strong peaks, and a smooth spectrum of continuum X-rays produced by coulombic interactions between the electron beam and the positive-charged nuclei (called *bremsstrahlung*, or *braking radiation*). In X-ray studies, characteristic K_α X-rays produced by transitions among inner shell electrons are typically used. These have a well-defined wavelength of about 1.5 Å, comparable to atom-atom bond lengths.

X-rays are scattered by the electron clouds of atoms, particularly by tightly bound electrons near the center of atoms, with no change in wavelength and no change in phase. Light scattered from single atoms is too weak to observe, but the amount of light can be amplified using purified crystals. In a crystal, large numbers of identical molecules are arranged in a regular lattice. Light passing through a crystal will be scattered in a variety of directions. Spherical wavelets scattered by different atoms will interfere, some constructively and some destructively. For certain wavelengths and scattering directions, the wavelets will be in phase to produce strong constructive interference. Constructive interference taking place between light waves as they are scattered off of the atoms serves to amplify the light, producing a characteristic pattern of light spots and dark areas, a diffraction pattern, that can be seen and analyzed to yield information on how the atoms are arranged.

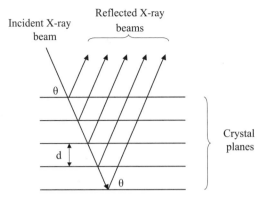

FIGURE 3.4. Scattering of X-rays from a series of planes in a crystal: A beam of X-rays impinges on a series of crystal planes. Some of the light is scattered back (reflected) from each of the planes while the remainder of the light is transmitted. The optimal situation where the angle of incidence equals the angle of reflection is depicted in the figure.

W.H. Bragg and his son W.L. Bragg formulated the basic theory of relating diffraction patterns to atomic positions in a crystal in 1912–1913. They noted that a three-dimensional crystal could be viewed as a set of equidistant parallel planes. The conditions for maximal constructive interference are twofold. First, the scattering from each plane must be a specular, or mirror, reflection in which the angle of incidence equals the angle of reflection. This situation is depicted in Figure 3.4. Second, the X-ray wavelength λ, distance between parallel planes d, and angle of reflection θ, obey the relationship known as *Bragg's law*:

$$2d \sin \theta = n\lambda. \tag{3.4}$$

In Eq. (3.4), n is an integer that can take on the values 1, 2, 3, and so on. When $n = 1$, the spots of light are known as *first order reflections*, and when $n = 2$, they are called *second order reflections*. First order reflections are more intense than second order reflections, and similarly for third and higher order contributions.

The light spots are produced when the light waves arrive at the detector in phase with one another and thus constructively interfere. In an X-ray diffraction experiment, the intensities and positions of the light spots are recorded. The diffraction patterns are converted into electron density maps through application of a mathematical operation known as a *Fourier transform*. Several tens of thousands of reflections are collected in a typical X-ray diffraction experiment. Computer programs, taking as input the resulting electron density map and knowledge of the primary sequence, are

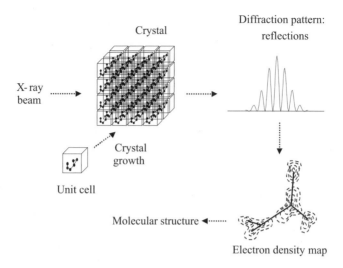

FIGURE 3.5. X-ray crystallography: A unit cell is shown containing the molecule to be studied. It is replicated many times in the crystal sample. X-rays scattered from the atoms in the crystal produce a complex three-dimensional pattern of light and dark spots in the detector called *reflections*. The patterns of light and dark spots are a consequence of the way the waves from different atoms in different locations in the unit cell interfere with one another. A typical diffraction pattern produced in one-dimension from a pair of atoms is depicted in the figure. As can be seen, there is a series of peaks where the waves constructively interfere, and each peak is separated from the next by a trough where the waves from the two atoms destructively interfere. These reflections are then converted into an electron density map. A typical portion of an electron density map is depicted. The map, a contour plot, shows peaks where the electron density is high and valleys or open areas where the electron density is low. In the model-building portion of the data analysis, the three-dimensional structure of the molecule is deduced using computer programs.

used deduce the three-dimensional arrangement of atoms in the protein (Figure 3.5).

X-ray crystallographic data have been used to deduce the atomic structure of more than 16,000 proteins to date. In 1953, Crick and Watson deduced the double helix structure of DNA using the X-ray crystallographic data of Franklin and Wilkins. This discovery was followed by those of Perutz and Kendrew, who used X-ray crystallography to deduce the atomic structure of hemoglobin and myoglobin in the period from 1953 to 1960. These were the first two proteins to be so described in atomic detail. Since that time a rapidly increasing number of proteins have had their atomic structure solved. The three-dimensional coordinates (x, y, z) for each atom in the protein are deposited in the Protein Data Bank (PDB) and made available to the research community. The PDB repository was established at Brookhaven National Laboratory in Long Island, and at several mirror sites throughout the world. Of the more than 16,000 structures for proteins and peptides that have been deposited in the PDB to date, 14,000 were determined using X-ray crystallography and 2000 using nuclear magnetic resonance (NMR) spectroscopy.

3.4 Membrane Protein 3-D Structure via Electron and Cryoelectron Crystallography

The overall limitation of X-ray crystallography is the growing of purified crystals with sufficient numbers of molecules. This outcome is exceptionally difficult to achieve in the case of large and complex protein molecules such as those embedded in membranes that pass back and forth through the lipid bilayer several times. These proteins have a hydrophobic band that tends to destabilize them when they are removed from the lipid bilayer and solubilized. In electron crystallography, both the native environment and the biological activity of the membrane proteins are preserved.

The central idea in electron crystallography is similar to that which drives X-ray crystallography. By growing a crystal containing a large number of the molecules of interest one can greatly amplify the crystal's weak signals. In this approach, two-dimensional membrane crystals are prepared. Electron diffraction patterns are then acquired using high-resolution cryoelectron microscopes. This method has been used to study light-harvesting complexes that function as antennae of solar energy in plants, and to investigate the atomic structure of porins of gram-negative bacteria. The method was first applied to the study of bacteriorhodopsin, the light-driven proton pump located in the purple membrane of *Halobacteria* and, more recently, to the analysis of the three-dimensional structure of the acetylcholine receptor, a neurotransmitter-gated ion channel located in nerve-muscle synaptic membranes.

3.5 Determining Protein Structure Through NMR

Nuclear magnetic resonance (NMR) spectroscopy is the second major experimental method used to determine the three-dimensional atomic structure of proteins. In this method, one utilizes pulses of electromagnetic radiation in the radiofrequency (RF) range of the EM spectrum. NMR pulse frequencies are typically in the range 300 to 600 MHz. These frequencies correspond to wavelengths of 100 to 50 cm, and the corresponding RF energies are 10 orders of magnitude weaker that the X-ray energies. Photons of these energies are used to induce transitions between nuclear (proton and neutron) spin states. The subsequent relaxation of the nuclear spin distributions back to their equilibrium distribution is sensitive both to the electron distribution surrounding the nucleus and to neighboring nuclear spins. As a consequence, one can deduce atomic structure of a protein from an analysis of its NMR spectra.

Electrons, protons, and neutrons have an intrinsic angular momentum, or *spin*. Because they have a spin angular momentum, they have a magnetic dipole moment and can interact with an external magnetic field as depicted in Figure 3.6. Like electrons, protons and neutrons have a spin of $1/2$. Nuclei with an odd number of neutrons and an even number of protons, or alternatively, an odd number of protons and an even number of neutrons will have a net spin of $1/2$. Spin $1/2$ nuclei encountered in biomolecules are ^1H, ^{13}C, ^{15}N, ^{19}F, and ^{31}P. This is not the only nonzero spin value possible. Nuclei with an odd number of proteins and also an odd number of neutrons, like ^2H, have a spin of 1. In Figure 3.6, spin is indicated by I_z.

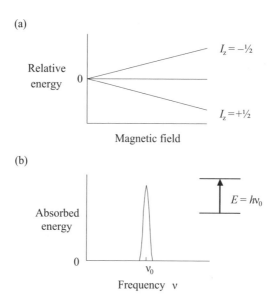

FIGURE 3.6. Splitting of energy levels of a spin $1/2$ particle by an external magnetic field: (a) In the absence of a magnetic field, the energies of the spin up and spin down states of the spin $1/2$ particle are the same. The energy levels are no longer the same in the presence of a magnetic field. The energy of the particle whose spin is aligned parallel to the magnetic field, the spin $+1/2$ particle, is lower than the particle whose spin is aligned against the external magnetic field, that is, the spin $-1/2$ particle. The energy levels are said to be split by the magnetic field. (b) If the spin $1/2$ particles are exposed to a source of electromagnetic energy, while in the magnetic field, there will be a sharp absorption peak when the frequency of the electromagnetic wave exactly matches the frequency of the photon needed to trigger a transition from the lower to the higher energy level.

The negatively charged electron clouds surrounding nuclei reduce the magnitude of the magnetic field experienced by the spin ½ protons and neutrons residing in the atomic nucleus. This effect is called *shielding*. Each atomic species has a uniquely different electron cloud, and thus different atoms will undergo different line splittings in the presence of the same external magnetic field. Nearby atoms will influence the energy splitting through the shielding effects, as well. Atoms such as oxygen that are strongly electronegative will have a far greater influence on neighboring atoms than will, for example, carbon atoms.

Shielding by the atom's electron cloud and those of its neighbors give rise to several different observable effects. The electron clouds will alter the locations of the peaks; they will create clusters of peaks in place of single isolated peaks and produce variations in the peak heights. The creation of clusters of peaks arises through interactions of the spins (magnetic fields) of the atoms in one group with the spins (magnetic fields) of atoms in neighboring groups. In more detail, interactions between spins split the energy levels, and the transitions between these energy levels appear as the distinct peaks in the plot of chemical shifts. An example of this, for the case of interactions between a methyl group and a methylene group, is presented in Figure 3.7. If the methyl group were isolated from its neighbors, there would be a single hydrogen peak, since each of the three hydrogen atoms sees the same environment. The presence of a nearby CH_2 group induces the splitting of the single peak into three peaks, and similarly, the presence

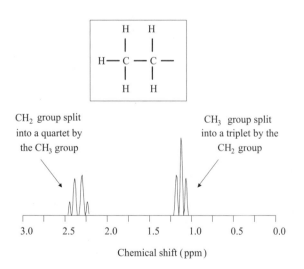

FIGURE 3.7. Chemical shifts of hydrogen atoms in a CH_3CH_2 molecular environment: The arrangement of atoms in the two groups is depicted in the insert. The chemical shifts are plotted with respect to a reference shift, taken as the zero of the axis, and are given in units of parts per million (ppm).

of the nearby CH_3 group induces the splitting of the single peak for CH_2 into four peaks. Variation in peak heights is not arbitrary, but instead is a direct consequence of the relative contribution to each peak from different spin-spin interactions (see Problem 3.4).

3.6 Intrinsic Magnetic Dipole Moment of Protons and Neutrons

Protons and neutrons, like electrons, have an intrinsic magnetic dipole moment. The magnitude of the magnetic moment for a proton is

$$\mu_p = 2.79\mu_N,$$

and for a neutron it is

$$\mu_n = -1.91\mu_N.$$

The quantity μ_N is known as the *nuclear magneton*. It is defined in a manner analogous to the electron dipole moment as

$$\mu_N = \frac{e\hbar}{2M_p}, \tag{3.5}$$

where e is the unit charge, $\hbar$ is Planck's constant divided by 2π, M_p is the proton rest mass, and c is the speed of light. Since the mass of a proton is 1837 times the mass of an electron, the magnetic dipole moment of a proton or neutron is $1/1000$ that of an electron.

A free proton at rest in a uniform static magnetic field B_0 oriented along the z-axis can occupy one of two spin states. It can either be in a spin up state ($I_z = +1/2$) or a spin down state ($I_z = -1/2$). The energy associated with lower energy spin up state is

$$E_+ = -\mu_p B_0,$$

and that of the higher energy spin down state is

$$E_- = \mu_p B_0.$$

If an RF pulse of electromagnetic energy is applied that matches the transition energy, or energy difference, between these two states, the spin can be flipped from the lower energy state to the higher. The proton will then relax back to its lower energy state. The resonant frequency of the RF pulse that triggers the spin transition is

$$\nu_0 = \frac{2\mu_p B_0}{h}. \tag{3.6}$$

Nuclear magnetic resonance of protons in bulk matter was first demonstrated in 1946 by Felix Bloch and, independently, by Edward M. Purcell.

Their efforts followed earlier work by Isidor I. Rabi, who discovered the NMR effect in molecular beam experiments in 1937. Norman F. Ramsey, working in Rabi's laboratory, acquired the first radiofrequency (RF) spectra the following year, and developed the chemical shift theory in 1949. These initial studies have evolved into a powerful technique used to study biomolecules in solution, and to a diagnostic tool, magnetic resonance imaging (MRI), used in medicine.

3.7 Using Protein Fluorescence to Probe Control Layer

Fluorescent proteins can be used as sensitive probes of the movements and interactions of control layer elements. Proteins that fluoresce emit light at a characteristic emission wavelength, λ_{em}, shortly after absorbing light at their characteristic excitation wavelength, λ_{ex}. The wavelengths at which both absorption and emission occur fall in the short to middle portion of the visual spectrum. An energy level diagram, a schematic depiction of the energies of the low-lying electron orbitals of the protein's fluorophore, is presented in Figure 3.3. As shown in the figure, there is a ground state band consisting of number of closely spaced energy levels, and an excited state band, also consisting of a cluster of energy levels. Levels within the excited state band are populated when light is absorbed. The emission of light corresponds to electronic transitions from the excited state band back down to the ground state band. The energies of absorbed and emitted photons differ slightly. The emitted photon has a greater wavelength, reflecting its slightly lower energy.

Green fluorescent protein (GFP) and other naturally fluorescent molecules are found in organisms ranging from the bioluminescent jellyfish *Aequorea victoria* to the non-bioluminescent coral *Discosoma striata*. The ability of GFP to fluoresce is due to the presence of a fluorophore consisting of a sequence of three amino acid residues—Ser 65-Tyr 66-Gly 67. This internal sequence is post-translationally modified to a ring structure and the tyrosine is oxidized. These two changes convert the trio of amino acid residues into a fluorescing center. The natural form, or wild type, of GFP has an excitation maximum at 395 nm, a secondary excitation maximum at 470 nm, and an emission peak at 509 nm. These peaks correspond to green light. The protein extracted from *Discosoma striata* fluoresces in the red range, and is named dsRed. In addition to these naturally occurring fluorescent proteins a number of variants have been created with altered properties that improve their utility as markers. The excitation and emission peaks for several fluorescent protein variants are presented in Table 3.3.

The DNA sequence for the 28 kDa green fluorescent protein has been determined. To make a fusion protein, the complementary DNA, or cDNA, of the protein of interest is inserted into a vector along with the DNA for the GFP. When the combined gene is expressed in a transfected cell (the

TABLE 3.3. Spectral properties of green fluorescent protein variants.

GFP variant	Color	λ_{ex} (nm)	λ_{em} (nm)
eBFP	blue	380	440
eCFP	cyan	434	476
eGFP	green	488	509
eYFP	yellow	514	527
dsRed	red	558	583

cell into which the vector was inserted) the resulting fusion protein will contain the GFP covalently attached to the study protein, in a way that does not interfere with the normal operation of that protein. The fusion protein functions as a fluorescent reporter. When excited either by a laser or by an incandescent lamp, the protein will emit light and thus report its presence. Fusion proteins made in this manner have been used to study how signaling proteins move about the living cell. They have been employed to visualize the subcellular compartmentalization of the signaling molecules, and to view how the proteins move from one locale to another, in and out of organelles, and back and forth to the plasma membrane.

In *Aequorea victoria*, GFP operates in close association with another protein, aquorin, a naturally luminescent protein that emits blue light. Energy is transferred in a radiationless manner from the fluorophore of aquorin to the fluorophore of GFP. The GFP absorbs the blue light and in turn emits green light. The process of radiationless energy transfer from one fluorophore to another is called *fluorescence resonance energy transfer* (FRET). This process can occur when the energy level differences involved in emission from one fluorophore overlap the energy level differences involved in excitation of the second fluorophore. That is, FRET can occur if the donor emission spectrum overlaps the acceptor excitation spectrum.

3.8 Exploring Signaling with FRET

Fluorescence resonance energy transfer can be used to explore signaling within the living cell. Fluorescence resonance energy transfer and related techniques can be used to study protein-protein interactions. In this approach, one fluorophore is fused to the first of two proteins and the other to the second protein. When the first protein is well separated from the second, the first protein will emit light at its characteristic emission wavelength upon stimulation. If the two fluorophores satisfy the conditions for FRET, there will be a shift in wavelength of the emitted light from that of the first protein (the donor) to that of the second (the acceptor) when the two proteins some into contact. These processes can then be studied with light microscopy. The nature of the emitted light is sensitive to relative orientation of the two fluorophores and to their spatial separation. These dependencies can be exploited in the design of the fusion proteins, so that the emission spectra reflect conformational changes and properties of the protein-protein interactions.

The principle behind the use of FRET for studying protein-protein interactions can be applied in a variety of ways to the study of signaling. As was the case for protein-protein interactions, two particles are brought into close proximity, one fused to a fluorescent donor and the other to a fluorescent acceptor. The sought after signaling interactions are then studied and quantified by measuring the changes in FRET efficiency. When the two fluorophores are far apart the efficiency is low, but this changes in the presence of the interaction of interest. This is the basis for studying post-translational modifications such as proteolytic cleavage and phosphorylation; the movement of small signaling intermediaries such as cyclic adenosine monophosphate (cAMP), Ca^{2+}, and cytochrome c; and it has been used to study the trafficking of proteins as they are secreted from the cell.

Several examples illustrating how the FRET principle is applied are presented in Figure 3.8. In the first example, that of phosphorylation by a protein kinase, a phosphorylation *reporter* is created. This (the reporter) protein contains an amino acid sequence that is typically phosphorylated by the kinase of interest. When the sequence is phosphorylated by the kinase a second domain in the reporter protein recognizes and binds to the phosphorylated sequence. The second domain contains a . fluorescent protein that acts as the acceptor, while the first domain contains the donor. When the protein is phosphorylated by the kinase and the second domain binds it, the fluorescent acceptor is brought closer to the donor and the FRET efficiency increases.

A similar procedure is followed for proteins called *cameleons,* .which report the presence of calcium, an exceptionally important small signaling molecule. In this case, calmodulin (CaM), a calcium-binding protein, is used in constructing the reporter. Again a donor and acceptor are appended; the donor is appended to CaM and the acceptor to a peptide known as M13 that binds tightly to CaM what the latter binds calcium. The third example presented in Figure 3.8 represents the converse situation, that of proteolysis. To study proteolysis, a protein is created with donor and acceptor fluorescent proteins. The two regions are connected to one another by an amino acid sequence that is characteristically cleaved by the proteolytic enzyme (protease) of interest. When the reporter protein is cleaved by the proteolytic protein, FRET ceases as the two fluorescent proteins drift apart and no longer interact.

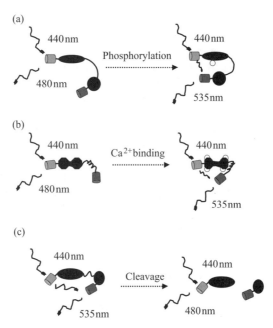

FIGURE 3.8. Fluorescence resonance energy transfer (FRET) used to explore intracellular signaling: (a) Phosphorylation—A laser emits radiation at 440 nm, triggering emission from a CFP at 480 nm in the left hand panel. The phosphorylation recognition domain connected to the first domain by a flexible linker binds the phosphorylated protein following kinase activity in the right hand panel. These changes result in FRET and an increased emission from the YFP at 535 nm. (b) Calcium binding—In the absence of calcium, a calcium-binding protein, calmodulin, is in a confirmation that loosely tethers an M13 peptide. When it binds four calcium ions, it assumes a dumbbell-like shape that induces a tight binding of the M13 peptide resulting in FRET and increased emission at 535 nm. (c) Proteolysis—In the absence of the protease, the two domains remain in close proximity and emit radiation at 535 nm. When the protease cleaves the linker, the two portions of the protein move apart and no longer exhibit FRET.

3.9 Exploring Protein Structure with Circular Dichroism

Another important property of biomolecules is their *chirality*, or *handedness*. All amino acids used to make proteins are left handed (L-amino acids) and all sugars used in DNA and RNA are right handed (D-sugars). This is not accidental. A consistent handedness on the part of protein-forming amino acids is essential for the proper folding of the polypeptide chains into compact three-dimensional shapes. Not all molecules possess a handedness. For this to occur a central atom surrounded by four different atoms or atomic groups must be present. The alpha carbon in the amino acids serves as the chiral center, and all amino acids with the exception of glycine satisfy the chirality condition. The left and right handed forms of chiral molecules are collectively referred to as *optical isomers* or as *enantiomers*—nonsuperimposable mirror images of one another.

Chiral biological materials are sensitive to the chiral behavior of left and right circularly polarized light and absorb the two light forms in different ways. In *circular dichroism* (CD), differences in absorption between left and right circularly polarized light are measured as a function of wavelength. These measurements provide information on the secondary structure content of the proteins. The measurements can be used to determine whether the protein is mostly alpha helical, or whether it is made up mostly of antiparallel beta sheets and turns. Circular dichroism utilizes the ultraviolet (UV) portion of the electromagnetic spectrum, typically in the range of 200 nm.

3.10 Infrared and Raman Spectroscopy to Probe Vibrational States

Transitions between rotational states in proteins are of fairly low energy and correspond to absorption of EM radiation in the microwave portion of the spectrum. Vibrational transitions are somewhat higher in energy. There are three kinds (modes) of vibrations—symmetric stretching, asymmetric stretching, and bending. In symmetric stretching, the stretching and contractions in bond length of a pair of bonds connecting a central atom to its two neighbors are of the same magnitude. In asymmetric stretching, the stretches and contractions are of unequal magnitude. The energies of the transitions between vibrational states lie in the infrared portion of the EM spectrum and can be explored using infrared (IR) spectroscopy and Raman spectroscopy.

In infrared spectroscopy, measurements are made of the frequencies of IR light that are being absorbed by the sample of interest. The IR wavelengths of most interest in explorations of biomolecules span the range from 2.5 to 15 microns, which is in the middle IR regime. The absorption spectra measured using IR spectrometers contain dips at wavelengths where the light has the correct energy to induce transitions between vibrational states and be absorbed. In Raman spectroscopy, inelastically scattered light is examined. The difference in energy between absorbed and emitted light is taken up by other vibrational transitions, and these transitions appear as spikes in the spectra.

3.11 A Genetic Method for Detecting Protein Interactions

The yeast two-hybrid system is a genetic method for detecting protein-protein interactions. There are a number of methods for detecting interactions between proteins. One of these, first introduced in 1989 and widely used since then, is the *yeast two-hybrid system*. This method involves the use of designed transcription factors (the hybrids) and reporter genes. Reporter genes are genes whose protein products have easily detected activities. Transcription factors are proteins that stimulate transcription. The transcription factors used in the yeast two-hybrid approach activate the reporters. The first of these to be used is the Ga l4 protein in the yeast *S. cerevisiae*. This protein stimulates transcription of the lacZ reporter gene that encodes the enzyme β-galactosidase, whose activity can be easily measured (assayed).

Ga l4, like many transcription factors, contains a DNA-binding domain and a transcription activation domain. (Transcription factors will be examined in detail in Chapter 16.) The modular organization of the Ga l4 protein is exploited to create a pair of hybrid proteins. One hybrid consists of the DNA-binding domain of Ga l4 fused to protein X. The other contains the activation domain of Ga l4 fused to protein Y. The two hybrids are reintroduced into the yeast cell. If protein X (the bait) interacts with protein Y (the prey) the Ga l4 DNA and activation domains will be brought into close proximity and the pair will be able to stimulate transcription of the reporter gene.

In the yeast two-hybrid method, different combinations of bait and prey are tested to identify interacting proteins. A common approach to carrying out large scale testing is to utilize two haploid yeast strains. One strain contains the bait and the other the prey. The strains are then mated and the interactions are determined by assaying the activity of the reporter gene in the diploid strain. This variant of the two-hybrid method is called *interaction mating* or *mating assaying*. It has been applied to several organisms whose genomes have been sequenced. For example, several hundred protein-protein interactions have been identified in this manner in *S. cerevisiae*.

3.12 DNA and Oligonucleotide Arrays Provide Information on Genes

One consequence of the work on complete genome sequences is the existence of libraries of genomes for many organisms. These libraries of DNA sequences form the basis for the newly developed *microarray technologies*. The physical basis for the microarray approaches is the complementary binding propensity of single-stranded nucleic acid sequences— nucleic acid bases preferentially bind to their complementary bases. A microarray is a glass slide containing single-stranded genetic material affixed in a regular grid. Each grid spot contains a single gene product that has been amplified (i.e., many copies made) in a polymerase chain reaction (PCR). A sample set of mRNAs from a cell of interest is then washed over the array. These nucleic acids will bind (hybridize) to the cDNA counterparts immobilized in the array yielding a profile of the particular gene being expressed in the cell.

In order to determine relative abundances the mRNA samples are labeled with a fluorescence dye. Typically, the sample of interest is labeled with a red fluorescent dye and there is a control sample that is labeled with a green fluorescent dye. By illuminating the spots with a laser and observing the color of the spot the relative abundances can be determined. Red would indicate a high abundance of the sample relative to the control, and green the opposite situation. Yellow would indicate equal abundances, and black a lack of hybridization by either sample or control at that spot.

DNA microarrays generate enormous amounts of data. They provide information on which sets of genes are expressed (and how strongly) at different times in the cell cycle and in response to which kinds of environmental stimuli. When interpreted and assembled into a regulatory model these kinds of data provide a molecular level explanation of how the cell or organism under study behaves.

3.13 Gel Electrophoresis of Proteins

Size and composition of protein complexes can be determined using one- and two-dimensional gel electrophoresis. In polyacrylamide gel electrophoresis, or SDS-PAGE, a polyacrylamide gel serves as an inert matrix over which proteins in a detergent solution migrate. Sodium dodecyl sulfate (SDS), the detergent, binds to hydrophobic portions of proteins causing then to unfold (denature) into extended chains, to dissociate from other proteins and lipid molecules, and for their subunits to unbind to each other. The large number of bound, negatively charged detergent molecules more than cancels any net positive charge on the protein. The overall negatively charged proteins migrate towards the positive electrode when a voltage is applied. The distance the protein moves is dependent upon the mass (size) of the protein, and so the proteins separate into bands arranged according to mass. The mass bands can then be analyzed to yield the protein masses and subunit composition.

In 2D gel electrophoresis, proteins are separated by both mass and charge. The proteins in the sample are again separated by a detergent, but this time the detergent is uncharged. The net charge on the now separated proteins is left unchanged. Recall that altering the pH of the solute will change the net charge on a protein, and proteins have an *isoelectric point*— the pH value at which the protein net change is zero. When electrophoresed in a gel in which a pH gradient has been applied the proteins will migrate to their individual isoelectric points and stop because at this point there is no force due to the electric field acting on the protein. When this procedure is followed with an SDS-PAGE procedure carried in a second direction orthogonal to that of the isoelectric focusing, the result is a set of proteins immobilized in a 2D gel separated by charge and mass.

3.14 Mass Spectroscopy of Proteins

Mass spectroscopy can be used to measure the masses of proteins, and to determine post-translational modifications. J.J. Thomson developed the mass spectrograph in the period from 1906 to 1913. He initially used his devices to study "canal rays," positively charged ions produced in a cathode ray tube. He later used the mass spectrograph to measure the masses of a variety of atomic species, establishing their discrete character, and then by this means established the existence of isotopes of stable elements. A number of researchers, most notably Aston, Dempster, Bainbridge, Mattuach, and Nier in the 1920s and 1930s further developed the discipline of mass spectrometry in which combinations of electric and/or magnetic fields are used to filter, disperse, and separate charged particles according to their charge (z) to mass (M) ratios.

All mass spectrometers have three main components: an ion source, a mass analyzer, and an ion detector. The source is responsible for producing a beam of ionized particles of the material to be mass analyzed. The analyzer separates the beam ions according to their M/z values, and the detector is responsible for their detection. The main breakthrough that lead to the use of mass spectrometry in the study of proteins was the development of techniques for producing beams of charged gaseous proteins. Two techniques are widely used. The first is *electrospray ionization* (ESI) and the second is *matrix-assisted laser desorption ionization* (MALDI). These methods were introduced during the 1980s, and can be employed for biomolecules with masses up to 50 kDa (ESI) and more than 300 kDa (MALDI).

In a mass spectrometer, a beam of particles with charge z moves with velocity **u**. The ions are first subjected to an accelerating voltage V in an ionization chamber, and, as a result, attain kinetic energies equal to zV, where z is the charge of the ion. They are then sent through a magnetic field that is applied in a direction perpendicular to the direction of motion of the particles. The particles experience a force equal to zuB, where B is the magnetic field strength, which is balanced by the centrifugal force. By subjecting the beam of charged biomolecules to a magnetic field, various M/z values are scanned. The basic expression is

$$M/z = r^2 B^2/2V. \qquad (3.7)$$

The voltage V is usually kept constant; the radius of curvature r is determined by the geometric properties of the magnet, and the user varies B to select different M/z ratios. In a further refinement of the technique, a pair of parallel electrodes can be added to the magnet. By adjusting the electric and magnetic fields according to the expression $r = 2V/E$, the velocity dependence can be removed so that all ionized proteins of a given M/z value fall on the same point in the focal plane of the detector.

In still another approach, a time-of-flight (TOF) instrument can be devised in which differences in arrival times are used to measure masses. In a TOF system all ions are accelerated to the same kinetic energy, and the time of flight t is measured over the flight length L. The M/z value is then determined according to the formula:

$$M/z = 2V(t/L)^2. \tag{3.8}$$

Books on Protein Structure, X-Ray Crystallography, and NMR

Brandon C, and Tooze J [1999]. *Introduction to Protein Structure*, 2nd edition. New York: Garland Science Publishing.

Drenth J [1998]. *Principles of Protein X-ray Crystallography*. New York: Springer-Verlag.

Hore PJ [1995]. *Nuclear Magnetic Resonance*. Oxford: Oxford University Press.

Levitt MH [2001]. *Spin Dynamics: Basics of Nuclear Magnetic Resonance*. New York: John Wiley and Sons.

Rhodes G [2000]. *Crystallography Made Crystal Clear: A Guide for Users of Macromolecular Models*, 2nd edition. San Diego: Academic Press.

Woolfson MM [2001]. *An Introduction to X-ray Crystallography*. Cambridge: Cambridge University Press.

References and Further Reading

X-Ray Crystallography, Electron Crystallography, and NMR

Campbell ID, and Downing AK [1998]. NMR of modular proteins. *Nat. Struct. Biol.* (Suppl.), 5: 496–499.

Kühlbrandt W, and Wang DN [1991]. Three-dimensional structure of plant light-harvesting complex determined by electron crystallography. *Nature*, 350: 130–134.

Unwin N [1993]. Nicotinic acetylcholine receptor at 9 Å resolution. *J. Mol. Biol.*, 229: 1101–1124.

Unwin N [1995]. Acetylcholine receptor channel imaged in the open state. *Nature*, 373: 37–43.

Wilson KS [1998]. Illuminating crystallography. *Nat. Struct. Biol.* (Suppl.), 5: 627–630.

Fluorescence Resonance Energy Transfer

Bastiaens PIH, and Pepperkok R [2000]. Observing proteins in their natural habitat: The living cell. *Trends Biochem. Sci.*, 25: 631–637.

Matz MV, et al. [1999]. Fluorescent proteins from nonbioluminescent *Anthozoa* species. *Nat. Biotechnol.*, 17: 969–973.

Pollok BA, and Heim R [1999]. Using GFP in FRET-based applications. *Trends in Cell Biol.*, 9: 57–60.

Weiss S [1999]. Fluorescence spectroscopy of single biomolecules. *Science*, 283: 1676–1683.

Circular Dichroism

Johnson WC [1990]. Protein secondary structure and circular dichroism: A practical guide. *Proteins*, 7: 205–214.

Woody RW [1995]. Circular dichroism. *Methods in Enzymology*, 246: 34–71.

Yeast Two-Hybrid System

Chien CT, et al. [1991]. The two-hybrid system: A method to identify and clone genes for proteins that interact with a protein of interest. *Proc. Natl. Acad. Sci. USA*, 88: 9578–9582.

Finley RL, Jr., and Brent R [1994]. Interaction mating reveals binary and ternary connections between *Drosophila* cell cycle regulators. *Proc. Natl. Acad. Sci. USA*, 91: 12980–12984.

Ito T, et al. [2001]. A comprehensive two-hybrid analysis to explore the yeast protein interactome. *Proc. Natl. Acad. Sci. USA*, 98: 4569–4574.

Uetz P, et al. [2000]. A comprehensive analysis of protein-protein interactions in *Saccharomyces cerevisiae*. *Nature*, 403: 623–627.

Walhout AJM, et al. [2000]. Protein interaction mapping in *C. elegans* using proteins involved in vulval development. *Science*, 287: 116–122.

2D Gel Electrophoresis

Gygi SP, et al. [1999]. Quantitative analysis of complex protein mixtures using isotope-coded affinity tags. *Nat. Biotechnol.*, 17: 994–999.

Mass Spectrometry

Link AJ, et al. [1999]. Direct analysis of protein complexes using mass spectrometry, *Nat. Biotechnol.*, 17: 676–682.

Yates JR, 3rd [1998]. Mass spectrometry and the age of the proteome. *J. Mass Spectrom.*, 33: 1–19.

Gene Fusion, Gene Expression Profiles, and Combined Methods

Enright AJ, et al. [1999]. Protein interaction maps for complete genomes based on gene fusion events. *Nature*, 402: 86–90.

Lockhart DJ, and Winzeler EA [2000]. Genomics, gene expression and DNA arrays. *Nature*, 405: 827–836.

Marcotte EM, et al. [1999a]. Detecting protein function and protein-protein interactions from genome sequences. *Science*, 285: 751–753.

Marcotte EM, et al. [1999b]. A combined algorithm for genome-wide prediction of protein function. *Nature*, 402: 83–86.

Pellegrini M, et al. [1999]. Assigning protein functions by comparative genome analysis: Protein phylogenetic profiles. *Proc. Natl. Acad. Sci. USA*, 96: 4285–4288.

Schwekowski B, Uetz P, and Fields S [2000]. A network of protein-protein interactions in yeast. *Nat. Biotechnol.*, 18: 1257–1261.

Young RA [2000]. Biomedical discovery with DNA arrays. *Cell*, 102: 9–15.

Problems

3.1 (a) The emission of laser light at 440 nm for use in FRET was discussed in Section 3.10. What is the energy of these photons in (i) eV, and (ii) kcal/mol? (b) What is the energy of a 1.5 Å X-ray in these two sets of units? (c) What are the corresponding frequencies of the photons at 440 nm and 1.5 Å.

The following list of physical constants and conversion factors may be helpful:

Physical constants and conversion factors
Avogardo's number $N = 6.022 \times 10^{23}$/mol
Velocity of light $c = 2.9979 \times 10^{10}$ cm/s
Planck's constant $h = 6.626 \times 10^{-27}$ erg sec
$1 \text{ eV} = 1.6 \times 10^{-12}$ erg
$1 \text{ cal} = 4.184$ joules .004 4.184×10^7 ergs

3.2 Using the figure shown below, derive Bragg's law, Eq. (3.4).

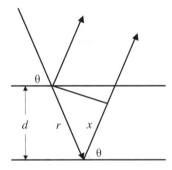

Start with the relationship that the increase in path length between the two scattered rays shown in figure must be an integral multiple of the wavelength, that is, the waves must be in phase with one another in order to constructively interfere. In terms of the notation of the figure this condition is

$$r + x = n\lambda.$$

Hint: Make use of the trigonometric identity:

$$2 \sin^2 \theta - 1 = \cos(180\,^\circ - 2\theta).$$

3.3 Show that the nuclear magneton

$$\mu_N = 5.05 \times 10^{-27} \text{ J/T}.$$

In the unit J/T, J denotes Joules and the symbol T denotes Tesla, the SI unit for B-fields. This quantity has dimensions of kilograms per second per second per ampere; that is $1\,T = 1\,kg\,s^{-2}\,amp^{-1}$. Use values for physical constants from the table in Problem 3.1 and the one shown below. (Note: The ampere is a unit of current; that is, 1 amp = 1 coulomb per second).

Physical constants and conversion factors

Charge on an electron $e = 1.6 \times 10^{-19}$ coulombs
Mass of a proton $M_p = 1.672 \times 10^{-27}$ kg

3.4 Recall from Figure 3.7 that the CH_2 group splits the single peak for the hydrogen atoms in CH_3 into three peaks, and the CH_3 group splits the single hydrogen peak for CH_2 into four peaks. In the figure shown below, the spin-spin effect of the CH_2 group on the CH_3 group is shown. The two hydrogen atoms in CH_2 can each have their spins either aligned with the external field (+) or have their spins aligned opposite (−). When opposite they shield the magnetic field and reduce the splitting, while aligned they add to the field and increase the splitting. There are three different spin-spin combinations: both negative, one positive and one negative, and both positive. This gives rise to the three peaks. Since there are two ways of arriving at a [+, −] combination and once each for the others the peak intensities are in a 1:2:1 ratio as illustrated in Figure 3.7.

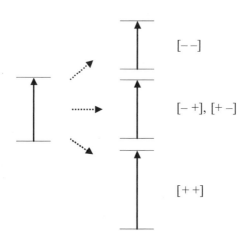

(a) In this problem, write down the possible spin combinations for the CH_3 group that splits the peak for CH_2 into four peaks shown in Figure 3.7, and give the ratios of the peak heights. (b) From the results for two and three hydrogens, infer the splittings of a hydrogen peak produced by four nearby hydrogen atoms. How would their peak heights vary?

3.5 Recall from the discussion on mass spectrometers that the beams of charged particles are subjected to first an accelerating voltage allowing the particle to attain a kinetic energy equal to

$$zV = mu^2/2.$$

When the particles feel the magnetic field the force zuB is exactly balanced by the centrifugal force:

$$zuB = mu^2/r.$$

(a) From these two relationships derive Eq. (3.7). (b) Then derive the time-of-flight expression given by Eq. (3.8).

4
Macromolecular Forces

Ordinary polymers such as rubber are constructed as repetitive sequences of a small number of basic units. In contrast, proteins are synthesized as non-repetitive sequences of the 20 different amino acids. Each kind of amino acid has a different side chain, and the variations in side chains endow each amino acid with a distinct set of physical and chemical properties. The amino acid compositions of proteins are not randomly selected. Instead, amino acid sequences are selected to allow the proteins to fold into compact three-dimensional forms in physiologically meaningful time periods, with specific binding properties that enable the proteins to carry out their cellular tasks.

Macromolecular forces—mixtures of covalent and noncovalent forces generated by the electron clouds surrounding the atomic nuclei—hold protein together. The starting point in understanding how proteins signal one another is to understand the physical properties of the amino acids and the macromolecular forces that hold biomolecules together. One question that one would like to answer is How do amino acids pack together? Another is How do the macromolecular forces and amino acid geometry shape the proteins? How do they guide their interactions with one another, and with lipids, carbohydrates, DNA, and RNA? The goal of this chapter is to begin to answer these questions and others like them. Amino acid composition and organization in proteins will be looked at along with macro-molecular forces holding the proteins together and mediating their interactions with one another through their interfaces. This exploration will continue into the next chapter, where the problem of how a protein folds, one of the most enduring and distinguished problems in all of science, will be explored.

4.1 Amino Acids Vary in Size and Shape

Amino acid side chains vary considerably in length and mass. The smallest amino acid is glycine. Its side chain consists of a single hydrogen atom. The next smallest amino acid is alanine with a CH_3 group as its side chain. The

TABLE 4.1. Physical properties of amino acid residues: Three- and one-letter abbreviations (codes) are listed in column 2. Masses are given in column 3 in Daltons. The volumes presented in column 4 are in cubic angstroms.

Amino acid	Code	Mass (Da)	Volume ($\mathring{A}^3$)
Alanine	Ala-A	71.08	89.3
Arginine	Arg-R	156.2	190.3
Asparagine	Asn-N	114.1	122.4
Aspartic acid	Asp-D	115.1	114.4
Cysteine	Cys-C	103.1	102.5
Glutamic acid	Glu-E	129.1	138.8
Glutamine	Gln-Q	128.1	146.9
Glycine	Gly-G	57.05	63.8
Histidine	His-H	137.1	157.5
Isoleucine	Ile-I	113.2	163.0
Leucine	Leu-L	113.2	163.1
Lysine	Lys-K	128.2	165.1
Methionine	Met-M	131.2	165.8
Phenylalanine	Phe-F	147.2	190.8
Proline	Pro-P	97.12	121.3
Serine	Ser-S	87.08	93.5
Threonine	The-T	101.1	119.6
Tryptophan	Trp-W	186.2	226.4
Tyrosine	Tyr-Y	163.2	194.6
Valine	Val-V	99.13	138.2

longest side chains belong to arginine and lysine, while the most massive are arginine, tyrosine and tryptophan. The lengths of the side chains vary by an order of magnitude and the masses of the amino acids, which are listed in Table 4.1, differ by more than a factor of three.

The amino acids used to make proteins vary not only in size, but also in shape. Some amino acids have open side chains (aliphatic); others contain closed rings (aromatic); and one amino acid, proline, has a side chain that closes back onto the main chain nitrogen. That the amino acids vary in size and shape is important for their packing. These variations make possible the close packing of amino acids in the core of the folded protein. The resulting interior packing densities approach those of organic solids.

4.2 Amino Acids Behavior in Aqueous Environments

Amino acids differ in their charge properties. Some are polar, charged, and either acidic or basic; others are polar and uncharged; and still others are nonpolar. Several side chains contain hydroxyl groups and are highly reactive, and two contain a sulfur atom. The physical properties of the amino acids—size, mass, shape, charge distribution, and bonding propensity—play major roles in determining how a protein folds, how a protein interacts

TABLE 4.2. Electrostatics and bonding propensities of the amino acids: The polar amino acids listed in column 3 are of three kinds. Some are basic (arginine, histidine, and lysine); others are acidic (aspartic and glutamic acids); and the remainder are uncharged polar molecules. Histidine is only weakly basic and is often considered a nonpolar amino acid, along with proline.

Amino acid	Nonpolar	Polar	H-bond	Salt Bridge	S Bridge
Alanine	*				
Arginine		*	*	*	
Asparagine		*			
Aspartic acid		*	*	*	
Cysteine	*				*
Glutamic acid		*	*	*	
Glutamine		*			
Glycine	*				
Histidine		*			
Isoleucine	*				
Leucine	*				
Lysine		*	*	*	
Methionine	*				*
Phenylalanine	*				
Proline	*				
Serine		*	*		
Threonine		*	*		
Tryptophan	*				
Tyrosine		*	*		
Valine	*				

with other biomolecules, and how proteins interact with their aqueous environments.

The results of grouping the 20 amino acids according to their charge and bonding properties are presented in Table 4.2. Columns 2 and 3 in the table group the amino acids according to whether they are polar or nonpolar. The significance of this partitioning is that amino acids with nonpolar side chains are hydrophobic, and those with polar side chains are hydroplilic. The distinction between the two kinds of interaction arises from the binding preference of water for the amino acid. In the case of a hydrophilic (water-loving) amino acid, a water molecule would rather bind to the amino acid than to another water molecule. In the case of a hydrophobic amino acid, a water molecule prefers another water molecule to the amino acid.

Nonpolar groups tend to come together in water, not because of an attraction for each other but rather because their clustering enables water molecules to make the maximum number of contacts with each another. The term *hydrophobic interactions* denotes the process whereby water molecules come together so that small nonpolar molecules and nonpolar portions of large molecules minimize their contacts with water. The propensity for water molecules to come together is a reflection of their hydrogen-bonding capabilities resulting in the formation of networks of hydrogen-bonded molecules.

4.3 Formation of H-Bonded Atom Networks

Extensive networks of hydrogen-bonded atoms are formed. In small hydrogen-bearing molecules such as water there is little screening of the positively charged hydrogen nuclei by electron clouds. The hydrogen atoms of these molecules easily form bonds with unshared electrons of electronegative atoms of nearby molecules. These *hydrogen bonds* form most often between *covalently bonded* O-H, N-H, and F-H groups and other O-H, N-H, and F-H groups. The tightness of the hydrogen bonding is limited by the mutual repulsion of the electron clouds of the two electronegative atoms. Typical bond lengths, representing the distances between centers of the two electronegative binding partners, range from 2.6 to 3.2 angstroms. Hydrogen bonding is dipole-dipole in character, and is therefore highly directional. The bond is strongest when the three atoms are colinear and rapidly diminishes with increasing bond angle. Typical maximum bond energies are 3 to 7 kcal/mol.

Water is a particularly important example of a molecule that forms hydrogen bonds with other molecules of the same kind. Each water molecule typically forms three or four hydrogen bonds with nearby water molecules. The result is a three-dimensional latticework of H_2O molecules. The ability of water molecules to form hydrogen bonds with other water molecules is responsible for water's cohesiveness. Most biomolecules are soluble in water and can move freely from one location to another without clumping together. But they are not so soluble that they are unable to expel intervening water molecules when coming together to form complexes and machines.

Hydrogen bonding is responsible for the formation and stability of alpha helices and beta sheets. The tendency is for the interior of proteins to be predominately hydrophobic, and in some studies of how proteins fold, hydrophobic and hydrophilic interactions of residues are found to have a dominating influence. However, polar amino acid residues do populate the interior of proteins, and hydrophobic residues populate the surface. In those instances where polar residues are located in the interior, hydrogen bonds alleviate the disruptive influences of charged groups on the stability of the protein.

4.4 Forces that Stabilize Proteins

Salt bridges, van der Waals forces, and disulfide bridges help stabilize proteins and their surface contacts. Salt bridges are formed by coulombic attractions between positively and negatively charged atoms. In the context of the amino acids, salt bridges are generated by electrostatic (coulombic) attractions between positively charged lysine or arginine residues and negatively charged aspartic acid or glutamic acid residues. These bonds are most often encountered on the surface of the protein.

Those atoms that do not form hydrogen bonds or salt bridges still attract one another through van der Waals attractions. The van der Waals attractions are dipole forces. They are shorter ranged than point coulomb forces and are weaker, as well. In a dipole-dipole interaction the region of negative charge of one molecule attracts the region of positive charge of another molecule. The strongest dipole-dipole interactions occur between molecules possessing permanent dipole moments. London (or London dispersion) forces arise from the motion of the electron clouds that surround the atomic nuclei. The electrons in any molecule are constantly in motion and undergo spontaneous distortions to form instantaneous dipoles where one portion of the molecule has a net positive charge and the other has a net negative one. These instantaneous dipoles induce matching dipoles on neighboring molecules. The result is a nonspecific and nondirectional attractive force between the two molecules. London forces are most pronounced when the molecules have large electron clouds that are easily polarized. It is the various dipole forces involving induced and permanent dipoles that are collectively referred to as *van der Waals forces*.

The last column in Table 4.2 lists two amino acids whose side chains contain sulfur atoms. Disulfide bonds depart from the aforementioned rule of weak bonding. The disulfide bonds are covalent in character. They are formed between sulfur atoms on cysteine side chains. These bonds make important contributions to the overall stability of the proteins.

4.5 Atomic Radii of Macromolecular Forces

Macromolecular forces have characteristic strength and distinct atomic radii. The different macromolecular forces vary considerably in strength from covalent bonds (the strongest) to van der Waals interactions (the weakest). Because the different kinds of interactions vary in their strengths, the amount of interpenetration of the electron clouds will vary depending on which forces are being experienced. This means that the radii of atoms bound inside proteins and other macromolecules will depend on the forces being experienced. This type of systematic behavior is quite useful, allowing investigators to deduce forces from observed bond lengths. Listed in Table 4.3 are a set of atomic radii for nitrogen and oxygen under the influence of different forces.

The average bond lengths and atomic radii are related to one another in a simple fashion. A bond length is just the sum of the two atomic radii for the bond partners. For bonds between like atoms, this means that the bond length is just twice the atomic radius for the interaction of interest. The

TABLE 4.3. Atomic radii: Listed are average values for atomic radii deduced from X-ray crystallography data for a variety of atomic groups in proteins. All atomic radii are in angstroms (Å). The radii listed in columns 2 through 5 are for covalent, coulombic, hydrogen-bonded and van der Waals interactions, respectively.

Atom	r_{cov}	r_{coul}	r_H	r_{vdW}
Nitrogen	0.70	1.45	1.55	1.70
Oxygen	0.65	1.40	1.40	1.50

atomic radii and bond energies behave in a systematic way with respect to one another. The stronger the bond the closer together the two atoms will be. Covalent bonding strengths can be as high as 100 kcal/mol or more. This is at least an order of magnitude greater than any of the noncovalent bond forms. As a result the bond lengths are about a factor of two shorter than those of any of the weaker bonding forms. The strongest of the noncovalent forces are the coulombic interactions. These can be as much as 5 kcal/mol. The hydrogen bonds are somewhat weaker yet, broadly distributed in the range of 2 kcal/mol. The weakest of the forces are the van der Waals interactions. These are not more than about 1 kcal/mol, barely above typical thermal energies of 0.6 kcal/mol.

The three kinds of interactions just discussed, hydrogen bonds, coulombic attractions and repulsions, and van der Waals forces, are all far weaker than the covalent bonds that underlie the protein backbone. Rather than forming a few strong covalent bonds proteins interact with one another by forming multiple weak bonds that can be easily broken and reestablished in the same or in different ways.

4.6 Osmophobic Forces Stabilize Stressed Cells

Osmophobic forces are an important stabilizing element when cells are stressed. There is one more class of forces that has an important bearing on protein-folding and stability in the cell. Cells are exposed to a variety of stressful conditions. Among these are thermal, osmotic, and salt stresses. Cells have developed a number of strategies that enable them to cope with stresses when they arise. One of these adaptive mechanisms is to build up intracellular concentrations of small organic metabolites called *osmolytes*. Free amino acids and amino acid derivatives are common osmolytes. These organic molecules protect the cell against denaturing effects on the proteins of abnormal cellular conditions. They help maintain the proteins in their folded conformations without interfering with the normal functioning of the proteins.

By analogy with the characterization of disfavored water-protein inter-actions as a hydrophobic effect, disfavored osmolyte-protein interactions are called *osmophobic interactions*. This effect manifests itself as a burial of the protein backbone to shield it from the osmolytes. The gain in stability from avoiding contacts between the backbone atoms and the osmolytes more than compensates for the attractive effects between the side chains and the osmolytes. By increasing the concentration of osmolytes when stressed, a cell supplies a driving force that promotes the folded state over the unfolded one.

4.7 Protein Interfaces Aid Intra- and Intermolecular Communication

Proteins possess interfaces that enable them to communicate with proteins, nucleotides, lipids, and carbohydrates. The region of contact between molec-ular surfaces is known as the *interface*. Interfaces belong to the *domain level* of protein organization: When a protein acquires a particular domain it inherits that domain's binding properties by possessing its interfaces. Protein interfaces recognize and bind other cellular components through the latter's interfaces. There are several different kinds of interfaces, listed in Table 4.4. The first entry is that of interfaces between adjacent domains located within a single protein. Communication across these interfaces allows the functionally separated parts of the protein to work together. The next two entries, *subunit interfaces*, also pertain to communication within a protein, but operate at the quaternary level of organization. One kind of subunit interface (tight) enables protein subunits that stay together for appreciable periods of time in membrane proteins to coordinate their activ-ities. The other (loose) permits subunits of cytosolic proteins that only come together for brief periods of time then immediately separate again to reg-ulate one another's actions.

TABLE 4.4. Classification of protein interfaces.

Type	Function
Domain-domain	Coordinates activities of adjacent domains within a protein
Subunit-subunit	
(a) Tight	Enables subunits on pore- and channel-forming proteins to work as a single unit
(b) Loose	Coordinates activities of cytosolic protein subunits that transiently associate
Protein-protein	Communication between proteins
Protein-DNA	Communication between proteins and regulatory sequences in DNA molecules
Protein-RNA	Communication between proteins and RNA molecules
Protein-lipid	Communication between proteins and membrane lipids
Protein-carbohydrate	Recognition of cellular and ECM carbohydrates

The other categories of interfaces operate between proteins and the different kinds of cellular biomolecules—proteins, DNA, RNA, lipids, and carbohydrates. These interfaces make possible the binding of proteins to their ligands, transcription factors to DNA, and splicing factors to RNA. They mediate interactions with the lipid membrane bilayer and are responsible for the coordinated activities of signal complexes. From the viewpoint of signaling the most widely encountered and studied interfaces are the protein-protein and protein-DNA interfaces. Carbohydrate interfaces are widely encountered in cells of the immune system, and these will be discussed in Chapter 10, which is devoted to cell adhesion and motility. Lipid interfaces are crucial for a variety of cellular processes, including signaling. These interfaces will be examined in detail in Chapter 8, where membrane lipid composition and lipid signaling are explored.

Recall from Chapter 2 that proteins involved in signaling are frequently post-translationally modified. Some amino acid residues acquire phosphoryl groups, while others gain acetyl groups or methyl groups. Some protein interfaces are sufficiently precise in their binding affinities that they can distinguish whether these groups are present or absent. As a result, an "upstream" signaling event, where a group is added, can be followed by a "downstream" event, which involves the recognition of that binding site.

4.8 Interfaces Utilize Shape and Electrostatic Complementarity

Surfaces establish contact with one another though their interfaces. The areas of surface contact may be planar, but most often they are irregular in shape. Interfaces utilize shape and electrostatic complementarity for recognition and binding. Shape *complementarity* denotes the propensity of the surfaces of two molecules to geometrically fit together so that multiple contacts can be established. Electrostatic complementarity denotes the matching of hydrophobic patches, the complementary pairing of hydrogen bond donors and acceptors, and the matching of positive and negative charges of basic and acidic polar residues from one surface to the other.

The amino acid residues that form the interfaces on the two complementary surfaces do not each contribute equally to the binding energy and specificity. Rather, some 5 to 10 amino acid residues in each complementary surface form energetic hot spots. These hot spots are responsible for most of the binding affinity of one surface for the other. This number may be compared to the 10 to 30 residues on each protein that form the interface. Interfaces are in general hydrophobic, but not overwhelmingly so. Hydrophilic residues assume a greater role in binding than in folding (to be discussed in the next chapter). For some interfaces electrostatic interactions help steer the ligand onto the correct docking orientation/location. Hot spots tend to be localized in the center of the interface, surrounded by hydrophobic rings containing energetically less important residues that shield the hot spot residues from the bulk solvent.

4.9 Macromolecular Forces Hold Macromolecules Together

The macromolecular forces that hold macromolecules together are of two kinds. One kind of force, the covalent bond, is generated when two atoms share electrons. The other kind of force operates between atoms that do not share electrons and are not covalently bonded. The interactions that take place in these situations are electrostatic in character consisting of combinations of point charge and dipole forces, and, as already discussed, are referred to as coulombic (salt bridges) and van der Waals interactions. Hydrogen bonds are included in this category. Unlike the electron-sharing mechanism underlying covalent-bonding interactions, hydrogen bonds are primarily electrostatic in nature. They arise from a balance between attractive and repulsive forces between partial charges.

Covalent-bonding forces can be thought of as operating in an elastic springlike manner. In an elastic spring, there is an equilibrium length where the spring is at rest. If compressed or stretched away from the equilibrium length, potential energy builds up, or is stored, in the spring. Once the perturbing force is removed from the spring, an elastic spring force, called a *Hooke's law force*, restores the spring to its equilibrium rest length. Hook'e law expresses the relationship between the displacement d from equilibrium position, the spring constant K_d, and the restoring force F:

$$F = -K_d \cdot d. \tag{4.1}$$

Several different kinds of motions of atoms about their bonds are possible. As depicted in Figure 4.1, there are stretching motions, bending motions, and torsions about the bond axis.

4.10 Motion Models of Covalently Bonded Atoms

Hooke's law and periodic potentials are often used to model how covalently bonded atoms move. A variety of conceptual approaches have been developed that enable researchers to study how proteins and other macromolecules move under the influence of bonding and nonbonding forces. These methods have been used singly and in combination. The most popular methods in use are:

- Continuum electrostatics formalism
- Molecular mechanics formalism
- Molecular dynamics method

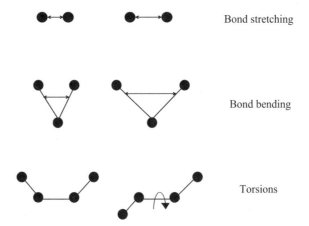

FIGURE 4.1. Motions about groups of two, three, and four covalently bonded atoms: Displayed in the upper portion of the figure are bond-stretching motions of a pair of covalently bonded atoms along their bond axis. Shown in the middle panel are bending motions involving three bonded atoms forming an angle θ. Displayed at the bottom are groups of four atoms, where the leftmost atom has rotated about the middle bond axis.

- Brownian dynamics method
- Quantum mechanics formalism

In continuum electrostatics, one replaces detailed models of interactions of the macromolecule atoms with their surrounding waters with a simplified treatment, one in which the molecule-solvent interactions are treated in an average way. The water is modeled at a macroscopic level while the atoms of the molecule of interest are still studied at an atomic level of detail. In this approach, a macroscopic expression is solved to give the electrostatic potential extending outward from the surface of the protein. When visualized, the potential highlights interesting regions of positive and negative charge and provides insight into the interface and binding properties of the protein.

The atomic level of detail is handled by the molecular mechanics (MM) formalism. In the MM formalism, a classical force field U is introduced consisting of a sum of bonding and nonbonding interactions. The three kinds of covalent bonds illustrated in Figure 4.1 are modeled using Hooke's law and periodic potentials. As indicated in expression (4.2), Hooke's law terms represent bond and angle interactions, and a periodic term represents the *dihedrals*, also referred to as *torsions*:

$$U_{\text{bonded}} = \sum_{\text{bonds}} K_b(b - b_{eq})^2 + \sum_{\text{angles}} K_\vartheta(\theta - \theta_{eq})^2$$
$$= + \sum_{\text{torsions}} K_\phi(1 - \cos(n\phi + \delta)). \tag{4.2}$$

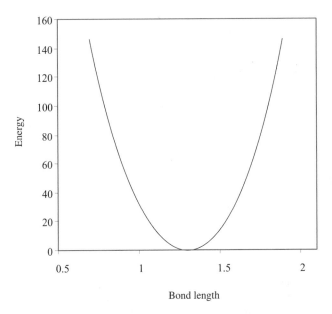

FIGURE 4.2. Bond stretching: Plotted are the potential energies associated with bond stretching. The harmonic potential has a minimum at the equilibrium bond length. It increases as the bonded atoms are pushed along the bond axis towards one another, so the bond is compressed, and as they are pulled apart, so the bond is stretched out. For the stretch spring constant, $K_b = 400 \, \text{kcal/(mol} \cdot \text{Å}^2)$, and for the equilibrium bond length, $b_{eq} = 1.3 \, \text{Å}$.

The total potential energy of the macromolecule U_{bonded} is equal to the sum (Σ) of the contributions from the individual bonds. All of the bond-stretching interactions are included along with the bond-rotations and bond-torsions. As the atoms forming these bonds move away from and towards their equilibrium configurations, the potential energies rise and fall. The shape of the potential energy curve for each of the covalent bond types is illustrated in Figures 4.2 to 4.4.

4.11 Modeling van der Waals Forces

The noncovalentbonded interactions are of two types—van der Waals forces and a coulombic term representing salt bridges and other point charge interactions. Because of the $1/r$ radius dependence the coulombic attractions and repulsions fall off with increasing separation of the inter-acting atoms very slowly, far more so than the contributions from the other terms in the nonbonded potential shown in Eq. (4.3).

Van der Waals forces are frequently modeled in terms of a Lennard–Jones, 6–12 potential. There are two terms in this potential. The term containing the sixth power of the radius is an attractive one while the term

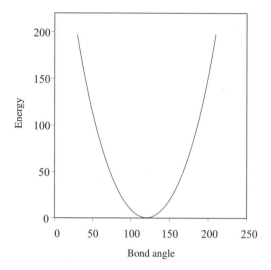

FIGURE 4.3. Bond bending: Plotted are the potential energies associated with rotations of the angle made by three atoms covalently bonded to one another. In a manner analogous to bond stretching, there is an equilibrium bond angle where the potential energy is at a minimum. As the rotation angles are either widened out or squeezed in, the potential energy rapidly increases. The bending spring constant $K_\theta = 40\,kcal/(mol \cdot rad^2)$, and equilibrium angle $\theta_{eq} = 120\,deg$.

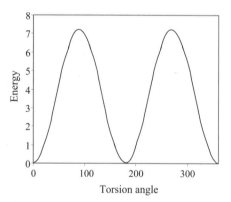

FIGURE 4.4. Bond torsions: Plotted are the periodic contribution to the potential energy from torsion (dihedrals) rotations of covalently bonded sets of four atoms. Calculations were done using a K_φ with four paths so that 3.625 kcal/mol is the constant in front of the expression. Parameters values used were $n = 2$ and $\delta = 180\,deg$.

involving the twelfth power of the radius is a repulsive one. As can be seen in the figure there are two distances of interest. One of these is the equilibrium radius r_0, the location where the potential has its minimum and where the force—the derivative of the potential—is zero. The other key dis-

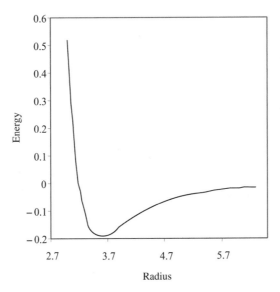

FIGURE 4.5. Van der Waals interactions: Plotted are the van der Waals potential energies in kcal/mol using a well depth $\varepsilon = 0.19$ kcal/mol and a radius $r_0 = 3.59$ Å, corresponding to nitrogen and oxygen contact radii of $1.7 + 1.5 = 3.2$ Å.

tance is the radius at which the repulsion exactly cancels out the attraction so that the net potential is zero. This repulsion is generated by the inter-penetrating electron clouds. This contact distance is the *van der Waals radius*. As can be seen in the plot, any further interpenetration is strongly resisted by the rapidly increasing repulsive force arising from the Pauli exclusion.

$$U_{\text{nonbonded}} = \sum_{i \neq j} \left\{ \varepsilon \left[\left(\frac{r_0}{r} \right)^{12} - 2 \left(\frac{r_0}{r} \right)^{6} \right] + \frac{q_i q_j}{4\pi\varepsilon_0 r} \right\} \qquad (4.3)$$

Hydrogen-bonding contributions to the total potential are treated in an approximate way in the MM formalism. They are modeled using Lennard–Jones and coulombic like terms suitably modified to match the characteristics of the hydrogen bonds.

4.12 Moleculer Dynamics in the Study of System Evolution

The evolution of macromolecular systems over time can be studied using the method of moleculer dynamics (MD). In molecular dynamics, the classical (Newtonian) equations of motion are solved through numerical integration. The forces F are the spatial derivatives of the potential energies U presented in Eqs. (4.2) and (4.3). These quantities are then converted to accelerations using Newton's second law, $F = ma$, where m is mass and a is acceleration. Once this is done the equations of motion are solved to give a picture of how the system evolves over time. Numerical methods are used to convert the equations of motion into a form suitable for numerical integration on a computer. A number of computer programs are in use that enable the researcher to carry out MD simulations.

The last entry in the list of methods is quantum mechanics (QM). The covalent bonds are only handled in an approximate way in the MM formalism. In a quantum mechanical approach, the empirical treatment of the covalent bonds through the use of a Hooke's law is replaced by an exact quantum mechanical treatment. The quantum mechanical method is exact and rigorous but is difficult to use in practice due to computational limitations. As computer technologies have advanced, quantum mechanical studies have become more numerous. An often-used approach is to combine the MM and QM formalisms, using one or the other of the two methods for different aspects of the system under study.

4.13 Importance of Water Molecules in Cellular Function

Water molecules are essential components of protein, DNA, and RNA function. Water plays a central role in protein structure and function. In the absence of water, a protein would not be able to fold into its native state, nor would it be able to catalyze reactions. Water is an essential component of many protein-protein, protein-DNA and protein-RNA interfaces, and without water proteins would not be able to recognize their substrates. Water surrounds a protein, fills in pockets and grooves on the surface, and occupies voids in the interior.

When a protein is placed into a water environment it alters the network of hydrogen bonds. The water molecules in the vicinity of the protein surface reorient themselves so that positive and negative regions of change are in opposition. The rotations of the water molecules disrupt the hydrogen bonds between adjacent water molecules and thus alter the network. The water molecules in the immediate vicinity of the protein surface that have reoriented themselves are referred to as the first hydration layer. The reorientation effect propagates outward from the protein and the next layer of disrupted hydrogen bonded water molecules is designated as the second hydration layer.

The same phenomena occur for DNA and RNA. Water molecules in the vicinity of the DNA and RNA form hydrogen bonds to the molecules. These bonds are not passive entities but instead contribute to the conformational stabilization and function of the macromolecules. Hydrogen bonding networks between water and DNA is essential for DNA stability. DNA denatures as it dehydrates. Hydrogen bonds between water and ssRNA are even more numerous than in the case of dsDNA because of the single strand character of the RNA leaves bases unpaired and there are additional ribose oxygen atoms available for bonding.

4.14 Essential Nature of Protein Dynamics

Macromolecules such as proteins are dynamic entities: their internal motions are essential. They are in continual thermal motion and through these motions are continually exploring different conformations. When especially stable states are encountered, the *dwell time*, the period of time that the protein remains in such states, increases; when highly unstable states are populated the dwell time decreases. This continual exploration of available conformations is central to binding and catalysis.

At the heart of the role of water in the activities of all macromolecules is its ability to promote rapid conformational changes. Water is a common element in the active site of enzymes, and is a key mediator of the catalytic activity of many, if not most, enzymes. When dehydrated these enzymes lose their catalytic abilities. Yet another place where water seems to be crucial is in regulation. Proteins such as hemoglobin that use allosteric mechanisms (this term will be defined and its properties explored in the next chapter) for regulation operate in a hydrated fashion, and loss of these water molecules impairs the regulatory functioning of the protein. Water, the hydrogen bonds that are continually being made and broken, and the underlying thermal agitation collectively serve as a "lubricant" that promotes conformation changes essential to the performance of protein, DNA, and RNA functions.

General Reference

Brandon C, and Tooze J [1999]. *Introduction to Protein Structure*, 2nd edition. New York: Garland Science Publishing.

References and Further Reading

Physical and Electrostatic Properties of Amino Acids and Proteins

Bolen DW, and Baskakov IV [2001]. The osmophobic effect: Natural selection of a thermodynamic force in protein folding. *J. Mol. Biol.*, 310: 955–963.

Dill KA [1990]. Dominant forces in protein folding. *Biochem.*, 29: 7133–7155.

Myers JK, and Pace CN [1996]. Hydrogen bonding stabilizes globular proteins. *Biophys. J.*, 71: 2033–2039.

Pace CN, et al. [1996]. Forces contributing to the conformational stability of proteins. *FASEB J.*, 10: 75–83.

Sheinerman FB, and Honig B [2002]. On the role of electrostatic interactions in the design of protein-protein interfaces. *J. Mol. Biol.*, 318: 161–177.

Tsai J, et al. [1999]. The packing density in proteins: Standard radii and volumes. *J. Mol. Biol.*, 290: 253–266.

Complementarity and Interfaces

Glaser F, et al. [2001]. Residue frequencies and pairing preferences at protein-protein interfaces. *Proteins*, 43: 89–102.

Lo Conte L, Chothia C, and Janin J [1999]. The atomic structure of protein-protein recognition sites. *J. Mol. Biol.*, 285: 2177–2198.

Jones S, and Thornton JM [1996]. Principles of protein-protein interactions. *Proc. Natl. Acad. Sci. USA*, 93: 13–20.

Jones S, et al. [1999]. Protein-DNA interactions: A structural analysis. *J. Mol. Biol.*, 287: 877–896.

Nadassy K, Wodak SJ, and Janin J [1999]. Structural features of protein-nucleic acid recognition sites. *Biochem.*, 38: 1999–2017.

Norel R, et al. [1999]. Examination of shape complementarity in docking of unbound proteins. *Proteins*, 36: 307–317.

Sheinerman FB, Norel R, and Honig B [2000]. Electrostatic aspects of protein-protein interactions. *Curr. Opin. Struct. Biol.*, 10: 153–159.

Hot Spots

Bogan AA, and Thorn KS [1998]. Anatomy of hot spots in protein interfaces. *J. Mol. Biol.*, 280: 1–9.

Hu ZJ, et al. [2000]. Conservation of polar residues as hot spots at protein interfaces. *Proteins*, 39: 331–342.

Theoretical Methods: Computer Modeling and Simulation

Cornell WD, et al. [1995]. A second generation force field for the simulation of proteins, nucleic acids, and organic molecules. *J. Am. Chem. Soc.*, 117: 5179–5197.

Elcock AH, Sept D, and McCammon JA [2001]. Computer simulation of protein-protein interactions. *J. Phys. Chem. B*, 105: 1504–1518.

Honig B, and Nicholls A [1995]. Classical electrostatics in biology and chemistry. *Science*, 268: 1144–1149.

Kollman PA, et al. [2000]. Calculating structures and free energies of complex molecules: Combining molecular mechanics and continuum models. *Acc. Chem. Res.*, 33: 889–897.

Problems

4.1 Atomic motions. Atoms in a protein are constantly in thermal motion. Assuming an average energy kT of 0.6 kcal/mol for each atom, how fast is a hydrogen atom moving? How fast is a carbon atom moving? How long will it take for each of these atoms to move 1 Å, roughly one bond length?

4.2 Numerical integration. Numerical techniques, known as *finite difference methods*, are used to convert the equations of motion into a form suitable for numerical integration on a computer. The basic idea is to take the position and momentum of each particle at a given time and compute how each changes over a small interval of time, the time step Δt, by calculating the accelerations from the forces, and these from the potentials of the form given in the chapter. In other words, for each particle i in the system

$$a_i = \frac{1}{m_i} F_i = \frac{1}{m_i} \frac{d}{dr_i} U_i.$$

A number of computer programs such as CHARMM, AMBER and GROMOS, are in use that enable a user to carry out MM simulations. A number of time-stepping algorithms are employed in determining the future positions from the past positions and forces. These algorithms are based on expansions such as

$$r(t + \Delta t) = r(t) + v(t)\Delta t + (1/2)a(t)(\Delta t)^2.$$

One of the most widely used stepping forms, known as the *Verlet algorithm*, is

$$r(t + \Delta t) = 2r(t) - r(t - \Delta t) + a(t)(\Delta t)^2.$$

Note that the velocities do not appear in this expression. The positions at time $t + \Delta t$ are computed from the positions at the present $(t + \Delta t)$ and previous (t) times and from the accelerations at the previous (t) time. Some algorithms use the velocities explicitly in the computations. One of these is the *velocity Verlet algorithm*. Its form is

$$v(t + \Delta t) = v(t) + (1/2)[a(t) + a(t + \Delta t)]\Delta t.$$

By making use of the appropriate expansions, and combining terms, derive both of these Verlet algorithms. What might be an appropriate time step size? (Hint: Think about the results from Problem 4.1.)

5
Protein Folding and Binding

The world contains a myriad of biological systems. All exhibit a considerable degree of order. They are organized in a hierarchical manner, and order is present at all levels of the hierarchy. From hydrogen, carbon, nitrogen, and oxygen, simple atomic groups are formed such as methyl (CH_3) and hydroxyl (OH). These groups are then used to form the basic building blocks of cells—sugars, fatty acids, nucleotides, and amino acids. Simple sugars (monosaccharides) are organized into short chains (oligosaccharides) or longer ones (polysaccharides). Fatty acids form complexes such as triglycerides and phospholipids. Nucleotides are used to make RNA and DNA, and the amino acids give rise to polypeptides, or proteins.

Biological order does not arise out of some mysterious "vital force," but rather is a consequence of the laws of thermodynamics and the character of the forces in our universe, their strength and their dependence on distance. At first glance the emergence of highly organized biological entities seems at odds with the second law of thermodynamics. This law establishes a thermodynamic arrow of time—the total disorder in the universe increases as the universe ages, until a terminal stage of disorder is reached in which the universe suffers a heat or *entropy* death. Yet, this first impression is wrong: Order comes about not in spite of the laws of physics but rather because of them.

Biological systems are open, continually exchanging matter and energy with their surroundings. They generate order by taking in energy and releasing heat to their surroundings. They absorb radiant energy from sunlight and from geothermal sources, and they take in foodstuffs that store energy in high energy chemical bonds. According to the second law of thermodynamics the amount of entropy, or disorder, in a cell and its surroundings must increase during any process. Thus, the production of order *within* a cell is accompanied by the creation of a greater amount of disorder *outside* a cell. This is accomplished through the release of heat from the cell at the same time that the order is produced. Biological entities are not only highly ordered, but actively generate these states in order to survive and propagate.

One of the central order-creating processes in a cell is the folding of a protein into its biologically active three-dimensional form. During folding, nascent proteins change their shape from a rather stretched out configuration to a highly compact form. They develop their secondary structure— alpha helices and beta sheets—and higher order structures with well-defined signaling roles such as binding motifs and functional domains. The folding process is the main focus of this chapter. Starting with a brief review of the thermodynamic conditions for order to emerge, the spontaneous folding of proteins will be examined. Large and complex proteins, especially those involved in signaling, often require the assistance of a class of molecules called *molecular chaperones* to fold and to maintain their correct form in the cell. Chaperone-assisted folding will be explored next. That topic will be followed by the third and final topic in this chapter, the relationship between the thermodynamic properties of the low-lying stable states of the folded proteins and their binding and signaling activities.

5.1 The First Law of Thermodynamics: Energy Is Conserved

The first law of thermodynamics is expressed in terms of three factors: the *internal energy* of a system, the *work* done by that system, and the *heat absorbed* by a system from its surroundings. The internal energy of a system is the sum of the kinetic and potential energies of its constituents. Work is done whenever a force is applied to an object to produce a displacement. Typical examples of systems doing mechanical work are pistons, levers, and pulleys. The most commonly encountered form of work in chemical systems is pressure-volume work. In these systems, pressure, or force per unit area, is usually held constant and there is a change in the volume occupied by the system doing the work. If two systems are in thermal contact with one another, energy will flow from the hotter (higher temperature) system to the colder (lower temperature) system. "Heat" is the designation given to the flow, or transfer, of (thermal) energy from one system to another due to a temperature gradient.

In a chemical reaction that takes place in a cell, or in any other system, the internal energy E is lowered by the amount of energy used to do useful work and increased by the amount of heat absorbed in the process. In more detail, as a system evolves from state a to state b, its internal energy will change by an amount $dE = E_b - E_a$. If the amount of work done by the system on its surroundings is written as W, and the amount of heat absorbed by the system from its surroundings is designated as Q, then the first law of thermodynamics states that

$$dE = Q - W. \tag{5.1}$$

Energy will be gained by a system whenever energy flows into the system due to temperature gradients and whenever the surroundings do work on the system. Energy will be lost from a system whenever the system does work on its surroundings and whenever energy is lost to the surroundings due to thermal gradients. If we consider pressure-volume work then we may write this as

$$dE = Q - PdV, \qquad (5.2)$$

where P is the pressure and $dV = V_b - V_a$ is the change in volume produced by the application of the (constant) pressure.

5.2 Heat Flows from a Hotter to a Cooler Body

Heat is associated with random molecular motion. If a hotter body is in contact with a cooler one in a way that allows matter and energy to flow from one to the other, temperature differences will be reduced and eventually vanish. The reason for this is a statistical one. There are many more ways that the contributions of energies of the randomly moving molecules can add up to a particular internal energy when the temperatures have equilibrated than when there are large temperature differences between parts of the whole. If, through random motions or perturbations of the individual molecules, one portion of the system gains an appreciable amount of thermal energy at the expense of the rest it will not retain it for long. Instead, the system will relax back to a thermally equilibrated distribution. The thermally equilibrated distribution thus has the property that it is the *stable* distribution. It is *also* the maximally disordered distribution, in contrast to highly ordered situations where each piece of a system is at some specific value of the temperature.

These everyday observations are depicted in Figure 5.1. Part (a) of the figure illustrates mass equilibration and part (b) the similar process of thermal equilibration. In the case of mass equilibration a system starts out with most of its mass concentrated in one of two interconnected compartments. The particles are free to diffuse, and over time the masses become far more evenly distributed between the two compartments. The statistical character of the process is easy to see. Situations (states) where all or mostly all of the mass is concentrated in one compartment are rare. In contrast, there are many ways of distributing half the mass in one compartment and half in the other, and so under the influence of random movements of mass these partitions will occur most, all the time. Situations where all the particles are in one compartment will rarely occur, and when they do the system will not remain so for long (these states are not stable ones).

The same reasoning applies to thermal equilibration. The rare velocity distribution, where all the fast particles are in one compartment and the slow ones in the other, is replaced over time by the usual one where both compartments containing similar mixes of fast and slow movers. Again, there are many ways of achieving this kind of distribution and few for the

(a)

(b)

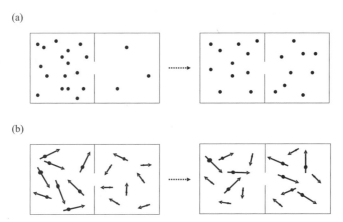

FIGURE 5.1. Mass and thermal equilibration: (a) Mass equilibration—In the left panel, most of the mass is concentrated in the left compartment. The particles are free to diffuse to and fro, and through the opening into the adjoining compartment. Over time, the system will mass equilibrate. In the right panel, roughly half of the particles are in each compartment. (b) Thermal equilibration—Arrows denote velocity; particles with longer arrows are moving with a greater velocity than those with shorter arrows. In the left panel, all of the fast particles are in the left compartment, and all the slow particles are in the right compartment. As a result, the temperature in the left compartment is higher than that in the right compartment. Again, there is an opening, and the particles can freely move about. Over time, the system equilibrates so that the temperature in each chamber becomes the same. In the right panel, each compartment contains a similar mix of slow and fast moving particles.

other. In the case of thermal equilibration, the temperatures in the two compartments initially differ. The compartment with the fast movers is at a higher temperature than that containing the slow movers, but over time the temperatures in the two compartments become the same.

5.3 Direction of Heat Flow: Second Law of Thermodynamics

The second law of thermodynamics formalizes the observation that heat flows from a hotter system to a cooler one, and not vice versa. The quantity called "entropy" gives a measure of the number of ways the molecular constituents can arrange themselves (i.e., it counts the number of microstates) to achieve a particular value of the macroscopic internal energy. The entropy increases as the particles approach the distribution that can be achieved in the largest number of ways, and entropy is maximal when the system equilibrates over time. The maximization process is interpreted as a flow of heat.

The second law of thermodynamics has three parts:

- Entropy, like the internal energy, is a property of a system.
- In an isolated system, all processes involving transitions from one internal state to another are accompanied by increases in entropy.
- In a system that is not isolated from its surroundings, any process occurring will increase the entropy S in the system by an amount dS proportional to the quantity of heat absorbed, and the constant of proportionality is the inverse of the temperature:

$$dS = Q/T. \tag{5.3}$$

If a living cell is to increase its internal order it must be in contact with its surroundings to allow for the exchange of energy. If it is isolated from its surroundings the amount of disorder *within* the cell can only increase since the second part of the second law asserts that

$$dS_{cell} > 0. \tag{5.4}$$

When a cell is in contact with its surroundings to allow for the flow of energy, there are two contributions to the total change in entropy:

$$dS_{total} = dS_{cell} + dS_{surr} > 0. \tag{5.5}$$

It is now possible for dS_{cell} to become negative so that the amount of order in the cell can increase. This can occur if the increase in disorder in the surroundings is greater than the production of order within the cell.

5.4 Order-Creating Processes Occur Spontaneously as Gibbs Free Energy Decreases

The first law of thermodynamics says that if there is no change in volume, then the change in internal energy of a system is equal to the amount of heat absorbed from the surroundings. That is, $Q = dE$. Conversely, if the pressure is held constant but the volume does change, then the heat absorbed can be written as $Q = (E_b + PV_b) - (E_a + PV_a)$. The enthalpy H of the system is defined as $H = E + PV$. Thus, at constant pressure and temperature, the heat absorbed is equal to the change in enthalpy, or $Q = dH$.

The change in entropy of the surroundings is proportional to the amount of heat released from the system. The inverse of the temperature T is the constant of proportionality. That is,

$$dS_{surr} = -Q/T. \tag{5.6}$$

In this expression a minus sign has been introduced since Q represents the amount of heat absorbed from the surroundings. Since at constant temperature and pressure $Q = dH$, the entropy of the surroundings is simply $dS_{surr} = -dH/T$, and

$$TdS_{total} = -dH + TdS_{cell}.$$ (5.7)

It is convenient to introduce another quantity, the *Gibbs free energy* of the system, G. This quantity represents the enthalpy minus the amount of energy tied up internally in random motion and thus not free to do useful work. It is defined as $G = H - TS_{cell}$ so that

$$dG = dH - TdS_{cell}.$$ (5.8)

Since the right hand side of this last equation is simply the negative of the right hand side of the previous equation, combining the two expressions yields the relation

$$TdS_{total} = -dG.$$ (5.9)

This is a remarkable result. It states that the net change in entropy can be determined from an examination of the change in the Gibbs free energy of the system alone. Thus, there is no need to evaluate the change in entropy in the surroundings in order to determine whether a process will occur spontaneously or not. For a process to occur spontaneously there must be a net increase in the entropy, or equivalently, the Gibbs free energy of the system must decrease:

$$dG < 0.$$ (5.10)

By considering the Gibbs free energy one can determine whether a process will occur spontaneously or not. There are two parts to consider—an energetic (enthalpic) piece and an entropic one. Crystalline solids are more ordered than liquids, and liquids, in turn, are more ordered than the gaseous phase of a given substance. A crystalline solid such as ice can spontaneously melt to become liquid decreasing the system order. In this process the energy of the system increases, but this increase is more than offset by the accompanying increase in entropy or disorder. Conversely, a process that increases the order within a system can occur spontaneously if the decrease in entropy is compensated for by a decrease in internal energy. The most favorable reactions are those where energetic and entropic changes are aligned, and spontaneous processes do not occur at all when entropic and energy changes both increase the Gibbs free energy.

5.5 Spontaneous Folding of New Proteins

Newly synthesized proteins spontaneously fold into their physiologically active three-dimensional shapes. Protein folding is the process whereby nascent proteins, newly synthesized linear polypeptide chains, spontaneously fold into functional three-dimensional forms. During this process they develop their secondary structures such as alpha helices and beta sheets. The set of similar states into which protein folds is collectively referred to as the *native state*. Conversely, the collection of states that a newly synthesized protein, or an unfolded protein, populates is called the *denatured state*. In a series of pioneering experiments in the 1950s and 1960s Ansfinsen showed that protein folding is a reversible process. By varying conditions in the aqueous environment, proteins were made to go back and forth between folded and unfolded configurations. Two main conclusions can be drawn from his experiments. First, all the information needed for folding is contained in the primary sequence. Second, the native and denatured states are thermodynamically stable states—they are states of minimum Gibbs free energy.

The most important environmental or physical parameter influencing protein stability is temperature. (Two others are pH and salt concentration.) The native state is a minimum in the Gibbs free energy at physiological temperatures (and conditions). However, if the temperatures are elevated above a critical temperature, the denatured state of a protein will lie at a lower Gibbs free energy than the native state. The reason for this can be discerned in an examination of the behavior of the two terms in Eq. (5.8) for the Gibbs free energy. As the temperature is raised, the entropic contribution gains in importance relative to the enthalpic term. The entropic term is a measure of the number of possible configurations for the main and side chains. The entropy favors the denatured state because there are many more ways for the side chains to arrange themselves when unfolded than when tightly folded into a globular form. The enthalpic contribution, on the other hand, strongly favors the native state. At low temperatures the entropic contribution is still appreciable, but the enthalpic term predominates in this regime.

The property of stability is an important one. To be useful a state must be stable long enough for the biological entity, whether it be a protein, a DNA molecule, or some larger structure, to carry out its biological function Such states must be stable in a thermodynamic sense. These states, once formed, do not change appreciably in time. The effects of small perturbations and of thermal fluctuations are rapidly damped out and the behavior of the system is not appreciably altered. These are equilibrium states in the language of thermodynamics.

While there are a multitude of states corresponding to the denatured state, there are usually only a few similar states corresponding to the native state and its conformation is essentially unique. This aspect is noted in Table 5.1. In order for the native state to be stable there must be an appreciable gap in energy between the native state and nearby nonnative ones. When the differences are appreciable, it is difficult for small perturbations and thermal fluctuations to induce transitions to the nearby higher energy states. Whenever the energy gaps are small, the proteins will be only marginally stable.

TABLE 5.1. Terms and concepts used to describe protein folding.

Terms and concepts	Meaning and significance
Denatured state	Name given to a large number of high energy configurations of a newly synthesized or an unfolded protein
Energy landscape	A graphical representation of the number of states available to a protein at each value of the potential energy as a function of a few significant degrees of freedom
Fast folding	Submillisecond folding of simple proteins, whose energy landscapes have few barriers and traps
Folding funnel	The overall shape of the potential energy landscape. With many high energy states and few low energy ones, the surface narrows as the potential energy (or enthalpy) is reduced
Kinetic trap	Any set of states forming a local minimum in the energy landscape that is enclosed by energy barriers large compared to the thermal energy
Native state	Name given to the small number of low energy configurations of a biologically active protein

5.6 The Folding Process: An Energy Landscape Picture

The process whereby a nascent protein folds into its physiologically viable 3D form can be envisioned in terms of an energy landscape. Each point in the landscape would represent a possible conformation of the protein. Similar conformations would be found near one another and dissimilar ones further apart. The vertical axis in this kind of description would represent the sum of all contributions to the free energy of the protein except for the configuration entropy. That is, it represents the enthalpy or potential energy of the protein. The horizontal axis of the landscape gives the values of the various degrees of freedom, coordinates such as the dihedral angle measures. Since there are too many coordinates to depict individually, one or two coordinates, or combinations thereof, are selected that capture the essential behavior of the protein as it folds. Two representative energy landscapes constructed in this manner are presented in Figure 5.2.

As can be seen in Figure 5.2, the energy landscapes are funnel shaped. They are broad at the top and narrow at the bottom. The reason for this is a general one. There are many energetically equivalent states at the top, but far fewer ones at the bottom. Since each point on the landscape represents a state, the width of the landscape is proportional to the number of states, that is, it is directly related to the entropy. This quantity is a rapidly increasing function of the (internal) energy.

The folding process can be depicted as a trajectory connecting many points on the landscape, denoting the sequence of small conformational changes that the protein undergoes as it folds. As shown in the figure, a folding trajectory starts out at a denatured state located at the top of the

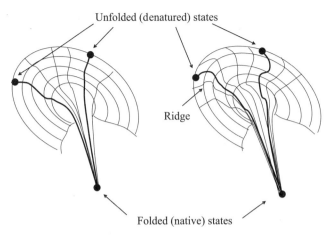

FIGURE 5.2. Energy landscapes: Each point in the cutaway views of the energy land-scapes represents a possible configuration of the protein. In these 3D depictions, the x, y-axes represent generalized coordinates indicative of the states of the protein. The vertical axis denotes the potential energy, or equivalently, the enthalpy. The overall shape—broad at the top and narrow at the bottom—gives rise to the description of these surfaces as folding funnels. The funnel on the left is smooth and the protein trajectories are straight, running down the funnel from the unfolded state to the native state. The funnel on the right is slightly more complex. A ridge is present which has to be surmounted before the protein can slide down to the native state. The trajectories are longer and convolute slightly as they pass over the ridge and then down the funnel. Even more complex funnels are possible, especially for multidomain signaling proteins, in which there are mountains and valleys that have to be traversed during folding.

landscape at a high potential energy and ends at the native state located at the bottom of the landscape at a low potential energy.

The amount of time required for a protein to fold into its native state is an important aspect of the process. This is referred to as a *kinetic requirement*. Not only must a protein fold into its native state, but also it must do so in a physiologically reasonable time interval. The speed depends critically on the topography of the potential energy surface. If the surface is studded with deep minima, and the folding trajectories pass close to them, the rate of folding will be slow. In these situations, the protein will fall into the minima and must escape before proceeding with its evolution towards its native state. The deep minima are called *kinetic traps* because of their slowing effects on the kinetics, or rates, of folding. Large and complex proteins, especially those involved in signaling pathways, tend to have rugged landscapes containing minima surrounded by high barriers. On the other hand, small single domain proteins often have landscapes that are fairly smooth and these proteins fold rapidly. The difference between smooth and rough funnels is highlighted in Figure 5.3.

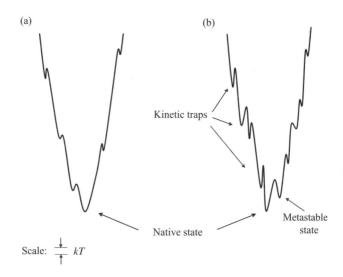

FIGURE 5.3. Smooth and rough folding funnels: (a) Smooth funnel in which there are few barriers and all of these are smaller in height than the thermal energy. (b) Rough funnel possessing several barriers that are difficult to surmount and serve as kinetic traps because their heights are much larger than the thermal energy.

One of the features present in the rough funnel depicted in Figure 5.3 is that of a low-lying metastable state. Recall from the discussion at the end of Section 5.5 that one of the conditions for stability of the native state is that there be an appreciable energy gap between the native state and those lying above it. This condition is violated in the rough landscape depicted in Figure 5.3. The protein will spend an appreciable time in the nearby excited state. Not only is that state not separated by a large energy gap, but also, once in that state, the protein must surmount a kinetic trap to get back out to the native state.

5.7 Misfolded Proteins Can Cause Disease

Not all polypeptide sequences are able to fold in any reasonable amount of time into a functionally meaningful native state. Rather, almost all randomly selected sequences will fail to do so. The replacement of even a single residue by another can sometimes convert a protein from a form that is folding-competent into one that is not. Failures of this kind often lead to disease. The cystic fibrosis transmembrane conductance regulator, or CFTR protein, serves as a prominent example of this sensitivity. Mutations to the gene encoding the CFTR protein lead to cystic fibrosis. The CFTR protein functions as a chloride channel, but when mutated it fails to fold properly and cannot insert in the membrane. The consequence is that the proper movement of chloride ions is impaired. Among the most prominent symptoms of the disease are the production of salty secretions by the sweat glands and thick mucus secretions by the lungs.

Another example is how point mutations can lead to impaired folding and disease is rhodopsin. This protein is found in rods in the retina, where it functions as the phototransducer. When this protein is mutated in regions away from the C-terminus it fails to fold properly and doesn't transduce light. In retinitis pigmentosa, the name for the resulting disorder, the rods die off; sight worsens progressively, and eventually the subject becomes blind.

One of the most striking examples of how even a single mutation can have disastrous consequences on protein folding is that of the blood oxygen carrier, hemoglobin. In this protein, a substitution of valine for glutamic acid in the beta chains leads to improper assembly of the four subunits. The result is sickle cell anemia. The hemoglobin molecules are misshapen and not only don't transport oxygen adequately, but tend to clump together causing further impairments. The clumping of misfolded proteins is not limited to hemoglobin. It is observed in a variety of neurological disorders, as well.

5.8 Protein Problems and Alzheimer's Disease

Alzheimer's disease is a neurological disorder that occurs with increasing frequency late in life. It is the leading cause of senile dementia. This neurodegenerative disorder, first described by Alois Alzheimer in 1906, can be identified by the presence of two kinds of lesions—amyliod plaques and neurofibrillary tangles. Amyloid plaques are deposits of the amyloid β (Aβ) protein. These proteins form filaments in the extracellular spaces, which are usually surrounded by abnormal cells and cell structures. The second kind of lesion occurs within cells. Neurofibrillary tangles are composed of a protein called *tau* that normally associates with microtubules. When this protein is hyperphosphorylated it dissociates from the microtubules and aggregates into insoluble filaments, the tangles.

The Aβ protein that forms the amyloid plaques is generated from a larger amyloid β protein precursor (APP) by means of a series of posttranslational cleavages. APP is a single-pass protein that resides in intracellular membranes. While membrane-bound it undergoes three proteolytic cleavage operations, first by an α-*secretase*, the second by a β-*secretase* and the third by a γ-*secretase*. The end product of these finishing operations is the Aβ protein. Mutations in ADD and in a family of proteins called *presenilins* lead to familial Alzheimer's disease. (The presenilins are 8-pass intramembrane proteins found in complexes that carry out the γ-secretase stage of proteolytic processing.)

5.9 Amyloid Buildup in Neurological Disorders

Amyloid buildups arising from misfolded proteins characterize many neurological disorders. As mentioned above, the amyloid deposits seen in Alzheimer's patients are composed of Aβ proteins. The underlying reason for their aggregation into insoluble clumps is that the proteins are misfolded. The formation of insoluble protein clumps in certain classes of cells in the brain is not restricted to Alzheimer's disease, but rather is encountered in a host of neurological disorders (Table 5.2). One of the most commonly encountered examples is Parkinson's disease. In this neurological disorder, aggregates of misfiled alpha synuclein proteins form clumps called *Lewy bodies* in dopamergic cells regulating motor function, leading to its impairment. Another entire set of examples is provided by the spongiform encephalopathies (SEs), where buildups of prion proteins occur. Mad cow disease in cattle and Creutzfeldt–Jakob disease in humans are two forms of SE. Finally, in Lou Gehrig's disease (amyotrophic lateral sclerosis, or ALS), CuZn superoxide dismutase (SOD1)-enriched inclusions develop in motor neurons possibly associated with the buildup of free radicals in the affected cells.

A clue as to what might be happening in all of these disorders is provided by the observation that the intracellular clumps of proteins that form in the neurons contain other proteins besides the misfolded ones. Proteins belonging to the ubiquitin-proteasome system are often found in these deposits. In Huntington's disease, the ubiquitin-proteasome system seems to have broken down, and the normal housekeeping function of removal of misfolded proteins does not occur. There is growing evidence for a breakdown in at least some forms of Parkinson's disease, as well. It may be that the ubiquitin-proteasome system, responsible for removal of misfolded proteins, is being overwhelmed, leading over time to cell death as more and more misfolded proteins accumulate in the long-lived neurons.

Even small clumps of misfolded proteins can be toxic. In Alzheimer's disease, there is either an excessive production of the Aβ protein, or the ratio of the amyloid-prone 42 amino acid residue forms over the less toxic

TABLE 5.2. Amyloid-producing neurological disorders.

Neurological disorder	Amyloid-forming protein
Alzheimer's disease	Amyloid β protein
Creutzfeldt–Jakob disease	Prion
Huntington's disease	Huntingtin
Lou Gehrig's disease	CuZn superoxide dismutase
Parkinson's disease	Alpha synuclein

40 amino acid residue forms is too high, or there is an alteration of the Aβ protein's biophysical properties promoting clumping. The proteins are partially folded resulting in the exposure of hydrophobic patches of residues that promote aggregation. The danger inherent in even small oligomeric assemblies of such proteins can be seen in the hippocampus where small soluble oligomers consisting of two or three Aβ proteins can impair normal physiological functions of the cells. Small oligomers of misfolded forms of even nominally harmless proteins can interact in an inappropriate fashion with other cellular proteins. In sum, small soluble aggregates of misfolded proteins can be highly toxic to the cells.

5.10 Molecular Chaperones Assist in Protein Folding in the Crowded Cell

The cell is a crowded place and as a result many polypeptides, especially large ones, require the assistance of helper molecules, the chaperones, to reach their native state. Because of crowding, proteins, especially those that are not completely folded and have exposed hydrophobic surfaces, tend to aggregate. One of the main functions of the molecular chaperones is to shuttle large polypeptides to their correct locations in the cell. In the process they prevent hydrophobic contact-driven aggregates from forming and assist in the formation of appropriate associations. Because of the crowding the energy landscape can sometimes be altered to the point where the proteins cannot reach their native state. Crowding also alters the kinetics, especially those involved in the assembly of multiprotein complexes. The chaperones perform several useful housekeeping functions. They assist in folding and in the rescue of proteins that have partially unfolded due to cellular stresses. In eukaryotes, which have a greater proportion of large polypeptides than prokaryotes, chaperones are among the most abundant proteins in the cell.

Folding in cells can be simple or complex depending on the nature of the protein and its cellular destination. Small, rapidly folding proteins can reach their native, necessary, functional three-dimensional conformation unaided. As noted by Anfinsen in his 1973 Nobel Prize lecture, all information required for folding is contained in the primary sequence. Because of cellular crowding, large and more complex polypeptides may require a protection from the immediate environment; that is, they may need a folding "cage" to provide a shielded folding environment. Large and complex polypeptides, especially those involved in signaling, destined for membrane insertion and/or attachments, may require the actions of chaperones to prevent aggregation. Some nascent proteins may be shuttled to their destinations in a form other than their native state and fold into the native state only upon arrival and insertion into functional multisubunit complexes.

5.11 Role of Chaperonins in Protein Folding

The energy landscapes for protein folding can be fairly smooth or they can be rough. The landscapes for the fast folding proteins correspond to the former situation. Large proteins may be able to reach their native state without any assistance, but may do so on a time scale that is too slow physiologically. These proteins tend to have rough energy landscapes and the proteins may easily be trapped in local minima. Molecular cheperones known as *chaperonins* do two things to alleviate these difficulties. First, they sequester the nascent protein away from other cellular proteins, forming folding cages. Second, they directly participate in the folding process, supplying energy through ATP hydrolysis to the substrate protein. The additional energy supplied to the protein enables that protein to surmount energy barriers and escape from kinetic traps. The protein is able to fold far more rapidly into its native state since it can avoid getting stuck in nonoptimal and nonfunctional states for long periods of time.

The subunits that form the folding cage repeatedly bind and release the substrate protein. They disrupt the misfolded structures that are kinetically trapped, and release the protein in a less folded configuration through the use of mechanical stretching forces. At the end of each round of binding and release, the protein is free to continue to fold and search for its low energy native state. The energy-dependent process of binding and release is called *annealing* because of its close resemblance to the metallurgical process of that name.

Chaperonins (Table 5.3) are large, 800- to 1000-kDa barrel-shaped structures that serve as folding cages. The group I GroEL chaperonin contains

TABLE 5.3. Molecular chaperones provide protected environments for protein folding, greatly accelerate the folding rate, and stabilize nascent chains.

Family	Distribution	Function
Group I Chaperonins		
Hsp60	GroEL in bacteria; Hsp60 in mitochondria and chloroplasts	Protected environment for folding, refolding, recovery from stress
Hsp10	GroES in bacteria; Hsp10 in mitochondria and chloroplasts	Co-chaperonin for Hsp60; assists in folding substrates bound to Hsp60
Group II Chaperonins	TriC/CCT chaperonins in archaea and eukaryotic cytosol	Protected environment for folding, refolding, recovery from stress
Hsp70 Chaperones		
Hsp70	DnaK in bacteria; Hsp70 in eukaryotes	Regulates heat shock response, stabilizes nascent chains, and prevents aggregation
Hsp40	DnaJ in bacteria; Hsp40 in eukaryotes	Co-chaperone for Hsp70; regulates activity of Hsp70
Hsp90 Chaperones	HtpG in bacteria; Hsp90 in eukaryotes	Maturation of signal transduction pathways

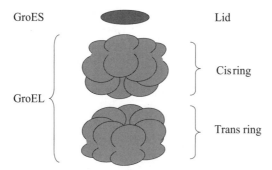

GroES   Lid

GroEL   Cis ring

                                               Trans ring

FIGURE 5.4. A GroEL-GroES Anfinsen cage for protein folding: Fourteen subunits
are arranged into two rinds of seven units each. Each chain contains equatorial,
intermediate, and apical domains. Polypeptide chains are inserted into the cage,
undergo a round of mechanical manipulations (i.e., the chains are "annealed" to
remove unfavorable bondings), and are ejected. Several rounds of these operations
may take place.

14 identical subunits; the eukaryotic Hsp 60 molecule has 16 identical
subunits. As shown in Figure 5.4 the subunits are arranged into a pair of
rings; in the case of GroEL each ring contains 7 identical subunits. Co-
chaperonins help in this activity. Group I co-chaperonins form a lid over
the barrel and assist in the folding.

5.12 Hsp 90 Chaperones Help Maintain Signal
Transduction Pathways

The second main group of molecular chaperones consists of the Hsp70
and Hsp90 families. The Hsp70 family of chaperones performs a variety of
tasks in the prokaryotes. DnaK is the outstanding bacterial Hsp70 family
member. It regulates the heat shock stress response and maintains proteins
in their physiologically viable configurations. Eukaryotic polypeptide chains
are some 30–40% larger on the average than their prokaryotic counterparts.
Eukarotic Hsp70 chaperones bind newly synthesized proteins. They assist
in transporting these proteins across organelle membranes and into the
endoplasmic reticulum (ER) and mitochondria. They also assist in the insertion of nascent membrane proteins into membranes.

Members of the Hsp90 family are involved in maintaining the functional
integrity of proteins involved in intracellular signaling. These chaperones
often work in association with Hsp70 family members and several other
small ancillary proteins to prevent aggregation and mediate refolding. The
Hsp90 chaperones form complexes with signal molecules, then help in their
translocation to the correct subcellular compartment and into association
with other elements of the signal pathway where they operate.

Members of the Hsp90 family of molecular chaperones are among the most abundant proteins in the cell. They account for 1–2% of the total cellular protein even under non-stressful conditions. Hsp90 acts later in the folding process than the other chaperones. In normal cells, Hsp90 family members bind to a selective set of proteins operating in the signaling pathways that regulate cell differentiation and embryonic development. The proteins targeted by Hsp90 typically have low energy conformations that are only marginally stable. These signaling proteins have difficulty remaining in their physiologically competent states and require the assistance of Hsp90 to stabilize them. For example, in the absence of Hsp90, steroid hormone receptors are unable to bind their ligands and DNA. Likewise, bereft of Hsp90, tyrosine kinases such as c-Src lose their ability to function as kinases and to be acted upon by their regulators.

5.13 Proteins: Dynamic, Flexible, and Ready to Change

Proteins are not static but instead are dynamic structures that exhibit considerable flexibility, and readily adopt new shapes. Proteins are dynamic entities. If one examines the energy landscape in the vicinity of the protein's native state one finds that there is an ensemble of low-lying states and the proteins is continually undergoing transitions from one state to another. This population of states and the barriers separating them play an important role in substrate recognition and catalysis. An example of such as ensemble of low-lying conformational states is presented in Figure 5.5. In this figure, none of the barriers are particularly high compared to the thermal energy kT. In this situation, the proteins will easily undergo transitions from one state to another, but will spend more time in the lowest energy state than in any of the others. The protein whose energy landscape is depicted in Figure 5.5 is quite flexible. It is able to move back and forth among a set of different shapes. The energy landscape of a protein that is completely rigid would be far sparser if it lacked low-lying states that were easy to get to.

Scale: kT

FIGURE 5.5. Low-lying ensemble of native and nearby state: These states determine the binding properties of the proteins. Different ligands acting as environmental factors select one or more of these preexisting states when they bind the protein.

The energy landscapes are influenced by environmental factors such as temperature, pH, ionic concentration, and binding events. Catalysts make use of protein motions to increase the rate of reactions. Environment factors, most notably charged groups such as acids, bases, metal ions, and dipoles belonging to the catalyst generate a shift in the energy landscape.

Two kinds of shifts are possible, *kinetic* and *thermodynamic*. In a kinetic shift, the energy barrier separating two conformational states is lowered, making possible transitions from a higher energy state to a lower energy state. The initial kinetic barrier is high compared to the thermal (kinetic) energy factor kT, and the system may be trapped in the higher, or metastable, state prior to the kinetic shift.

In a thermodynamic shift (the situation illustrated in Figure 5.5) the barrier between two states remains the same but the relative energies are modified so that a formerly higher energy state is now a lower energy state. The barriers in this scenario low so there are no kinetic barriers, and transitions among the many states can be frequent. In either case, a catalyst generates a shift in the population of conformations towards those that favor the reaction being catalyzed. This latter utilization of protein flexibility is often referred to in the literature as *stabilizing the transition state* in the case of catalysis, and as *the induced fit model of surface complementarity* in the case of binding. The key point is that protein flexibility underlies the ability of a protein to bind to another molecule—and to entire groups of proteins of differing shape and size—with the appropriate degree of specificity. In all of these situations the ligand may be thought of as selecting out one or more states from the ensemble of preexisting low-lying states.

References and Further Reading

Anfinsen Nobel Prize Lecture

Anfinsen CB [1973]. Principles that guide the folding of protein chains. *Science*, 181: 223–230.

Motions of Proteins

Cavanagh J, and Akke M [2000]. May the driving force be with you—Whatever it is. *Nature Struct. Biol.*, 7: 11–13.

Feher VA, and Cavanagh J [1999]. Millisecond-timescale motions contribute to the function of the bacterial response regulator protein SpoOF. *Nature*, 400: 289–293.

Forman-Kay JD [1999]. The "dynamics" in the thermodynamics of binding. *Nature Struct. Biol.*, 6: 1086–1087.

Frauenfelder H, Sligar SG, and Wolynes PG [1991]. The energy landscapes and motions of proteins. *Science*, 254: 1598–1603.

Kay LE, et al. [1996]. Correlation between dynamics and high affinity binding in an SH2 domain interaction. *Biochem.*, 35: 361–368.

Kern D, et al. [1999]. Structure of a transiently phosphorylated switch in bacterial signal transduction. *Nature*, 402: 894–898.

Lee AL, Kinnear SA, and Wand AJ [2000]. Redistribution and loss of side chain entropy upon formation of a calmodulin-peptide complex. *Nature Struct. Biol.*, 7: 72–77.

Stock A [1999]. Relating dynamics to function. *Nature*, 400: 221–222.

Zidek L, Novotny MV, and Stone MJ [1999]. Increased protein backbone conformational entropy upon hydrophobic ligand binding. *Nature Struct. Biol.*, 6: 1118–1121.

Protein Folding: The Energy Landscape Picture

Bryngelson JD, et al. [1995]. Funnels, pathways, and the energy landscape of protein folding: A synthesis. *Proteins: Structure, Function and Genetics*, 21: 167–195.

Chan HS, and Dill KA [1998]. Protein folding in the landscape perspective: Chevron plots and non-Arrhenius kinetics. *Proteins: Structure, Function and Genetics*, 30: 2–33.

Dill KA, and Chan HS [1997]. From Levinthal to pathways to funnels, *Nature Structure Biology*, 4: 10–19.

Leopold PE, Montal M, and Onuchic JN [1992]. Protein folding funnels: A kinetic approach to the sequence-structure relationship. *Proc. Natl. Acad. Sci. USA*, 89: 8721–8725.

Onuchic JN, et al. [1995]. Toward an outline of the topography of a realistic protein-folding funnel. *Proc. Natl. Acad. Sci. USA*, 92: 3626–3630.

Sali A, Shakhnovich E, and Karplus M [1994]. How does a protein fold? *Nature*, 369: 248–251.

Molecular Chaperones and Protein Folding in the Cell

Hartl FU, and Hayer-Hartl M [2002]. Molecular chaperones in the cytosol: From nascent chain to folded proteins. *Science*, 295: 1852–1858.

Pratt WB [1998]. The Hsp90-based chaperone system: Involvement in signal transduction from a variety of hormone and growth factor receptors. *Proc. Soc. Exp. Biol. Med.*, 217: 420–434.

Rutherford SL, and Lindquist S [1998]. Hsp90 as a capacitor for morphological evolution. *Nature*, 396: 226–342.

Sauer FG, et al. [2000]. Chaperone-assisted pilus assembly and bacterial attachment. *Curr. Opin. Struct. Biol.*, 10: 548–556.

Binding Mechanisms

DeLano WL, et al. [2000]. Convergent solutions to binding at a protein-protein interface. *Science*, 287: 1279–1283.

Freire E [1999]. The propagation of binding interactions to remote sites in proteins: Analysis of the binding of the monoclonal antibody D1.3 to lysozyme. *Proc. Natl. Acad. Sci. USA*, 96: 10118–10122.

Hilser VJ, et al. [1998]. The structural distribution of cooperative interactions in proteins: Analysis of the native state ensemble. *Proc. Natl. Acad. Sci. USA*, 95: 9903–9908.

Kumar S, et al. [2000]. Folding and binding cassettes: Dynamic landscapes and population shifts. *Protein Sci.*, 9: 10–19.

Ma B, et al. [2002]. Multiple diverse ligands binding at a single protein site: A matter of pre-existing populations. *Protein Sci.*, 11: 184–197.

Teague S [2003]. Implications of protein flexibility for drug design. *Nature Rev. Drug Dis.*, 2: 527–541.

Problems

5.1 Entropy, density of states, and probabilities. Consider a system of N atoms, each of which is placed in one of M bins. The atoms are labeled by the bins into which they are placed, but are otherwise indistinguishable from one another. That is, n_1 atoms are placed in bin 1; n_2 atoms are placed in bin 2, and so on up to n_M atoms in bin M. The quantities

$$p_i = n_i/N$$

represent the probabilities of finding an atom in a particular bin. The number of ways a specific configuration can be realized (where "configuration" means finding n_1 atoms in bin 1, n_2 atoms in bin 2, and so on) is, from elementary probability theory, given by the multinomial coefficient

$$W(\{n_1, n_2, \ldots, n_M\}) = \frac{N!}{n_1 n_2 \cdots n_M}.$$

The relationship between number of states k, actually coeffs, and entropy S was first established by Ludwig Boltzmann and Max Planck. Their results are usually presented in the form:

$$S = \frac{k}{N} \ln W(\{n_i\}),$$

and sometimes in the form

$$S = -k \sum_{i=1}^{M} p_i \ln p_i.$$

Show that the two forms are equivalent; that is,

$$\left(\frac{k}{N}\right) \ln W(\{n_i\}) = -k \sum_{i=1}^{M} p_i \ln p_i.$$

5.2 Frustration and rugged energy landscapes. Rugged energy landscapes are formed whenever there is a lack of good low energy conformations. Situations of this type arise when the system of atoms

(a)   (b)

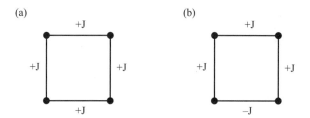

FIGURE for Problem 5.2. Nonfrustrated and frustrated squares: Atoms (vertices) are connected by bonds (sides) with bond (coupling) strengths as indicated. (a) Nonfrustrated square in which all couplings are positive. (b) Frustrated square in which there is one negative coupling and three positive couplings.

and bonds connecting them becomes frustrated. The term *frustration* is intended to convey the inability of the atomic system to fold itself in such as way that all bond orientations lead to good low energy states. Instead, for any spatial orientation of bonds, some interactions will be positive and others will be negative, and in place of a few deep low energy states there will be a large number of less deep low energy states.

The notion of frustration can be illustrated using a mini system consisting of four spin $1/2$ atoms arranged in a pair of squares as shown above. Each vertex in a square contains an atom whose spin value is either $+1$ or -1. The energy of the systems is the sum of four interactions of the form $Js_i s_j$, where s_i and s_j are the spin values of atoms at neighboring vertices connected by bonds, and J is the bond strength. In other words, the energy E is given by the expression

$$E = -\sum J_{ij} s_i s_j,$$

and the sum is over the four pairs of neighboring vertices. There are 16 possible states of this system: $2 \times 2 \times 2 \times 2$. The nonfrustrated square on the left possesses two good (deep) low-energy states—one where all spins are $+1$ and one where all spins are -1, each with energy $-4J$. All other states have higher energies. (a) Tabulate the distribution of states and then compare this distribution to that for the frustrated square shown in the right hand part of the figure. (b) What happens if the top and bottom couplings are negative while the two side couplings remain positive?

5.3 Transition rates and their dependence on barrier height. The transition rate for passage over a barrier from state A to state B is given by the Arrhenius formula:

$$v = v_0 e^{-E_0/kT}.$$

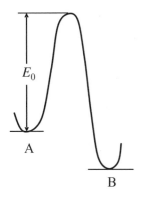

FIGURE for Problem 5.3. Transition over a barrier from one state to another.

In this expression, v is the transition rate, which is equal to the product of a factor v_0 that gives the number of attempts being made on the barrier (the attempt frequency) and the Boltzmann factor (i.e., the exponential term) describing the probability for success for any given attempt. Assuming no change in the attempt frequency, how much less likely is a transition for the case where barrier height is a factor of 3 greater than the thermal energy kT as compared to a situation where the barrier height is half that of the thermal energy?

6
Stress and Pheromone Responses in Yeast

Unicellular organisms such as the budding yeast *Saccharomyces cerevisiae* have to cope with continually changing environments and are subject to a variety of environmental and physiological stresses. These include unfavorable temperatures that produce heat shock, osmotic stresses that disrupt the water balance, carbon source deprivation that leads to starvation, and oxidative stresses. Yeast cells respond to these and other unfavorable conditions by altering their patterns of gene expression and by adjusting their rates of protein synthesis. They may respond in a rather general way to a stressful situation, or they may respond in a far more specific manner to a particular stress condition.

The signaling routes that ultimately produce the alterations in gene expression and protein synthesis begin at the plasma membrane with signal reception. Signal receptors embedded in the plasma membrane sense changes in the local environment and receive chemical messages sent by other cells. The end points of the signaling routes are control points in the fixed infrastructure. It is at these control points that the signals are converted into cellular response. Some signaling routes, especially those found in bacteria, are fairly short with at most a few elements lying between sensing and contact elements. Other signal routes are far longer. In these situations, the routes from cell membrane to the contact points in the cell interior often involve several intermediate signaling elements and control points where several signals converge. Whether short or long, a set of signaling elements is activated in a sequential fashion, from cell surface molecules to intracellular signal elements to one or more final contact elements. The ensemble of signaling elements activated in this way is referred to as *forming a signaling pathway.*

The different kinds of signaling elements that make up the typical eukaryotic signaling pathway will be introduced in the first part of this chapter. Examples will then be given in the second part of the chapter of how these elements come together to form stress and pheromone signaling pathways in yeast. The short signaling routes found in bacteria will be explored in Chapter 7, and an important set of signaling intermediaries, commonly referred to as *second messengers*, will be explored in the chapter after that, Chapter 8.

6.1 How Signaling Begins

Signaling begins at the plasma membrane with receptor-ligand binding and signal transduction. Transmembrane receptors function as sensors of environmental stresses and as receivers of chemical messages. They transmit, and, in the process, convert the signals from an external, outside-the-cell form to an internal, inside-the-cell one that can be understood and further processed. This process is called *signal transduction*. Several kinds of transmembrane receptors are presented in Figure 6.1. In each example, part of the receptor lies in the extracellular spaces enabling it to bind to ligands. There are also one or more transmembrane segments, portions of the polypeptide chain that either pass once through the plasma membrane or wind back and forth several times. Lastly, there is a cytoplasmic segment that permits the receptor to make contact with signaling molecules inside the cell, thereby allowing the receptor to transduce a signal.

The notion of signal transduction is a central one. The ligand itself is not passed into the cell. Rather, a different molecule becomes activated and conveys the message inside the cell. This conversion process can happen several times inside the cell with one protein contacting a second one, which then contacts and activates a third molecule, and so on until the final signaling element reaches a control point where the signal is converted into a cellular response. In each of these transfers, one signal molecule replaces another in a sequential manner to carry the signal. Thus, each step involves transduction, the conversion of a message from one form to another.

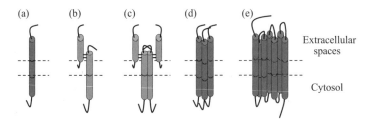

FIGURE 6.1. Signal receptors: Each receptor has an extracellular region, a transmembrane segment, and an intracellular (cytosol) portion. (a) Single-chain, single-pass receptor; (b) two-chain, single-pass receptor; (c) dimeric arrangement of two-chain, single-pass receptors; (d) trimeric arrangement of single-chain, single-pass receptors; and (e) single-chain, seven-pass receptor. The N-terminal is usually located on the extracellular side and the C-terminal on the cytoplasmic side in the single-chain and seven-pass receptors. In the double-chain receptors, the uppermost side is N-terminal and the lower side is C-terminal.

Receptors can be grouped into families according to their topology, structure, and the kinds of ligands they bind. The first two examples presented in Figure 6.1 are single-pass receptors. In Figure 6.1(b), two chains are used rather than a single chain as in Figure 6.1(a). One chain lies entirely in the extracellular space and is responsible for ligand binding. The other chain is covalently linked to the first (through disulfide bonds). It contains transmembrane and cytoplasmic segments and transduces the signal into the cell.

Receptors form associations with other receptors. One way for a signal to be conveyed into the cell is through the formation and stabilization of receptor complexes. The complex may consist of a pair of receptors, as in Figure 6.1(c), simultaneously bound to a single ligand (1:2 stoichiometry) or to a pair of ligands (2:2 stoichiometry). Alternatively a receptor trimer may form as in Figure 6.1(d) bound to, for example, three ligands. Higher-order complexes may form and these complexes may even involve a mix of different receptors. The last example, presented in Figure 6.1(e), is for a seven-pass, single chain receptor. This is the topology of the largest family of receptors found in the body—the G protein-coupled receptors (GPCRs). They sense and transduce sensations such as light and odors, and cell-to-cell hormonal and neuromodulatory signals. They are also the targets of about half of all the drugs commercially produced.

6.2 Signaling Complexes Form in Response to Receptor-Ligand Binding

Receptor-ligand binding stimulates the assembly of signaling complexes at and just below the plasma membrane. Ligand binding provokes changes in the environment around the cytoplasmic portion of the receptor(s) leading to formation of these complexes. These changes may take one or more forms. In some receptor families, ligand binding stabilizes the formation of receptor dimers. The receptors possess an intrinsic kinase activity, which is turned on when the two chains are brought into close proximity to one another. The ensuing phosphorylation opens up docking sites for cytoplasmic signaling proteins that, in turn, seed the formation of the signaling complex. In families such as the GPCRs, ligand binding triggers a changes in conformation that are sufficient to activate nearby cytoplasmic G-proteins leading to activation of signaling intermediates—second messengers—and the formation of signaling complexes.

Proteins functioning as anchor, scaffold, and adapter proteins help organize these signaling complexes. As depicted in Figure 6.2, *anchor proteins* provide platforms for proteins to attach in close proximity to receptors and other upstream signaling elements. Anchor proteins attach to the plasma membrane and to membranes of organelles. They allow two or more sig-

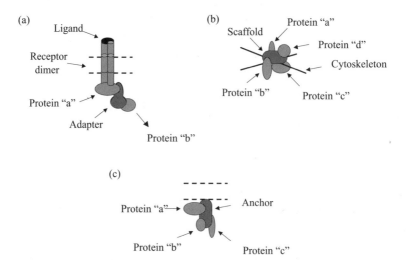

FIGURE 6.2. Assembly of signaling complexes: (a) Signal complex organized by exposure of a docking site by a receptor aided by the utilization of an adapter protein; (b) signal complex organized by a cytoskeleton-associated scaffold protein; and (c) signal complex organized by a membrane-associated anchor protein.

naling proteins to position themselves in close proximity to one another and to their membrane-associated substrates. *Scaffold proteins* operate in a similar fashion to anchor proteins. They too provide platforms for attachment of multiple proteins. They attach to components of the actin cytoskeleton, or alternatively, to microtubules. These attachment sites are usually located near the plasma membrane and play an important role in relaying signals from the plasma membrane to downstream control sites. The third group of nonenzymatic intermediaries is the *adapter proteins*. These proteins contain protein-protein interaction domains and serve as intermediaries that allow two proteins that would otherwise not be able to communicate to do so.

In order to form a complex, proteins diffuse over to and collect at the docking sites. The proteins are said to be "recruited" to the collection point. The process of recruitment is mostly a passive one driven by diffusion of proteins localized in the vicinity of docking sites opened up by receptor-ligand binding. The diffusion is largely planar; that is, it is a two-dimensional rather than a three-dimensional process because all signaling elements are located at and just below the plasma membrane. The adapters, anchors, and scaffolds localize the necessary signaling elements close to one another at and just below the plasma membrane. Because the proteins are localized nearby and because the process is a two-dimensional one, it is fairly rapid.

6.3 Role of Protein Kinases, Phosphatases, and GTPases

Protein kinases, phosphatases, and GTPases convey signals from the cell surface to downstream cytoplasmic effectors. The organization of signaling complexes by anchors, scaffolds, and adapters is illustrated in Figure 6.2. In each example, proteins designated in a generic sense as protein "a," or protein "b," and so on are recruited to the complex. The proteins so indicated are mostly enzymes. Two different kinds of enzymes predominate—*protein kinases* and *protein phosphatases*, and these are the primary intracellular signal transducers. Protein kinases are large proteins that catalyze the transfer of phosphoryl groups to target proteins using ATP as a donor. Protein phosphatases do the opposite: They catalyze the removal of phosphoryl groups using ADT as an acceptor. These two operations are illustrated in Figure 6.3. The substrates that accept and lose phosphoryl groups are often kinases themselves, so that one kinase activates another, which then activates a third kinase, and so on.

Protein kinases and phosphatases are not the only categories of enzymes that participate in signaling. Enzymes called *GTPases* also contribute to signaling, as do a variety of proteolytic enzymes, or *proteases*. GTPases are enzymes that are activated when they bind guanosine triphosphate (GTP) and are deactivated when they bind guanosine diphosphate (GDP). Two sets of ancillary enzymes catalyse GDP/GTP binding and release from the GTPases. As illustrated in Figure 6.4, a guanine nucleotide exchange factor (GEF) catalyzes the dissociation of GDP from the GTPase. GTP is an abundant cytosolic protein and in response to the actions of the GEF it binds the GTPase, thereby activating it. The GTPase-activating protein (GAP) accelerates the enzymatic actions of the GTPase. It also speeds up the hydrolysis of the bound GTP molecule, leaving a bound GDP molecule in its place, resulting in the inactivation of the GTPase.

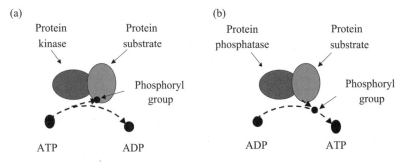

FIGURE 6.3. Enzymatic actions of protein kinases and protein phosphatases: (a) A protein kinase catalyzes the transfer of a phosphoryl group to its substrate using ATP as a donor. (b) A protein phosphatase catalyzes the removal of a phosphoryl group from its substrate using ADP as an acceptor.

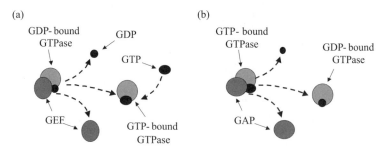

FIGURE 6.4. Catalytic activities of GTPase-associated factors: (a) A GEF catalyzes the removal of GDP from a GTPase, which then binds to a GTP molecule. (b) A GAP catalyzes the hydrolysis of the GTP molecule bound to the GTPase, leaving a bound GDP molecule.

6.4 Role of Proteolytic Enzymes

Proteolytic enzymes transform, activate, and deactivate signaling elements. *Proteolytic processing* is yet another kind of posttranslational modification that has an important role in signaling. Several different kinds of proteolytic enzymes, or proteases, are active in the cell. Some of the them catalyze the cleavage of membrane-associated proteins either on the extracellular side or the intracellular side of the plasma membrane. The cleavage of a tether that anchors a signaling protein on the outside of the cell converts that protein into a diffusible ligand, thereby greatly extending its range of influence. The intracellular cleavage of the protein either anchored to or embedded in the plasma membrane converts that protein into a messenger that can relay messages into the cell interior, allowing it to function not only as an upstream signaling element but also as a downstream one. These operations are illustrated in Figure 6.5(a) and (b).

The enzymes involved in transducing signals are localized where they are needed in inactive forms. When an initiating event takes place, such as the arrival of an extracellular ligand that binds a receptor located in the plasma membrane, the proteins that the signaling pathway comprises are activated. A series of events take place with elements upstream activating substrates downstream until the last signaling elements contact the fixed infrastructure at a control point.

The signaling elements must be carefully controlled to prevent unwanted signaling. The signals that convert the signaling molecules from an inactive to an active form are of several kinds, and these often work in concert to "turn on" signaling when and only when appropriate activating signals are present. One way of accomplishing this is to immobilize a signaling protein through binding it to a scaffolding protein. If the scaffold protein is proteolytically processed, the signaling protein will be freed up and can convey

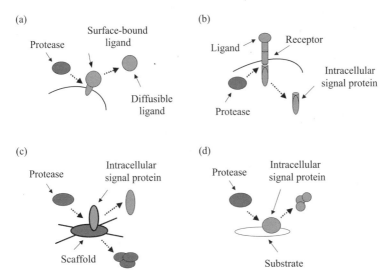

FIGURE 6.5. Signaling activities of proteolytic enzymes: (a) A protease cleaves the tether, detaching the ligand from the outer leaflet of the plasma membrane allowing it to function as a diffusible ligand. (b) A protease cleaves the polypeptide chain just below the plasma membrane, allowing the cytoplasmic portion of the protein to function as a downstream signaling element. (c) A protease degrades a scaffold protein, freeing up the signaling protein to convey a signal to its downstream substrate. (d) A protease degrades a signaling protein shortening its lifetime and terminating signaling.

a message as shown in Figure 6.5(c). Proteolytic processing can do the opposite and turn off signaling. If a signaling protein rather than a scaffold is tagged for proteolytic processing as indicated in Figure 6.5(d), the signal protein's lifetime will be reduced and its signaling activity will be damped down.

6.5 End Points Are Contact Points to Fixed Infrastructure

The end points of signaling pathways are control points that make contact with the fixed infrastructure. The furthest downstream signaling elements in the signaling pathways make contact with one or more components of the fixed infrastructure. The components being contacted may be a receptor or ion channel embedded in a membrane, or they may belong to the cytoskeleton. They may belong to an intracellular transport organelle, or to the mitochondria, or to the transcription apparatus in the nucleus, or to the translation machinery situated in the endoplasmic reticulum.

The most commonly occurring end points are those that contact and regulate the transcription machinery. Proteins that convey regulatory instructions to the transcription machinery are called *transcription factors*. These are the last or furthest downstream signaling elements in the various stress and pheromone pathways. These proteins bind to specific DNA sequences located in noncoding regions of genes and as a consequence influence transcription either positively or negatively. In the former case by makig it easier to carry out transcription and in the latter case by making it more difficult to do so. The noncoding gene regulatory regions contain binding sites for elements of the fixed infrastructure to bind and also sites for regulatory proteins to attach. The control regions are known as *promoters* and the sequence-specific control points where transcription factors bind and come together to regulate transcription are known as *responsive elements* (REs).

6.6 Transcription Factors Combine to Alter Genes

In the budding yeast, several types of stress response can be evoked. Those elements of a stress response that are common to all stresses are handled by the yeast general stress response, while additional elements needed to treat certain kinds of stresses are handled by a set of specific stress responses. Organisms such as the yeast respond to stresses by remodeling their pattern of gene expression, upregulating some genes and downregulating others. Experiments performed using DNA microarrays show that changes in the patterns of gene expression are global, often involving substantial fractions of the entire genome. The control points are organized in a way that makes possible these stress responses. Genes that require coexpression because their products must all be present at the same time and work together possess a common set of responsive elements to which the transcription factors may bind. That is, multiple copies of each of these responsive elements are distributed among promoters throughout the genome. Certain combinations of responsive elements are encountered in genes that are cotranscribed in response to specific stress and pheremone signals; other combinations are often encountered in genes coexpressed in response to a different set of stress signals and some are encountered in almost all stress responses. By this means a small number of transcription factors can coordinate and control a major remodeling of the cellular infrastructure.

Transcription factors activated by stresses act in a combinatorial fashion to trigger genome wide alterations in gene expression. In eukaryotes, these noncoding DNA sequences almost always contain binding sites for more than one transcription factor. The transcription factors jointly determine whether and how strongly to initiate transcription. The process whereby several transcription factors regulate transcription is known as *combinatorial control*. This term signifies one of the key aspects of the process. Transcription factors can come together in a variety of ways, acting in different combinations of twos, threes and so on. A small number of transcription factors can therefore generate many different patterns of gene expression in response to stresses and other environmental changes.

Combinatorial control is a synergistic process. In combinatorial control the effect produced by two positively acting transcription factors is far greater than the sum of the effects of each individual in the pair. This can happen in several ways. One way of producing an effect of this sort is for one of the partners to relieve an autoinhibitory interaction present in its partner that prevents the partner from acting. Alternatively, one of the partners may act on the DNA substrate to expose or better prepare a binding site for the other partner. These interactions are all cooperative with the first interaction making it easier for the second interaction to occur.

In the remainder of this chapter, the operation of signaling pathways activated by yeast cells in response to stresses will be examined. The yeast stress pathways illustrate how the various kinds of signaling proteins come together to form a pathway. Before examining these pathways the different categories of signaling elements will be looked at in more detail.

6.7 Protein Kinases Are Key Signal Transducers

Protein kinases are central regulators of cellular physiology. These proteins catalyze the transfer of a gamma-phosphoryl group from an ATP molecule (usually bound to Mg^{2+} ion) to a hydroxyl group on the side chain of specific residues in target proteins. In other words they are phosphotransferases that use ATP (and sometimes GTP) as a donor and specific residues in target proteins as acceptors. Their ability to function as enzymes is due to the presence of a catalytic domain of some 200 to 300 amino acids. The kinase domain, along with one or more regulatory domains, comprise the kinase. Some protein kinases encode their catalytic domains in separate subunits; in others, the catalytic and regulatory domains are all part of a single chain molecule.

The protein kinases can be divided into two large groups. In the first group, are the protein kinases that target hydroxyl groups on serine/threonine residues. These are called *serine/threonine (ser/thr) kinases*. In the second group are the protein kinases that target hydroxyl groups on tyrosine residues, and these are called *tyrosine (tyr) kinases*. The proteins in each major grouping can be further placed into a number of families that are highly conserved across all eukaryotes. Listed in Table 6.1 are five prominent groups of serine/threonine kinases, organized according to similarities in kinase domain structure, substrate specificity, and method of regulation/activation. These kinases are encountered in eukaryotes ranging from yeast to man. Tyrosine kinases are more restricted in their distribution. They are largely absent from unicellular organisms and appear to have emerged

TABLE 6.1. Serine/threonine kinases of eukaryotes.

Group	Family	Comments
AGC	Protein kinase A (PKA)/Protein kinase G (PKG)	Cyclic nucleotide-regulated; basic amino acid-directed
	Protein kinase C (PKC)	Lipid-, calcium-regulated
	Protein kinase B (PKB), Akt, RAC	Lipid-regulated
	G protein-coupled receptor kinase (GRK)	Lipid-, calcium-regulated
	Ribosomal S6 kinase (S6K)	Lipid-regulated
	Phosphoinositide-dependent protein kinase-1 (PDK1)	Lipid-regulated
CaMK	CaM kinase (CaMK)	Ca^{2+}/calmodulin-regulated; Basic amino acid-directed
	AMP-dependent protein kinase (AMPK)	Activated when AMP:ATP is elevated; basic amino acid-directed
CMGC	Cyclin-dependent kinase (CDK)	Proline-directed
	Mitogen-activated protein kinase (MAPK)	Substrate of MEKs; proline-directed
	Glycogen synthase kinase-3 (GSK-3)	Principal substrate of PKB; dual specificity; proline-directed
	Casein kinase-2 (CK2)	Uses both ATP, GTP; Acidic amino acid-directed; dual specificity
MEK	MAP/ERK kinase (MEK)	Dual specificity; highly specific
	MAP/ERK kinase kinase (MEKK)	Highly specific
	MAP/ERK kinase kinase kinase (MEKKK)	Highly specific
PIKK	ATM/ATR; DNA-PKcs	Glutamine-directed
	TOR	Proline-directed

in response to the need for increased cell-to-cell signaling and control in multicellular organisms. Tyrosine kinases will be examined in detail in later chapters.

Serine/threonine kinases have many features in common. The most important of these similarities from the signaling perspective is their activation through phosphorylation. The kinases become catalytically active only when crucial residues in a region of their catalytic domain known as the *activation loop* becomes phosphorylated. In many, if not most instances one kinase activates a second kinase, and these kinases, activated sequentially in an upstream to downstream direction, form the core of the signaling pathway. Many of the entries in Table 6.1 can be arranged into pathways according to their actions on one another. That is, these kinases catalyze the transfer of phosphoryl groups to other kinases. Many of these are called by the same name as the downstream kinase, with the addition of another kinase in the name. An illustration of a kinase signaling cascade whose components are named in this manner is presented in Figure 6.6. Some common examples of this naming convention are *ERK kinase kinases* (ERKKs), *AMP kinase kinases* (AMPKKs), and *CaM kinase kinases* (CAMKKs). Other kinases that act on kinases have their own distinct names. These

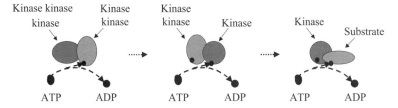

FIGURE 6.6. A kinase signaling cascade: A kinase kinase kinase phosphorylates a kinase kinase, which in turn phosphorylates a kinase, which then phosphorylates its substrate. In some cascades, the kinases are localized in close proximity to one another by scaffolding proteins; in others, the upstream element must diffuse to where the downstream element is tethered.

kinases typically can act on more than one kind of target kinase; that is, they have broader substrate specificity than the first group of rather narrowly acting kinase kinases, and consequently have their own name. An example of this kind of broader acting kinase kinase is phosphoinositide dependent kinase (PDK). This signaling molecule acts upstream of many, if not most members of the Protein kinase A, G, and C (AGC) group.

6.8 Kinases Often Require Second Messenger Costimulation

Phosphorylation by itself may not be sufficient to activate a kinase. Many of the serine/threonine kinases require the presence of a *second messenger molecule*. Second messengers are signaling intermediates connecting events taking place at the plasma membrane with the intracellular signaling that eventually converts the signal into a cellular response. Second messengers are small molecules. There are three main kinds—cAMP, lipids, and calcium. These intermediates are sent into the cytoplasm in response to receptor binding. Members of the AGC group and the CaM kinase family members are regulated in this way. Binding of a second messenger to the kinase, and also to the kinase kinase, precedes activation of the signaling pathways by phosphorylation.

The first of the second messengers mentioned above, cyclic adenosine monophosphate (cAMP), is generated from ATP by the enzyme adenylate cyclase embedded in the plasma membrane. Catalytic and regulatory sites are located within the cytoplasmic loops and termini of this double-clustered, multipass molecule (depicted in Figure 8.9). The production of cAMP is triggered by signals relayed from the cytosolic surface of plasma membrane-bound receptors, and other intracellular signaling elements located in its near vicinity. Cyclic AMP binds to protein kinase A and this kinase is the sole cellular mediator of cAMP stimulated signaling.

Lipid second messengers regulate the activity of a number of kinase families belonging to the AGC group. The kinases of the AGC group are activated in a multistep manner. First, there is an initial phosphotransfer process that induces the migration of the protein kinases to the plasma membrane where subsequent phosphotransfers serve to activate the catalytic properties of the kinase. Prominent agents of the second messenger system are lipid kinases and phospholipase C that generate the second messengers, such as diacylglycerol (DAG) that activates protein kinase C, and other lipid second messengers that trigger the release of Ca^{2+} from intracellular stores. The lipid second messenger generating system and AGC kinase activation will be examined in detail in Chapter 8.

The third major kind of second messenger is intracellular calcium. As noted in Table 6.1 calcium is a regulator of the CaM kinases. It activates the kinase by first binding to calmodulin; then the calcium-calmodulin complex binds and activates the protein kinase and its upstream kinase kinases. Calcium is not only stored and released from intracellular stores located in the endoplasmic reticulum, but also enters the cells through ion channels found in excitable cells such as neurons. Calcium signaling and CaM kinases, along with many of the AGC kinases, are key agents of neural signaling and for these reasons the CaM and associated signaling pathways are called *learning pathways*.

6.9 Flanking Residues Direct Phosphorylation of Target Residues

A third common theme in kinase actions pertains to the establishment of substrate specificity. This issue was briefly touched upon in the naming of kinases depending on whether they had a narrow or broad specificity. The key notion is that specificity is determined not only by the presence or absence of serine and threonine residues in a potential target protein, but also by the identity of the residues that flank the potential acceptor of the phosphoryl group. The flanking residues, that is, the amino acid residues immediately preceding or following the target site, help determine whether a particular target residue can be phosphorylated or not.

The most general observation that can be made is that kinases belonging to the AGC and CaMK groups tend to be basic amino acid-directed while those belonging to the CDK, MAPK, GSK3, CK2, and cyclin dependent kinase-like kinase (CMGC) group are mostly proline directed. In more detail, kinases belonging to the AGC group are more likely to catalyze the transfer of phosphoryl groups to S/T residues lying near basic amino acids such as Lys and Arg. The GRKs are an exception to this preference, as they prefer flanking acidic residues. A similar observation holds for the CaMK group. These kinases too prefer basic amino acid environments. In contrast, members of the CMGC group are mostly proline-directed, phosphorylating S/T residues lying within proline-rich regions. The main exceptions to this rule are the CK2s. These kinases prefer the flanking amino acids to be acidic. Finally, phosphoinositide 3-kinase related kinases (PIKKs) are either glutamine or proline directed, as listed in Table 6.1.

6.10 Docking Sites and Substrate Specificity

One of the challenges to the reliable operation of cellular control layers is how to ensure that the kinases phosphorylate the correct substrate at the right time. The cell is crowded with many proteins, and each of these proteins likely to contain many potential phosphorylation sites. This selection problem is solved, but only in part, by the presence of specific flanking residues. It turns out that the amino acids lying either just before or just after the target serines and threonines are not the only amino acid residues involved in kinase-substrate recognition. Additional residues located well away from the target phosphorylation sites confer additional substrate specificity. These sites are referred to as *docking motifs* or *docking domains*.

The presence of docking sites on kinase and substrate increases the substrate specificity. These sites are typically located far from the catalytic site of the kinase and far from the phosphoacceptor site of the substrate. An example of a kinase family that uses this form of matching is the MAP kinase family. In this family, the docking sites are located in a C-terminal domain of the kinase. The docking site is characterized by the presence of several acidic amino acids, and is matched to a cluster of basic amino acids in the matching docking sites used by upstream activators, deactivators, and downstream substrates. Another family of kinases, the glycogen synthase kinase-3s (GSK-3s), uses a different docking strategy. In this family the presence or absence of a phosphorylated serine located four residues from the substrate target serine is a crucial determinant of whether the kinase can dock at the substrate. A crucial arginine residue serves as the kinase docking site, and the complementary matching of the arginine residue in the kinase to the phosphorylated, or "primed," site in the substrate stabilizes the kinase in a catalytically active conformation.

6.11 Protein Phosphatases Are Prominent Components of Signaling Pathways

Reversible phosphorylation, the covalent attachment and removal of phosphoryl groups to and from proteins, is the most widespread method of regulation protein activity in the cell. In addition to a various families of protein kinases there are several families of protein phosphatases. Protein kinases catalyze the transfer of phosphoryl groups to target proteins using

ATP as a donor. Protein phosphatases do the opposite. They catalyze the removal of phosphoryl groups using ADP as an acceptor.

Like the protein kinases, the protein phosphatases fall into two main groups—those that catalze the removal of phosphoryl groups from serine/threonine residues, and those that do the same on tyrosine residues. Eukaryotic genomes encode a smaller number of protein phosphatases than protein kinases. The serine/threonine phosphatases can be placed into two families—PPP and PPM—and the tyrosine phosphatases are placed into one large familty designated as PTP. The best studied of the serine/threonine phosphatases are the members of the PP1 and PP2 families. These proteins possess nearly identical catalytic subunits, and it is the regulatory units that provide substrate specificity. Each phosphatase family member can associate with any of a number of different regulatory units.

6.12 Scaffold and Anchor Protein Role in Signaling and Specificity

Scaffold and anchor proteins help organize the signaling pathway and confer specificity. Spatial localization is a powerful way of ensuring that protein kinases only phosphorylate the correct substrates. While cytosolic proteins are, in principle, free to diffuse about in the cytosol, they generally do not do so. Instead, proteins are restricted to specific locations in the cell, either in particular organelles or in specific subcellular compartments, that is, in restricted regions of space in the cytosol. The restriction of signaling molecules to specific locations in the cell is known as *spatial localization* or *compartmentalization*.

Scaffold and anchor proteins promote spatial localization and substrate specificity. These proteins do not have any enzymatic activity but rather serve as platforms that help organize the signaling pathways. The scaffold and anchor proteins (Figure 6.2) provide binding sites for attachment of the kinases and their substrates at specific locales in the cell. Scaffolding proteins were first discovered in the MAP kinase pathways of yeast cells, and then in the mammalian MAP kinase pathways as well. Since the initial discoveries, several different kinds of scaffold proteins have been identified in nerve cells, where they play a prominent role in organizing the signaling machinery.

Anchor proteins are similar to scaffold proteins in that they too help organize signaling routes, but unlike scaffolds are attached, or anchored, to the cytoplasmic face of the plasma membrane. They sequester signaling proteins in either active or inactive states, ready for activation or deactivation, positioned close to their substrates that are usually other plasma membrane-associated signaling proteins. These too are prominent components of the signaling machinery in nerve cells.

6.13 GTPases Regulate Protein Trafficking in the Cell

The transcription machinery resides in the cell nucleus while the translation machinery is located in the cytosol. Among the major consequences of this compartmentalization is the addition of control points that regulate nucleocytoplasmic transport. In order for a transcription factor to influence transcription it must be located in the nucleus. If it is kept out of the nucleus then it cannot act. Similarly, a signaling molecule whose target lies in the cytosol cannot act on that molecule if its location is restricted to the nucleus. Spatial localization and sequestering is another way of regulating kinase actions so that they only act on the appropriate targets at the right time. A hallmark of this sort of regulation is its dynamic character. In response to regulatory signals, some kinases may migrate from the nucleus into the cytosol while others translocate in the opposite direction from the cytosol into the nucleus.

GTPases play a crucial role in regulating protein traffic in a cell. These are a large family of GTP-binding proteins that function as molecular switches, operating at key control points in the cell. Members of this family of control elements are involved in shuttling proteins in and out of the nucleus. Other family members regulate the movement of cargo throughout the cell, and especially from organelles in the interior to the surface and back, and still others regulate the organization of the actin cytoskeleton. GTPases along with a variety of small adapter molecules are often concentrated at and just below the plasma membrane, where they help organize the intracellular signal-transducing pathways leading to the various control points that serve as interfaces to the fixed infrastructure and mediate the cellular responses to the signals.

6.14 Pheromone Response Pathway Is Activated by Pheromones

Yeasts respond to mating pheromones and to stresses by altering their patterns of gene expression. Environmental and pheromone signals are sent from the plasma membrane to the nucleus through a set of parallel pathways named for the last in a series of protein kinases that serve as the central signal transducers. The kinases that lend their name to the pathways are mitogen-activated protein (MAP) kinases, members of the CMGC group of serine/threonine kinases. Six of these pathways have been found in yeasts and three in mammals.

One of the best-characterized pathways, the pheromone response pathway, is diagrammed in Figure 6.7. As depicted in the figure, this pathway contains a kinase core that sits between the receptor and associated signal transduction machinery positioned at the plasma membrane and down-

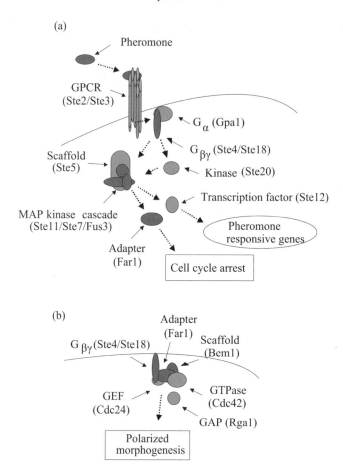

FIGURE 6.7. Pheromone response pathway: (a) Pheromones initiate signaling when they attach to the extracellular ligand-binding portion of the G protein-coupled receptor (GPCR). Receptor-binding triggers dissociation and activation of the G_α and $G_{\beta\gamma}$ subunits of the G-protein. The $G_{\beta\gamma}$ subunit activates the Ste20 kinase and interacts with the Ste5 scaffold protein resulting in activation of signaling through a MAP kinase cascade, leading to activation of Ste12 and the subsequent transcription of genes responsive to pheromone signals, and of Far1, a mediator of cell cycle arrest (in G1). (b) Far1 is an adapter protein and serves as an adapter between $G_{\beta\gamma}$ and the Cdc24 GEF at the plasma membrane.

stream elements that make contact with the transcription machinery to influence the program of gene transcription. The transmission of a signal from the extracellular side of the plasma membrane into the cell interior and from there to the cell nucleus involves the execution of several activities, or processes, each designed to enhance the specificity, or fidelity, of the signaling. These are: transmembrane signal transduction; nucleocytoplasmic shuttling; and recruitment to the cytosolic face, leading to assembly and activation of the kinase core unit.

Membrane-spanning receptors bind ligands and send signals indicative of the binding event across the plasma membrane to the cytoplasmic surface. In the pheromone pathway, members of the G protein-coupled receptor (GPCR) family perform this task. The GPCR family is a prominent component of the human genome with roughly 600 members identified so far. They serve as sensors for physical and chemical signals such as light and odors, and for a variety of cell-to-cell, hormonal and neural signals. These receptors and their methods of action will be explored in detail in Chapter 12. In brief, when a ligand binds a G protein-coupled receptor, it produces changes in the electrostatic environment of the cytoplasmic portions of the receptor. As its name indicates, a GPCR acts through a G protein to which it is coupled, or tethered. The G proteins are built from three distinct subunits, called *alpha*, *beta*, and *gamma*, and so are referred to as being *heterotrimeric*. The changes in the cytoplasmic portions of the GPCR cause the dissociation of the G protein alpha subunit from the beta and gamma subunits, allowing the subunits to move about and activate other signaling molecules.

The set of actions that follow signal reception and conveyance across the plasma membrane may be collectively termed *recruitment and pathway formation*. These activities occur at and just below the plasma membrane. In the yeast pheromone pathway, the recruitment of two proteins, Ste20 and Ste5, is promoted primarily by the activated G protein beta subunit. Ste5 continually shuttles between the cytoplasm and nucleus, and in response to activation of G protein beta subunits, Ste5 starts to accumulate just below the cell surface. The binding by the G protein beta subunit to Ste5 and Ste20, the latter a protein concentrated just below the cell surface, is the crucial step in forming the kinase core. ("Ste" is an abbreviation for the term "sterile," assigned to proteins whose mutated forms produce nonmating, i.e., sterile phenotypes in the yeast.)

In the pheromone response pathway the kinase core consists of three kinases, Ste11, Ste7, and Fus3, organized in a linear fashion with one signaling to the next in the chain. The Ste5 protein functions as a molecule scaffold for the formation of the kinase module. The protein organizes the kinases into a signaling pathway leading to the activation of the last member of the module. This assembly step takes place subsequent to binding to the G protein beta subunit of Ste20, which phosphorylates and activates Ste11, and to activation of Ste5, which interacts with all members of the MAP kinase module.

The Fus3 MAP kinase functions as the output unit from the kinase module. In the absence of pheromone signals, this protein, like the Ste5 scaffold, continually shuttles between nucleus and cytoplasm. (The other members of the module, Ste7 and Ste11, are cytoplasmic proteins.) In the presence of pheromone signals, the Fus3 protein assembles into the MAP kinase module, along with Ste11 and Ste7, and is activated. It then translocates back to the nucleus. There it activates the transcription factors Ste12.

The Fus3 MAP kinase also phosphorylates and thus activates the Far1 adapter protein. This protein has several functions. It is a mediator of cell cycle arrest (in G1) as indicated in Figure 5.7. Like Ste5, it shuttles between different subcellular locations. It shuttles the Cdc24 GEF to the plasma membrane and mediates formation of a complex containing several proteins at the $G_{\beta\gamma}$ location where the pheromone signals are strongest. This action makes the place wherein the cytoskeleton reorients itself in preparation for bud formation.

6.15 Osmotic Stresses Activate Glycerol Response Pathway

The *high osmolarity glycerol (HOG) response pathway*, activated by osmotic stresses, is shown in Figure 6.8. This is another of the yeast stress pathways. It enables the yeast cells to adapt to changes in external osmolarity by altering their patterns of gene expression. In response to elevated

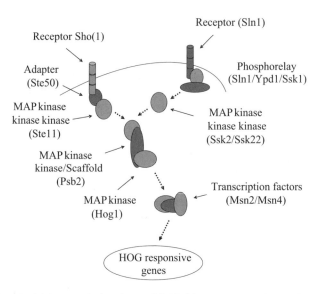

FIGURE 6.8. The high osmolarity glycerol (HOG) response pathway: Two receptors, Sln1 and Sho1, signal through a common MAP kinase cascade and downstream Hog1 and Msn2/Msn4 transcription factors to influence the expression of HOG-responsive genes.

external osmolarity, the yeast cells increase glycerol synthesis and decrease glycerol permeability, thereby increasing their internal osmolarity and restoring a correct osmotic gradient for water uptake.

The HOG pathway has two distinct *osmosensors*, Sho1 and Sln1. Sho1 and Sln1 (along with Ypd1 and Ssk1) operate in place of the GPCR and associated G proteins that sense and transduce pheromone signals. Both Sho and Sln1 signal through components of the HOG MAP kinase core module resulting in the activation of the Hog1 MAP kinase. The central element in the kinase core, Psb2, doubles both as the kinase kinase and as the scaffold protein for the module. Signals conveyed into the cell through the Sln1 receptor utilize Ssk2 or Ssk22 as the kinase kinase kinase, while Sho1 signals are routed via the Ste11 kinase kinase kinase. The last element in the MAP kinase core, Hog1, like Fus3, shuttles back and forth between the cytoplasm and nucleus. In response to elevated osmotic stress, the kinase is phosphorylated by Psb2 and becomes concentrated in the nucleus along with the Msn2 and Msn4 transcription factors.

The second osmosensor, Sln1, is unusual for a eukaryotic sensor. It is a histidine kinase, a kind of signal protein that is common in bacteria, but is not only uncommon in eukaryotes, but also seems to be absent all together in animals. These systems are often called *histidine-aspartate* (His-Asp) *phosphorelays* because they involve the transfer, or relay, of phosphoryl groups between histidine and aspartate residues. The mechanistic details of how His-Asp systems operate in bacterial cells will be a main focus of the next chapter.

6.16 Yeasts Have a General Stress Response

The term *general stress response* was coined to describe the expression of a common set of genes by a number of different stress conditions. Roughly speaking, the same set of genes was observed to be upregulated in response to unfavorable temperature shifts, osmotic shock, starvation conditions, and DNA damage. Among the proteins upregulated in response to these conditions are those such as heat shock proteins that confer general protection against the stresses. The stimulation of gene transcription of a specific set of genes is associated with the presence in the promoters of each of these genes of a *stress responsive element*, or STRE.

The pathways leading to the transcription initiation sites located in the nucleus start with a sensing activity taking place in the outer reaches of the yeast cell. These activities are most often performed by receptors located in the plasma membrane, but may also be initiated by proteins functioning as sensors located near or at the cytoplasmic face of the plasma membrane. As indicated in Figure 6.9 two signaling routes are involved in signaling to the STREs. One involves glucose sensing by a transmembrane receptor, and the other involves sensing of internal stresses such as low pH. The crucial

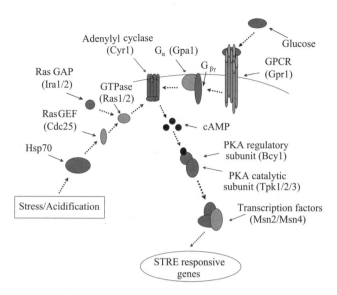

FIGURE 6.9. General stress response pathway: Protein kinase A, the core signaling element, is stimulated by Ras-mediated internal stress signals and by GPCR/G protein transduced glucose signals, both acting through adenylyl cyclase/cAMP intermediaries.

element in the internal pathway is the heat shock protein Hsp70 introduced in the last chapter. Under low pH conditions the number of denatured proteins increases. The Hsp70 proteins must deal with this situation and are not available to signal to the Ras GEF, Cdc25.

The G protein-coupled receptor (GPCR) functions as a sensor of glucose, the preferred sugar for yeast cells. In response to the presence of glucose it stimulates the production of cAMP by adenylate cyclase. These molecules function as second messengers and have as their cellular role the activation of protein kinase A (PKA). In the yeast this protein has a pair of regulatory subunits and a pair of catalytic subunits. It is activated when two cAMP molecules bind to each regulatory subunit. In response, the catalytic subunits dissociate from the regulatory subunits and become activated. The glucose-activated pathway leading from a GPCR is not the only path to activation of PKA. Alternatively, Ras may stimulate cAMP production in response to stress conditions, again leading to activation of PKA.

Activated PKA is a negative regulator of Msn2p and Msn4p, transcription factors (TFs) that bind to the STRE. Under good (logarithmic) growth conditions these TFs are localized in the cytoplasm, but migrate to the nucleus when glucose conditions are degraded. The import and export of Msn2p and Msn4p from the nucleus is regulated by PKA activity and also independently by signals from the TOR pathway (to be discussed next). The overall mechanism is that when growth conditions are poor and the cell is subjected to stressful conditions the Msn2p and Msn4p proteins accumulate in the nucleus, bind to the STREs, and stimulate transcription of genes whose protein products protect the cell.

6.17 Target of Rapamycin (TOR) Adjusts Protein Synthesis

TOR adjusts protein synthesis in accordance with growth conditions. The process of growth is distinct from that of proliferation, although the latter often follows from the former. Cell growth refers to the increase in cell size and mass and is driven by rapid protein synthesis. Cell proliferation is the result of the progression through the cell cycle resulting in cell division. Under good growth conditions yeast cells will grow rapidly. The average lifetime of a ribosome is about 100 minutes, and to maintain an optimal rate of growth, yeast cells produce 2000 ribosomes per minute. The rRNA genes account for more than 60% of the total cellular transcription activity. The yeast genome contains 137 ribosomal protein (RP) encoding genes; there are some gene duplications and these genes encode 78 different RPs. Their transcription yields 25% of the total number of mRNA molecules in the cell.

As a large expenditure of resources is needed to maintain this growth machinery, when conditions are no longer favorable to growth, cellular resources must be directed elsewhere. *Targets of rapamycin* (TOR) proteins are central controllers of cellular resources. They regulate the balance between protein synthesis and protein degradation, throttling back the program of cellular growth in response to reductions in the quality of nitrogen and carbon nutrients. In *S. cerevisiae* this means shutting down budding, the growth at a particular location on the cell surface that lends the name "budding yeast" to the organism. TOR proteins carry out their resource management functions by sending out regulatory signals to the translation apparatus and the transcription machinery.

TOR is a serine/threonine kinase belonging to the PIKK family (Table 6.1). It remains in an active state as long as intracellular amino acid concentrations, most notably, that of leucine, are adequate. Under plentiful conditions it phosphorylates and thus activates a protein called Tap42 (Figure 6.10). This protein is a regulator of phosphatase PP2A activity, functioning as a subunit of PP2A and PP2A-like phosphatases. The PP2A phosphatase is composed of three subunits that must come together for the phosphatase to carry out its dephosphorylation actions on its cognate target proteins. When Tap42 is activated it binds the catalytic subunit of the phosphatase PP2A. This binding event prevents the catalytic subunit from associating with the two other subunits of the phosphatase PP2A and by this means immobilizes the phosphatase in an inactive form. Whenever the quality of

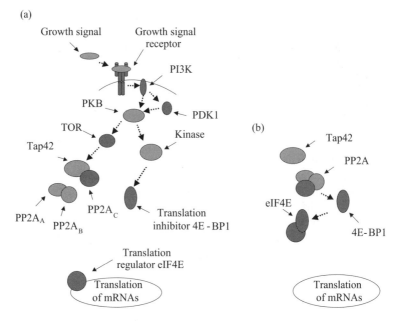

FIGURE 6.10. Regulation of translation initiation through the TOR signaling pathway: (a) When growth signals are present (good growth conditions) TOR is phosphorylated by upstream kinases such as PKB, and TOR then phosphorylates Tap42. In response Tap42 binds the catalytic PP2A subunit, keeping it away from the other two subunits. The translation inhibitor 4E-BP1 remains in a phosphorylated state and cannot prevent the translation regulator eIF4E from stimulating translation of mRNAs. (b) In the absence of growth signals, the PP2A phosphatase is active and dephosphorylates 4E-BP1, which in turn inhibits eIF4E translation initiation.

carbon and nitrogen nutrients degrades, TOR becomes inactive. It no longer activates Tap42, which, in turn, no longer prevents activation of PP2A. The result of turning off TOR and turning on PP2A is that downstream targets involved in translation initiation and elongation are dephosphorylated and, by these actions, turns down translation of resident mRNAs.

In yeast, TOR proteins match ribosome biogenesis to nutrient availability. Not only do the TOR proteins regulate genes that encode metabolic proteins, but also they regulate genes that encode components of the ribosomes such as the ribosomal proteins mentioned at the beginning of this section. TOR signaling regulates the transcription of RPs by poly II and also regulates the machinery responsible for transcribing rRNA subunits and tRNAs. End points of TOR signaling include sites at poly I responsible for transcribing the 35S rRNA subunit, and poly III that transcribes the 5S rRNA precursor and tRNAs.

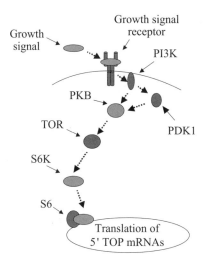

FIGURE 6.11. Regulation of translation of 5'TOP genes through the TOR signaling pathway: When growth conditions are favorable TOR is active and phosphorylates the S6K kinase, which in turn phosphorylates the ribosomal factor S6 thereby stimulating transcription of mRNAs bearing a 5'TOP sequence in their upstream untranslated (regulatory) region (UTR).

When growth conditions are favorable, TOR phosphorylates a serine/threonine kinase called S6K (Figure 6.11). This kinase, a member of the AGC family (Table 6.1), is a key regulator of ribosomal biogenesis. When activated it phosphorylates S6 ribosomal protein resulting in increased translation of mRNAs that encode components of the translation apparatus. The genes encoding these ribosomal proteins are referred to as 5'TOP genes because they contain a characteristic 5'TOP (terminal oligopyrimidine tract) sequence in their mRNA regulatory region.

6.18 TOR Adjusts Gene Transcription

TOR adjusts gene transcription in accordance with nutrient conditions. Under poor nutrient conditions, transcription factors are activated that regulate genes encoding proteins involved in metabolism. The specific transcription factors involved in adjustments to poor quality nitrogen supply are the Gln3p and Gat1p, and for poor carbon conditions the transcription factors are Msn2p and Msn4p. The other set of TOR end points, besides beings elements of the translation initiation and elongation regulatory apparatus, are regulators of these transcription factors. In a manner analogous to the immobilization of the catalytic subunit of PP2A, these transcription factors are kept out of the nucleus when growth conditions are

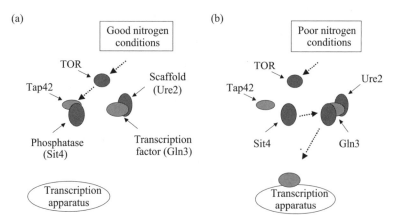

FIGURE 6.12. Regulation of transcription in response to nitrogen conditions through the TOR signaling pathway: (a) Under good nitrogen nutrient conditions, TOR catalyzed phosphorylation of Tap42 leads to the immobilization of Sit4, a PP2A-like phosphatase. The transcription factor Gln3 is sequestered in the cytoplasm by the Ure2 scaffold and does not stimulate transcription. (b) Under poor nitrogen conditions, TOR does not phosphorylate Tap42. Sit2 is able to dephosphorylate Gln3, which then decouples from Ure2 and enters the nucleus, where it stimulates transcription of genes in response to the poor nutrient conditions.

good, and allowed to translocate to the nucleus when the conditions no longer favor growth.

Two anchor proteins, Ure2p and a 14-3-3 family member, are involved in this regulation. Under good conditions (Figure 6.12) TOR stimulates Tap42p, which inhibits Gln3p and Gat1p translocation. TOR also directly stimulates Ure2p, which inhibits the migration of the two transcription factors into the nucleus. When conditions are poor these multiple inhibitory influences are lifted and the transcription factors are able to migrate into the nucleus and activate gene transcription. Similarly, the Msn2p and Msn4p proteins are inhibited by 14-3-3 anchor proteins (not shown) in a nutrient supply-dependent way.

6.19 Signaling Proteins Move by Diffusion

The proteins responsible for signaling in the cell are mobile; they move from one cellular location to another. For example, the Fus3 protein (Figure 6.7 and Section 6.13) involved in the pheromone response not only translocates from the plasma membrane and the nucleus, but also shuttles back and forth between the cytoplasm and nucleus. Some of the pathways discussed in this chapter end at transcription sites in the nucleus, while other pathways terminate at the translation machinery in the cytoplasm. Mobility is central to the signaling activities of many, if not most, signaling proteins in the cell. The predominant means of movement for signaling proteins is *passive diffusion*.

Particles diffuse from one location to another through random movements known as Brownian motion. Brownian motion is named after the English botanist Robert Browning who in 1826 studied and puzzled over the zig-zag movements of dust particles suspended in a fluid that he observed using a light microscope. Albert Einstein provided an explanation of this phenomenon in 1905. In his analysis, Einstein showed that Brownian motion arises from random collisions of the dust particles with the molecules of the fluid.

The cell interior is an aqueous medium filled with ions, biomolecules, a cytoskeleton, and organelles. The speed at which a particular signaling molecule can diffuse depends on three factors:

- viscosity of the medium,
- binding, and
- crowding effects.

Diffusive motion is slower than inertial motion. In diffusion, the medium exerts a drag or frictional force on the molecules that slow them down. The difference between inertial and diffusive motion is reflected in the mean square displacements $\langle x^2 \rangle$ produced by each kind of motion, namely,

$$\langle x^2 \rangle = \left(\frac{kT}{m} \right) t^2, \quad t \ll \frac{m}{\alpha}, \tag{6.1}$$

and

$$\langle x^2 \rangle = \left(\frac{2kT}{\alpha} \right) t, \quad t \gg \frac{m}{\alpha}. \tag{6.2}$$

In the first expression, corresponding to inertial motion, the mean square displacement grows as the square of the time, while in the second, that of diffusive motion, the mean square displacement increases only linearly with time. The overall scale is established by the ratio of the particle mass m to the friction coefficient α. At small times the distances traveled by the particles are so small and they have not yet collided with any other particles. At large times the particles have undergone a number of random collisions with particles in their surroundings, changing direction each time. This type of random movement, Brownian motion, underlies the diffusion process as first noted by Einstein.

It is customary to discuss diffusion in terms of a diffusion coefficient rather than a friction coefficient. The diffusion coefficient D represents the ratio of thermal energy to the friction coefficient:

$$D = \frac{kT}{\alpha}. \tag{6.3}$$

This expression formalizes the notion that as the viscosity (friction) goes up the mobility represented by the quantity D goes down. The cellular medium is viscous but not excessively so. Small and moderately sized biomolecules diffuse about four times slower in the cytosol as they would in a pure water medium.

As might be expected the larger the molecule the slower it will diffuse. This aspect of diffusive motion is encapsulated in the Stokes–Einstein relationship:

$$\alpha = 6\pi\eta r_0, \tag{6.4}$$

where η is the viscosity and r_0 is the hydrodynamic (particle) radius.

Biomolecules can be slowed down by binding effects and by crowding, beyond the factor of 4 reduction in rate mentioned above and the hydrodynamic dependence on the radius. Organelles and the cytoskeleton can impede rectilinear motion, as can other biomoelcules through steric hindrance effects. Large macromolecules, especially those with masses greater than about 200 kDa, might be slowed down considerably. Binding effects can come into play not only to slow down movement but also to halt it entirely. For example, calcium ions are rapidly bound by calmodulin, which acts as a buffering agent to prevent large-scale diffusion of the ions. Binding reactions taking place before the biomolecules reach their targets can also hinder if not stop the biomolecules entirely. And of course, signaling proteins become immobilized when they reach their cellular targets and bind them.

References and Further Reading

Global Patterns of Gene Transcription and Combinatorial Control

Chu S, et al. [1998]. The transcriptional program of sporulation in budding yeast. *Science*, 282: 699–705.

Causton HC, et al. [2001]. Remodeling of yeast genome expression in response to environmental changes. *Mol. Cell Biol.*, 12: 323–337.

Gasch AP, et al. [2000]. Genomic expression programs in the response of yeast cells to environmental changes. *Mol. Cell Biol.*, 11: 4241–4257.

Holstege FCP, et al. [1999]. Dissecting the regulatory circuitry of a eukaryotic genome. *Cell*, 95: 717–728.

Pilpel Y, Sudarsanam P, and Church GM [2001]. Identifying regulatory networks by combinatorial analysis of promoter elements. *Nature Genet.*, 29: 153–159.

Spellman PT, et al. [1998]. Comprehensive identification of cell cycle-regulated genes of the yeast *Saccharomyces cerevisiae* by microarray hybridization. *Mol. Cell Biol.*, 9: 3273–3297.

Protein Phosphatases

Cohen PTW [2002]. Protein phosphatase 1—Targeted in many directions. *J. Cell Sci.*, 115: 241–256.

Janssens V, and Goris J [2001]. Protein phosphatase 2A: A highly regulated family of serine/threonine phosphatases implicated in cell growth and signaling. *Biochem. J.*, 353: 417–439.

Mitogen-Activated Protein Kinases

Banuett F [1998]. Signalling in the yeasts: An informational cascade with links to filamentous fungi. *Microbiol. Mol. Biol. Rev.*, 62: 249–274.

Estruch F [2000]. Stress-controlled transcription factors, stress-induced genes and stress tolerance in budding yeast. *FEMS Microbiol. Rev.*, 24: 469–486.

Görner W, et al. [1998]. Nuclear localization of the C_2H_2 zinc finger protein Msn2p is regulated by stress and protein kinase A activity. *Genes Dev.*, 12: 586–597.

Gustin MC, et al. [1998]. MAP kinase pathways in the yeast *Saccharomyces cerevisiae*. *Microbiol. Mol. Biol. Rev.*, 62: 1264–1300.

Herskowitz I [1995]. MAP kinase pathways in yeast: For mating and more. *Cell*, 80: 187–197.

Widmann C, et al. [1999]. Mitogen-activated protein kinase: Conservation of a three-kinase module from yeast to human. *Physiol. Rev.*, 79: 143–180.

Scaffold Proteins and MAP Kinase Cascades

Garrington TP, and Johnson GL [1999]. Organization and regulation of mitogen-activated protein kinase signaling pathways. *Curr. Opin. Cell Biol.*, 11: 211–218.

Madhani HD, and Fink GR [1998]. The riddle of MAP kinase signaling specificity. *Trends Genet.*, 14: 151–155.

Schaeffer HJ, and Weber MJ [1999]. Mitogen-activated protein kinases: Specific messages from ubiquitous messengers. *Mol. Cell. Biol.*, 19: 2435–2444.

Whitmarsh J, and Davis RJ [1998]. Structural organization of MAP-kinase signaling modules by scaffold proteins in yeast and mammals. *Trends Biochem. Sci.*, 23: 481–485.

Shuttling

Elion EA [2001]. The Ste5p scaffold. *J. Cell Sci.*, 114: 3967–3978.

Mahanty SK, et al. [1999]. Nuclear shuttling of yeast scaffold Ste5 is required for its recruitment to the plasma membrane and activation of the mating MAPK cascade. *Cell*, 98: 501–512.

Van Drogen F, et al. [2001]. MAP kinase dynamics in response to pheromones in budding yeast. *Nature Cell Biol.*, 3: 1051–1059.

TOR Central Controller

Gingras AC, Raught B, and Sonenberg N [2001]. Regulation of translation initiation by FRAP/mTOR. *Genes Dev.*, 15: 807–826.

Schmelzle T, and Hall MN [2000]. TOR, a central controller of cell growth. *Cell*, 103: 253–262.

Shamji AF, Kuruvilla FG, and Schreiber SL [2000]. Partitioning the transcriptional program among the effectors of the TOR proteins. *Curr. Biol.*, 10: 1574–1581.

Diffusion

Luby-Phelps K [2000]. Cytoarchitecture and physical properties of cytoplasm: Volume, viscosity, diffusion, intracellular surface area. *Int. Rev. Cytol.*, 192: 189–221.
Verkman AS [2002]. Solute and macromolecular diffusion in cellular aqueous compartments. *Trends Biochem. Sci.*, 27: 27–33.

FRAP

Lippincott-Schwartz J, Altan-Bonnet N, and Patterson GH [2003]. Photobleaching and photoactivation: Following protein dynamics in living cells. *Nature Cell Biol.*, 5: S7–S14.

Problems

6.1 Diffusion rates of various biomolecules can be measured inside the cell using GFPs in a technique called *fluorescence recovery following photobleaching* (FRAP). In this approach, a region inside the cell is subjected to intense laser light, which photobleaches any fluorescence in that region. GFPs located outside the exposed region then diffuse back into the bleached region, and this movement is studied using low intensity laser light.

What is the hydration radius of a GFP molecule in water at a thermal energy $kT = 0.6\,\text{kcal/mol}$? Use the Stokes–Einstein relationship, Eq. (6.4) together with Eq. (6.3) assuming viscosity $\eta = 0.01\,\text{P} = 0.01\,\text{g/cm s}$, and a GFP diffusion coefficient $D = 87\,\mu\text{m}^2/\text{s}$. How long will it take for the GFP proteins to diffuse 10 microns? Hint: Note from Eqs. (6.2) and (6.3) that

$$\langle x^2 \rangle = 2Dt.$$

(This result holds in one dimension. In two dimensions, a factor of 4 appears in place of the factor of 2, and in three dimensions a factor of 6 appears.)

7
Two-Component Signaling Systems

Bacteria such as *Escherichia coli* are able to sense their external environments and, in response to changes in these conditions, alter their physiology. Lifetimes of bacterial proteins are generally short, and bacteria alter their metabolism and reproductive strategies by turning on some genes and turning off others. The corresponding signal pathways start at the plasma membrane, where sensing takes place, and terminate at DNA control points, where transcription is initiated. The pathways in bacteria are generally short and typically involve two core elements, a membrane-bound sensor and signal transmitter, and a receiver that establishes contact with the transcription apparatus.

Several representative examples of bacterial two-component signaling pathways are listed in Table 7.1. Two kinds of signaling pathways are presented. The first kind of pathway is represented in the table by the chemotaxis pathway. This pathway controls a piece of fixed infrastructure in the cell directly. In the case of chemotaxis, it controls the operation of the flagellar motor and terminates at a motor control point. The second kind of pathway, represented by all other entries in the table, regulates gene expression and by that means controls the fixed infrastructure too. These pathways terminate at DNA promoters, and the second components, the response regulators, serve as transcription factors. Like the yeasts discussed in the last chapter, bacteria cannot control their environments and so must continually sense and respond to changes in their surroundings by altering their patterns of gene expression.

Bacteria are not the only organisms that use two-component systems. Plants use these systems to respond to hormones and to light. In the first part of this chapter, the bacterial chemotactic system will be examined in detail. Binding and catalysis will be discussed and the notion of a linear pathway, introduced in the last chapter, will be broadened to include feedback. These explorations will be followed by a discussion of two-component signaling in plants.

TABLE 7.1. Representative two-component signaling systems and their cellular roles.

Sensor	Response regulator	Function
Tar/CheA	CheY, CheB	Regulates chemotaxis
EnvZ	OmpR	Osmoregulation; controls porin gene expression
FixL	FixJ	Regulates nitrogen fixation genes in symbiotic bacteria
KinA, KinB	Spo0A, Spo0F	Regulates stress-induced sporulation gene expression
NarX, NarQ	NarL, NarP	Regulates nitrate/nitrite metabolism gene expression
NtrB	NtrC	Nitrogen assimilation; controls glutamine synthetase gene expression
PhoR	PhoB	Phosphate uptake gene expression

7.1 Prokaryotic Signaling Pathways

Two-component signaling and phosphorelays form the core of prokaryotic signaling pathways. The signaling pathways used by the bacteria to relay signals to the transcription machinery and to morphological structures such as flagellar motors during chemotaxis are built in a fashion similar to those found in eukaryotes. Like the eukaryotic pathways, protein kinases and phosphotransfer processes lie at the heart of the signaling routes. In place of the serine, threonine, and tyrosine residues used by eukaryotes as attachment sites for phosphoryl groups, two other amino acids, histidine and aspartate, are used. The basic signaling unit has a histidine kinase sensor and an aspartate-based response regulator that mediates the cellular response to the signal event. Like all signaling systems the design is a modular one; as a result many different combinations of the two components and their modular domains can be formed. The more elaborate versions of the basic two-component system are called *phosphorelays*.

The arrangements of elements of a basic two-component signaling system and a typical phosphorelay are presented in Figure 7.1. The first signaling element in the two-component system is the *sensor/histidine kinase*. It has a receptor (input) unit and a transmitter. The receptor spans the plasma membrane, and in response to ligand binding, transduces a signal across the membrane from the extracellular side to the cytoplasmic side. Some receptors respond to osmolarity conditions. Others sense nutrient conditions or respond to harmful chemicals such as heavy metals or respond to signals sent by other bacteria. The transmitter unit may be part of the same polypeptide chain as the receptor, or it may be encoded as a separate protein. When signaled to do so by the receptor, the transmitter autophosphorylates itself on a conserved histidine residue. The phosphoryl group is then transferred to the *response regulator*, the second element in the two-component signaling system. The response regulator, like the sensor/histidine kinase, has two functional units, and these usually belong to a single

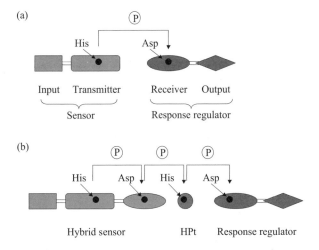

FIGURE 7.1. Phosphotransfer systems: (a) Two-component system; and (b) phosphorelay. The dark circles in the units symbolize histidine and aspartate residues serving as binding sites for phosphoryl groups. Phosphotransfer steps are denoted by squared-off arrows accompanied by circled P's.

polypeptide chain. The first is the *receiver*. It possesses a conserved aspartate residue that binds, or receives, the transferred phosphoryl group. The second part of the response regulator chain is the *output unit*. Once fully activated, the response regulator diffuses to, and establishes contact with, the transcription machinery.

The second part of Figure 7.1 illustrates the operation of a typical phosphorelay. There are two additions to the system depicted in the first part of the figure. The sensor unit is more complex; it now becomes a hybrid sensor, containing an extra module, an aspartate-bearing receiver functionally identical to the receiver unit in a response regulator. Sandwiched between the sensor unit and the response regulator is a small histidine phosphotransfer (HPt) protein. It is functionally identical to the catalytic portion of the transmitter unit and contains the conserved histidine. As a result of these two additions, the two-step His-Asp phosphotransfer becomes a four-step His-Asp-His-Asp phosphorelay.

7.2 Catalytic Action by Histidine Kinases

Histidine kinases catalyze the transfer of a phosphoryl group first to its own histidine residue and then to a conserved aspartate residue in a response regulator. Histidine kinases (HKs) must carry out a number of binding operations. These operations are sequestered in specific domains resulting

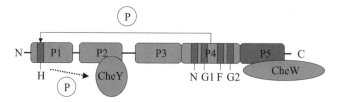

FIGURE 7.2. Domain organization of the CheA histidine kinase: A number of conserved residues characterize the histidine kinases. The conserved residues and their surrounding 5 to 12 amino acid residues are known as *homology boxes*. The conserved residues are used, in one-letter amino acid code form, to name the boxes. Four of these, the N, G1, F, and G2 homology boxes are involved in forming the cleft that holds the ATP molecule in position. The fifth, the H box, contains the conserved histidine residue involved in phosphotransfer.

TABLE 7.2. Domain organization of the CheA histidine kinase.

Domain	Function
P1	H domain; contains the conserved histidine (H), the site of autophosphorylation
P2	Regulatory domain; binds CheY and CheB
P3	Dimerization domain; binds CheA
P4	Kinase domain; contains the asparagine (N), phenylalanine (F), and glycine-rich (G1 and G2) boxes; binds ATP
P5	Interaction domain; binds receptors and CheW

in a highly modular protein organization. The domain organization of several histidine kinases has been determined using X-ray crystallography and NMR. Among the proteins studied are CheA, used in chemotaxis, and EnvZ, used in osmosensing. The CheA and EnvZ proteins are representative members of two classes of histidine kinases. In one class (CheA) the histidine residue that functions as the site for phosphotransfer is located in the C-terminal region; in the other class (EnvZ) it is found in the N-terminal region.

There are five CheA domains (Figure 7.2 and Table 7.2). These are named in order from the N-terminal to the C-terminal as P1 to P5. The P1 domain contains the conserved histidine (H box) that is the site for the phosphotransfer. The P2 domain is a regulatory domain and binds the CheY and CheB response regulators. The P3 domain is a dimerization domain. This domain is present because the autophosphorylation process requires two HK proteins. One HK molecule catalyzes the phosphorylation at the appropriate histidine residue in the second HK molecule (Figure 7.3). The P4 domain contains the homology boxes that facilitate nucleotide (ATP) binding and catalysis of the phosphotransfer from ATP to the histidine residue with the exception of the H box. The last domain, P5, mediates interactions with the receptor and the CheW scaffold (Figure 7.2).

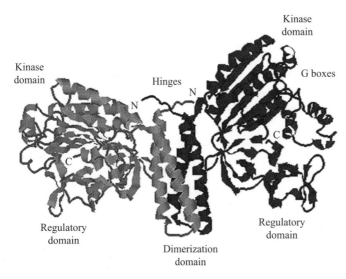

FIGURE 7.3. Structure of the CheA dimer: Shown are the P3 dimerization domain, P4 kinase domain and P5 regulatory domains arranged in a dimer. The figure was prepared using the *Protein Explorer* with atomic coordinates as determined by X-ray crystallography and deposited in the PDB file under accession code 1b3q.

As noted above, the CheA domain organization is not the only one found in histidine kinases. A second, rather common organization is for the H box to be located in the C-terminal region rather than the N terminal one. In this mode of organization, the H box containing the conserved histidine residue is found in close proximity to the dimerization domain and the N, G1, F, and G2 homology boxes to form a catalytic core.

7.3 The Catalytic Activity of HK Occurs at the Active Site

The part of a histidine kinase responsible for catalyzing the transfer of phosphoryl groups is referred to as the *active site*. The active site of the histidine kinase is centered about the residue that receives the phosphoryl group. This residue, along with a small number of other highly conserved residues that surround the phosphoacceptor and assist in binding, defines the active site. In CheA, the active site includes residues comprising the boxes of the P4 kinase domain. These form a cleft that holds the ATP molecule in the correct orientation for transfer of the γ-phosphoryl group from the ATP donor to the His48 acceptor located in the P1 domain.

There are several different kinds of catalysis. The type of catalysis carried out by the histidine kinases, called *covalent catalysis*, requires the presence of a nucleophile to break bonds. A nucleophile is an atom or molecule that can donate an electron pair. A nucleophilic (nucleus-loving) process is one in which an electron pair is donated to another atom or molecule. There are a number of common nucleophiles. Deprotonated water molecules can serve as a nucleophiles. Alternatively, solvent-exposed oxygen atoms in hydroxyl groups in amino acid residue side chains can function as nucleophiles. The nucleophile in histidine kinase action is the imidazole ring structure on the histidine side chain. In this instance, the imidazole ring is said to attack the terminal phosphoryl group of the ATP molecule.

A key feature in the catalytic actions of the kinases is the presence of Mg^{2+} ions that facilitates the breaking of the bond holding terminal phosphoryl groups in place in ATPs. The Mg^{2+} ion binds the ATP molecule prior to catalysis. During catalysis, the Mg^{2+}-bound ATP and the substrate come together in a small region of space, usually the groove, pocket, or cleft formed by the kinase that was mentioned in the last paragraph. The combined actions of physical positioning and electrostatic shaping greatly accelerate the phosphotransfer reaction. These actions are said to stabilize the transition state, lowering the barrier for transfer of the phosphoryl group from donor to acceptor. (See Problem 7.1 for a further discussion of catalysis.)

7.4 The GHKL Superfamily

Histidine kinases along with a large number of ATPases form the GHKL superfamily that is quite distinct from the eukatyotic kinase STTE superfamily. The phosphoaccepting histidine, His48, is surrounded by glutamate, and lysine residues. The histidine, glutamate, and lysine residues form a hydrogen-bonding network. This arrangement and that of the four-helix bundle (Figure 7.4) also characterize the histidine-containing phosphotransfer (HPt) domains that are key components of phosphotransfer systems. The amino acid residues that make up the HPt domains fold in an up-down, up-down manner into four-helix bundles. The second helix in the bundle contains a highly conserved histidine residue that is exposed to the solvent and serves as the acceptor site for transfer of the phosphoryl group. The phosphoryl group covalently attached to the histidine residue is subsequently transferred to an aspartate residue located in a response regulator protein.

The ensemble of alpha helices and beta sheets that forms the three-dimensional structure of the kinase core differs markedly from that of the serine/threonine and tyrosine kinases (STTKs). The histidine kinases closely resemble a number of ATPases. ATPases are proteins that couple metabolic energy stored in the high-energy bonds of phosphoryl groups to operate as pumps and motor proteins, all of which require that work be done. These proteins, including the histidine kinases, form the GHKL *super-*

FIGURE 7.4. The CheA P1 domain: The structure revealed by X-ray crystallography is that of a four-helix bundle (A through D), plus a flanking helix (E). A box shows the location of the conserved His48 residue plus conserved flanking residues on helix B. The figure was prepared using *Protein Explorer* with atomic coordinates from PDB file under accession code 1i5n.

TABLE 7.3. Domain structures of the response regulators.

Family	Domain structures
CheY	Receiver domain
OmpR	Receiver domain, winged helix-turn-helix output domain
FixJ	Receiver domain, four-helix bundle output domain
NtrC	Receiver domain, ATPase domain, DNA-binding domain

family. They each have a characteristic fold consisting of five helices and three sheets, and possess the conserved residues that define the N, G1, F, and G2 boxes. Crystal structures for ATPases such as DNA gyrase B, Hsp90, and MutL closely resemble those for CheA and EnvZ.

7.5 Activation of Response Regulators by Phosphorylation

Response regulators can be placed into families according to their domain structure. Most transcription factors contain at least two domains. There is an N-terminal domain that functions as a receiver, and there is a C-terminal domain that operates as an output unit (Table 7.3). Receivers function as protein kinases that catalyze the transfer of the phosphoryl groups from the conserved histidines on histidine kinases to one of their own conserved aspartate residues. Output domains typically contain DNA-binding and regulatory sequences that enable the response regulators to function as transcription factors. Some response regulators, such as the chemotactic CheY and CheB proteins (to be discussed later in this chapter) contain only the receiver domain and do not bind DNA. However, most proteins diffuse to DNA control points once the phosphoryl group is transferred to the receiver.

Proteins are continually in motion and undergoing shape (conformational) changes. There are atomic, group, and residue movements, backbone motions, side-chain motions, shifts in secondary structure elements, and domains movements. These motions enable a protein to continually sample a population of states, and this ability is important for binding and release, catalysis, and signaling. Rapid motions occur on picosecod-nanosecond time scale, involving only a few atoms and small energy changes. Slower motions occurring on longer microsecond to millimicrosecond time scales, involving appreciable numbers of amino acid residues and far larger energy changes.

NMR spectroscopy has been used to explore protein motions and conformational changes. Using these techniques, it has been observed that the active site of an enzyme undergoes conformation changes taking place on a time scale of microseconds to milliseconds. The regions that undergo these motions correlate with the region involved in protein-protein interactions. In examining what happens to the NtrC protein, it was found that NtrC dimers form in response to phosphorylation and then oligomers form that activate gene transcription. The picture that emerges from studies of NtrC and several other response regulators is one in which response regulators and activated in an allosteric manner by phosphorylation.

There are two *populations of states*. One population of states corresponds to the inactive protein, the other to the active form. Phosphorylation shifts the equilibrium between the two populations of states. Both populations preexist and are continually sampled, but phosphorylation shifts the balance in favor of the active form. Shifts in equilibria between two preexisting populations of conformational states, produced either by ligand binding or by covalent attachment of phosphoryl groups, are known as *allosteric modifications*. In an allosteric modification, binding at one location in the molecule alters how other portions of the molecule respond to their binding partners. These alterations may be thought of as a consequence of the conformational changes accompanying the shifts in equilibrium.

7.6 Response Regulators Are Switches Thrown at Transcriptional Control Points

The commonality with pumps and other ATPases and GTPases seen with histidine kinases extends to response regulators. One of the key observations of how the pumps work is that they utilize a highly conserved aspartate residue in the active site along with several other residues. Like the pump ATPases, an aspartate functioning as the nucleophile, working along with two other aspartate residues, a lysine, a threonine or serine, and a pair of water molecules, form a dense network of bonds in this region of the molecule. Phosphorylation of the conserved aspartate is a key intermediate step in the energy-generating hydrolysis process carried out by the pumps. As they did for the histidine kinases, the common properties of the residues provide useful insights into how the response regulators function.

Response regulators are switches. The energy stored in the high-energy acyl phosphate bond is released when the switch is thrown. The response regulators involved in transcription activation form complexes that jointly drive a series of conformational changes leading to transcription activation. The throwing of the switch is achieved through hydrolysis of the bond, that is, the response regulators catalyze their own dephosphorylation. This occurs when response regulators and their target proteins come together so that the energy released through hydrolysis is used to drive conformational changes in the complexes that activate transcription.

7.7 Structure and Domain Organization of Bacterial Receptors

Bacterial receptors have a stereotypic structure and domain organization. Most bacterial sensor units are constructed from a single polypeptide chain that passes through the plasma membrane twice. The N-terminal lies in the cytoplasm at the end of a short segment. The chain threads out and then back through the plasma membrane as shown in Figure 7.5. The extra-

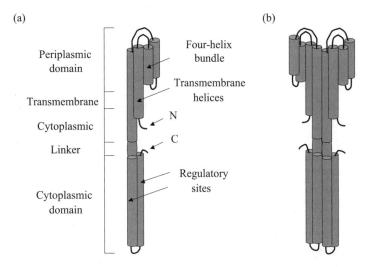

FIGURE 7.5. Tar receptor: (a) Tar monomer with labels denoting the various regions—The presence of several regulatory (methylation) sites in the cytoplasmic domain is denoted. (b) Tar dimer—The basic structure is that of a homodimer that appears as a 35-nm long helical bundle oriented perpendicular to, and passing through, the plasma membrane. The outer, ligand-binding, portion has the form of two four-helix bundles, one per dimer subunit. Two helices from each bundle span the plasma membrane. The cytoplasmic portion is arranged into a four-helix bundle, formed by two helical hairpins, one from each subunit.

cytoplasmic looping portion serves as the ligand-binding region. The portion that is C-terminal to the transmembrane segments contains a linker followed by a long signaling domain. While most receptors conform to this plan, there are some receptors, most notably those involved in cell-to-cell signaling, such as the quorum-sensing (AgrC) receptor and the competence (ComD) receptor, that pass through the plasma membrane six times. The two- and six-pass receptors are bacterial counterparts to the one- and seven-pass receptors found in eukaryotes.

The EnvZ sensor unit is a representative example of the single chain, two-pass membrane topology. This form differs from the bacterial chemotaxis receptor *Tar*, which signals to a separate histidine kinase CheA protein. Tar is a two-pass receptor of the form described above, but some of the signaling pieces such as the histidine kinase domain are sequestered on a separate protein. NtrB represents a third kind of sensor unit; it is a soluble protein and does not attach to the plasma membrane at all.

7.8 Bacterial Receptors Form Signaling Clusters

Bacterial receptors form dimers, trimers, and higher order oligomers. The two subunits of the Tar dimer are linked through multiple bonds, and form a tightly packed and relatively inflexible set of parallel helices. The multiple bonds limit the possible relative movements of the signal helix in response to ligand binding, especially since the energy effects of ligand binding are small. The movements used to transmit a signal over a distance of 35 nm from the outside to the histidine kinase on the inside are modest. In response to ligand binding, the signal helix undergoes a 0.1 to 0.2 nm sliding, or piston-like, displacement towards the cytoplasm. The mechanism is an allosteric one in which ligand binding shifts the equilibrium towards a population of displaced conformations.

The cytoplasmic signaling module of the Tar receptor forms associations with two kinds of proteins. One of these is CheW that functions as a molecular scaffold. The other is CheA, the histidine kinase. CheA proteins are linked to the Tar signal helix through the CheW scaffold. Both CheW and CheA contain modules termed *SH3 domains*. These modules serve as interfaces for the protein-protein interactions leading to the assembly of not just a pair of receptors, scaffolds, and histidine kinases, but rather for the formation of an extended network of such units. The Tar receptors are not the only ones in these signaling clusters. Tar receptors are intermingled with another major receptor, the *Tsr serine receptor*, and with three less prominent chemotactic sensors. The signals sent through the receptors converge upon and are integrated by the CheA histidine kinase.

The net result of the interactions between receptors, scaffolds, and protein kinases is the formation at one pole of the bacterium of a signaling mesh. In this mesh, the extracellular portions of the receptors bind to ligands, while the cytoplasmic portions of the receptors composed of sets of helical coils form a set of signaling "whiskers." These whiskers bind to the scaffolds and through them to the protein kinases. The overall system functions much like a primitive nose, exhibiting considerable sensitivity to external stimuli over a broad range of concentrations.

The signaling complex operates through ligand-induced expansions and contractions. In the absence of ligand binding, the receptors bind CheA and thus stimulate the kinase (autophosphorylation) activity leading to structural changes within the array. Ligand binding inhibits these interactions. When ligands bind the receptors, the pistonlike movements of the signal helices cause the elements of the array to move apart, or disperse. This expansion is sufficient to shut down the autophosphorylation of histidine residues in CheA proteins and the attendant downstream signaling. The formation of large signaling complexes consisting of multiple transmembrane receptors is not restricted to bacteria. Lymphocytes belonging to the vertebrate immune system and neurons utilize extensive signaling complexes and meshes on their surfaces, too.

7.9 Bacteria with High Sensitivity and Mobility

Bacteria such as *Escherichia coli* and *Salmonella typhimurium* are highly mobile, free-swimming organisms. Bacteria such as *Escherichia coli* and *Salmonella typhimurium* are able to sense nutrients and noxious substances in their local environments, and, in response, swim towards the nutrients and away from dangerous chemicals. The essential components of the system that makes possible these chemotactic responses lie at the poles of the cell. A flagellar motor complex resides at one pole, and a sensor complex lies at the other. About 50 genes are involved in chemotaxis. Roughly 10 genes are needed to encode the sensor and signaling complex, and about 40 genes are needed to encode assembly and structural proteins of the flagellar motor. These two systems are coupled to one another. Signals are sent from the sensor system to control point at the flagellar motor switch.

The flagellar motor converts an electrochemical gradient into a mechanical torque that drives the bacterium forward at speeds up to $25\,\mu/s$. The propulsive force is provided by 6 to 10 flagella organized into a bundle. Each flagellum can either rotate clockwise (CW) or counterclockwise (CCW). A flagellum has an intrinsic handedness. When all flagella undergo CCW rotation, the bundle movement is concerted and the cell moves uniformly. If one or more flagella rotate in a CW fashion, the bundle flies apart, the movement is disorganized, and the cell tumbles. Because of its small size, Brownian effects limit the effective straight-line distance that can be traversed without readjustments to the trajectory. The movement of a swimming bacterium consists of a series of smooth runs punctuated by periods of tumbling in which a new direction for running is chosen randomly. The seemingly random movements of the bacterium are biased towards attractants or away from repellents through modifications in the tumbling frequency arising from the sensory input signals.

7.10 Feedback Loop in the Chemotactic Pathway

The chemotactic pathway contains a feedback loop that promotes robust behavior. The Tar, CheW, and CheA proteins constitute the sensor/transmitter unit of the two-component bacterial chemotaxis system. The CheY protein functions as the response regulator. The overall arrangement of the various units in the chemotactic system is depicted in Figure 7.6, and their functions are summarized in Table 7.4. CheY is a *diffusible messenger molecule*. Upon activation through phosphorylation it diffuses to the motor complex where it binds to motor switch protein FliM to promote CW motion and tumbling. CheY is phosphorylated by activated CheA, and dephosphorylated by the protein phosphatase CheZ. In the absence of

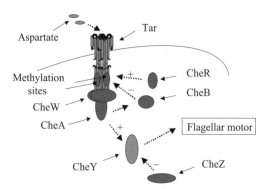

FIGURE 7.6. Tar chemotactic signaling pathway: The receptor (Tar) and histidine kinase (Che A) are encoded on separate chains. CheW is a scaffold that mediates the interactions between sensor and histidine kinase. CheY is the aspartate-bearing response regulator that mediates the cellular response (contact with flagellar motor).

TABLE 7.4. Tar chemotactic pathway.

Component	Function
Tar	Receptor
CheW	Scaffold
CheA	Histidine kinase transmitter
CheY	Aspartyl response regulator
CheZ	Regulatory (phosphatase)
CheR	Regulatory (methyltransferase)
CheB	Regulatory (methylesterase)

CheZ, CheY dephosphorylation would take about 10 s, a period of time that is too long for rapid adaptation to changes in nutrient concentration. When CheZ is present it binds to CheY and reduces the time for dephosphorylation from 10 s to 1 s.

The remaining two proteins, CheB and CheR regulate the activity of the Tar receptor. Tar, like many receptors, is a conduit for two-way communication between outside and inside. Ligand binding on the outside alters the binding properties and catalytic actions of the cytoplasmic part of the protein. Binding events taking place in the cytoplasmic portion of the molecule influence the receptor's outside ligand binding properties. The cytoplasmic part of the Tar signal helix contains binding sites for attachment of methyl groups. As a consequence, Tar and receptors like it are known as *methyl-accepting chemotactic proteins* (MCPs). CheR catalyzes the transfer of methyl groups to glutamate residues on the Tar signal helix using S-adenosyl methionine as the methyl donor. This activity is continual, and the addition of these methyl groups on the helix progressively reduces the binding affinity of the Tar receptors for their ligand. CheB does the opposite. It removes methyl groups from Tar and restores a high binding affinity.

CheB is activated by phosphorylated CheA, thereby forming a feedback loop—Tar signals to CheA, which signals to CheB, which signals back to Tar. As the ligand concentration builds up, and more and more Tar receptors are bound, CheA is less and less phosphorylated, CheB remains inactive, and only a few methyl groups are removed from Tar. Thus, at high ligand concentrations, the sensitivity of Tar to its ligand is reduced. In the absence of ligand binding, that is, at low ligand concentrations, CheA is phosphorylated and can activate CheB. CheB removes methyl groups from Tar, and returns Tar to a condition of high binding affinity and sensitivity to its ligand. By this means, the methylation level tracks the concentration. At low concentrations, Tar has few attached methyl groups, and it has a high affinity for its ligand. At high concentrations, many methyl groups are added, and Tar has a low binding affinity.

Tar can maintain its sensitivity to concentration gradients over some five orders of magnitude through this feedback mechanism. This process is called *exact adaptation*; the steady state tumbling frequency in a homogeneous ligand environment (no signal) is insensitive to the ligand concentration over a broad range of attractant or repellent concentrations. The chemotactic process is independent of specific values of the concentrations over a broad range. As a result, the bacterium can sense small spatial changes in concentration, i.e., it can detect small concentration gradients over many orders of magnitude in mean concentration. This ability is referred to as *robustness*. It is a consequence of the network connectivity, particularly that of the crucial feedback loop. This feedback loop compensates for the effect of ligand binding, thereby maintaining a robust performance regardless of the ligand concentration.

A bacterium is too small to be able to detect concentration gradients by comparing concentrations at different points along its body. Instead of using a spatial strategy, bacteria adopt a temporal one: They determine concentration gradients by comparing concentrations at slightly different *times*. This is equivalent to determining a *spatial* concentration gradient, since there is translational motion of the bacterium during the measurement time interval. The key to the temporal or time derivative method is to compare two different quantities, each measuring in some way concentration. One quantity is the concentration at an instant in time, represented as a percent of receptor occupancy. The second quantity is the receptor concentration a few seconds earlier, represented by the level of receptor methylation. There is a lag between receptor binding and the adjustment in methylation— CheA must be phosphorylated, and then CheB must be phosphorylated before there is an adjustment in receptor methylation. The lag between receptor binding and methylation of a few seconds serves as a memory that enables determination of temporal gradients.

7.11 How Plants Sense and Respond to Hormones

Plants use two-component systems and phosphorelays for sensing and responding to hormones. At certain times in its life a plant will undergo *senescence*, the state in which somatic tissues no longer required are broken down and their components reused in younger tissues. At other times a plant will undergo *organ abscission*, in which organs such as seeds and fruit are detached from the main body of the plant in a developmentally regulated manner. Plants, like animals, undergo wound-healing and mount defenses against pathogens. These developmental and defense processes are regulated by plant hormones such as ethylene (C_2H_4) and cytokinins.

Arabidopsis is a small weed, and it was the first plant species to have its genome sequenced. The *Arabidopsis* genome encodes 12 histidine kinases, 22 response regulators, and 5 HPt proteins. Histidine kinases used for sensing and responding to ethylene, cytokinins, and for osmosensing are listed in Table 7.5. All of these with the exception of ERS1 are hybrids. They contain a receiver domain and signal through HPt proteins.

The Arabidopsis response regulators (RRs) can be partitioned into two groups. A-type ARRs contain the required Asp phosphorylation site, as well

TABLE 7.5. Hormone and osmosensing in plants.

Histidine kinase	Sensing function
ETR1	ethylene
ERS1	ethylene
CRE1/AHK4/WOL	cytokinins
AHK2	cytokinins
AHK3	cytokinins
AtHK1	osmosensor

as several other required residues, but lack the DNA-binding domain needed for operation as transcription factors. A-type ARRs are found in increased numbers shortly after a cell is exposed by cytokinins. B-type ARRs contain a DNA-binding domain in their C-terminal regions; nuclear localization signals (NLSs) needed for entry into the nucleus; and modules associated with Asp response regulation and signaling. Thus, the type-B ARRs contain all the components needed for operation as transcription factors.

Cytokinin ligands are recognized by the sensor histidine kinases CRE1, AKH2 and AKH3. Receptor-binding activates a phosphorelay from CRE1 to an HPt protein called AHP, which translocates to the nucleus and transfers its phosphoryl group to a type B response regulator. The type B RRs activate transcription of genes encoding Type A response regulators. The type A RRs operating in conjunction with AHPs and histidine kinases modulate the activities in other signaling pathways (Figure 7.7).

The ethylene-signaling pathway is erected in a way that combines histidine kinase action with MAP signaling cascades. Ethylene receptors are histidine kinases, but these proteins do not appear to transfer phosphoryl groups to aspartyl response regulators, but rather signal through one or more MAP-like serine/threonine kinases. The signaling mechanisms appearing in the *Arabidopsis* ethylene pathway bear a striking resemblance to those of the yeast high osmolarity (HOG) pathway discussed in the last chapter. In the ETR1 pathway, ethylene is a negative regulator of its recep-

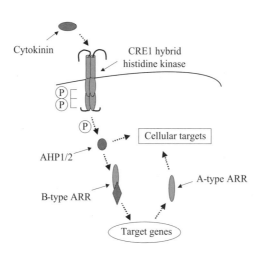

FIGURE 7.7. Cytokinin signal transduction in *Arabidopsis thaliana*: Binding of the cytokinin ligand to the Cre1 receptor/histidine kinase activates a His-Asp phosphorelay. Signals are relayed through AHP1/2 types of HPt intermediates, and a B-type *Arabidopsis* response regulator (ARR), resulting in the expression of cytokinin-responsive genes including A-type of ARRs. These together with B-type ARRs and AHPs regulate downstream cellular targets.

TABLE 7.6. Light-sensitive proteins and their distribution.

Sensor protein	Chromophore	λ_{max} (nm)	Distribution
Rhodopsins	Retinal	500	Animals
Phytochromes	Tetrapyrrole	665, 730	Bacteria, protists, plants
Cryptochromes	Flavins	400–500	Plants, insects, mammals

tor. If the ligand is not present, the receptor is in an "on" state and sends signals through the MAP3 kinase CTR1. The activated CTR1 proteins negatively regulate transcription of ethylene response genes. When ethylene is present in the environment, the receptor state is "off", and it can no longer activate CTR1. Turning off CTR1 relieves the repression by CTR1 of transcription of ethylene response genes.

7.12 Role of Growth Plasticity in Plants

Plants retain a great deal of plasticity in their growth and developmental programs and tie these programs to environmental cues such as lighting. In plants, light sensing is used to guide developmental, morphological, and physiological responses, whereas in animals light sensing guides behavioral responses. Plants are responsive not only to light intensity, but also to its orientation and duration, and its spectral properties. Light sensing enables plants to project growth into well-illuminated spaces and away from regions shaded by other plants. Light sensing allows plants to synchronize, or *entrain*, their growth and developmental rhythms with daily and seasonal cycles. It enables plants to arrive at decisions on when to germinate and when to flower, for example.

Light sensing by plants is an activity distinct from photosynthesis. Sensing of light in plants is achieved using two kinds of photoreceptors—phytochromes and cryptochromes (Table 7.6). These molecules, like the rhodopsin molecules found in the rods of the animal eye, contain a light absorbing group, or chromophore, that alters its conformation upon absorbing light of the appropriate wavelength. The conformational changes are sent on to other signaling elements to complete the transduction of the light signal into intracellular signal events.

7.13 Role of Phytochromes in Plant Cell Growth

Phytochromes are red/far-red photoreceptors that initiate developmental and proximity responses to light. Phytochromes exist in two forms. One form operates as a red photoreceptor with peak absorption at 665 nm, and the other as a far-red photoreceptor with peak absorption at 730 nm. Light absorption triggers the conversion of one form to the other. The red light absorbing form is designated as Pr and the far red light form as Pfr. When the red form (Pr) absorbs light it is converted into the far-red form Pfr, and similarly, when the far-red form (Pfr) absorbs light it is converted to the red form Pr. The ratio of these two forms is a measure of the amount of red light incident on the plant compared to the amount of far red light. Direct sunlight provides more red than far red light, but the light characteristics change when surrounding foliage shades the plant. Then the light shifts towards the far red. Thus, during exposure to direct sunlight Pfr will predominate in the leaves and when shaded Pr will be more abundant.

The signaling pathway leading to alterations in gene expression has been explored in *Arabidopsis*. The pathway is direct and short. Using green fluorescent protein it has been found that the Pfr form, but not the Pr form, translocates to the nucleus. Upon arrival, the Pfr form interacts with a transcription factor PIF3. The interaction of Pfr with PIF3 triggered by the red light influences a large number of cellular signaling and metabolic pathways. As observed using microarrays, some pathways are turned on while others are turned off (Figure 7.8a).

Phytochromes are not limited to plants. They are found in photosynthetic bacteria and in protists (algae). Cyanobacterial phytochromes are histidine

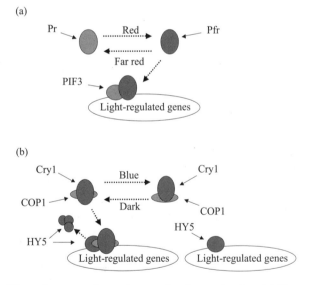

FIGURE 7.8. Phytochrome and cytochrome signal transduction: (a) Phytochrome signaling, in which the active form of the light receptor Pfr acts in conjunction with the transcription factor PIF3 to regulate the expression of light-regulated genes. (b) Cytochrome signaling, in which expression of (blue) light-sensitive genes is turned off under dark conditions. The blue light receptor in conjunction with the regulator COP1 translocates to the nucleus where COP1 stimulates the proteolysis of the transcription factor HY5. Under blue light conditions the Cry1/COP1 complex remains in the cytosol and COP1 cannot inhibit HY5-mediated transcription.

kinases that relay signals to aspartyl response regulators. The plant phytochromes appear to have replaced a histidine kinase activity with a serine/threonine kinase one. The transformed kinases still exhibit many of the characteristics of histidine kinases, supporting the notion of a bacterial origin. These signaling proteins are thus regarded as divergent histidine kinases that along with some of the ethylene receptors lack some of the amino acid residues necessary for histidine kinase activity.

7.14 Cryptochromes Help Regulate Circadian Rhythms

Cryptochromes are blue/ultraviolet-A photoreceptors that help regulate circadian rhythms. They have a peak absorption in the range 400 to 500 nm. They are found not only in plants, but also in insects and mammals. The cryptochrome signaling pathways regulate physiological processes such as the setting of period and phase of an organism's internal circadian clock so that internal body rhythms match those present in the external environment (daily and seasonal). This process is called *entrainment*.

Like the phytochrome signaling pathway, the cryptochrome signaling pathway is direct and short. The control point at the end of the signaling pathway is one that regulates transcription, but it is not a transcription factor at the promoter site as in the case of the phytochromes. Instead blue light photoreceptors remain bound to a transcription regulator called COP1. The COP1 protein regulates the activity of transcription factor HY5 that stimulates transcription of a large number of genes. When it is activated, COP1 tags HY5 for degradation by proteolytic proteins in the nucleus, and thus COP1 inhibits the ability of the HY5 protein to stimulate transcription.

The C-terminal domain of Cry1, Ctt1 binds to COP1 under both dark and light conditions. In the dark COP1 is localized predominantely in the nucleus and in light it becomes mostly cytosolic. The blue-light transcription factor HY5 is localized in the nucleus under both kinds of illumination. When blue light strikes the Cry1 photocenter, conformational changes occur in COP1 altering its cellular location and turning off its inhibition of HY5. In blue light conditions, HY5 escapes being tagged for destruction, and is then able to stimulate transcription (Figure 7.8(b)).

In this form of regulation, the presence or absence of blue light modulates the effective lifetime of the HY5 protein. This form of regulation is a fairly common one. It is fast because no lengthy protein synthesis steps are required. Rapid modulatory actions in a signaling pathway can take one of several forms. They can involve immobilizing a signal protein at a particular location, or they may involve proteolytic degradation of a signaling element as in the case of cryptochrome signaling. One form of regulation by localization already encountered in this and the previous chapter is the sequestering of transcription factors in the cytosol and away from the nucleus until the appropriate signal is received.

References and Further Reading

Histidine Kinases

Bilwes AM, et al. [1999]. Structure of CheA, a signal-transducing histidine kinase. *Cell*, 96: 131–141.

Mourey L, et al. [2001]. Crystal structure of the CheA histidine phosphotransfer domain that mediates response regulator phosphorylation in bacterial chemotaxis. *J. Biol. Chem.*, 276: 31074–31082.

Response Regulators

Baikalov I, et al. [1996]. Structure of the *Escherichia coli* response regulator NarL. *Biochem.*, 35: 11053–11061.

Birch C, et al. [1999]. Conformational changes induced by phosphorylation of the FixJ receiver domain. *Structure*, 7: 1505–1515.

Feher VA, and Cavanagh J [1999]. Millisecond-timescale motions contribute to the function of the bacterial response regulator protein SpoOF. *Nature*, 400: 289–293.

Kern D, et al. [1999]. Structure of a transiently phosphorylated switch in bacterial signal transduction. *Nature*, 402: 894–898.

Lewis RJ, et al. [1999]. Phosphorylated aspartate in the structure of a response regulator protein. *J. Mol. Biol.*, 294: 9–15.

Stock J, and Da Re S [2000]. Signal transduction: Response regulators on and off. *Curr. Biol.*, 10: R420–R424.

Volkman BF, et al. [2001]. Two-state allosteric behavior in a single-domain signaling protein. *Science*, 291: 2429–2433.

Chemotaxis Receptors

Falke JJ, and Hazelbauer GL [2001]. Transmembrane signaling in bacterial chemoreceptors. *Trends Biochem. Sci.*, 26: 257–265.

Mowbray SL, and Sandgren MOJ [1998]. Chemotaxis receptors: A progress report on structure and function. *J. Struct. Biol.*, 124: 257–275.

Ottemann KM, et al. [1999]. A piston model for transmembrane signaling of the aspartate receptor. *Science*, 285: 1751–1754.

West AH, and Stock AM [2001]. Histidine kinases and response regulator proteins in two-component signaling systems. *Trends Biochem. Sci.*, 26: 369–376.

Volz K [1993]. Structural conservation in the CheY superfamily. *Biochem.*, 32: 11741–11753.

Feedback, Methylation, and Robust Behavior

Barkai N, and Leibler S [1997]. Robustness in simple chemical networks. *Nature*, 387: 913–917.

Djordjevic S, et al. [1998]. Structural basis for methylesterase CheB regulation by a phosphorylation-activated domain. *Proc. Natl. Acad. Sci. USA*, 95: 1381–1386.

Djordjevic S, and Stock AM [1997]. Crystal structure of the chemotaxis receptor methyltransferase CheR suggests a conserved structural motif for binding S-adenosylmethionine. *Structure*, 5: 545–558.

Yi TM, Huang Y, Simon MI, and Doyle J [2000]. Robust perfect adaptation in bacterial chemotaxis through integral feedback control. *Proc. Natl. Acad. Sci. USA*, 97: 4649–4653.

Receptor Clusters and Formation of a Primitive Nose

Liu Y, et al. [1997]. Receptor-mediated protein kinase activation and mechanism of transmembrane signaling in bacterial chemotaxis. *EMBO J.*, 16: 7231–7240.

Levit MN, Liu Y, and Stock JB [1998]. Stimulus response coupling in bacterial chemotaxis: Receptor dimers in signaling arrays. *Mol. Microbiol.*, 30: 459–466.

Duke TAJ, and Bray D [1999]. Heightened sensitivity of a lattice of membrane receptors. *Proc. Natl. Acad. Sci. USA*, 96: 10104–10108.

Shimizu TS, et al. [2000]. Molecular model of a lattice of signaling proteins involved in bacterial chemotaxis. *Nature Cell Biol.*, 2: 792–796.

Plant His-Asp Signaling

Cashmore AR, et al. [1999]. Cryptochromes: Blue light receptors for plants and animals. *Science*, 284: 760–765.

Lohrmann J, and Harter K [2002]. Plant two-component signaling systems and the role of response regulators. *Plant Physiol.*, 128: 363–369.

Neff MM, Fankhauser C, and Chory J [2000]. Light: An indicator of time and place. *Genes Dev.*, 14: 257–271.

Smith H [2000]. Phytochromes and light signal perception by plants—An emerging synthesis. *Nature*, 407: 585–591.

Young MW, and Kay SA [2001]. Time zones: A comparative genetics of circadian clocks. *Nat. Rev. Genet.*, 2: 702–715.

Problems

7.1 A significant number of the proteins discussed in the last two chapters—kinases, phosphatases, GTPases, and proteases—are enzymes. Catalysts increase the rate of the reactions by many orders of magnitude, but are not themselves altered during catalysis. In their absence the biochemical reactions to be catalyzed are too slow because of unfavorable mixes of positive and negative charges that have to be brought into close proximity at the transition state long enough for the reaction to occur. As was discussed in the chapter, the histidine kinases not only grip and position the ATP in a favorable orientation for transfer of the gamma phosphoryl group, they also modify the electrostatic environment. Negative charges of the phosphoryl group are countered by the positive charges of the divalent magnesium cation; a nucleophile is present to break bonds, and the cleft at the active site in laced with charged residues. These activities help stabilize the transition state, and in the process lower the activation barrier for the transition to the final state. This lowering is depicted schematically in the figure shown below.

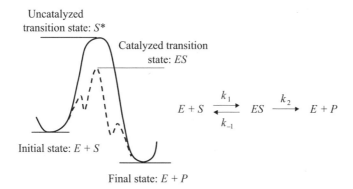

Uncatalyzed
transition state: $S*$

Catalyzed transition
state: ES

$$E + S \; \underset{k_{-1}}{\overset{k_1}{\rightleftarrows}} \; ES \; \xrightarrow{k_2} \; E + P$$

Initial state: $E + S$

Final state: $E + P$

FIGURE for Problem 7.1. Uncatalyzed and catalyzed transition states.

The equation shown to the right of the figure defines the three rates involved in the catalytic process. Kinetic equations governing the catalysis can be constructed by applying the law of mass action. The resulting expressions in terms of enzyme and substrate molar concentrations are

$$\frac{d[E]}{dt} = -k_1[E][S] + k_{-1}[ES] + k_2[ES],$$

$$\frac{d[S]}{dt} = -k_1[E][S] + k_{-1}[ES],$$

$$\frac{d[ES]}{dt} = k_1[E][S] - k_{-1}[ES] - k_2[ES],$$

where the k's are the reaction rates.

The amount of enzyme is usually far less than the amount of substrate and is completely bound up by the substrate. As a result one can make the steady state assumption that

$$\frac{d[ES]}{dt} = 0.$$

Using the above formulas, derive the Michaelis–Menton equation

$$V_0 = \frac{V_{\max}[S]}{K_M + [S]},$$

where $V_{\max}$ is equal to $k_2 \cdot ([E] + [ES])$, the Michaelis constant $K_M = (k_1 + k_{-1})/k_2$, and the steady state velocity V_0 is $k_2[ES]$.

Plot V_0 versus $[S]$. What happens at high $[S]$? It is customary to determine the constants appearing in the Michaelis–Menton equation by making measurements of V_0 values at different substrate concentrations $[S]$ and arranging the data in the form $1/V_0$ versus $1/[S]$. What does a plot of $1/V_0$ versus $1/[S]$ look like? What is the slope and what is the intercept?

8
Organization of Signal Complexes by Lipids, Calcium, and Cyclic AMP

Cytoplasmic signaling proteins are recruited to the cell plasma membrane in response to signals sent from other cells. The recruited signal molecules are organized into modules and complexes, and the first steps in converting the extracellular signals into cellular responses are taken. Several kinds of small signaling molecules, most notably, lipids, calcium, and cyclic adenosine monophosphate (cyclic AMP, or cAMP), act as signaling intermediaries. They tie together events taking place subsequent to ligand binding by helping to recruit and organize the proteins that function as the primary intracellular signal transducers. Because of their role as signaling intermediaries, the small molecules are commonly termed "second messengers," the first messengers being the extracellular signal molecules, and the third messengers being the large protein kinases and phosphatases that are recruited to the plasma membrane.

Localization plays an important role in the organization of signaling pathways, especially those that lead from the cell surface and end at control points at the nucleus, cytoskeleton, and elsewhere. First, second, third and higher order messengers do not diffuse about the cell but rather act in restricted portions of space. The plasma membrane contributes to the localization. It does far more than simply separate outside from inside and provide a fluid medium for the insertion, uniform lateral movement, and tethering of proteins; it assumes an active role in the signaling processes taking place.

In the first part of this chapter, the composition and organization of the plasma membrane will be examined. This exploration will be followed by a discussion of how lipid, calcium, and cAMP second messengers are generated. Second messengers stimulate the activities of serine /threonine kinases belonging to the AGC family. Signaling by these kinases will be explored in the third part of the chapter.

8.1 Composition of Biological Membranes

Biological membranes are composed of phospholipids, glycolipids and cholesterol. Membrane lipids are linked together through the cooperative effects of multiple weak noncovalent interactions such as van der Waals forces and hydrogen bonds, and as a result there is considerable fluidity of movement within the membrane—the constituents are free to diffuse laterally and rotationally. The overall structure is that of a fluid of lipids and membrane-associated proteins undergoing Brownian motion. The motions of the molecules are not completely unrestricted, but instead are limited to specific regions of the membrane, giving rise to a mosaic of membrane compartments.

Three classes of lipids are found in biological membranes—phospholipids, glycolipids, and cholesterol. Phospholipids are the primary constituents. The four most common phospholipids are listed in Table 8.1. Phosphoglycerides contain a glycerol backbone that is linked to a phosphoryl group bonded to a phosphorylated alcohol group. A different backbone component, sphingosine, is used in the sphingolipids. Of the four commonly occurring phospholipids, all except phosphatidylserine have uncharged head groups. Phosphatidylserine has a negatively charged head group and is found exclusively in the cytoplasmic leaflet. Phosphatidylinositol is of special importance in metazoans. It is reversibly phosphorylated at one or more OH sites on the inositol ring by lipid kinases to generate lipid signal molecules that coordinate a number of cellular processes including cytoskeleton control and motility, insulin signaling (glucose and lipid metabolism), and growth factor-promoted cell survival.

The glycolipids, like the phospholipids, have a backbone connected to fatty acyl chains. They differ from the phospholipids in that they contain one or more sugar groups in place of the phosphoryl-alcohol bearing head-group of the phospholipids (Figure 8.1). Glycolipids are found in the exoplasmic leaflet of the plasma membrane, and are believed to promote cell-to-cell recognition. The sugar residues that form the hydrophilic head extend out from the cell surface.

TABLE 8.1. Lipid constituents of the plasma membrane: Exoplasmic—Outer; Cytoplasmic—Inner.

Type	Lipid	Comments
Phospholipids	Phosphatidylcholine (PC)	Exoplasmic leaflet
	Phosphatidylethanolamine	Cytoplasmic leaflet
	Phosphatidylserine	Cytoplasmic leaflet, negatively charged head group
	Sphingomyelin	
	Phosphatidylinositol	Exoplasmic leaflet, sphingosine backbone
		Major role in signaling
Glycolipids		Exoplasmic leaflet, cell-to-cell recognition
Cholesterol		Influences fluidity and membrane organization

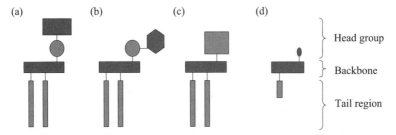

FIGURE 8.1. Schematic representations of phospho- and glycolipids and cholesterol: (a) A phospholipid such as PC consisting of a pair of acyl chains in the tail region, a backbone, which in this case is glycerol, and a head region consisting of a phosphoryl group (circle) plus an alcohol group (rectangle). (b) A phosphatidylinositol molecule consisting of a tail region, a glycerol backbone, and a phosphoryl group coupled to a hexagonal inositol ring in the head region. (c) A glycolipid in which the head group consists of one or more sugar groups (square). (d) A cholesterol molecule is composed of a fatty acyl tail connected to a rigid steroid ring assembly with an OH group at the terminus that serves as its polar head.

8.2 Microdomains and Caveolae in Membranes

Biological membranes contain microdomains and caveolae specialized for signal transduction. Lipids found in biological membranes vary in chain length and degree of saturation. Chains vary in length, having an even number of carbons typically between 14 and 24, with 16, 18, and 20 most common in phospholipids and glycolipids. Chains with one or more double bonds are unsaturated. These bonds are rigid and introduce kinks in the chain. In a fully saturated acyl chain the carbon-carbon atoms are covalently linked by single bonds. Each carbon atom in such a chain can establish a maximum possible number of bonds with hydrogen atoms, hence the term "saturated." Such chains are free to rotate about their carbon-carbon bonds, and can be packed tightly. In contrast, the kinks present in an array of unsaturated lipids cause irregularities or voids to appear in the array; these molecules cannot be packed as tightly.

The degree of saturation of the acyl chains and the cholesterol content influence the melting point and fluidity of the lipids in the membrane. This point can be illustrated by some everyday examples. Fats, oils, and waxes are examples of lipids. Butter, a saturated lipid, is a solid gel at room temperature, while corn oil, an unsaturated lipid, is a liquid at the same temperature. Cholesterol plays an important role in determining the fluidity of the membrane compartments. It is smaller than the phospho- and glycolipids and is distributed between both leaflets. As the concentration of cholesterol increases, the lipid membrane becomes less disordered, gel-like and more like an ordered liquid in which the lipids are more tightly packed together, especially when saturated sphingolipids are present (Figure 8.2).

(a)

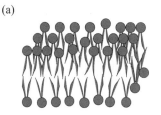

(b)

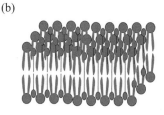

FIGURE 8.2. Lipids and the lipid bilayer: (a) Membrane bilayers contain a mixture of amphipathic lipids. Each lipid molecule has a hydrophilic (polar) head region and a two-pronged hydrophobic tail region oriented as shown. Tails are fatty acyl chains, hydrocarbon chains with a carboxylic acid (COOH) group at one end and (usually) a methane group at the other terminus. In cells, the polar head of each lipid molecule is surrounded by water molecules and thus is hydrated. The density of water molecules drops off rapidly in the hydrophobic interior. (b) Small cholesterol molecules are situated in between the larger lipid molecules. Lipids differ from one another in the number, length, and degree of saturation of the acyl chains, and in the composition of their head groups. The overall packing density is greater in (b) than in (a) due to the presence of cholesterol and of lipids with straighter saturated acyl chains. Signaling proteins (not shown) carrying a GPI anchor attach to the exoplasmic leaflet while signaling proteins bearing acetyl and other kinds of anchors attach to the cytoplasmic leaflet. These proteins congregate in cholesterol and sphingolipid-enriched membrane compartments.

The plasma membranes of eukaryotes are not uniform, but rather contain several kinds of lipid domains, each varying somewhat in its lipid composition. Compartments enriched in cholesterol and/or sphingolipids contain high concentrations of signaling molecules: GPI-anchored proteins in their exoplasmic leaflet and a variety of anchored proteins in their cytoplasmic leaflet. Two kinds of compartments—*caveolae* and *lipid rafts*—enriched in cholesterol and glycosphingolipids, are specialized for signaling.

Caveolae (little caves) are detergent-insoluble membrane domains enriched in glycosphingolipids, cholesterol, and lipid-anchored proteins. Caveolae are tiny flask-shaped invaginations in the outer leaflet of the plasma membrane. They play an important role in signaling as well as in transport. Caveolae may be flat, vesicular, or even tubular in shape, and may be either open or closed off from the cell surface. They are detergent insoluble and are enriched in coatlike materials, caveolins, which bind to cholesterol. Cholesterol- and sphingolipid-enriched microdomains can float within the more diffuse lipid bilayer.

The second kind of cholesterol- and sphingolipid-enriched compartment is a *lipid raft*. It does not have a cave-like shape and does not contain caveolins, but instead is rather flat in shape. The fluid and detergent-insoluble properties of both the rafts and the caveolae arise from the tight packing of the acyl chains of the sphingolipids and from the high cholesterol content. The cholesterol molecules not only rigidify the compartment but

TABLE 8.2. Phosphoinositide nomenclature.

Phosphoinositide	Members	Designation
D3	Phosphatidylinositol-3 phosphate	PtdIns(3)P
	Phosphatidylinositol-3,4 biphosphate	PtdIns(3,4)P$_2$
	Phosphatidylinositol-3,5 biphosphate	PtdIns(3,5)P$_2$
	Phosphatidylinositol-3,4,5 triphosphate	PtdIns(3,4,5)P$_3$ or PIP$_3$
D4	Phosphatidylinositol-4 phosphate	PtdIns(4)P
	Phosphatidylinositol-4,5 biphosphate	PtdIns(4,5)P$_2$ or PIP$_2$
D5	Phosphatidylinositol-5 phosphate	PtdIns(5)P

also facilitate the formation of signaling complexes and the initiation of signaling by them.

8.3 Lipid Kinases Phosphorylate Plasma Membrane Phosphoglycerides

The plasma membrane phosphoglyceride known as phosphatidylinositol plays an important role in signaling, cytoskeleton regulation, and membrane trafficking. The inositol ring of the phosphatidylinositol molecule contains a phosphoryl group at position 1 that is tied to the glycerol backbone. All other OH groups of the inositol ring can be phosplorylated except those at positions 2 and 6. Just as protein kinases catalyze the transfer of phosphoryl groups to selected amino acid residues, lipid kinases catalyze the transfer of phosphoryl groups to specific sites on lipids. Several lipid kinases catalyze the phosphorylation of phosphatidylinositol.

The lipid kinase phosphoinositide-3-OH kinase (PI3K) catalyzes the transfer of a phosphoryl group from an ATP molecule to the OH group at position 3 of the inositol ring of the lipid. Other lipid kinases, PI4K and PI5K, catalyze the transfer of phosphoryl groups to the other available sites, positions 4 and 5, on the ring. An entire ensemble of phosphoinositides can be produced through the addition and subtraction of phosploryl groups from positions 3, 4, and 5 of the inositol rings. These phosphorylated lipid products, their placement into D3, D4, and D5 phosphoinositide groups, and their common abbreviations are listed in Table 8.2.

8.4 Generation of Lipid Second Messengers from PIP$_2$

Two lipid second messengers are generated from PIP$_2$ by phospholipase C. Phosphatidylinositol 4,5 biphosphate (PIP$_2$) serves as the source of two lipid second messengers: diacylglycerol (DAG) and inositol 1,4,5-triphosphate (designated Ins(1,4,5)P$_3$ or IP$_3$). The plasma membrane functions as a cellular repository for the PIP$_2$ and other phosphoinositides.

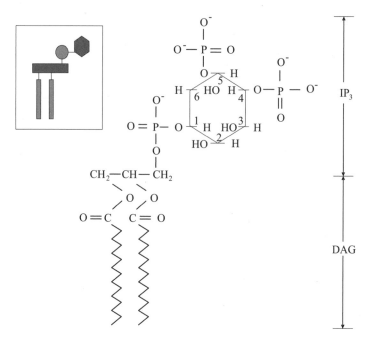

FIGURE 8.3. Structure of phosphatidylinositol 4,5 biphosphate (PIP₂): Cleavage of this molecule by PLC generates the lipid second messengers diacylglycerol (DAG) and inositol 1,4,5-triphosphate (IP₃). The insert shows in block form the organization of the molecule into tail, backbone, linker phosphoryl group, and inositol head groups.

Phospholipase C (PLC) cleaves the membrane-situated PIP₂ into DAG—which contains the acyl chains plus the glycerol backbone—and IP₃—which contains the rest of the head group (Figure 8.3).

A large number of plasma membrane receptors use PLC as an intermediary to signal and activate downstream kinases. Prominent among these are G protein-coupled receptors (GPCRs) and growth factor receptors. There are three PLC subtypes, designed as PLCβ, PLCγ, and PLCδ. These enzymes have a modular organization that supports their (a) localization at the plasma membrane, (b) activation by upstream receptor signals, and (c) catalytic activities (Figure 8.4). The Pleckstrin homology domain and the C-terminus SH2 domain mediate binding to the plasma membrane PtdIns. The upstream, signaling elements such as G protein subunits bind to and activate PLC, and the SH3 and N-terminal SH2 domain mediates interactions with upstream growth factor receptors. Once formed by PLC acting on PIP₂, IP₃ diffuses to intracellular stores located in the endoplasmic reticulum (ER) and triggers the release of Ca^{2+}. The calcium ions together with DAG activate protein kinase C, the main downstream target of PLC signaling (Figure 8.5).

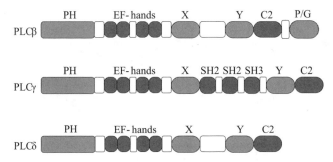

FIGURE 8.4. Organization of phospholipase C: The domain structure of the three classes of PLC isozymes is shown. Each type of PLC has a Pleckstrin homology (PH) domain in its N-terminus. PH domains bind to plasma membrane PtdIns proteins. The PH domains are not all the same; they vary their binding affinities among the three classes. The X and Y domains form the catalytic domain of the enzyme.

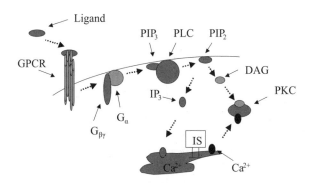

FIGURE 8.5. Signaling through PLC: Activation of PLC by ligand-GPCR binding stimulates dissociation of the G protein subunits, which then activate PLC. The PLC proteins tether to the plasma membrane by binding PIP$_3$ lipids, and cleave PIP$_2$ into DAG and IP$_3$. The latter translocates to the intracellular stores (IS) triggering release of calcium ions, which along with DAG bind to and stimulate protein kinase C activity.

8.5 Regulation of Cellular Processes by PI3K

Phosphoinositide-3-OH kinase (PI3K) helps regulate a variety of cellular processes. PI3Ks mediate cellular responses to GPCRs, growth factors and insulin, activation by cell adhesion molecules called *integrins*, and leukocyte (white blood cell) function. An important characteristic of the proteins that interact with lipid second messengers is the presence of one or more lipid-

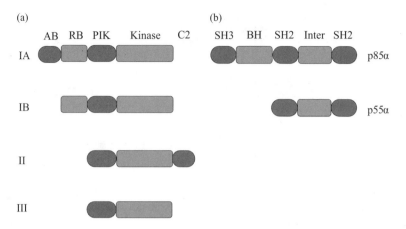

FIGURE 8.6. PI3 kinase domain structure: (a) Classes of catalytic subunits. (b) Examples of regulatory subunits that associate with Class IA subunits of PI3 kinases. Abbreviations: Adapter-binding (AB) domain; Ras-binding (RB) domain; Src homology domain-3 (SH3); Src homology domain-2 (SH2); BCR-homology GTPase activating (BH) domain.

binding domains in their regulatory regions. Several kinds of lipid-binding modules—PH domains, C2 domains, and FYVE domains—mediate the binding of lipid second messengers to their downstream targets.

There are three classes of mammalian PI3Ks (Figure 8.6). Class I PI3Ks are heterodimers composed of a 100-kDa catalytic subunit and an 85-kDa or 55-kDa joint regulatory/adapter subunit. There are two kinds of Class I PI3Ks, determined by the presence or absence of an adapter-binding (AB) domain in its N-terminal region (Figure 8.4a) and by kinds of receptor binding events that activate them. The adapter for the Class IA PI3Ks binds to growth factor receptors, while Class IB PI3Ks are activated primarily by G protein coupled receptors operating through the associated Gβγ subunits. Class II PI3Ks contain a C2 domain in their C-terminal region. Class III PI3Ks may be constitutively active in the cell and help regulate membrane trafficking and vesicle formation, two housekeeping activities carried out all the time. As shown in Figure 8.7, PI3K functions as a key intermediary to activate protein kinase B.

8.6 PIPs Regulate Lipid Signaling

There is a corresponding set of lipid phosphatases that catalyzes the removal of phosphoryl groups from inositol rings. Perhaps the most prominent of these is PTEN (phosphatase and tensin homolog deleted on chromosome 10). PTEN acts in opposition to PI3K and catalyzes the removal

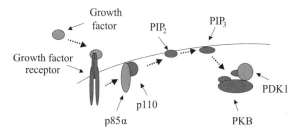

FIGURE 8.7. Signaling through PI3K: Ligand growth factor receptor binding stimulates the dissociation of the regulatory and catalytic subunits of PI3K from each other. The catalytic subunit phosphorylates the PIP_2 proteins at the 3′ position, thereby making PIP_3, which then diffuses to and binds PDK1 and protein kinase B.

of phosphoryl groups from position 3 on inositol rings. It acts on PIP_3 to return it to a PIP_2 form, thereby reversing the catalytic effects of PI3K.

The importance of dephosphorylating actions is made apparent by the high frequency of either mutated or missing forms of PTEN in at least one kind of brain cancer (glioblastoma), in prostate cancer, and in endometrial (uterine) cancer. The major downstream target of the PIP_3 lipids is protein kinase B (Figure 8.7). This kinase supplies what may best be termed a "survival" signal in response to growth factor-binding to receptors such as the insulin receptor. In the absence of growth signals such as insulin, platelet-derived growth factor, and neural growth factor, the levels of activated protein kinase B remain low. They increase in response to the just mentioned growth signals. When PTEN does not throttle back the survival signaling to a baseline level by dephosphorylating the lipid second messengers, the cells undergo uncontrolled growth and proliferation. The actions taken by PTEN tend to suppress the cancer-promoting actions of overly active protein kinase B. For this reason PTEN is referred to as a *tumor suppressor*.

8.7 Role of Lipid-Binding Domains

Lipid-binding domains facilitate the interactions between proteins and lipids. As has been discussed in the last few sections, a number of different lipid-binding domains mediate the interactions between proteins and lipid bilayers. Four of these domains—the PH, C1, C2, and FYVE domains—are especially prominent. They are found in hundreds of proteins and mediate protein recruitment to lipid membranes. These domains and their properties are summarized in Table 8.3. As indicated in the table, the PH domain also mediates protein-protein interactions. Two modules that appear in Table 8.3—the SH2 and PTB domains—are primarily known as

TABLE 8.3. Lipid-binding domains found on proteins.

Designation	Domain name	Description
C1	Protein kinase C homology-1	~50 amino acid residues; binds DAG
C2	Protein kinase C homology-2	~130 amino acid residues; binds acidic lipids
FYVE	Fab1p, YOTB, Vac1p, Eea1	~70 amino acid residues; binds PI3P
PH	Pleckstrin homology	~120 amino acid residues, binds PIP_3 headgroup, PIP_2 and its headgroup; binds proteins
PTB	Phosphotyrosine-binding	~100 amino acid residues; binds phosphorylated tyrosine residues; binds phospholipids in a weak nonspecific manner
SH2	Src homology-2	~100 amino acid residues; binds phosphorylated tyrosine residues; binds phospholipids

phosphotyrosine binding domains. These domains possess phospholipid-binding properties, and for that reason have been included in Table 8.3.

8.8 Role of Intracellular Calcium Level Elevations

Transient elevations in intracellular calcium levels serve as a second messenger. As is characteristic of a second messenger, local increases in intracellular calcium concentration are triggered by the binding of signal proteins to receptors embedded in the plasma membrane. In the absence of triggering signals, intracellular calcium levels are maintained at a low level, no more than $0.1\,\mu M$ in some cell types. Unlike cAMP and cellular lipids, calcium is not synthesizesd by cells. Instead, there are two reservoirs of calcium—the extracellular spaces outside the cell and the calcium stores located within the cell. The calcium concentration in the extracellular spaces is on the order of $2\,mM$, some 20,000 times greater than the resting levels within the cell. Extracellular calcium enters a cell through ion channels located in the plasma membrane. Calcium is sequestered within the cell in intracellular stores (IS), regions enriched in calcium buffers located in the lumen of the endoplasmic reticulum, the matrix of mitochondria, and in the Golgi. In response to the appropriate signals, calcium is released into the cytosol from the stores.

Signals that trigger the entry of extracellular calcium through ion channels and the release of intracellular calcium from stores are sent through two kinds of receptors embedded in the plasma membrane. The first kind of receptor is the *voltage-gated ion channel*. These are opened and closed, or gated, through changes in membrane voltage. These channels are found in cells whose membranes are excitable. Whenever the membrane is depolarized the ion channels open allowing calcium ions from the extracellular spaces to diffuse through and enter the cell. The second kind of membrane signal molecule is the *ligand-gated receptor* such as the G protein-coupled receptor. When a ligand binds the GPCR receptors, phospholipase C is activated. As discussed earlier in this chapter, PLC hydrolyzes PIP_2 to IP_3 and DAG. IP_3 diffuses over to, and binds to, IP_3 receptors located in the ER. This event serves as a release signal, resulting in the movement of Ca^{2+} out of the stores and into the cytosol.

The duration of a calcium signal is short. Intracellular calcium levels are restored to their base values fairly rapidly. Buffering agents bind calcium ions before they can diffuse appreciably from their entry point. Free calcium path lengths, the distance traveled by calcium ions before being bound, average less than 0.5μ, which is far smaller than the linear dimensions, 10 to 30μ, of typical eukaryotic cells. In addition to being buffered, ATP-driven calcium pumps located in the plasma membrane rapidly remove calcium ions from the cell, and other ATP-driven pumps transport calcium back into the intracellular stores. The take-up of calcium by buffers along with its rapid pumping out of the cytosol and into the stores produces a sharp localization of the signaling both in space and time.

8.9 Role of Calmodulin in Signaling

Calmodulin is a calcium sensor involved in activating many signaling pathways. Calmodulin is an abundant protein, consisting of 0.1% of all the protein present at any given time in the cell. It functions as a calcium sensor, and to carry out this role it is distributed throughout the cytosol and nucleus. Calcium is an extremely important regulator of cells in the brain, and the cytosolic concentrations of calmodulin in neurons may reach 2%. Calmodulin is a small protein, consisting of only 148 amino acid residues. It is organized into two lobes connected by a flexible helix giving it a fairly elongated dumbbell shape. As shown in Figure 8.8, calmodulin has four calcium-binding sites; two are in the N-terminal lobe and two are in the C-terminal lobe. Calcium binding produces a shift in the population of equilibrium states from a fairly closed to a more open elongated structure. The shift in population exposes a number of hydrophobic patches that serve as attachment sites to downstream signaling partners.

Calmodulin serves as a key intermediary in a number of signaling pathways. When bound to calcium it promotes the activity of PI3K, nucleotide phosphodiesterases and adenylyl cyclases (both to be discussed shortly), protein kinases such as multifunctional CaM-dependent protein kinase II (CaMKII), protein phosphatases such as CaM-dependent protein phosphatase 2B(PP2B), and a number of cytoskeleton regulators.

(a)   (b)

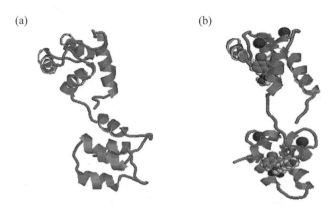

FIGURE 8.8. Solution NMR structure of calmodulin, free and bound to calcium: (a) Calcium-free calmodulin consisting of an (upper) N-terminal domain and a (lower) C-terminal domain. (b) Ca^{2+}_4-bound calmodulin. The four calcium ions are depicted as dark gray spheres. The two prominent hydrophobic patches that are exposed in this more open conformation are bound by W7 molecules shown in a space-filled model. The figure was prepared using Protein Explorer with atomic coordinates deposited in the PDB under accession numbers 1dmo (a) and 1mux (b).

8.10 Adenylyl Cyclases and Phosphodiesterases Produce and Regulate cAMP Second Messengers

Adenylyl cyclase is an integral membrane enzyme that catalyzes the conversion of intracellular ATP into cyclic adenosine monophosphate (cyclic AMP or cAMP). The organization of the adenylyl cyclase molecule is depicted in Figure 8.9. As can be seen there are two transmembrane (TM) regions (M_1 and M_2) and two large cytoplasmic regions (C_1 and C_2). Each transmembrane region consists of six highly hydrophobic membrane-spanning helices connected by short loops. One of the cytoplasmic regions lies topologically in-between the TM regions. The other cytoplasmic region is a situated C-terminal to the second membrane-spanning region. The overall structure of the adenylyl cyclase molecule resembles a dimer, each unit consisting of a TM region followed by a cytoplasmic region. However, monomer-like structures are not functional. The cytoplasmic regions together form the catalytic core of the molecule. The relative orientation of C_1 relative to C_2 is important, and both C_1 and C_2 are required for binding and catalysis.

The magnitude and duration of cyclic nucleotide second messenger signaling is regulated by another class of enzymes, namely, nucleotide phosphodiesterases (PDEs). As shown in Figure 8.10, cAMP has a phosphate group attached to both the 3′ carbon and 5′ carbon of the ribose. PDEs are enzymes that catalyze the hydrolytic cleavage of 3′ phosphodiester bonds

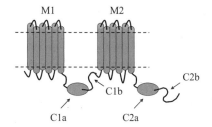

FIGURE 8.9. Organization of adenylyl cyclase: The cylinders denote transmembrane segments. These are organized into a repeated set of six segments (M1 and M2). The cytoplasmic C1 and C2 catalytic domains consist of a compact region (C1a and C1b) and a broad loop (C1b and C2b).

(a)

(b)

FIGURE 8.10. Adenosine triphosphate and cyclic adenosine monophosphate: (a) ATP molecule consisting of an adenine joined to a ribose to which are attached three phosphoryl groups, named in the manner shown. (b) cAMP showing the cyclic structure.

in cAMP resulting in its degradation to inert 5'AMP. They also carry out the same operation in cGMP to yield inert 5'GMP. The PDEs terminate second messenger signaling. They modulate these signals with regard to their amplitude and duration, and through rapid degradation restrict the spread of cAMP to other compartments in the cell.

8.11 Second Messengers Activate Certain Serine/Threonine Kinases

Second messengers acting in the vicinity of the plasma membrane help organize the signaling pathways. They exert their influences by activating and regulating a large number of serine/threonine kinases, among which are

TABLE 8.4. Members of the AGC family of serine/threonine kinases: Different gene-encoded isozymes and alternatively spliced isoforms are listed in column 3. Second messengers required for the activation of the kinases are listed in column 4.

AGC kinase family	Structure	Forms	Regulation
Protein kinase A	2 regulatory subunits; 2 catalytic subunits	RIα, RIβ, RIIα, RIIβ Cα, Cβ, Cγ	cAMP
Protein kinase B	Single chain; PH, catalytic, regulatory domains	α, β, γ1, γ2	PI3K lipid products
Protein kinase C Classical	Single chain; catalytic, regulatory domains	PKC-α, PKC-β1, PKC-β2, PKC-γ	Ca^{2+}, DAG, phosphatidylserine
Novel	Single chain; catalytic, regulatory domains	PKC-δ, PKC-ε, PKC-η, PKC-θ	DAG, phosphatidylserine
Atypical	Single chain; catalytic, regulatory domains	PKC-ζ, PKC-ι, PKC-λ	

those belonging to the AGC family. Three subfamilies of AGC kinases are included in Table 8.4 along with the second messengers involved in their activation. The kinases sequester their catalytic activities within a catalytic domain (or subunit) and similarly combine their regulatory activities into one or more regulatory domains (or subunits). Some of the kinases possess a separate lipid-binding PH (Pleckstrin homology) domain, while others incorporate lipid-binding structures such as C1 and C2 domains into their regulatory regions.

The kinases all have a common structure and similar modes of activation. Second messengers and upstream kinases activate them. Binding of the second messengers to PH domains and regulatory motifs induces the movements of the kinases to the plasma membrane near the sites of second messenger release. This step is followed by phosphorylation by an upstream kinase. In the case of protein kinase B and protein kinase C the upstream kinase kinase has been identified. It is called phosphoinositide-dependent kinase-1 (PDK1). This enzyme may also be the one responsible for activating protein kinase A. In all cases, the role of the upstream kinase is to catalyze the transfer of a phosphoryl group to a crucial residue situated in the activation loop of the AGC kinase. When this occurs the amino acid residues involved in catalysis are unblocked and can carry out their functions.

8.12 Lipids and Upstream Kinases Activate PKB

Protein kinase B is the primary target of signals relayed from membrane-bound signal receptors via lipid second messengers. It is activated in two stages. In the first stage the PI3K product PIP$_3$ binds to PKB through its

PH domain. In response the kinase migrates to the plasma membrane where it is phosphorylated by 3-phosphoinositide-dependent kinase-1 (PDK-1) and then again at a second location either by another kinase or by itself. Once it is recruited to the plasma membrane and phosphorylated, the PKB enzyme is fully activated. As discussed earlier PIP_3-mediated signaling is terminated by a protein phosphatase PTEN that converts PIP_3 to PIP_2.

Protein kinase B is centrally involved in insulin signaling. One of its immediate downstream targets is glycogen synthase kinase 3 (GSK3). In the absence of insulin signaling, GSK3 is active, and when it phosphorylates glycogen synthase it inhibits its enzymatic stimulation of glycogen synthesis. Signaling is fairly rapid. Within a few minutes of insulin binding, PKB is activated and phosphorylates GSK3, thereby inactivating it so that it cannot phosphorylate glycogen synthase. The latter becomes dephosphorylated and consequently is better able to stimulate glycogen production. Another effect of insulin binding leading to GSK3 inactivation is that of an increase in protein translation. This, too, is slowed down by GSK3, but insulin signaling frees the protein translation initiation regulator eIF2B from its inhibition by GSK3 (Figure 8.11).

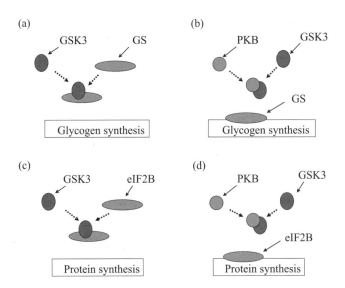

FIGURE 8.11. Insulin signaling through protein kinase B: (a) In the absence of PKB activity GSK3 inhibits the stimulation of glycogen synthesis by glycogen synthase (GS). (b) When PKB is activated it binds to and prevents GSK3 from inhibiting GS. (c) In the absence of PKB activity GSK3 inhibits the initiation of protein synthesis by eIF2B. (d) When PKB is activated it binds to and prevents GSK3 from inhibiting eIF2B.

8.13 PKB Supplies a Signal Necessary for Cell Survival

Protein kinase B is activated by a number of signals, not just insulin. It is activated through PI3K when growth factors bind to members of the receptor tyrosine family to which the insulin receptor belongs. The effect on the cell of protein kinase B signaling is to enhance cell survival. It mainly does so by inhibiting proteins that promote cell suicide, or apoptosis. Two examples of this kind of action are presented in Figure 8.12. The first example is inhibition of *Bad signaling*. Bad is a member of a group of apoptosis regulators known as the Bcl-2 family that will be explored in Chapter 15. Some Bcl-2 family members promote apoptosis while others inhibit it. These proteins are regulated by phosphorylation. The Bad protein, in particular, promotes apoptosis, but this activity is turned off by phosphorylation. When PKB is fully activated, it diffuses over to and phosphorylates Bad, thereby preventing its proapoptosis actions.

The second example presented in Figure 8.12 is inhibition of *Forkhead-mediated transcription*. The Forkhead protein is a transcription factor. Protein kinase B catalyzes the phosphorylation of substrates at sites characterized by the consensus sequence RXRXXS/T. Among the PKB

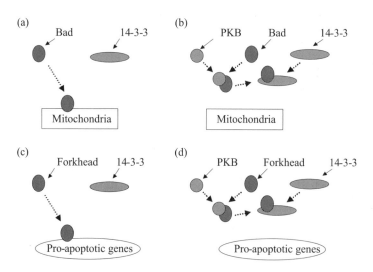

FIGURE 8.12. Survival signaling through protein kinase B: (a) In the absence of PKB signaling Bad translocates from the cytoplasm to the mitochondria where it promotes apoptotic responses. (b) When PKB is activated it phosphorylates Bad, which then binds to a 14-3-3 protein and becomes immobilized in the cytoplasm. (c) In the absence of PKB-signaling, the Forkhead (FH) transcription factor translocates from the cytoplasm to the nucleus where it promotes the expression of proapoptotic genes. (d) When PKB is activated it phosphorylates FH, which then binds to a 14-3-3 protein and becomes immobilized in the cytoplasm.

substrates possessing this consensus sequence are the transcription factors belonging to the Forkhead (FH) family. As was the case for Bad, phosphorylating these proteins inactivates them. In the absence of PKB signaling, the FH proteins translocate from the cytoplasm to the nucleus where they stimulate transcription of apoptosis-promoting genes. When protein kinase B is actived it can phosphorylate FH resulting in its binding to, and immobilization by, cytoplasmic 14-3-3 proteins. Binding to the 14-3-3 proteins prevents FH from carrying out its anticell-cycle progression actions, and thus again promotes cell survival. In both of these examples, the cellular response to lipid-mediated protein kinase B signaling is survival—the suppression of the cell suicide program, and continued progression through the cell cycle.

8.14 Phospholipids and Ca^{2+} Activate Protein Kinase C

There are several subfamilies of protein kinase C (PKC), each characterized by a slightly different domain structure. As depicted in Figure 8.13, the different isozymes of PKC belong to either the *classical, novel,* or *atypical* subfamilies. The pseudosubstrate (PS) is a portion of the chain that interacts with and blocks the activity of the catalytic domain, but activation relieves the block. There are three requirements for full activation of protein kinase C. The first is phosphorylation. Protein kinase C, like protein kinase B, is activated by phosphorylation. PKC is phosphorylated by upstream kinases such as PDK1. The second requirement is presence of subfamily-specific second messengers acting as coactivators. The various isozymes of protein kinase C are listed in Table 8.4 along with the associated coactivators.

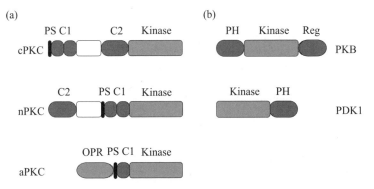

FIGURE 8.13. Structure of protein kinase C and protein kinase B: (a) Protein kinase C families—*Conventional* (c), *novel* (n), and *atypical* (a) protein kinase C proteins. (b) Protein kinase B and PDK1 proteins. Abreviations: pseudosubstrate (PS); Pleckstrin homology (PH); octicopeptide repeat (OPR).

The best-characterized PKC proteins are the classical (conventional) isozymes. As indicated in Table 8.4, these enzymes are activated by calcium along with DAG. Their C1 domain binds DAG while their C2 domain binds Ca^{2+}. Two other parts of the protein, C3 and C4, form the ATP and substrate-binding lobes of the catalytic core. The steps leading to second messenger signaling have been discussed. Binding of a ligand to a receptor activates PLC, which then splits PIP_2 to create DAG and IP_3. The latter stimulates the release of Ca^{2+} from intracellular stores and together the Ca^{2+} and DAG activate protein kinase C. Binding of DAG to the C1 domain and Ca^{2+} to the C2 domain facilitates the recruitment of the protein to the plasma membrane, where it binds to the negatively charged phosphatidylserine molecules concentrated in the cytoplasmic leaflet. In this process the C1 domain is mostly responsible for the recognition of phosphatidylserine while the C2 domain confers a more general specificity for lipid membranes.

8.15 Anchoring Proteins Help Localize PKA and PKC Near Substrates

Anchoring proteins were introduced in Chapter 6. These proteins do not function as enzymes but instead help localize the kinases near their substrates. They provide sites for the tethering of protein kinase A and protein kinase C, and protein phosphatases PP1 and PP2B, to the cytoplasmic face of the plasma membrane and similarly to organelle membranes. There are several prominent families of kinase-anchoring proteins, among which are the A-kinase anchoring proteins (AKAPs) and receptors for activated C kinase (RACKs).

The AKAPs localize protein kinase A close to its substrates. The AKAPs contain an amphipathic helix that binds to specific regulatory subunits of protein kinase A, and the AKAPs have a targeting sequence that directs the anchor protein to a specific subcellular location. Sites of AKAP attachment include cell membranes, cytoskeleton, nuclear matrix, and endoplasmic reticulum. The AKAPs serve as control points where the primary protein kinase-signaling elements, and their regulators and modulators, are brought together near or at their fixed infrastructure targets. The AKAPs serve as platforms for assembly and integration of several signaling proteins, thereby serving as points of control of their substrates.

The RACKs bind protein kinase C proteins once they have been partially activated by their upstream kinases and coactivators. Different isozymes of the PKCs localize to different subcellular locations. In response to production of cofactors such as Ca^{2+} and DAG, different PKC isozymes translocate from the plasma membrane to distinct subcellular locations, aided by the RACKs. Some PKCs are localized by their RACKs to the plasma membrane, where they phosphorylate L-type calcium channels, ligand-gated ion channels (discussed in Chapter 18), and other membrane-bound signaling proteins. Other PKCs are localized to the nucleus or to the mitochondria or to other cellular compartments where they phosphorylate their substrates.

8.16 PKC Regulates Response of Cardiac Cells to Oxygen Deprivation

Cells in any tissue of the body will not survive prolonged periods of oxygen deprivation. In oxygen deprived heart muscle (myocardium) cells shift from oxidative phosphorylation to anaerobic glycolysis. The amount of available ATP drops, and cellular pH rises. Contractile forces are reduced, and energy dependent ion pumps are impaired. Myocardial cells (and others, too) respond to brief periods of oxygen deprivation by adjusting their cellular processes, and as a result are be able to tolerate longer periods of poor oxygen conditions. This kind of response is known as *ischemic preconditioning* (IP). The key signaling events responsible for IP are sketched in Figure 8.14. As can be seen in the figure, signaling starts with release of adenosine by stressed cells resulting in their binding to adenosine receptors, members of the G-protein coupled receptor (GPCR) family. The heterotrimeric G proteins activated by these receptors stimulate the hydrolysis of $PtdIns(4,5)P_2$ to DAG and IP_3; the DAG acting as a cofactor for ε and δ members of the nPKC subfamily, which are anchored nearby the receptors along the plasma membrane, stimulates their activation.

Once activated the PKC ε and δ proteins change their cellular location, translocating from the plasma membrane to the mitochondria. They become anchored there, and associate with and activate members of the Src family of protein tyrosine kinases such as Lck leading to activation of a MAP kinase cascade. The downstream PKC ε-signaling targets are the mitochondrial ATP-dependent potassium channel proteins, which are phosphorylated by the last MAP kinase in the cascade. The PKC ε and δ have opposing effects. While PKC ε promotes preconditioning responses, PKC δ stimulates proapoptotic responses. The cellular response to oxygen deprivation will depend at least in part on the balance between these two signals.

Anchoring proteins such as the RACKs help localize the PKC isozymes in their proper locations. This aspect is illustrated in Figure 8.14 where several RACKs are depicted localizing the PKCs to the plasma membrane, mitochondria, and nucleus. At the plasma membrane the RACKs position the kinase near its substrate—cardiac L-type calcium channels. In the nucleus, the RACKs serve a similar role, placing the kinase near transcription factors that it regulates by means of phosphorylation. The RACKs not only help localize the proteins near their substrates, but also help organize signaling complexes. This aspect is represented in the figure by the binding of Src/Lck along with PKC ε to the mitochondrial RACK.

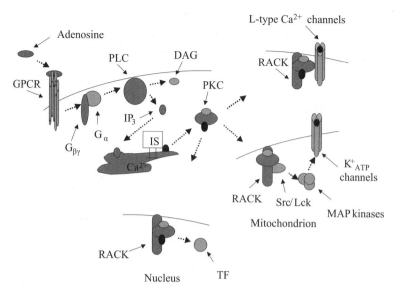

FIGURE 8.14. Protein kinase C signaling in cardiac cells: Signaling is initiated by binding of adenosine to a G protein-coupled receptor (GPCR) resulting in dissociation of the G protein tethered nearby into its Gα and Gβγ subunits. The Gα subunit activates phospholipase C, leading to formation of IP_3 and DAG second messengers. IP_3 triggers the release of calcium from intracellular stores (IS). Calcium and DAG activate cPKCs and DAG activates nPKCs. The RACKs position these protein kinases near their substrates along the plasma membrane and in the mitochondria and nucleus. (To keep the figure from getting too cluttered, phosphorylation of the PKCs by upstream kinases is not shown.)

8.17 cAMP Activates PKA, Which Regulates Ion Channel Activities

Cyclic AMP is a second messenger that acts through cAMP-dependent protein kinase A (PKA) to form the cAMP signaling pathway. This pathway is used to relay a variety of hormonal and neuromodulatory signals to the transcription machinery in the nucleus, to ion channels embedded in the plasma membrane, and to other targets such as the actin cytoskeleton. The cAMP signaling pathway consists of a receptor such as a GPCR, an intermediary such as a G protein, adenylyl cyclases and phosphodiesterases that form and regulate cAMP production, and protein kinase A. In more detail, ligand binding to a GPCR activates a G protein (the intermediary) tethered to the cytoplasmic region of the plasma membrane close enough to the receptor to be activated by its cytoplasmic region. Once activated, the G protein translocates to and stimulates the production of cAMP by adenylyl cyclases. The cAMP molecule, in turn, binds to and activates protein kinase A.

Protein kinase A contains a pair of catalytic (C) subuints and a pair of regulatory (R) subunits. In its inactive form, the regulatory units bind to the catalytic site and inhibit its activity. Binding of cAMP to the regulatory subunits results in a conformational change that permits the dissociation of the regulatory units form the complex. Once freed of its regulatory subunits, the catalytic subunits are able to catalyze the transfer of phosphoryl groups to their targets using ATP as the phosphoryl group donor. Protein kinase A is able to phosphorylate a number of different proteins. The choice of target is dictated by the choice of regulatory units and by its subcellular location. The specific location in the cell is determined by an AKAP.

The AKAP complexes incorporate not only protein kinases but also protein phosphatases that after a time turn off what the kinases turn on. In Figure 8.15, PKA together with protein phosphatases PP1 or PP2A regulate the opening of an ion channel embedded in the plasma membrane. An AKAP called *Yotiao* helps regulate the opening and closing of ion channels called *NMDA receptors* in neurons in the brain. The manner of this is depicted in the figure.

AKAP79 is a well-studied member of the A-kinase anchoring protein family. This protein binds two AGC kinases, protein kinase A and protein kinase C along with a prominent serine/threonine protein phosphatase called *calcineurin* (CaN), or as it is sometimes known, protein phosphatase 2B (PP2B). These enzymes are capable of binding many different substrates

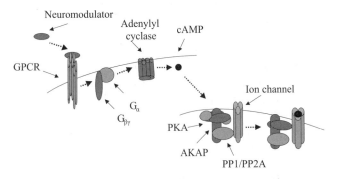

FIGURE 8.15. PKA signaling through AKAPs regulating ion channel openings: Binding of a neuromodulator to a G protein-coupled receptor initiates signaling in the neuron. The activated Gα subunit stimulates production of cAMP by adenylyl cyclase. The cAMP molecules bind to the regulatory subunits of protein kinase A. In response, the regulatory subunits dissociate from the catalytic subunits, which then phosphorylate the ion channel, overcoming its inhibition (closed state) by the protein phosphatases. After a time, the protein kinases are no longer able to maintain the channel in its phosphorylated state and the ion channel closes again (not shown).

with high affinity. The desirable property of high affinity is combined with another important property, substrate specificity, through use of an AKAP. The AKAP79 situates the three signaling proteins near their targets: plasma membrane receptors and ion channels. The enzymatic functions of these proteins are disabled by their association with the anchoring protein. Ca^{2+}/CaM supplies the "on" signal for these enzymes. When this complex binds protein kinase C and calcineurin, the enzymes detach, and diffuse to and phosphorylate/dephosphorylate their nearby targets. Similarly, cAMP binding will enable protein kinase A to detach from the AKAP79 and phosphorylate its targets.

In the brain, protein kinase A participates in the "learning pathway" along with several other protein kinases. The main endpoint of the learning pathway is the nucleus where kinases functioning as transcription factors regulate gene expression. Signaling interactions of a short term nature involving the regulation of ion channels, and long term nature-producing changes in gene expression, are central to brain function. They will be examined in Chapter 21.

8.18 PKs Facilitate the Transfer of Phosphoryl Groups from ATPs to Substrates

The core unit of protein kinase A is representative of the core units of all serine, threonine, and tyrosine kinases. As shown in Figure 8.16, it consists of a small lobe, a linker, and a large lobe. A cleft formed by elements of the small and large lobes operates as the catalytic site. The large lobe binds the substrate peptide or protein, while the small lobe supplies the main site for attachment of the ATP molecule. The ATP molecule sits at the base of the cleft and provides structural support, helping to fix and maintain the orientations of the large and small lobes with respect to one another. The cleft is bordered by the glycine-rich loop and the C helix from the small lobe and by a beta sheet from the large lobe. As shown in Figure 8.16, the phosphoryl groups attached to the adenosine are oriented towards the cleft with the gamma phosphoryl group at the end. The loops and beta sheet position this group for transfer to the substrate.

A common feature in protein kinases is the presence of metal ions. As was the case for the histidine kinases discussed in the last chapter, these positively charged ions help position and stabilize the cell's assembly. The Mg ion-positioning loop highlighted in the figure assists in positioning the magnesium ions.

There are two phosphorylation sites on the core unit—Thr197 and Ser338. In the active state conformation, the cleft is opened up and the core unit is catalytically competent. Phosphorylation at Thr197 and Ser338 helps stabilize the open conformation.

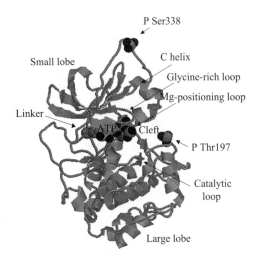

P Ser338

Small lobe

C helix

Glycine-rich loop

Mg-positioning loop

Linker

ATP

Cleft

P Thr197

Catalytic
loop

Large lobe

FIGURE 8.16. Crystal structure of the conserved core of the catalytic subunit of protein kinase A: The core unit consists of a small lobe, a linker and a large lobe. The small lobe contains a five-stranded antiparallel beta sheet and a conserved helix (the C helix). The large lobe consists mainly of helices plus a grouping of four beta strands at the bottom of the active site cleft. The phosphoryl groups covalently attached at Thr197 and Ser338 are depicted as space-filled model. The ATP molecule is also shown with a space-filled representation. The figure was prepared using Protein Explorer with atomic coordinates deposited in the Protein Data Bank (PDB) under accession code 1ATP.

References and Further Reading

Plasma Membrane Organization

Brown DA, and London E [1998]. Structure and origin of ordered lipid domains in biological membranes. *J. Mem. Biol.*, 164: 103–114.

Sheets ED, Holowka D, and Baird B [1999]. Critical role for cholesterol in Lyn-mediated tyrosine phosphorylation of FcεRI and their association with detergent-resistant membranes. *J. Cell Biol.*, 145: 877–887.

Simons K, and Ikonen E [2000]. How cells handle cholesterol. *Science*, 290: 1721–1726.

Simons K, and Toomre D [2000]. Lipid rafts and signal transduction. *Nature Rev. Mol. Cell Biol.*, 1: 31–41.

Smart EJ, et al. [1999]. Caveolins, liquid-ordered domains, and signal transduction. *Mol. Cell Biol.*, 19: 7289–7304.

Vereb G, et al. [2003]. Dynamic, yet structured: The cell membrane three decades after the Singer–Nicolson model. *Proc. Natl. Acad. Sci. USA* 100: 80553–80558.

Lipid Signaling, Phosphoinositide-3-OH Kinase, and PIP₂ Signaling

Honda A, et al. [1999]. Phosphatidylinositol 4-biphosphate kinase α is a downstream effector of the small G protein ARF6 in membrane ruffle formation. *Cell*, 99: 521–532.

Raucher D, et al. [2000]. Phosphatidylinositol 4,5-biphosphate functions as a second messenger that regulates cytoskeleton-plasma membrane adhesion. *Cell*, 100: 221–228.

Toker A, and Cantley LC [1997]. Signaling through the lipid products of phosphoinositide-3-OH-kinase. *Nature*, 387: 673–676.

Vanhaesebroeck B, and Waterfield MD [1999]. Signaling by distinct classes of phosphoinositide 3-kinases. *Exp. Cell Res.*, 253: 239–254.

Lipids and Lipid-Binding Domains

Czech MP [2000]. PIP$_2$ and PIP$_3$: Complex roles at the cell surface. *Cell*, 100: 603–606.

Hurley JH, and Meyer T [2001]. Subcellular targeting by membrane lipids. *Curr. Opin. Cell Biol.*, 13: 146–152.

Lemmon MA, and Ferguson KM [2000]. Signal-dependent membrane targeting by Pleckstrin homology domains. *Biochem. J.*, 350: 1–18.

Calcium/Calmodulin

Berridge MJ, Lipp P, and Bootman MD [2000]. The versatility and universality of calcium signaling. *Nature Revs. Mol. Cell Biol.*, 1: 11–21.

Carafoli E [2002]. Calcium signaling: A tale for all seasons. *Proc. Natl. Acad. Sci. USA*, 99: 1115–1122.

Chin D, and Means AR [2000]. Calmodulin: A prototypic calcium sensor. *Trends Cell Biol.*, 10: 322–328.

Adenylyl Cyclases

Cooper DMF, Mons N, and Karpen JW [1995]. Adenylyl cyclases and the interaction between calcium and cAMP signaling. *Nature*, 374: 421–424.

Hurley JH [1999]. Structure, mechanism, and regulation of mammalian adenylyl cyclase. *J. Biol. Chem.*, 274: 7599–7602.

Taussig R, and Gilman AG [1995]. Mammalian membrane-bound adenylyl cyclases, *J. Biol. Chem.*, 270: 1–4.

Cyclic AMP

Houslay MD, and Milligan G [1997]. Tailoring cAMP-signaling responses through isoform multiplicity. *Trends Biochem. Sci.*, 22: 217–224.

Rich TC, et al. [2001]. A uniform extracellular stimulus triggers distinct cAMP signals in different compartments of a simple cell. *Proc. Natl. Acad. Sci. USA*, 98: 13049–13054.

Schwartz JH [2001]. The many dimensions of cAMP signaling. *Proc. Natl. Acad. Sci. USA*, 98: 13482–13484.

Cyclic Nucleotide Phosphodiesterases

Beavo JA [1995]. Cyclic nucleotide phosphodiesterases: Functional implications of multiple isoforms. *Physiol. Rev.*, 75: 725–748.

Francis SH, Turko IV, and Corbin JD [2001]. Cyclic nucleotide phosphodiesterases: Relating structure and function. *Prog. Nucl. Acid Res. Mol. Biol.*, 65: 1–52.

PTEN

Yamada KM, and Araki M [2001]. Tumor suppressor PTEN: Modulator of cell signaling, growth, migration and apoptosis. *J. Cell. Sci.*, 114: 2375–2382.

Anchoring Proteins

Colledge M, and Scott JD [1999]. AKAPs: From structure to function. *Trends Cell Biol.*, 9: 216–221.

Feliciello A, Gottesman ME, and Avvedimento EV [2001]. The biological function of A-kinase anchoring proteins. *J. Mol. Biol.*, 308: 99–114.

Mochly-Rosen D, and Gordon AS [1998]. Anchoring proteins for protein kinase C: A means for isozyme selectivity. *FASEB J.*, 12: 35–42.

Protein Kinases B and C Signaling

Dudek H, et al. [1997]. Regulation of neuronal survival by the serine-threonine protein kinase Akt. *Science*, 275: 661–665.

Newton AC, and Johnson JE [1998]. Protein kinase C: A paradigm for regulation of protein function by two membrane-targeting modules. *Biochim. Biophys. Acta*, 1376: 155–172.

Ron D, and Kazanietz MG [1999]. New insights into the regulation of protein kinase C and novel phorbel ester receptors. *FASEB J.*, 13: 1658–1676.

Shepherd PR, Withers DJ, and Siddle K [1998]. Phosphoinositide 3-kinase: The key switch mechanism in insulin signaling. *Biochem. J.*, 333: 471–490.

Vanhaesebroeck B, and Alessi DR [2000]. The PI3K-PDK1 connection: More than just a road to PKB. *Biochem. J.*, 346: 561–576.

Protein Kinases

Huse M, and Kuriyan J [2002]. The conformational plasticity of protein kinases. *Cell*, 109: 275–282.

Taylor SS, et al. [1999]. Catalytic subunit of cyclic AMP-dependent protein kinase: Structure and dynamics of the active site cleft. *Pharmacol. Ther.*, 82: 133–141.

Problems

8.1 As discussed in the chapter, biological membranes can be described as fluids in which lipids and proteins freely diffuse within the plane of the membrane. Singer and Nicholson [Singer SJ and Nicholson GL [1972]. The fluid mosaic model of the structure of cell membranes. *Science*, 175: 720–731] presented the basic features of this picture in a 1972 paper that appeared in *Science*. The membranes are not homogeneous structures but rather are organized into domains, each characterized by a somewhat different mix of lipids and proteins. A typical value for the diffusion coefficient for lateral diffusion in the plane of the lipid bilayer in the fluid phase is $D = 3 \times 10^{-8}\,\text{cm}^2/\text{s}$. When the cholesterol content is increased to high values the diffusion coefficient decreases to $D = 2 \times 10^{-9}\,\text{cm}^2/\text{s}$. How far will the lipids diffuse in 1 s in the fluid phase?

8.2 Proteins diffuse more slowly than lipids. Some proteins, especially those involved in signaling, are not free to diffuse at all. These proteins are immobilized through interactions with each other and with the cytoskeleton elements. Protein diffusion coefficients are consequently quite variable, but in those instances where the proteins can diffuse freely within a domain, diffusion coefficients ranging from $D = 4 \times 10^{-9}\,\text{cm}^2/\text{s}$ to $D = 2 \times 10^{-10}\,\text{cm}^2/\text{s}$ are observed. Given these numbers, how does the viscosity of the lipid bilayer compare to the viscosity of water?

9

Signaling by Cells of the Immune System

The human body has three super signaling and control systems—the immune system, the endocrine system, and the nervous system. Each system has a myriad of cells extending throughout the body and specialized for signaling. Cells of the immune system communicate using cytokines; cells of the endocrine system send out hormones and growth factors, and the cells of the nervous system utilize neurotransmitters and neuromodulators. The immune system, the subject of this chapter, consists of several organs, colonies of leukocytes, and large numbers of extracellular messengers. The immune system's job is to identify and destroy pathogens, entities that enter the body, establish themselves in a specific locale, or niche, multiply, cause damage, and exit. Leukocytes, the white blood cells, are highly motile, short-lived cells that move through the cardiovascular and lymphatic systems into damaged tissues. The leukocytes kill bacterial, protozoan, fungal, and multi-cellular pathogens; they destroy cells infected with viruses and bacteria, and eliminate tumor cells.

Leukocytes are highly mobile cells that can migrate from blood into tissues and back again into blood. They respond to infections by attacking and destroying the causative agents, or pathogens. They carry out inflammatory responses, innate immune responses, and adaptive immune responses. The inflammatory response to an infection involves the triggering of physiological responses such as fever and pain, redness and swelling, and a buildup of white blood cells, the leukocytes, at the infection site. A local environment is formed that promotes migration of leukocytes to the infection site, the destruction of the invasive agents, and the repair of damaged tissues. It takes several days for the adaptive immune response to develop, and during that the inflammatory response contains the infection.

The innate immune response is a phylogenetically ancient form of defense by multicellular organisms against pathogens. It involves recognition by host cell surface receptors of molecules situated on the outer surface of pathogens. Such molecules are characteristic of the pathogen. Because the receptors are encoded by the host genome and thus innate, the recognition response is termed an *innate immune response*. Lipopolysaccharides

(LPS) are an example of molecules that are recognized by receptors expressed on cells of host multicellular organisms. They are prominent components of the outer membrane of gram-negative bacteria such as *Escherichia coli*.

The *adaptive immune response* is unique to vertebrates. It involves the production of antibodies and enables the host to respond to pathogens that have eluded the innate immune response, and to pathogens have not been encountered before. There are two basic situations: Pathogens may reside outside of the host cells, or they may hide within cells of the body. *Antibodies* are receptors that recognize and bind *antigens*, foreign substances uniquely derived from the pathogens. Antigens (antibody generators) may be molecules or structures located on the surface of pathogens, toxins secreted by a pathogen, foreign RNA, or any other molecule identified by the host as not belonging to the host (not self). In situations where the pathogens hide within the cell the solution used by the immune system is to present peptide fragments of the pathogen, i.e., antigens, on the surface of the host cells where they be seen and can trigger immune responses. Antigens that are encountered extracellularly are taken up by leukocytes specialized for their ingestion, and these are presented on the cell surface of the leukocytes.

This chapter will begin with a review of the different kinds of leukocytes found in the body and the signaling proteins—the cytokines—used by them to communicate with one another. The signaling pathways used by each of the five classes of cytokines will then be examined. The chapter will conclude with an examination of signaling through the T cell receptor.

9.1 Leukocytes Mediate Immune Responses

An immune response involves detection, the marshalling and movement of resources (cells) between tissue and blood, creation of a protected environment, production of leukocytes needed for fighting the infection, and destruction of the invaders and diseased cells. Colonies of leukocytes are formed from hematopoietic (blood-forming) stem cells in several developmental stages in a number of locations in the body—in bone marrow, in lymphoid tissue, in the circulating blood, and in body tissues. There are two main categories of leukocytes: Cells that migrate in and out of the lymphatic system and mature and differentiate there are categorized as *lymphocytes*. These cells are the key players in the adaptive immune response. Members of the second group of cells differentiate and mature in bone marrow. These *myeloid cells* are key mediators of inflammation and innate immune responses and are categorized as in Table 9.1 as *myeloid-derived phagocytes/ granulocytes*.

Dendritic cells and *mast cells* are distributed throughout the body, serving as sentinels whose purpose is to detect the presence of pathogens. As

TABLE 9.1. Leukocytes: Cells of the immune system.

Leukocyte	Function
Lymphocytes	
Dendritic cells	Antigen-presenting cells (APCs); sentinels distributed throughout the body
B cells	Produce antibodies, mediate adaptive immune responses; antigen-presenting cells
T cells	Mediate adaptive immune responses; help B cells respond to antigens
NK cells and cytotoxic T cells	Kill virally infected and cancerous cells
Myeloid-derived phagocytes/granulocytes	
Basophils	Circulate in the bloodstream; mediate inflammatory responses and recruit leukocytes
Eosinophils	Reside in submucosal tissues; kill multicellular parasites
Mast cells	Sentinels broadly distributed throughout the body; initiate inflammatory and allergic responses and recruit leukocytes
Macrophages	Antigen-presenting cells; engulf and digest bacteria, fungi, dead/dying cells
Neutrophils	Engulf and digest viruses, bacteria, protozoa, fungi, viral infected cells, and tumor cells

already noted, antigen presentation is the way cells respond to pathogens such as viruses that hide within host cells, thereby eluding direct detection by the sentinels. The immune system deals with these pathogens by shipping and displaying antigens on the external surfaces of antigen-presenting cells (APCs), where T cells can recognize them through ligand-receptor binding. In their role as sentinels, dendritic cells are able to take foreign materials from the extracellular spaces and display them on their surfaces leading to activation of T cells. They along with B cells and macrophages are called *professional APCs*. All nucleated cells in the body are capable of displaying peptides derived internally in the cytosol from invading pathogens. This kind of activity is constitutive so that in the absence of pathogens only self-molecules are displayed on the surface. These surface display and recognition processes are the essence of self- versus non-self recognition. If not dealt with in the treatment of patients, they lead to the rejection of tissue grafts and organ transplants.

Many of the leukocytes listed in Table 9.1 contain vesicles filled with histamines, hormones, and other inflammatory agents; with cytokines for signaling; and with enzymes that mediate the destruction of pathogens. They release the contents of their vesicles (granules) in response to the appropriate signals, and as a result they are called *granulocytes*, while leukocytes that are able to engulf pathogens are referred to in Table 9.1 as *phagocytes*.

9.2 Leukocytes Signal One Another Using Cytokines

Leukocytes continually send and receive messages from one another, acting in many ways like a highly mobile nervous system whose job it is to identify and destroy pathogens. Cytokines are small, secreted molecules, usually less than 30 kDa in mass. They are synthesized and secreted by leukocytes, most commonly macrophages and T cells. They convey messages to other leukocytes and to nonhematopoietic cells such as neurons. The cytokine signals instruct leukocytes to grow, differentiate, mature, migrate, and die. The signals stimulate antiviral and antitumor activities, and stimulate and regulate the three kinds of immune responses.

Cytokines, like many proteins belonging to the control layer, are *pleiotropic*, or multifunctional, in their actions. The specific physiological response elicited by a given cytokine is dependent upon cellular context. Different responses are produced in different cellular environments, that is, in cells expressing different mixes of proteins. Furthermore, the specific effect of a cytokine on a leukocyte not only depends upon the type of leukocyte but also upon the leukocyte's physiological state. For example, it will depend upon the cell's state of maturation.

Cytokines often act in a redundant manner in which several different cytokines produce the same physiological response in a cell. Several different cytokines are usually produced at the same time and they act synergistically. The receipt of messages conveyed by cytokines often triggers another burst of cytokine signals by the recipient thereby creating a cascade of signaling events. Cytokines convey signals in a targeted fashion. They usually operate over short distances and have short half-lives. However, they can convey signals over long distances too. For example, they can send messages to bone marrow to instruct cells to make more leukocytes. The overall result of the cytokine signaling is to rapidly activate and recruit cells of the immune system in response to the onset of an infection.

Interleukins are cytokines secreted mostly by T cells. After T cells, monocytes (and macrophages) are the most prolific interleukin conversationalists. The messages are received by other leukocytes, but especially by B and T cells, and instruct the recipients to growth and proliferate and to differentiate. A representative sampling of the interleukins is presented in Table 9.2, while a more extensive but compressed listing of cytokines is presented Table 9.3. As can be seen from an examination of the entries in Table 9.2, T cells send out multiple cytokines. These signaling proteins work synergistically with one another in cell-dependent ways to induce a variety of changes in the recipients.

Although colonies of different leukocytes are present at all times, their numbers are increased when an infection occurs, and then their numbers are reduced once the infection has been treated. Many of these cells are maintained in an immature state awaiting signals that instruct them to differentiate into specific, more mature forms. This aspect is reflected in the

TABLE 9.2. A sampling of leukocyte-to-leukocyte signaling proteins (interleukins).

Interleukin	Sending cell	Receiving cell	Instructions
IL-1	Monocytes and DCs	Th cells	Proliferation
IL-2	Th1 cells	Activated B, T cells	Growth and proliferation; Ig production
IL-3	Th cells	Stem cells	Growth and differentiation
IL-4	Th2 cells	Activated B, T cells	Proliferation and differentiation; promotes Th2 differentiation; Ig production
IL-5	Th2 cells	Activated B cells	Maturation, proliferation and differentiation; Ig production
IL-6	Monocytes and other cells	Activated B cells	Proliferation and differentiation; Ig production
IL-7	Stromal cells	Stem cells	Growth factor for pre-T, B cells
IL-9	T cells	T cells	Growth
IL-10	B, T cells, monocytes	Monocytes, Th cells	Inhibits cytokine production
IL-11	Bone marrow stromal cells	B cells	Growth and proliferation
IL-12	Monocytes and other APCs	Th1 cells	Promotes Th1 differentiation

TABLE 9.3. Signaling molecules of the immune system.

Family and representative members	Main role(s)
Toll/IL-1: IL-1, TLR1-10, Toll	Mediates the innate immune response to bacterial pathogens; mediates inflammatory responses and stimulates lymphocytes
Tumor necrosis factor (TNF): FasL, TNFα, TRAIL, Apo3L	Mediates adaptive immune responses of growth, proliferation, and death (apoptosis)
Hematopoietic (Class I cytokine receptors): Prolactin, EPO, GH, GM-CSF, TPO, IL-2, IL-3, IL-4, IL-5, IL-6, IL-7, IL-9, IL-11, IL-12	Key mediators of the adaptive immune response; regulates and coordinates leukocyte activities; functions as leukocyte-specific growth hormones, promoting growth and differentiation
Interferon/IL-10 (Class II cytokine receptors): IFNα, IFNβ, IFNγ, IL-10	Mediates antiviral and antitumor responses, and promotes the adaptive immune response
Chemokine: IL-8, Rantes, MIP-1	Leukocyte chemoattractants

table by the frequent presence of differentiation in the list of instructions conveyed by the interleukins. Yet another kind of instruction conveyed by interleukins is to express a specific set of cell surface molecules, for example, those belonging to the immunoglobulin (Ig) family of antigen-binding receptors.

9.3 APC and Naïve T Cell Signals Guide Differentiation into Helper T Cells

Interactions between dendritic cells and naïve T cells leading to differentiation into helper T cells is illustrative of interleukin signaling. Immature dendritic cells serve as sentinels in the body. Their morphology is specialized for capturing antigens from their surrounding. When they encounter an antigen denditic cells migrate to the secondary lymphoid organs. They differentiate into antigen-presenting dendritic cells, and interact with naïve T cells thereby stimulating their differentiation into either T1 helper cells or T2 helper cells (Figure 9.1). T1 and T2 helper cells perform different immune functions. T1 helper cells secrete cytokines that stimulate actions by macrophages in inflammatory responses. T2 helper cells secrete a mix of cytokines that triggers the differentiation of B cells into antibody-releasing plasma cells.

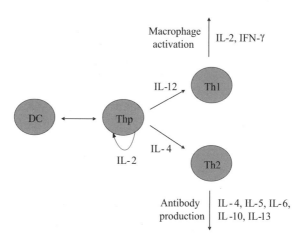

FIGURE 9.1. Helper T cell development: The first steps in the specification involve interactions between the immature dendritic cells and the pathogen they have encountered. These interactions determine the precise nature of the receptor complexes expressed on the surface of the antigen-presenting dendritic cells and the type of cytokines secreted when they interact with the naïve T cells in the secondary lymphoid tissue. Differentiation is triggered by interactions between the dendritic cell (DC) and a naïve T cell (Thp). The Thp expresses IL-2, which, acting in an autocrine manner on itself, leads to its differentiation into either a Th1 cell or a Th2 cell. IL-12 production stimulates differentiation into Th1 cells and inhibits formation of Th2 cells. IL-4 production stimulates differentiation into Th2 cells and inhibits formation of Th1 cells.

9.4 Five Families of Cytokines and Cytokine Receptors

There are five families of cytokines and cytokine receptors. The five families, representative members, and their most prominent activities are listed in Table 9.3. The first family in the table is the *Toll/IL-1 family*, named for the interleukin-1 (IL-1) cytokine, the family's founding member. IL-1 is primarily sent from macrophages to lymphocytes stimulating lymphocyte activation. When received by granulocytes, IL-1 promotes inflammatory responses. The Toll receptors are key mediators of the innate immune response. Their ligands are not cytokines but instead are molecular markers found on the surfaces of pathogens.

The next family of cytokines is the *tumor necrosis factor (TNF) family*. The tumor necrosis factor superfamily consists of a group of ligands and receptors that coordinate and regulate growth, proliferation, and survival of leukocytes. They coordinate the development of lymphoid organs and temporary inflammatory structures, an essential part of the adaptive immune response. One group of these proteins conveys death signals that instruct cells to suicide, or die by apoptosis (discussed in Chapter 15). A key function of these signals is homeostasis, the maintenance of stable (baseline) populations and the preservation of a balance between different subpopulations of cells ready to respond to future infections.

Hematopoietin receptors are sometimes referred to as *Class I cytokine receptors* and the *interferon/IL-10 receptors* as *Class II cytokine receptors*. The hematopoietins are a family of more than 20 different signal molecules that are especially prominent in lymphocyte signaling. As already mentioned, interleukins, secreted by macrophages, T cells, and bone marrow stromal cells stimulate growth, proliferation, and differentiation of several cell types. In response to these signals, stem cells differentiate into progenitor B and T cells; B cells differentiate into antibody-secreting plasma cells and B cells and macrophages express core components of their antigen-specific responses.

Interferons mediate antiviral responses by alerting other leukocytes that a virus attack is underway and by upregulating genes whose products have antiviral activities. In addition, interferons act as antitumor agents by down-regulating genes important for cell proliferation, and they regulate the expression of many genes essential for the adaptive immune response.

Lastly, leukocytes are guided to the site of an infection by *chemokines*, a family of small, 7- to 15-kDa protein chemoattractants. These molecules are secreted by several different kinds of cells at an infection site in order to recruit leukocytes from blood to sites of infection in tissue. Neutrophils, monocytes, fibroblasts, endothelial cells, and epithelial cells secrete chemokines. The chemokines form chemical concentration gradients that guide the migration of monocytes, neutrophils, eosinophils, and lymphocytes, and they arrest their movement at the appropriate location.

9.5 Role of NF-κB/Rel in Adaptive Immune Responses

The NF-κB/Rel signaling module plays an important role in the inflammatory, innate, and adaptive immune responses. The signaling pathway activated in response to Toll receptor binding has several features in common with the pathway activated in response to tumor necrosis factor receptor binding. Both utilize signaling through the NF-κB signaling module; both convey signals through MAP kinase modules, and both use adapters belonging to the TRAF family to link upstream receptor binding to downstream intracellular signaling.

NF-κB proteins are transcription factors that regulate genes for cytokines, chemokines, adhesion molecules, antimicrobial peptides, and other factors important for immune responses. In the absence of activating signals these proteins form an inactive cytosolic module with their IκB inhibitors, and a set of IKK regulatory kinases. The IKK proteins consist of two catalytic subunits, IKKα and IKKβ, and a regulatory subunit, IKKγ, also called NF-κB essential modulator (NEMO). Both of the IKK catalytic subunits are competent to phosphorylate IκB proteins. The IKKs are themselves activated by upstream kinases, and are the key point of convergence of a variety of regulatory events triggered by extracellular signals and intracellular stresses.

There are five members of the NF-κB family of proteins: NF-κB1 (p50/p105), NF-κB2 (p52/p100), c-Rel, RelB, and RelA (p65). In the absence of activating signals, these proteins are sequestered in inactive forms in the cytosol by inhibitory IκB factors. When the repressive effects of the IκBs are lifted, the NF-κBs form homo- and heterodimers, the most common combination being p65/p50, and translocate to the nucleus where they function as transcription factors (Figure 9.2).

The NF-κB1 and NF-κB2 proteins are formed and sequestered in the cytosol as p105 and p100 precursors. They are proteolytically processed to make the functional p50 and p52 forms. The C-terminus portion that is

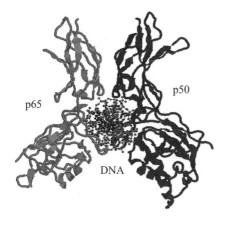

p65   p50   DNA

FIGURE 9.2. Binding of a NF-κB heterodimer to DNA: Shown is a p65/p50 NF-κB heterodimer (ribbon model) bound to an IFNβ enhancer (ball-and-stick representation) viewed looking down along the DNA axis. The figure was prepared using Protein Explorer with atomic coordinates deposited in the PDB under accession numbers 1le5.

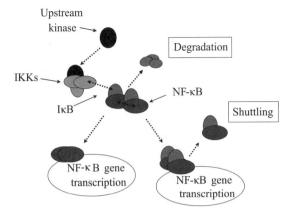

Upstream
kinase

Degradation

IKKs

NF-κB

IκB

Shuttling

NF-κB gene
transcription

NF-κB gene
transcription

FIGURE 9.3. The NF-κB node: Extracellular stimuli and intracellular stresses activate upstream protein kinases, which phosphorylate the IKKs. In response, the IKKs phosphorylate the IκB proteins, triggering either ubiquitination/degradation and/or nucleocytoplasmic shuttling, depending on the upstream signals and subunits present. The NF-κB proteins form dimers that translocate to the nucleus where they stimulate transcription of NF-κB responsive genes.

removed from p105 and p100 contains a series of *ankyrin repeats* that are also present in the IκB proteins and mediates their sequestration. Ankyrin repeats are 33 amino acid residues, protein-protein interaction modules. They are found in many signaling proteins, usually as four or more tandem-arranged repeated copies. The chain of steps leading to mobilization of the NF-κB proteins is depicted schematically in Figure 9.3.

In the absence of activating signals, the IκB proteins bind to sites in the Rel homology domain (RHD) of the NF-κBs, thereby masking their nuclear localization sequences (NLSs) and forcing their retention in the cytosol. The IκBα proteins contain nuclear export sequences (NESs) and, when they are bound to nuclear NF-κB dimers, they promote their export to the cytosol. The inhibition on NF-κB by the IκB proteins is relived by phosphorylation, which induces the ubiquitin-mediated degradation of the IκBs by the 26S proteasome (Figure 9.3). IκB proteins contain N-terminal serines that are the substrates of IκB kinases (IKKs).

The NF-κB module turns on and then turns off. One of the genes upregulated by the NF-κB proteins encodes IκBα. Signaling through the TNF receptor leads to increased migration of NF-κB proteins to the nucleus where they stimulate transcription of the gene for IκBα, which once synthesized binds to NF-κB, thereby shutting down its response to TNF. Because of the time delays inherent in the signaling pathway the NF-κB protein activity levels move up and down over time. The other IκB subunits do not depend on the NF-κB proteins for their transcription. The subunits' activity remains fairly constant over time, enabling them to smooth out the oscillations in NF-κB activity brought on by the negative feedback.

9.6 Role of MAP Kinase Modules in Immune Responses

MAP kinase modules convey cytokine and stress signals in the inflammatory, innate, and adaptive immune responses. Mitogen-activated protein (MAP) kinase modules convey cytokine, stress, and growth signals that are sent to them from the plasma membrane to the nucleus where they influence transcription of target genes. There are three mammalian MAP kinase pathways. One of these, the extracellular signal-regulated kinase (ERK) pathway, primarily carries growth signals. The other two, the c-Jun NH$_2$-terminal kinase (JNK) and p38 MAP kinase pathways, relay inflammatory cytokine and stress signals.

As was the case for the yeast MAP kinases discussed in Chapter 6, there are three kinases in a MAP kinase module. The first of these is a serine/threonine kinase. It phosphorylates the second kinase in the module, which is a dual specificity kinase. The middle kinase phosphorylates the third kinase in the module at threonine and tyrosine residues, hence the name *dual specificity*. The target residues are arranged as Thr-X-Tyr, where X is either Glu, Pro, or Lys for the ERK, JNK, and p38 pathways, respectively. Once they are phosphorylated, the third and last kinases in the module typically stimulate the transcriptional activity of members of several families of transcription factors. The three families, their upstream activating signals, and their downstream targets are depicted schematically in Figure 9.4.

The Ras and Rho GTPases function as molecular switches that relay growth, cytokine, and stress signals from receptors and their adapters and other intermediaries to the MAP kinase modules. Ras is a crucial regulator of growth signals. When its gene is mutated at certain residues it can get stuck in the "on" position and will continually send inappropriate growth messages to the ERK MAP kinase module. Ras will be discussed in more detail along with the other GTPases and their upstream growth factor receptors in Chapter 11 and again in Chapter 14, on cancer.

9.7 Role of TRAF and DD Adapters

TRAF and DD adapters transduce signals from TNF receptors into the cell. Scaffolds, anchors and adapters were introduced in Chapter 6. These proteins are not enzymes, but rather they organize the signaling nodes and control points where the enzymes can interact with one another and with the fixed infrastructure. Scaffolds that help organize MAP kinase cascades were examined in Chapter 6; anchors that tether serine/threonine kinases to membranes were discussed in the last chapter, and now an

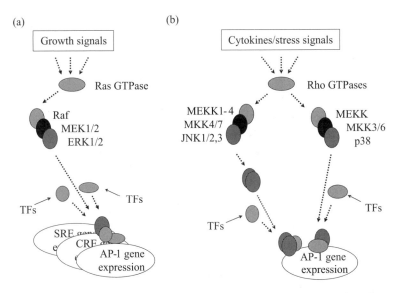

FIGURE 9.4. Mammalian MAP kinase modules: The upstream activators fall into two groups. (a) The ERK MAP kinase module is activated by growth and mitogenic signals relayed to it by the Ras GTPase. (b) The JNK and p38 MAP kinase modules are activated by proinflammatory cytokines and stress signals relayed to them by the Rho family of GTPases.

important class of adapters—the TRAF proteins—will be introduced. The tumor necrosis factor (TNF) receptor-associated factors, or TRAFs, help organize signaling nodes in pathways that mediate innate and adaptive immune responses and stress responses. These proteins contain a TRAF domain in their C-terminus. The TRAF domain consists of two portions: a TRAF-C domain that interfaces to the receptors and a coiled-coil domain that mediates interactions with other TRAFs. These are shown in Figure 9.5.

The TRAF adapters act as key intermediaries between the receptors and downstream signaling elements. The N-terminal domains of the TRAFs contain motifs called *zinc* and *RING fingers*. These motifs consist of arrangements of four cysteine and/or histidine residues that bind zinc ions, facilitating the formation of a compact domain that, along with the C-terminus coiled-coil domain, mediates downstream signaling. There are six mammalian TRAFs, named TRAF1 through TRAF6. The TRAF2 and TRAF6 proteins are crucially involved in TNF and Toll signaling, where they function as adapters that link the receptors to downstream NF-κBs, AP-1 signaling, and MAP kinase cascades.

(a)   (b)

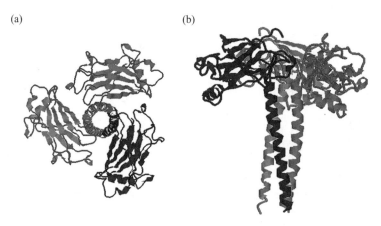

FIGURE 9.5. Structure of the TRAF domain of the TRAF2 adapter determined by X-ray crystallography: (a) Top view—Shown is a trimeric arrangement of TRAF domains that transduce signals from a 3:3 arrangement of TNF receptors and ligands. (b) Side view—The overall shape of the signaling unit resembles a mushroom with the C-TRAF domains forming the cap and the coiled-coil domains making up the stalk. The figure was prepared using Protein Explorer with atomic coordinates deposited in the PDB under accession number 1ca9.

9.8 Toll/IL-1R Pathway Mediates Innate Immune Responses

A group of plasma membrane-bound receptors called the Toll/IL-1R family plays a key role in leukocyte responses to bacterial lipopolysaccharide (LPS). The Toll/IL-1R signaling pathway activated in response to ligand binding by receptors in this family is an ancient one. It has been identified in plants, in insects (*Drosophila*), where it is known as the Toll/Dorsal pathway, and in vertebrates where it is referred to as the IL-1R/NF-κB pathway.

LPS is a prominent component of the outer membrane of gram-negative bacteria such as *E. coli*. The sensing of the presence of LPS by leukocytes, most notably, macrophages, activates the Toll/IL-1R signaling pathway. The focus of this pathway is the nuclear factor-κB (NF-κB) and a MAP kinase cascade leading to activation of members of the AP-1 family of transcription factors such as c-Jun (Figure 9.6). Ligand binding results in the rapid activation and subsequent translocation of NF-κB to the nucleus, where it and c-Jun upregulate the expression of genes for IL-1, IL-6, interferons, TNF, the cell adhesion molecules ICAM-1 and E-selectin, and the chemokine IL-8 (not shown). Signaling through this pathway not only starts an infection-containing innate immune response, but also launches the adaptive immune response.

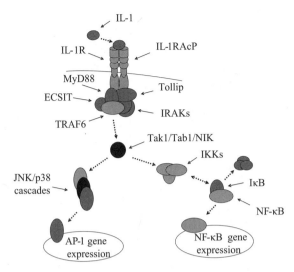

FIGURE 9.6. Signaling through the IL-1/Toll pathway: Shown are the IL-1R and its associated IL-1RAcP receptor along with several adapter proteins, namely, MyD88, Tollip, and ECSIT, which assist in recruiting kinases and other signaling elements to the plasma membrane. The first step in their activation is the recruitment of the adapter proteins to the plasma membrane. The serine/threonine kinase IRAK (IL-1R-associated kinase) and the TRAF6 adapter are then assembled at the site. These proteins activate a number of upstream kinases collectively designated in the figure as Tak1/Tab1/NIK, which then activate the NF-κB pathway and the JNK/p38 MAP kinase pathways.

9.9 TNF Family Mediates Homeostasis, Death, and Survival

Cell suicide, or apoptosis, instructions are an important part of the immune response. Immature T lymphocytes that do not respond properly to self-antigens are eliminated in the thymus through apoptosis. Other thymocytes that do not have the correct arrangement of their T cell receptors (TCRs) are disposed of in this way, too. At the end of an immune response, super-fluous, mature, activated T cells are removed, and the immune system is returned to its basal level ready to respond to the next infection. The maintenance of basal levels of the different kinds of leukocytes when there are no infections are present is referred to as *cellular homeostasis*. It relies on apoptosis to remove the superfluous leukocytes.

Apoptosis signaling mechanisms are not only used for elimination of incorrectly functioning and superfluous cells. It provides a mechanism for ensuring that cells do not survive outside of the tissue to which they belong. It is used by cytotoxic T cells to kill virus-infected cells and cancerous cells.

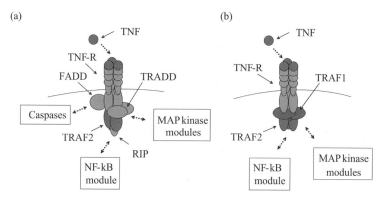

FIGURE 9.7. Upstream signaling in the TNF pathways: Trimeric arrangements of ligands, receptors, and adapters convey messages to signal enzymes. (a) Apoptosis signals are conveyed by death domain-bearing receptors and adapters among which are Fas-associated death domain (FADD) proteins, TNF-R-associated death domain (TRADD) proteins and receptor-interacting proteins (RIPs). (b) Inflammatory cytokine- (e.g., IL-6 and IFNγ) promoting signals.

Some tissues such as the eye, testis, and parts of the central nervous system cannot tolerate inflammation. These are privileged sites that insulate themselves from immune responses. If a lymphocyte comes into contact with a cell in any of these sites it will rapidly undergo apoptosis. Tumor cells trying to evade the immune response use this mechanism, as well.

The TNF receptors can be divided into two groups according to the kind of adapter proteins they use. Members of one group of TNF receptors have TRAF-binding motifs in their cytoplasmic domain, which mediate binding to TRAFs, while members of the second group have death domain (DD) motifs and bind to DD-bearing adapters. Receptors belonging to the second group are called *death receptors* because they convey cell suicide (apoptosis) instructions to the recipient. Upstream signaling events in these two TNF pathways are depicted in Figure 9.7. Most TNF ligands are single-pass transmembrane proteins with their C-terminus outside the cell and their N-terminus in the cytoplasm. They contact TNF receptors expressed on the surface of an opposing cell. The arrangement of receptors and ligands is 3:3. Three receptors arranged symmetrically contact three ligands. This signaling arrangement is facilitated by a trimeric arrangement of adapters of the form illustrated in Figure 9.5.

9.10 Role of Hematopoietin and Related Receptors

Hematopoietin receptors bind most of the interleukins, while a closely related set of receptors binds the interferons. Members of the hematopoietin family of cytokines are leukocyte hormones and growth factors. Although all members of this family of ligands and receptors have similar

TABLE 9.4. Hematopoietic and interferon receptors, Jaks and STATs: Abbreviations—erythropoietin (EPO); prolactin (PRL); thrombopoietin (TPO); growth hormone (GH); granulocyte macrophage colony stimulating factor (GM-CSF).

Receptor	Jak	STAT
Class I-Hematopoietin cytokines:		
Homodimeric receptors:		
EPO, PRL, TPO	Jak2	STAT5a, STAT5b
GH	Jak2	STAT5a, STAT5b, STAT3
Receptors that share γ_c:		
IL-2, IL-7, IL-9, IL-15	Jak1, Jak3	STAT5a, STAT5b, STAT3
IL-4	Jak1, Jak3	STAT6
Receptors that share β_c:		
GM-CSF, IL-3, IL-5	Jak2	STAT5a, STAT5b
Receptors that share gp130:		
IL-6, IL-11	Jak1, Jak2, Tyk2	STAT3
IL-12:		
IL-12Rβ1, IL-12Rβ2	Jak2, Tyk2	STAT4
Class II-Interferon/IL-10:		
IFNα, IFNβ	Jak1, Tyk2	STAT1, STAT2
IFNγ	Jak1, Jak2	STAT1
IL-10	Jak1, Tyk2	STAT1, STAT3, STAT5

structures, there are some differences in structure and subunit composition that allow grouping the family members into subfamilies. This breakdown is presented in Table 9.4. Hematopoietic receptors are assembled from two or three subunits, each a single-pass glycoprotein. Each receptor contains a ligand-binding subunit that is unique for that receptor type, plus one or more subunits that are common, or shared, by several other receptor types. A number of cytokines have a common beta chain (β_c); others share a gamma chain (γ_c); still others have a common gp130 subunit. The IL-12 group of cytokines listed in the table departs somewhat from the overall pattern. This group consists of cytokines that bind to homodimeric chains. Interferon and IL-10 receptors (Class II) are constructed from multiple unique subunits and in this way differ from the other grouping in the table, which are collectively called *Class I receptors*.

Cells regulate the binding activity of their plasma membrane receptors by varying the subunit composition. Cells can switch between low, medium, and high affinity-binding by differentially expressing different receptor subunits. Several examples of this form of control are supplied by the cytokine receptors. IL-2 is composed of IL-2Rα, IL-2β, and IL-2γ subunits. The IL-2γ chain is expressed widely while the other two subunits are expressed in a restricted fashion. By varying the subunit composition the receptor affinity for its cognate ligand can change from high affinity to low affinity with one or more intermediate affinity complexes. A similar dependence of binding affinity upon receptor subunit composition occurs in the IL-12 and IL-6 receptor systems.

Cytokine ligands are typified by their structure, a four-helix bundle fold consisting of two pairs of antiparallel α-helices linked together by loops. Some are short chain cytokines; IL-2, 3, 4 are constructed from short α-helices, 8 to 10 residues in length. Long chain cytokines such as GH, EPO, G-CSF, and the gp130 cytokines are built from chains that are longer, 10 to 20 residues in length. A third group of cytokine ligands, notably IL-5 and IFN-γ, are formed as a doubled four-α-helix fold, and contain eight α-helices.

9.11 Role of Human Growth Hormone Cytokine

The human growth hormone (hGH) cytokine and its receptor serve as a model system for cytokine receptor-ligand recognition. The hematopoietin chains bind their ligands in characteristic ways. In Figure 9.8, one molecule of the growth hormone ligand simultaneously binds two growth hormone receptor chains. This 1:2 binding is fairly typical of the entire group. An examination of the human growth hormone provides some insights into how this happens. Each hGH molecule contains two receptor-binding sites. One interface is located in helix 4 and the other is formed from helices 1 and 3. Thus, a single molecule of hGH forms a homodimeric complex with a pair of hGH receptor molecules.

Not all residues in the hGH ligand-hGH receptor interface contribute equally to the binding energy. Instead, a few residues located in the vicinity

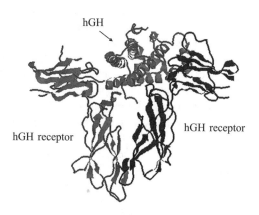

FIGURE 9.8. Structure of the human growth hormone (hGH) ligand bound to two hGH receptors: Shown are two extracellular domains of the hGH receptors bound to a single hGH ligand. The hGH ligand binds in the cleft between the two extracellular domains of the receptors. The figure was prepared using Protein Explorer with atomic coordinates deposited in the PDB under accession number 1hwg.

of the center of the interface contribute most of the binding energy. These residues form a "functional epitope," or "hot spot," that mediates ligand-receptor binding. Ligand engagement occurs when residues belonging to ligand and receptor portions of the hot spot come into close proximity. During the binding process an initial contact between the ligand and receptor surfaces is established. The contact is followed by a random diffusion stage where one surface moves (rolls) relative to the other until the motion is stabilized. This happens when the two portions of the functional epitope make contact. The maintenance of close surface contact through rolling diffusion greatly accelerates the association rate over that which would result from a single collision followed by a separation and then another elastic collision, and so on, until the correctly oriented surfaces are engaged.

The 1:2 association of the hGH ligand with its receptors is more efficient than a 2:2 association. The 1:2 complex is formed in a stepwise fashion. In the first step, a ligand molecule finds and establishes contact with one receptor molecule. The 1:1 complex then makes contact with the second (unbound) receptor molecule to create a stable 1:2 signaling unit. The advantage to this mechanism is that the second step involves diffusion in two dimensions—the second receptor molecule moves along membrane surface to contact the bound pair—rather than three. This may be compared to the case where the second step is another single ligand-single receptor binding event followed by association of the pairs. In this latter 2:2 scenario a second three-dimensional diffusion process would be required.

Many of the cytokines bind their receptors in a manner similar to that of the human growth hormone. Like hGH, the ligands for the β_c family of receptors have two binding sites. Members of the β_c family form high affinity complexes stepwise in the following manner. The ligand first binds the Rα subunit. The β_c chain then binds to the bound pair to form a 1:1:1 complex. Two complexes then come together to form a dimerized complex that activates the intracellular components of the signaling machinery. Hematopoietin receptors such as IL-2R that belong to the γ_c family are composed of three distinct subunits. A single ligand molecule possesses three binding sites and so is able to bind the three subunits when forming the high affinity 1:1:1:1 complex. A similar process is used by the gp130 cytokines to form stable complexes that are signaling-competent.

9.12 Signal-Transducing Jaks and STATs

Jaks and STATs transduce signals into the cell from the plasma membrane. The reason for receptor dimerization can be understood by noting that there is considerable flexibility in the membrane-spanning polypeptide chains. Under these conditions the binding of a ligand to the extracellular portion of a chain will not produce a large and long-lasting (stable) shift in conformation of the cytoplasmic portion of the chain needed to serve as a

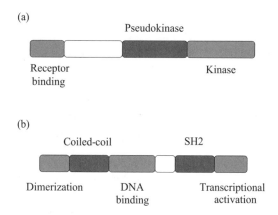

FIGURE 9.9. Domain organization of Jaks and STATs: (a) Janus tyrosine kinases, or Jak proteins. (b) Signal transducer and activator of transcription (STAT) proteins.

signal. The solution to this problem of how to transduce a signal into the cell is solved by dimerization. The combined effect of ligand binding and dimerization alters the intracellular electrostatic environment sufficiently to stabilize a different mix of conformational states that will activate the Jaks, triggering phosphorylation and the subsequent recruitment of the STATs and other signaling elements to the receptors. The Jaks (tyrosine kinases) and STATs (transcription factors) are then able to transduce the hematopoietin signals into the cell responses.

The domain organization of the Jak and STAT proteins is presented in Figure 9.9. As shown in part (a) of the figure, the Janus kinases possess kinase and pseudokinase domains that face one another, hence the name *Janus kinase* (Jak), named after the Roman god of gates and doorways. They also possess a receptor-binding region that mediates their binding to recognition motifs in the cytoplasmic terminal of the cytokine receptors. Once recruited to the receptors, autophosphorylation occurs, resulting in their activation. The activated kinases then phosphorylate the receptors, thereby exposing docking sites for the STATs. As depicted in part (b) of the figure, the STATs possess a dimerization domain, an SH2 domain, a tyrosine phosphorylation site, and a DNA-binding domain. The signal transducer and activator of transcription proteins are transcription factors, and they are recruited to the activated signaling complex formed by the phosphorylated cytoplasmic receptor terminals and Jaks. They bind the phosphorylated tyrosine residues in the receptors through their SH2 domains, and they are phosphorylated on their tyrosine residues by the Jaks. Once these recruitment and activation steps take place, the STATs are able to form heterodimer and homodimers, and translocate to the nucleus where they stimulate transcription of their target genes.

9.13 Interferon System: First Line of Host Defense in Mammals Against Virus Attacks

Interferons are secreted into the bloodstream where they circulate to all cells in the body to alert them that a virus attack is underway. They trigger an organism-wide, or global, response to the virus attack in cells that have not been attacked along with those that have been invaded. Invasion of a virus and the resulting viral replication stage produce dsRNA (double-stranded RNA) molecules. These are present in the cell only when a virus has invaded and started replication. The presence of dsRNA sets off a series of regulatory events. A resident cellular sensor of dsRNA called *protein kinase R* (PKR) is activated by the dsRNAs. The PKR proteins activate a NF-κB, and IFN regulatory factor (IRF), which binds within the IFN stimulated responsive element (ISRE) to promote transcription of the antiviral proteins including the interferons. In addition, PKR signals other control proteins to halt all protein synthesis (Figure 9.10). This action blocks the virus' ability to replicate and make proteins that it needs. The overall process operates though a negative feedback loop. The presence of dsRNA is required for action by PKR, and when this quantity is reduced the block on protein synthesis is relieved.

Interferon signal transduction is initiated when a ligand binds to an interferon receptor dimer (Figure 9.11). As is typical of the hematopoietins this event stimulates the recruitment to the plasma membrane and activation of members of the Janus family of protein tyrosine kinases. Phosphorylation stimulates the further recruitment of the STAT proteins, which are phosphorylated by the Jaks. In the case of IFNα/β-mediated signaling, Jak1 and Tyk2 phosphorylate STAT1 and STAT2, which then form heterodimers that further associate with an IRF protein.

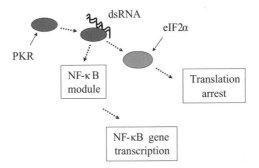

FIGURE 9.10. Antiviral activities of PKR: Protein kinase R, a sensor of dsRNAs, stimulates production of interferons by activating NF-κB and IRF (not shown) transcription factors, and halts protein synthesis by activating eIF2α, a negative regulator of protein synthesis.

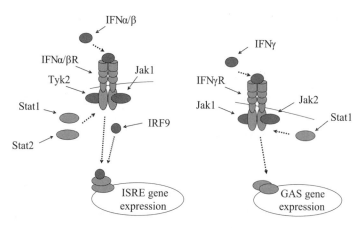

FIGURE 9.11. Interferon signaling: Displayed in the figure are the main components of the IFNα/β and IFNγ pathways from the plasma membrane to the nucleus. Distinct α and β chains of each receptor jointly bind a ligand. Janus family kinases are recruited and are activated. They create docking sites for the STATs by phosphorylating the receptors. The STATs are recruited and are phosphorylated by the Jaks. STATs form dimers (trimers) and translocate to the nucleus where they bind to interferon-responsive DNA transcriptional sites, the ISREs and interferm-gamma activated sites (GASs).

The trimers translocate to the nucleus where they stimulate transcription of IFNα/β-responsive genes IFNγ utilizes a different set of signal transducers. Jak1 and Jak2 are recruited to the plasma membrane and phosphorylate STAT1s, which form homodimers that translocate to the nucleus where they stimulate transcription of IFNγ-responsive genes. Cytokines such as the interferons mount a multilayered response. Besides arrest of protein synthesis and cell multiplication, they stimulate the production of enzymes that kill (degrade) the viruses, recruit cells that attack virus-containing cells, and organize other elements of the adaptive immune response.

9.14 Chemokines Provide Navigational Cues for Leukocytes

Chemokines are chemoattractants that guide the movement of leukocytes. Some chemokines are inflammatory chemokines while others are lymphoid chemokines. *Inflammatory chemokines* attract neutrophils, macrophages, and other leukocytes central to the innate immune response. Inflammatory chemokines are secreted by several different kinds of cells including leukocytes and endothelial cells that line the blood vessels. *Lymphoid chemokines* help regulate leukocyte traffic and cell compartmentalization within lymphoid tissues. Thus, they act to maintain homeostasis among the various populations of lymphocytes. Stromal cells, endothelial cells, and other cells residing in the lymphoid organs produce these signaling molecules.

Chemokines have a positive charge and bind to heparin and heparin sulfate, negatively charged polysaccharides found within the extracellular matrix and on the surface of endothelial cells. The heparin and heparin sulfate molecule act as receptors for the chemokines. The chemokines become immobilized when bound to the polysaccharide-covered substrates and form stable gradients of chemokine concentration. The migratory leukocytes navigate up the gradients to the target sites.

Chemokines are bound at the cell surface by G protein-coupled receptors (GPCRs), seven-pass transmembrane receptors coupled to G proteins. Their method of transducing signals into the cell was examined briefly in the last chapter and will be discussed in detail in Chapter 12. There are four families of chemokine receptors. These families are distinguished from one another by their structure and by their chromosomal locations. The distinguishing structural feature is the presence of four conserved cysteine (C) residues in specific positions. The CXC and CC families are fairly large with several members each, while the two small families, designated as CX3C and C, have a single known member each. The CXC family has an amino acid located between the first and second cysteines; in the CC family the two cysteines are directly adjacent to one another. There is considerable redundancy; within each ligand-receptor family, the chemokines can bind to more than one type of receptor, and the receptors bind to more than one type of chemokine.

Lymphoid chemokines form chemical highways within lymphoid organs that guide the migration of lymphocytes from site to site bringing together lymphocytes that must interact with one another in the adaptive immune response. The B-cell attracting chemokine (BCA-1), and secondary lymphoid chemokine (SLC) are representative lymphoid chemokines. They bind the CXCR5 and CCR7 receptors, respectively. Gradients of BCA-1 chemokines guide the entry of T cells and dendritic cells into lymphoid organs. Similarly, gradients of SLC are crucial for compartmental homing of B cells and T cells, helping to establish distinct territories within the secondary lymphoid organs.

9.15 B and T Cell Receptors Recognize Antigens

The first stage of the lymphocyte immune response is the production by B cells of antibodies, receptors that bind to specific antigens. There are two classes of antigen-recognizing receptors, one set expressed on B cells and referred to as *B cell receptors* (BCRs), and the other set expressed on T cells and referred to as *T cell receptors* (TCRs). *Major histocompatability complexes* (MHCs) also recognize antigens, and these may be regarded as a third class of antigen-recognizing receptors. All are members of a large family of glycoproteins known as *immunoglobulins* (Igs) that share a similar structure and fold.

Each B cell expresses on its surface a large number of a single kind of Ig molecule. When a B cell encounters an antigen that it recognizes (i.e., that it binds), it matures and differentiates into a plasma cell that produces many more antibodies of that specific type. These antibodies are no longer membrane-bound receptors but instead are secreted and become free to diffuse and bind to their specific antigen whenever that antigen is exposed on the surface of the bacterium or other pathogen. This binding event not only tags that object for destruction by other leukocytes, but also inactivates viruses and bacterial toxins by impeding their ability to attach to their cellular targets.

At least two distinct signals are needed to elicit cellular responsiveness. Antigen binding to the BCR is the first signaling step. The B cell ingests the antigen-antibody complex, processes the antigen, and presents it bound to an MHC Class II molecule on its cell surface. Activated helper T (Th) cells possessing receptors that recognize that specific antigen convey the second signal. Signals relayed into the Th cell trigger the production and release of cytokines by the Th cell. The cytokines are bound to cytokine receptors on the B cell. This second signal triggers B cell differentiation and proliferation and the release of antibodies from the resulting plasma cells.

T cells target pathogens such as viruses and small bacteria that hide inside other cells. Antigen receptors on TCRs recognize short peptide sequences, typically 8 to 15 residues long, belonging to an antigen that has been processed and bound to MHC molecules expressed on the surface of MHC-presenting cells. The TCR recognition process differs from BCR antigen recognition since the latter targets entire molecules, either denatured or native forms of proteins or cell-bound carbohydrates. Both the BCR and TCR molecules have distinct antigen-binding and signaling units. Unlike the BCR, TCRs are not later made in quantity and secreted but instead remain membrane-bound. The signaling tail of the T cell receptor associates with a set of chains collectively termed CD3. The CD3 chains, and a pair of disulfide-linked ζ chains, are immunoglobulin family members that assist the TCR molecule in transducing signals into the cell.

9.16 MHCs Present Antigens on the Cell Surface

Almost all cells in the human body express MHC proteins on their surface. Between 10^4 and 10^6 Class I MHCs are present on the surface of a typical nucleated cell, and similar numbers of Class II MHC proteins are expressed on the surfaces of MHC Class II-expressing cells. Each MHC allele can bind some 1000 to 2000 different self-peptides represented at greater than 10 copies per cell and up to several hundred in some cases. Pathogen proteins are encountered at far smaller numbers. 10^2 to 10^4 copies accumulate on the cell surface. Pathogen peptides trigger responses even though they are only a small fraction of the total peptide fragment population—from 0.01 to 0.1%. Thus, the MHC must exhibit high sensitivity and selectivity of foreign peptides while maintaining low responsiveness to self-peptides.

Different individuals express different mixes of these molecules. These proteins lie at the core of self versus not-self recognition. The job of an MHC is to present antigens to T cells and to NK cells. Class I MHCs interact with CD8 receptors on cytotoxic T cells and with KIR receptors on NK cells. Class II MHCs interact with CD4 receptors on helper T cells. Class I members bind peptides derived from molecules encountered in the cytosol while Class II members bind peptides from molecules broken down in the endosomes.

What the MHCs of both classes have to do is present the broadest range of peptides possible while simultaneously satisfying the requirement that these peptides be presented in a way that maximally stimulates TCR responses. The minimal size of a peptide fragment needed to discriminate between self and foreign is about 8 or 9 amino acid residues. One way of satisfying the broadness requirement is to have several binding sites along the MCH. However, this would violate the TCR stimulation requirement. To maximally induce a TCR response, there should be a single peptide-binding site with each peptide of a given sequence binding to a given MHC allele in exactly the same way. The solution is to have a single binding pocket and a way to generate a spectrum of different MHCs utilizing a combination of conserved and nonconserved (variable region) residues in the binding pocket.

9.17 Antigen-Recognizing Receptors Form Signaling Complexes with Coreceptors

The mixes of proteins expressed on the surfaces of leukocytes are unique markers of the population or subpopulation to which the leukocytes belong. The different ensembles of proteins reflect not only the kind of leukocyte but also characterize the different stages of leukocyte development and activation/inactivation. The proteins that can be so used are given a "cluster of differentiation," or CD, designation because of their association with specific lineages or stages of differentiation. CD proteins function as receptors, coreceptors, and as cell adhesion molecules.

Transmembrane-signaling proteins that form signaling complexes with the BCRs and TCRs are well represented in the CD listing. The two main classes of T cells are distinguished from one another by the presence of either CD4 or CD8 molecules on their surfaces. CD4 cells such as helper T cells signal other leukocytes to converge on the site of an infection. The CD4 molecula acts in concert with T cell receptors to bind peptides derived from antigens bound to MHC-II molecules. (The TCR binds the peptide, while the CD4 molecule binds MHC-II). The T8 group of T cells, such as

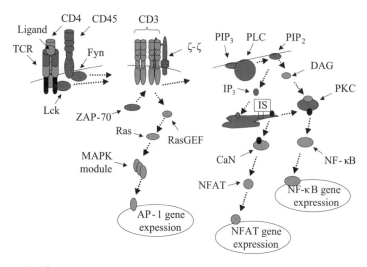

FIGURE 9.12. T cell receptor (TCR) signal transduction pathways leading to expression of genes activated by AP-1, NF-κB, and NFAT transcription factors working in concert: In response to ligand binding at the TCR, CD45 (a receptor tyrosine phosphatase) and CD4 activate Fyn and Lck. Fyn and Lck phosphorylate the γ, δ, and ε chains of the CD3 complex and the ζ dimer. After phosphorylation, ZAP-70 containing two SH2 domains is recruited and binds to the phosphatyrosines and is then activated through phosphorylation by Lck. ZAP-70 then interacts with PLC-γ, which subsequently cleaves PIP2 to produce IP3 leading to the aforementioned increased concentration of intracellular calcium. DAG activates PKC, which serves as the upstream kinase for activation of NF-κB, while ZAP-70 acts through a RasGEF, Ras, and a MAP kinase cascade leading to activation of AP-1. Intermediate steps involved in activation of NF-κB by PKC have been omitted for simplicity.

cytotoxic T cells, express CD8 proteins on their surfaces. The CD8 proteins act in concert with the T cell receptors to bind peptides derived from antigens bound to MHC-I molecules.

Several transmembrane proteins must associate with one another for efficient signaling into the cell. The upstream signaling starts with ligand binding and association of the TCRs with their CD4 or CD8 coreceptors and with CD3 chains (Figure 9.12). CD3 consists of several separate chains, denoted as δ, γ, and ε. In addition, there is a fourth kind of chain, called ζ. These chains form homo- and heterodimers that associate with the α/β chains of the TCR. The CD3 and ζ chains possess immunoreceptor tyrosine-based activation motifs, or ITAMs, in their cytoplasmic regions. These enable the CD3 chains to recruit and bind Fyn and Lck protein tyrosine kinases (protein tyrosine kinases will be discussed in detail in Chapter 11), triggering a series of steps leading to activation of AP-1, NF-κB, and NFAT-responsive genes.

Signaling through the TCR and its associated proteins culminates in the transcription of genes-encoding receptors, signal transducers, cell adhesion molecules, and cytokines. One of the most prominent of the cytokines upregulated through TCR signaling is IL-2. Compared to the straightforward signal transduction pathways utilized by cytokines, signaling through the TCR is far more complex. The steps leading to IL-2 production, along with many other gene products, involve activation of the lipid second messenger pathways discussed in the last chapter. Intracellular calcium levels are increased through stimulation of phospholipase C activity leading to a greater IP3 production, which increases the concentration of intracellular calcium. The calcium ions along with calmodulin activate the protein phosphatase, calcineurin, which, in turn, dephosphorylates and thus activates the nuclear factor of activated T cells (NFAT) transcription factors. The NFATs work in cooperation with members of the AP-1 family of transcription factors and the NF-κB transcription factor. These transcription factors stimulate the transcription of genes leading to the differentiation and proliferation of the T cells.

9.18 Costimulatory Signals Between APCs and T Cells

The task of discriminating between self and non-self is a difficult one. Several failsafe mechanisms operate to ensure that inappropriate actions are not taken by B cells and T cells. One of the mechanisms is the requirement that there be two independent activating signals before T and B cells are fully activated. The first signals are conveyed by the BCRs and TCRs in association with their CD4 and CD8 coreceptors. The second set of signals is sent into the cell via *costimulatory pathways*. The most prominent of the costimulatory pathways are the CD28/B7 system and the CD40/CD40L signaling pathway (Table 9.5).

The first entry in Table 9.5 is the CD40 receptor and its ligand. They are members of the TNF superfamily. Costimulatory signals supplied via CD40 are an important part of the help supplied by helper T cells. The CD40 ligand is found on CD4+ and CD8+ T cells, while the CD40 receptor is found on APCs such as B cells and dendritic cells. Signaling through

TABLE 9.5. T lymphocyte costimulators: Abbreviations—Inducible costimulator (ICOS); programmed death-1 (PD-1).

Receptor	Ligand	Comments
CD40	CD40L	TNF superfamily
CD28/B7		Ig superfamily
CD28	B7-1, B7-2	Positive regulator
CTLA-4	B7-1, B7-2	Negative regulator
ICOS	ICOSL	Positive regulator
PD-1	B7-L1, B7-L2	Negative regulator

CD40/CD40L activates resting B cells and primes the dendritic cells to adequately stimulate the T cells.

The remaining receptors and ligands all belong to the CD28/B7 branch of the immunoglobulin (Ig) superfamily. The first of these, CD28, is constitutively expressed on the surface of most T cells. Signaling through CD28 helps determine how the T cells will differentiate. In the absence of the costimulatory signals the T cells undergo *anergy*, becoming nonresponsive, or tolerant, of antigens presented on the surface of the antigen-presenting cells (APCs). In this state the T cells are inactive. They do not proliferate nor do they send out cytokines signals. The co-stimulator signals not only prevent anergy but also help determine whether the activated T cell will become a helper T cell or an effector T cell.

CD28 and CTLA-4 both bind to a pair of ligands, B7-1 and B7-2. The CTLA-4 receptor is transiently expressed. It has a higher affinity for the B7 ligands than does CD28, and when it is expressed it shuts down the signaling. A second pair of positive and negative regulators, ICOS and PD-1, also appears in the table. These glycoproteins are not constitutively expressed but are inducible through CD28 signaling and act later during immune responses. The positive regulator, ICOS, helps promote B-T cell interactions while PD-1, like CTLA-4, halts cytokine production and the cell cycle progression.

9.19 Role of Lymphocyte-Signaling Molecules

Lymphocyte-signaling molecules form immunological synapses. In order to carry out the kind of signaling required for an immune response several kinds of molecules must come together at and just below the proximal cell surfaces of the T cell and the APC. Signaling into the leukocyte cell interior is triggered at the plasma membrane by surface contact (cell adhesion), by antigen-signaling through the T cell receptors and coreceptors, and by costimulatory signals. Cell adhesion molecules important for this process include integrins such as LFA-1, and cell adhesion molecules belonging to the Ig superfamily, such as the ICAMs. Cell adhesion molecules play a crucial role in leukocyte motility and will be examined in detail in the next chapter.

The transmission of messages across the plasma membrane and into the cell interior is best thought of as a process. It is carried out in several distinct steps. It takes a certain amount of time and covers a region of space. Depending on the character of the message, the transmission can be fairly straightforward or can require several intermediate steps in order to elicit the appropriate cellular response. In either case signal transduction is not carried out by a single gene product but rather by multiple gene products working in close physical proximity and contact with one another. In the immune response the receptors themselves are composed of a number of polypeptide chains. Some of the chains are involved in recognition, others with transducing a signal from the extracellular face to the cytoplasmic face. The receptors work together with coreceptors and with cell adhesion molecules to relay information into the cell.

In the immune systems receptors cluster and rearrange themselves. Signals received at the cell surface help shape the clusters; the character of the cluster in turn shapes the signal transduced into the cell interior. The lipid bilayer is not passive but instead plays an active role in these adaptive rearrangements. These procedures are utilized by leukocytes, by nerve cells, and, as was discussed in Chapter 7, in the bacterial chemotactic system.

The proteins required for signaling between T cells and APCs are organized into signaling structures called immunological synapses (ISs). The term "synapse" denotes a junction between cells where information is transmitted. Synapses are highly enriched in signaling molecules that not only transmit signals but also adaptively control the properties of the relayed messages such as their strength. The term "synapse" originates in studies of the transmission of signals between nerve cells. The literal meaning of the term, coined by Sir Charles S. Sherrington in 1897, is "to clasp together." Synapses between nerve cells will be explored in Chapter 21.

Immunological synapses bring together a host of proteins required to support sustained signaling. Among the components of the synapse are receptors, cytoplasmic kinases, and adapters and anchors that link events at the cell surface to the cytoskeleton. The TCRs and MHCs lie in the middle of immunological synapse. As discussed earlier, the TCRs associate with cytoplasmic CD3 complexes, protein tyrosine kinases such as Lck and Fyn, and costimulatory molecules such as CD28 and CD40. Other proteins belonging to the immunological synapse include ICAM-1 immunoglobulins and LFA-1 intergrins, which bind one another, and members of the CD2 grouping of Ig adhesive molecules, which include CD48 (rat) and CD58 (human). This second group of mainly adhesive proteins surrounds the central portion of the immunological synapse.

9.20 Kinetic Proofreading and Serial Triggering of TCRs

The TCR utilizes kinetic proofreading and serial triggering to discriminate between self and non-self antigens. The TCR is remarkable. It can tell self from non-self antigens even when the latter are buried in a sea of self antigens. At first glance, the TCR assemblage looks unwieldy, if not outright baroque. A whole series of diffusion, phosphorylation, and binding events are required to elicit a proper cellular response to antigen binding. Involved are several cytoplasmic proteins, namely, Lck, Fyn, and ZAP-70, a number of adapters, a cluster of transmembrane chains, some of which, most notably the ζ-ζ chains associated with CD3, are subjected to multiple phosphorylations, and finally PLC and the RasGTPase. It turns out that these two aspects of TCR signaling are tied together.

Completion of the series of phosphorylation steps, including those of the ζ-ζ chains, depends on the nature of the peptide gripped by the MHC molecule and bound to the TCR. If the ligand is not a non-self antigen, the activation series will not run to completion before the ligand dissociates from the receptor. The dissociation will abrogate the sequence of diffusion and phosphorylation events before a strong (optimal) signal can be generated. By this means differences in lifetime of the receptor-ligand binding are converted to differences in signaling because of the presence of many intermediate time- and energy-consuming steps.

This process is called *kinetic proofreading*. This term was first introduced in the 1970s to explain how high levels of fidelity could be achieved in operations such as transcription and translation. In these processes, the error rates are exceedingly low, on the order of 10^{-4} to 10^{-9}. These low error rates are hard to understand using thermodynamic arguments alone because the energy differences between correct and incorrect steps are miniscule. This situation is comparable to that of the TCR, which can tell self from non-self antigens when the latter are present in as few as 1 part in 10,000.

Kinetic proofreading occurs along with another chain of steps called *serial triggering*. The term "triggering" in this case refers to the generation of a signal by the TCR at the end of the series of internal steps. In serial triggering, a single peptide-bound to an MHC molecule (pMHC) is able to bind to several TCRs, one after the other in a serial fashion. Once the series of diffusions and phosphorylation is completed the TCR is internalized and the pMHC is free to engage another TCR. By this means the effect of a small number of non-self peptides is amplified. *Serial engagement favors short lifetimes* so that many receptors can be engaged, while *kinetic proofreading favors longer ones* to increase the fidelity. The result is a balance between the two processes and a dissociation rate optimized for TCR signaling.

The formation of an immunological synapse permits optimal signaling through the TCR. As noted above, in an immunological synapse, a ring of adhesive molecules called the *peripheral supramolecular activation cluster* (pSMAC) surrounds a central ring of TCRs and associated signaling proteins, collectively referred to as the *central supramolecular activation cluster* (cSMAC). This clustering together of TCRs into a cSMAC accomplishes two things. It promotes serial triggering, and it helps to rapidly turn off signaling through receptor internalization once the signaling has reached its appropriate level.

General References

Benjamini E, Coico R, and Sunshine G [2000]. *Immunology* (4th edition) New York: John Wiley and Sons.

Janeway CA, Jr., Travers P, Walport M, and Capra JD [1999]. *Immuno Biology* (4th edition) London: Current Biology Publications.

References and Further Reading

Lymphocytes

Abbas AK, Murphy KM, and Sher A [1996]. Functional diversity of helper T lymphocytes. *Nature*, 383: 787–793.

Fearon DT, and Locksley RM [1996]. The instructive role of innate immunity in the acquired immune response. *Science*, 272: 50–54.

Glimcher LH, and Murphy KM [2000]. Lineage commitment in the immune system: The T helper lymphocyte grows up. *Genes Dev.*, 14: 1693–1711.

Goldrath AW, and Bevan MJ [1999]. Selecting and maintaining a diverse T-cell repertoire. *Nature*, 402: 255–262.

Pulendran B, Palucka K, and Bancherseau J [2001]. Sensing pathogens and tuning immune responses. *Science*, 293: 253–256.

Rissoan MC, et al. [1999]. Reciprocal control of T helper cell and dendritic cell differentiation. *Science*, 283: 1183–1186.

NF-κB Signaling Node

Hoffmann A, et al. [2002]. The IκB-NF-κB signaling module: Temporal control and selective gene activation. *Science*, 298: 1241–1245.

Li QT, and Verma IM [2002]. NF-κB regulation in the immune system. *Nature Rev. Immunol.*, 2: 725–734.

Senftleben U, et al. [2001]. Activation by IKKα of a second, evolutionary conserved, NF-κB signaling pathway. *Science*, 293: 1495–1499.

Silverman N, and Maniatis T [2001]. NF-κB signaling pathways in mammalian and insect innate immunity. *Genes Dev.*, 15: 2321–2342.

MAP Kinase Modules

Cowan KJ, and Storey KB [2003]. Mitogen-activated protein kinases: New signaling pathways functioning in cellular response to environmental stress. *J. Exp. Biol.*, 206: 1107–1115.

English J, et al. [1999]. New insights into the control of MAP kinase pathways. *Exp. Cell Res.*, 253: 255–270.

TRAFs

Arch RH, Gedrich RW, and Thompson CB [1998]. Tumor necrosis factor receptor-associated factors (TRAFs)—A family of adapter proteins that regulate life and death. *Genes Dev.*, 12: 2821–2830.

Chung JY, et al. [2002]. All TRAFs are not created equal: Common and distinct molecular mechanisms of TRAF-mediated signal transduction. *J. Cell Sci.*, 115: 679–688.

Toll and Toll-Like Receptors

Aderem A, and Ulevitch RJ [2000]. Toll-like receptors in the induction of the innate immune response. *Nature*, 406: 782–787.

Hemmi H, et al. [2000]. A Toll-like receptor recognizes bacterial DNA. *Nature*, 408: 740–745.

Medzhitov R, and Janeway C, Jr. [1997]. Innate immunity: The virtues of a nonclonal system of recognition *Cell*, 91: 295–298.

Medzhitov R, Preston-Hulburt P, and Janeway CA, Jr. [1998]. A human homologue of the *Drosophila* Toll protein signals activation of adaptive immunity. *Nature*, 388: 394–397.

TNFs

Bodmer JL, Schneider P, and Tschopp J [2002]. The molecular architecture of the TNF superfamily. *Trends Biochem. Sci.*, 27: 19–26.

Locksley RM, Killeen N, and Lenardo MJ [2001]. The TNF and TNF receptor super-families: Integrating mammalian biology. *Cell*, 104: 487–501.

Jaks, STATS, and Hematopoietins

Guthridge MA, et al. [1998]. Mechanism of activation of GM-CSF, IL-3, and IL-5 family of receptors. *Stem Cells*, 16: 301–313.

Heinrich PC, et al. [2003]. Principles of interleukin (IL)-6-type cytokine signaling and its regulation. *Biochem. J.*, 374: 1–20.

Levy DE, and Darnell JE, Jr. [2002]. STATs: Transcripional control and biological impact. *Nature Rev. Mol. Cell Biol.*, 3: 651–662.

Ortmann RA, et al. [2000]. Janus kinases and signal transducers and activators of transcription: Their roles in cytokine signaling, development and immunoregulation. *Arthritis Res.*, 2: 16–32.

The Interferon System

Goodbourn S, Didcock L, and Randall RE [2000]. Interferons: Cell signalling, immune modulation, antiviral responses and viral countermeasures. *J. Gen. Virol.*, 81: 2341–2364.

Chemokines

Baggiolini M [1998]. Chemokines and leukocyte traffic. *Nature*, 392: 565–568.

Cyster JG [1999]. Chemokines and cell migration in secondary lymphoid organs. *Science*, 286: 2098–2102.

Rollins BJ [1997]. Chemokines. *Blood*, 90: 909–928.

T Cell Costimulation

Lanzavecchia A [1998]. License to kill. *Nature*, 393: 413–414.

Schwartz RH [2001]. It takes more than two to tango. *Nature*, 409: 31–32.

Sharpe AH, and Freeman GJ [2002]. The B7-CD28 superfamily. *Nature Rev. Immunol.*, 2: 116–126.

Viola A, et al. [1999]. T lymphocyte costimulation mediated by reorganization of membrane microdomains. *Science*, 283: 680–682.

Immunological Synapses

Coombs D, et al. [2002]. Activated TCRs remain marked for internalization after dissociation from pMHC. *Nature Immunol.*, 3: 926–931.

Dustin ML, and Cooper JA [2000]. The immunological synapse and the actin cytoskeleton: Molecular hardware for T cell signaling. *Nature Immunol.*, 1: 23–29.

Grakoui A, et al. [1999]. The immunological synapse: A molecular machine controlling T cell activation. *Science*, 285: 221–227.

Krummel MF, et al. [2000]. Differential clustering of CD4 and CD3ζ during T cell recognition. *Science*, 289: 1349–1352.

Lee KH, et al. [2003]. The immunological synapse balances T cell receptor signaling and degradation. *Science*, 302: 1218–1222.

Monks CRF, et al. [1998]. Three-dimensional segregation of supramolecular activation clusters in T cells. *Nature*, 395: 82–86.

Problems

9.1 There are two basic approaches used in theoretical studies of how signaling pathways operate: macroscopic and microscopic. The starting point for the macroscopic approaches is the *law of mass action*, and for the microscopic approach it is the *principle of detailed balance*. In both classes of approaches one constructs a set of equations that captures the salient features of the signaling process in terms of rates and other quantities that can be related to experiment. One then follows the evolution of the system on a computer to answer questions posed by the experimental data. In many situations, modeling a system and then simulating its behavior on a computer is the only way to see just what the biomolecules are doing over time.

The law of mass action is a statement that the rate of a chemical reaction is proportional to the concentrations of the reactants. To illustrate how this rule works, consider a reaction of the form

$$[A] + [B] \underset{k_{-1}}{\overset{k_1}{\rightleftarrows}} [C] + [D].$$

The rate of change of the product [C] is

$$\frac{d[C]}{dt} = k_1 [A][B] - k_{-1}[C][D]. \tag{9.1}$$

In the above, the rate of change of a reactant is expressed as the difference between its rate accumulation and its rate of loss through degradation and additional/reverse reactions. Expressions of this form can be simplified if some of the concentrations do not change appreciably over time, if some intermediates are formed rather transiently compared to the other steps, and if the reaction only proceeds in one direction or the other. For example, if the concentration of D does not change over time the second term in Eq. (9.1) becomes a linear one in which [D] is absorbed into the constant. An intermediate step, where a complex between A and B is formed, has been omitted in writing the above under the transient formation assumption.

One can use the law of mass action to construct models of signaling pathways. Consider a set of kinases arranged in a pathway so that one kinase activates a second kinase and so. Let X_i^P denote the ith activated kinase in the pathway. Let X_{i+1} be its substrate kinase, and let Y denote a protein phosphatase that acts on X_{i+1}^P to return it to an unphosphorylated state. The following pair of expressions can describe the joint · actions of X_i^P and Y on X_{i+1}:

$$[X_i^P] + [X_{i+1}] \xrightarrow{k_{i+1}} [X_i^P] + [X_{i+1}^P],$$

$$[Y] + [X_{i+1}^P] \xrightarrow{j_{i+1}} [Y] + [X_{i+1}].$$

The rate equation governing the buildup of the phosphorylated $(i+1)$th kinase is

$$\frac{d[X_{i+1}^P]}{dt} = k_{i+1}[X_i^P][X_{i+1}] - j_{i+1}'[X_{i+1}^P], \tag{9.2}$$

where [Y] has been assumed to be constant and was absorbed into the phosphatase rate constant j'. Other equations can be constructed that describe interactions between ligands and receptors, between receptors and kinases/adapters, and between adapters and kinases. The result is a model of each of the steps in the signaling pathway.

(a) Derive the above expression, Eq. (9.2), for the rate of change in $[X_{i+1}^P]$. This equation can be solved to determine the steady-state (ss) concentration of X_{i+1}^P.

(b) Show that the result, taking the concentration of the unphosporylated plus phosphorylated forms of X_{i+1} to be a constant T, is

$$[X_{i+1}^P]_{ss} = \frac{[X_i^P]T}{(j_{i+1}'/k_{i+1}) + [X_i^P]}. \tag{9.3}$$

This steady-state formula, Eq. (9.3), relates the stimulus $[X_i^P]$ to the response $[X_{i+1}^P]$. What is the behavior of the resulting stimulus-response curve at low $[X_i^P]$, and at high $[X_i^P]$? How does this compare to Michaelis–Menten kinetics discussed in the problems for Chapter 7?

9.2 The reaction kinetics models based on the law of mass action make up a macroscopic description of the biochemical processes taking place. Alternatively one could employ a microscopic description that deals with the movement and interactions at the molecular and atomic levels rather than dealing with macroscopic quantities such as the concentrations. The starting place for the microscopic approaches is the principle of detailed balance, which states that the rate at which a system makes a transition into a particular microstate from a preceding state is equal to the rate that the system will transition back out of that particular microstate to its predecessor. In mathematical terms the principle is

$$p_i P_{ij} = p_j P_{ji}, \tag{9.4}$$

where p_i is the probability that the ith microstate is occupied (i.e., the *occupation probability*), and P_{ij} is the transition probability from microstate i to microstate j. The above expression is a statement that the laws of physics are the same going forward or backward. In particular, the expression states that the probability that a state is occupied times the rate at which there is a transition out of that state to another is equal to the probability that the entered state is occupied times the rate at which that state is exited back to its predecessor. The occupation probabilities can be defined in terms of the Boltzmann factor β:

$$p_i = \frac{1}{Z} e^{-\beta E_i}, \tag{9.5}$$

where $\beta = 1/k_B T$, E_i is an energy that characterizes the microstate i, and Z is a normalization factor. With this definition one can introduce a sampling algorithm that allows a researcher using a computer to simulate the evolution of a biochemical system over time in order to study its properties and see how it works. The sampling algorithm for the transitions is

$$P_{ij} = \begin{cases} e^{-\beta \Delta E_{ij}}, & E_j > E_i, \\ 1, & E_j \le E_i. \end{cases} \tag{9.6}$$

In the simulations, one starts out with the system in some initial state, then randomly picks a new microstate and calculates the energy difference $\Delta E_{ij} = E_j - E_i$. If the energy difference is negative then the transition leads to a state of lower energy and the move is allowed. If the energy difference is positive, then the transition increases the energy of the system. One then selects a random number between 0 and 1. If that random number is less than the calculated P_{ij}, the transition is allowed. Otherwise the move is rejected and another proposed transition is selected. Note that in this approach moves that increase the energy become possible allowing the system to transition over barriers and out of pockets in the energy landscape. This would not be possible in situations where only energy decreasing steps were allowed.

This algorithm, known as the *Metropolis algorithm*, was introduced in a landmark paper in 1953. The general method of randomly sampling a set of moves of a microsystem according to a probability distribution is known as the *Monte Carlo method* and is extensively used in several different forms to study how complex physical and chemical systems evolve over time.

Show that the Metropolis algorithm, Eq. (9.6), satisfies the detailed balance condition, Eq. (9.4), using Eq. (9.5). What are the results of this sampling procedure when the energy of state j just barely exceeds that of state i? What happens when the energy differences between the two states are large? The quantity Z appearing in the definition of the occupation probabilities is known as the *partition function* in statistical physics. Write an expression for it in terms of the Boltzmann factor β.

Theoretical (Modeling) Studies

Asthagiri AR, and Lauffenburger DA [2001]. A computational study of feedback effects on signal dynamics in a mitogen-activated protein kinase (MAPK) pathway model. *Biotechnol. Prog.*, 17: 227–239.

Heinrich R, Neel BG, and Rapoport TA [2002]. Mathematical models of protein kinase signal transduction. *Mol. Cell*, 9: 957–970.

Tyson JJ, Chen KC, and Novak B [2003]. Sniffers, buzzers, toggles and blinkers: Dynamics of regulatory and signaling pathways in the cell. *Curr. Opin. Cell Biol.*, 15: 221–231.

10
Cell Adhesion and Motility

Movement and adhesive contacts between cells and opposing surfaces are necessary for embryonic development and normal adult functioning of cells in the body. During development, cells do more than grow, multiply, and differentiate. They also segregate, migrate, and aggregate in forming tissues and organs. In adult life, vascular endothelial cells and fibroblasts migrate during wound healing, leukocytes migrate in response to infections, and cancerous cells migrate during metastasis.

Cells establish and maintain contacts with other cells and with the extracellular matrix. These contacts stabilize the aggregates of cells and enable them to work together to carry out their functions. Adhesive contacts underlie axon outgrowth during development of the nervous system. Paralleling the steps taken by migrating cells, growth cones, motile sensory structures enriched in adhesion molecules, guide the formation of connections in the nervous system.

Several different families of cell adhesion molecules—integrins, cadherins, immunoglobulin superfamily cell adhesion molecules (IgCAMs), and selectins—mediate adhesive contacts between surfaces in the body. These molecules, acting as receptors and counterreceptors/ligands on opposing surfaces, are responsible for establishing and maintaining physical contact and communication between the extracellular matrix, the cell surface, and the cytoskeleton. These families of adhesion molecules, along with several different kinds of diffusible molecules and their receptors, also mediate axon outgrowth in the nervous system. Leukocyte adhesion and mobility will be examined in the first part of this chapter and axon outgrowth in the second.

10.1 Cell Adhesion Receptors: Long Highly Modular Glycoproteins

The extracellular regions of cell adhesion receptors are mosaics of domains, strung together in a linear fashion. The domains may be composed of a number of short identical subdomains, referred to as *repeats*, or they may be erected from a mixture of two or more different domains, or they may be built from both repeats and single domains. Some of the domains in the extracellular regions are specific to a particular family and define membership in that family. Examples are cadherin domains in cadherins, and lectin domains in selectins. Other domains, most notably the immunoglobulin-(Ig) like domains and fibronectin type III repeats, occur in many different proteins. A representative sampling of extracellular regions of adhesion-promoting proteins is presented in Figure 10.1.

The extracellular regions of these proteins are attached to several different kinds of membane-associated segments. The most commonly encountered form of attachment is to a single-pass transmembrane segment that is followed by a short cytoplasmic tail. All of the proteins appearing is Figure 10.1 attach in this manner. There are three families of selectins—vascular endothelial (E), leukocyte (L), and platelet (P). All of these attach by means of transmembrane segments. So do 5/2 IgCAMs such as NCAM, 6/5 IgCAMs such as L1, and 4/6 IgCAMS such as deleted in colorectal cancer (DCC), but IgCAMS such as Contactin attach by means of a GPI anchor. Most cadherins attach through a single-pass TM segment, but cadherins such as Flamingo attach by means of a seven-pass transmebrane chain and may signal through G proteins.

During an inflammatory/immune response, leukocytes emigrate out of the bloodstream and converge upon the injured tissue. The emigration process occurs in several stages. Leukocytes first establish a loose contact with the endothelial cells lining the vessel walls and begin rolling. Next, a hard adhesive contact is established, and finally the leukocytes crawl out of the blood vessel and migrate into the injured tissue. Adhesion receptors

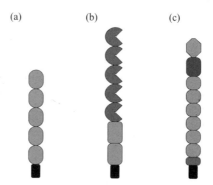

(a)   (b)   (c)

FIGURE 10.1. Representative extracellular regions of proteins involved in cell adhesion: (a) Classical cadherins containing 5 cadherin domains N-terminal to the transmembrane segment shown in black. (b) Neural cell adhesion molecules (NCAMs) containing 5 Ig-like domains and 2 fibronectin type III (FNIII) repeats. (c) E-selectins containing an amino terminal lectin domain, an EGF domain, 6 short consensus repeats (SCRs), and a membrane proximal cleavage region.

work synergistically with receptors for cytokines and soluble growth factor. They signal through many of the same pathways as growth factors and together promote cell survival, proliferation, and differentiation. The cytoplasmic domains of cell adhesion receptors make contact with the cytoskeleton, and the receptors help regulate cell shape and polarization, and cytoskeleton organization and motility. During metastasis cancer cells utilize a similar ensemble of mechanical and adhesion receptor signaling processes. In metastasis, cells detach from the surrounding tissue, migrate to remote sites elsewhere in the body using the lymphatic and circulatory systems, reattach, and then establish a colony. Signaling between the cell surface and the nucleus coordinates the expression of specific cell adhesion molecules at different stages in metastasis.

Cell adhesion receptors are both large and flexible and because of these properties present challenges in their study using high resolution NMR and X-ray crystallography methods. The solution to this technical challenge is to take advantage of their inherent modularity and characterize portions—fragments and domains—of these molecules. The resulting NMR and X-ray crystallography data together with electron microscopy results provide a core body of structural information on these essential proteins.

10.2 Integrins as Bidirectional Signaling Receptors

Integrins are membrane-spanning glycoproteins composed of noncovalently attached α and β subunits. In vertebrates, 18 distinct alpha subunits and 8 different beta subunits have been identified so far. Not all of the 18 × 8 = 144 combinations of alpha and beta subunits can be formed. Instead, a far smaller number, namely, 24, αβ heterodimers can occur. Each integrin subunit contains a large extracellular domain, a single transmembrane segment, and a short cytoplasmic domain. There is one exception to this rule: The β_4 subunit has a large cytoplasmic domain. The α subunits are larger than the β subunits. The α subunits vary in size from 120 to 170 kDa (up to 1114 amino acid residues) while β subunits range in size from 90 to 100 kDa (up to 678 residues).

Integrins are multidomain proteins. The extracellular parts of the alpha and beta chains each contain five or more domains and some of these domains may consist of multiple subdomains. The domain organization of a representative integrin alpha chain and of a typical beta chain are depicted in Figure 10.2. In part (c) the two chains of the integrin molecule are bent in a V-shape that supports low affinity ligand binding. Not all integrin molecules contain inserted (I) or I-like domains. These ligand-binding domains are included in the model integrin depicted in the figure. The most important feature of this figure is the bent shape. The integrin molecule undergoes massive rearrangements in response to allosteric signals, and the far more open shapes that result mediate intermediate and high affinity

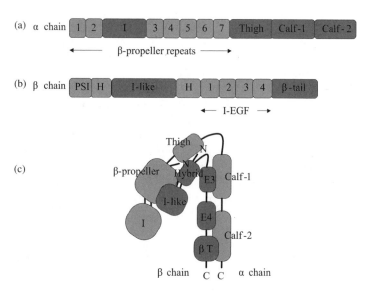

FIGURE 10.2. Domain organization and assembly of the extracellular portions of an integrin containing inserted (I) and I-like ligand-binding domains: (a) Domain organization of the alpha subunit. (b) Domain organization of the beta subunit. (c) Assembly of the alpha and beta chains in a V-shaped conformation showing the exposed ligand binding I and I-like domains at the end of the molecule. Abbreviations: β-tail (βT); hybrid (H); integrin-epidermal growth factor (I-EGF, or E); plexin/semaphorin/integrin (PSI).

ligand binding. X-ray crystallography data revealing this bent shape are presented in Figure 10.2.

Integrins and cadherins differ from most transmembrane signal transducers that transmit signals in one direction, from outside the cell inward. Integrin and cadherin receptors transmit signals in both directions—from outside the cell inward (outside-in) and from inside the cell outward (inside-out). In outside-in signaling, binding of the integrins to the ECM triggers changes in the pattern of gene expression. Inside-outside signals produce changes in the integrin conformation resulting in changes in adhesiveness. Integrins tie the ECM to the cellular cytoskeleton and anchor cells in a fixed position, while cadherins are a primary element of cell-to-cell junctions.

10.3 Role of Leukocyte-Specific Integrin

The leukocyte-specific integrin LFA-1 mediates the migration of T cells. The migration of T cells occurs in several stages. In the first stage, lamellipodia, broad flat structures, form and extend out from the leading edge of the cell. This stage is followed by a step in which new adhesive contacts are established and stabilized. The cell body then contracts, and the following edge, or tail, detaches. T cells use the $\alpha_L\beta_2$ integrin, also called the leukocyte

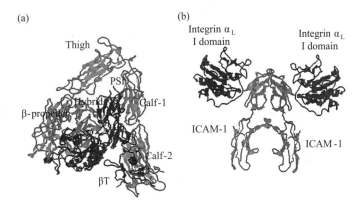

FIGURE 10.3. Integrin crystal structure: (a) V-shaped, or bent, closed conformation of the $\alpha_V\beta_3$ integrin. The alpha chain is shown in light gray and the beta chain in black. (b) Complex formed between two α_L inserted (I) domains (black) and a pair of ICAM-1 ligands (light gray). The figure was prepared using Protein Explorer with atomic coordinates deposited in the PDB under accession numbers 1jv2 (a) and 1mq8 (b).

function-associated antigen-1 (LFA-1) integrin, to contact ICAM-1 proteins on opposing endothelial cells (Figure 10.3). LFA-1 is also a main component of the pSMAC rings of immunological synapses.

Contacts between LFA-1 integrins and ICAM-1s trigger the formation of signaling complexes and signaling through Rho GTPases to downstream kinases, most notably myosin light chain kinase (MLCK) and Rho kinase (ROCK). These kinases are restricted to specific locations in the cell. MLCK is concentrated near the leading edge of the cell, and ROCK is localized to the tailing edge. These serine/threonine kinases act on myosin light chains (MLCs) which when phosphorylated promote actin-myosin-fiber contractions. When coordinated over time, signaling through the adhesion receptors produces leading edge attachments, cell body contractions, tailing edge detachments, and T cell movement.

10.4 Most Integrins Bind to Proteins Belonging to the ECM

The extracellular matrix (ECM) is intercellular material, mostly glycoproteins and proteoglycans, that surrounds cells and forms the connective tissue in the body. Examples of ECM connective tissue include teeth and bone, cartilage and tendons, and skin. The ECM encompasses the various basement membranes that provide structural support for tissues and organs. A typical basement membrane has a basal lamina formed by glycoproteins secreted by the attached cells and a reticular lamina formed by protein fibers from the underlying connective tissue. Collagens, glycine-rich glycoproteins, are the primary ECM elements. Collagens are long ropelike linear molecules composed of three chains arranged in a triple helix. Typical lengths of collagens are 300 to 400 nm. Two families of adapter molecules are present in the basal lamina—laminins and fibronectins. These molecules connect ECM collagens to adhesive cellular proteins.

The extracellular matrix has considerable influence over the metazoan cell. Both mechanical and chemical signals are sent from the ECM to the cell surface, and from there they are transduced into the cell interior. Integrins have as their ligands the aforementioned ECM adapter molecules laminin and fibronectin. Binding to these proteins triggers the formation of adhesive contacts at specific locations along the plasma membrane. Binding to the matrix proteins stimulates integrin clustering, signaling to proteins embedded in the plasma membrane, and to proteins located in the cytoplasm.

The cytoplasmic tails of integrin beta chains contain motifs of the form NPxY (arginine-proline-X-tyrosine) that are recognized by phosphotyrosine-binding (PTB) domains of cytoplasmic adapter molecules, nonreceptor tyrosine kinases such as Src, focal adhesion kinase (FAK), and the actin cytoskeleton-binding protein talin. The binding of the beta chains to talin and other proteins establishes mechanical linkages called *focal adhesions* between the ECM, the cell surface, and the cytoskeleton that facilitates clustering of integrins and the onset of signaling. Once the focal adhesions are formed positive feedback signals to the integrin receptors stimulate their further clustering and additional signaling from them to the focal adhesions. The result of this two-way signaling is the tying together, or integration, of the ECM with the cytoskeleton of the cell. The integrative properties of these receptors led to their being given the name "integrins."

10.5 Cadherins Are Present on Most Cells of the Body

Cadherins (calcium dependent adherins) are a large family of transmembrane proteins. In mammalian species there are at least 80 family members. A large number of these proteins is expressed preferentially in the nervous system. Cadherins contain an extracellular domain, a transmembrane segment, and a cytoplasmic tail. The defining characteristic of this family is the presence of a *cadherin motif*, or *EC domain*, in the extracellular domain. These motifs are tandemly repeated anywhere from 2 times to 30 or more times. The most N-terminal of these domains is the adhesive site, while other portions of the extracellular domain serve as spacers and supply multiple binding sites for calcium ions. Like integrins, cadherins form clusters when they bind their ligands, in this case, counterreceptors on opposing surfaces. The juxtamembrane portion of the cytoplasmic domain—the part of the cytoplasmic domain nearest the plasma membrane—interacts and contributes to the clustering and adhesive strengthening taking place upon ligand binding.

FIGURE 10.4. Ectodomain of a classical cadherin: Shown are the five domains, labeled EC1 through EC5, from the N-terminus towards the C-terminus, which is nearest plasma membrane. Small single and paired spheres represent calcium ions, which are necessary for proper functioning of the receptor. Clusters of spheres denote the locations of N-linked and O-linked sugars. The figure was prepared using Protein Explorer with atomic coordinates deposited in the PDB under accession number 1l3w.

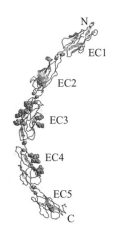

Classical cadherins such as E-cadherin (epithelial cadherin) and N-cadherin (neural cadherin), contain about 750 amino acid residues. Their extracellular domain consists of five repeats of a 100-amino acid EC domain, which extends out 250 Å from the cell surface (Figure 10.4). These cadherins attach to the actin cytoskeleton by means of a linker protein called *catenin*. The cytoplasmic domain of these cadherins is approximately 150 amino acid residues in size, and contains a catenin-binding site. There are two catenin subunits. The β subunit binds to the catenin domain of the cadherin and to the α subunit of the catenin. The α subunit, in turn, connects to the actin cytoskeleton. This cadherin-catenin complex is referred to as a *zonula adherens junction in epithelial cells* and as an *adherens junction in neurons*. In epithelia, these junctions link cells together laterally to permit them to form stable sheets, and they separate the apical (top) region from the basolateral (bottom) region of each of the cells. In neurons, the adherens junctions connect the pre- and postsynaptic cells, thereby providing mechanical stability and a means of conveying signals in both directs across the synapse.

Cadherins operate in a homophilic manner, and have a role in maintaining tissue boundaries and stabilizing synapses. Cells that segregate into distinct tissues express different cadherins, so that a cadherin receptor on the surface of one cell binds to a cadherin receptor of the same type on the surface of an opposing cell. Calcium binding is necessary for clustering, or oligomerization, of cadherin receptors as well as for ligand binding. An example of how this requirement might serve a useful function is provided by N-cadherin binding in synaptic junctions. Intense synaptic activity generates long-lasting changes in synaptic transmission. Remodeling of the synapse underlies these changes. These changes might occur through the following sequence of steps: Intense activity depletes the supply of calcium in the vicinity of the synapse. This lowering in the local calcium concentration weakens the links between cadherin molecules allowing remodeling activities to proceed. This step is followed by reestablishement of firm contact when the calcium concentration returns to its basal levels.

10.6 IgCAMs Mediate Cell–Cell and Cell–ECM Adhesion

Cell adhesion molecules (CAMs) of the immunogloulin superfamily (Ig-SF) mediate cell–cell and cell–ECM adhesion. Cell surface receptors belonging to this superfamily, the IgCAMs, are characterized by the presence of one or more immunoglobulin-like domains in their extracellular region. These adhesion molecules mediate cell-to-cell contact by binding to cell surface counterreceptors, and help establish and maintain contact with the extracellular matrix by binding to ECM constituents. Unlike cadherins, calcium binding is not required either for ligand binding or for clustering to occur. Some IgCAMs bind in a homophilic manner while others bind in a heterophilic way.

One of the main kinds of IgCAM is the neural IgCAM, or NCAM, expressed on neurons (Table 10.1). These receptors contain one or more repeats of the Ig fold plus a number of fibronectin type III folds, in their extracellular domain as depicted in Figure 10.1. The NCAMs are grouped into families according to the organization of the extracellular domains. NCAM contains five immunoglobulin-fold repeats plus a pair of fibronectin type III (FnIII) repeats proximal to the plasma membrane. It is thus classified as a 5/2 IgCAM. Another important and well-studied neural IgCAM is L1. This cell adhesion molecule is a 6/5 IgCAM family member. A third

TABLE 10.1. Members of the IgCAM group of cell adhesion receptors: Abbreviations—ICAM (intercellular cell adhesion molecule); LFA (lymphocyte function-associated antigen), NCAM (neural cell adhesion molecule), PECAM (platelet endothelial call adhesion molecule), VCAM (vascular cell adhesion molecule), CD (cluster of differentiation).

Ig-SF CAM	CD designation	Ligand	Distribution/role
ICAM-1	CD54	LFA-1, Mac-1, fibrinogen	Lymphocytes, endothelial cells when inflammation is present
ICAM-2	CD102	LFA-1, Mac-1	Recirculating leukocytes
ICAM-3	CD50	LFA-1	Leukocytes
LFA-2	CD2	LFA-3	Lymphocytes
LFA-3	CD58	LFA-2	Broad distribution
NCAM	CD56	NCAM, collagen, heparin	Neural tissue
PECAM-1	CD31	PECAM-1	Platelets, leukocytes
VCAM-1	CD106	$\alpha_4\beta_1$ and $\alpha_4\beta_7$ integrins	Lymphocytes, endothelial cells when inflammation is present

group of IgCAMs are the DCC family members. Named for the protein DCC (*deleted in colorectal cancer*) these are 4/6 IgCAMs. Molecules closely related to DCC have a prominent role in the development of the nervous system. They are found on growth cones where they bind a family of diffusible chemoattractants called *netrins* that guide growing axons during embryonic development, as will be discussed later in this chapter.

The two most prominent families of IgCAMs are the NCAMs and ICAMs (intercellular cell adhesion molecules). The ICAMs bind to integrins and thus mediate heterophilic binding. NCAMs bind to identical receptors on opposing cells and thus mediate homophilic binding. Besides their role in the immune/inflammatory system, these proteins play an important role during embryonic development and in the nervous system both during development and in adult life. NCAMs are also expressed in the heart, gonads, and skeletal muscle. Because of their role in signaling and maintaining tissue integrity they have a prominent role in several forms of cancer.

10.7 Selectins Are CAMs Involved in Leukocyte Motility

Leukocytes circulate in blood vessels and the lymphatic system, and when an infection is detected they converge to infection sites. Selectins mediate the movement of leukoctyes within blood vessels to sites of infection, and out of the blood vessel into the surrounding inflamed tissue. They are expressed on the surface of leukoctes, the endothelium lining the blood vessels, and platelets that form adhesive plugs, or clots, at sites of wounds. There are three kinds of selectins: *L-selectins* are expressed on leukocytes, *E-selectins* are expressed on the endothelial cells, and *P-selectins* are expressed on platelets on the endothelium. These adhesion molecules mediate the capture of the circulating leukocytes by the walls of the blood vessels and make possible their subsequent rolling.

Selectins are mosaic proteins with a common structural organization. They each have an NH_2 terminal lectin domain followed by an epidermal growth factor (EGF) domain followed by a number of consensus repeats (CRs) of a complement-like binding sequence, a single transmembrane segment, and a short cytoplasmic tail. The lectin domain is a carbohydrate-binding domain that enables the selectin to bind to carbohydrate structures on its ligand. The CRs are thought to function as spacers that extend the molecule a distance that supports optimal rolling. Deletion of portions of this segment impairs rolling. The three selectins vary in the number of CRs. Human P-selectin has nine CRs, while E-selectin has six and L-selectin just two.

The lectin domain situated near the NH_2 terminus is a carbohydrate-binding domain. Selectin ligands such as GlyCAM-1 and CD34 are members of the *mucin family*. These are long rodlike glycoproteins with multiple serine/threonine residues and are heavily glycosylated (O-linked).

TABLE 10.2. Selectins, cells, and structures that express them, their functions, and ligands: Ligand abbreviations are as follows: PSGL-1 (P-selectin glycoprotein ligand-1); MAdCAM-1 (mucosal addressin cellular adhesion molecule-1); GlyCAM-1 (glycosylation-dependent cell adhesion molecule-1).

Selectin	Expression	Functions	Ligands
L-selectin	Leukocytes	Leukocyte trafficking, rolling adhesion	PSGL-1, GlyCAM-1, MAdCAM-1, CD34
P-selectin	Platelets, endothelium	Rolling adhesion	PSGL-1
E-selectin	Endothelium	Rolling adhesion induced by inflammation	PSGL-1

Another ligand, MAdCAM-1, is an IgCAM and directs selectins to mucosal tissues. The primary P-selectin ligand is PSGL-1. Both ligands and selectins support rapid bond formation and dissociation.

Selectins are the first cell adhesion receptors to be activated in an inflammatory response. In the absence of an injury E-selectins and P-selectins are not expressed on the cell surface while L-selectins are constitutively active in order to promote continual leukocyte homing. This activity is guided by the appearance of its ligand in the vicinity of an inflammation. The expression of E-selectin on the surface of endothelial cells is triggered by cytokines such as TNF-α, IL-1, and lipopolysaccharide (LPS). P-selectin is stored in granules inside platelets (α-granules) and endothelial cells (Weibel–Palade bodies). P-selectin can be shipped to the outer surface in minutes in response to histamines and other triggering signals. Cytokines induce synthesis of P-selectin mRNAs.

Platelets are cytoplasmic fragments of bone marrow cells called *megakaryocytes*. Platelets are released from the bone marrow and circulate in the bloodstream. Their role in the inflammatory response is to form clots or plugs that block blood flow at sites of injury. Adhesion and aggregation are central to platelet function, and they express integrins, selectins, and IgCAM receptors. *Microvilli* are adhesion molecule-rich extensions of the cell surface. They are formed at the outside facing, or apical, surface in a variety of different cell types in order to increase the effective surface area. The primary platelet selectin ligand is the P-selectin glycoprotein-1 (PSG1). It is constitutively expressed on the tips of microvilli of leukocytes such as neutrophils and monocytes.

10.8 Leukocytes Roll, Adhere, and Crawl to Reach the Site of an Infection

Leukocytes are highly mobile cells that migrate from blood in and out of lymphatic organs and into tissues, and then back out again into circulation. They respond to infections by attacking and destroying the causative agents.

The leukocytes do not passively float along the blood vessels, but instead remain in contact with the cells lining the walls of the blood vessels. They move along the walls by rolling in a controlled way. This mode of locomotion enables them to stop at the right location and then exit by crawling out between the cells. Adhesion is critically important, enabling the leukocytes to first roll, and then crawl, and finally to kill the pathogens.

In an inflammatory response, the physiology of the blood vessels is altered to better enable the leukocytes to reach the site of infection. Postcapillary *venules* are the main locus of vascular inflammatory activity during an inflammatory response. These structures, some 30 to 40 microns in diameter, consist of a layer of endothelial cells on the inside and a supporting basement membrane on the outside. Blood flow in small vessels such as the postcapillary venules takes place under low flow rates, reasonably high viscosities, and small vessel diameter. Under these conditions the velocity of the blood flow is greatest in the center of the vessel and decreases to zero as the vessel walls are approached. During inflammation the blood vessels dilate and the overall flow rate is slowed. Red blood cells aggregate into large assemblies called *rouleaux* that collect into the center of the vessels displacing white blood cells. In response the leukocytes move from the center of the blood vessel to the vicinity of the vessel walls. When this happens cell surface receptors expressed on the plasma membranes of the endothelial cells and leukocytes can engage one another to promote rolling and then hard adhesion.

10.9 Bonds Form and Break During Leukocyte Rolling

Shear stresses are forces generated when adjacent layers of a fluid slip over each other. Because of the shear stresses that are present, leukocytes will either tumble or roll in the direction of the blood flow. Tumbling motion is a rotary movement done without any contact with the cell wall. Rolling, in contrast, is a rotary movement carried out while in contact with the endothelial cells of the vessel wall. The observed motions of leukocytes are jerky. Periods of rapid motion are interrupted by periods in which the motions are far slower. The periods of slow motion correspond to tethering of the leukocytes to the membranes of the endothelium and the periods of rapid motion to contact-free flow.

Bonds form and break during leukocyte rolling (Figure 10.5). Bond lifetimes must be long enough to permit formation of multiple bonds. If the off (bond dissociation) rates are too high multiple bonds cannot form. One bond will dissociate before another can be formed. At the other extreme, too tight a bonding will either immobilize the cell or will lead to situations where large forces can pull a receptor molecule out of the membrane. The association and dissociation rates should be rapid, but not too rapid. In leukocyte rolling, the bonds that tether leukocytes to the endothelial wall

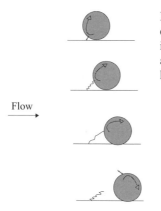

FIGURE 10.5. Bond under shear forces during leukocyte rolling: Depicted are the compression, stretching, and finally dissociation of bonds between a cell adhesion molecule and its ligand during rolling of a leukocyte along the inner wall of a blood vessel.

Flow

are subjected to stretching, or tensile, forces as a result of the shear stresses. Forces have an important effect on the bond lifetimes. They shorten the lifetimes of the bonds. The enhancements can be appreciable—the off-rate rises exponentially with increasing force. The enhanced dissociation rates arising from the tensile forces shift the bond lifetimes into the range needed to ensure proper rolling.

The off-rate amplification is a consequence of the lowering of the energy barriers by the applied forces. The dependence of bond lifetimes on the presence and strength of applied forces is not specific to rolling or leukocytes. Instead, these considerations apply equally well to other bonds between receptors and ligands. The acceleration in dissociation rates due to the presence of applied forces provides a general mechanism for transducing mechanical stresses into signaling responses. Membrane, cytoskeletal, and signaling elements coming together at control points will sense and respond to mechanical stresses by dissociating far more rapidly than would be the case in the absence of stresses.

10.10 Bond Dissociation of Rolling Leukocyte as Seen in Microscopy

Rugged energy landscapes describe bond dissociation during leukocyte rolling. The stretching properties of bonds between cell adhesion molecules and their ligands have been explored using atomic force microscopy (AFM) and optical tweezers. In a typical AFM experiment, the adhesion molecules of interest are attached to a sharp tip, one having a diameter of no more than a few microns. The tip is mounted on a flexible cantilever. When the adhesion molecules are brought into close proximity to their ligands, mounted on a separate surface, the cantilever will bend and undergo a displacement. These displacements are measured by sweeping a laser beam

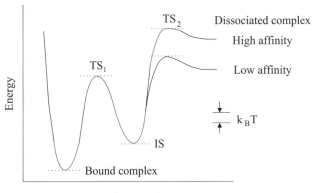

FIGURE 10.6. Energy landscape for dissociation of bonds between adhesive molecules and their ligands: Abbreviations—transition state 1 (TS_1); transition state 2 (TS_2); intermediate state (IS).

across the cantilever and measuring the angle and orientation of the reflected light. In an optical tweezers experiment, one uses a laser beam to trap dielectric particles. External forces applied to the trapped particles can be determined by monitoring the angular deflection of the laser light. The adhesive molecules are mounted on a pedestal attached to a surface and the ligands are attached to optically trapped beads.

Experiments of several different adhesion molecule/ligand combinations have been studied, and the results placed within an energy-landscape perspective. A representative landscape deduced from these experiments is presented in Figure 10.6. This landscape has two prominent features. First, it is a rugged landscape. The pockets are deep, several times the thermal energy, resulting in the presence of at least one metastable intermediate state. The situation shown in the figure is that of two transition states. Second, two different outer barriers heights are present—one of these corresponds to low affinity binding and the other to high affinity binding. The high inner barrier is responsible for high strengths seen at short time scales and the major portion of the activation energy. The lower barriers located further out extend bond lifetimes when forces are absent.

10.11 Slip and Catch Bonds Between Selectins and Their Carbohydrate Ligands

The thrust of the previous section was that externally applied forces reduce the lifetimes of receptor-ligand bonds by lowering the energies of the transition barriers. The force-driven reductions in lifetime allow rolling leukocytes to detach from surface tethers at the right time while maintaining good adhesion to that place on the surface at earlier times. The corresponding bonds are called *slip bonds* because they allow the rolling cells to slip away. These are not the only kinds of bonds that form. An observation often made not only of rolling leukocytes, but also migrating bacteria, is that shear stresses promote good surface contact. The physical mechanism underlying this phenomenon is captured by the notion of a *catch bond*—a bond that is strengthened by the external forces rather than weakened. The selectins that mediate leukocyte rolling exhibit both kinds of bonds. The catch bonds are formed first. They enable the leukocytes to be caught by the surface. Once the leukocytes are caught and begin rolling, a transition takes place from a catch to a slip bond regime so that the latter can mediate proper detachment from the surface.

Selectins are not only expressed by migrating leukocytes. They are expressed early during embryonic development and enable embryos to attach to the lining of the uterus. Failure to properly lodge in the uterus is a prominent cause of fertilization and implantation failures. In the attachment process, carbohydrate ligands upregulated and expressed on the surface of the uterus bind the L-selectin molecules expressed on the surface of the embryo. The adhesion takes place under shear stresses created by mucin secretions and, once established, sets off a chain of signaling events.

10.12 Development in Central Nervous System

Development of the central nervous system involves many of the same operations as used in body development. The development of the central nervous system, with its laminar structure and organization into dozens of anatomically and functionally distinct areas, is one of the most remarkable and dramatic processes in nature. It takes place in several stages. It begins with the production of immature neurons, or *neurites*, and *glia*, cells that eventually supply neurons with growth factors, nutrients, protection (astrocytes); and electrical insulation (Schwann cells). The neurites multiply, grow, differentiate, and migrate to various sites where they aggregate into distinct cortical layers and regions. During this period support structures such as the ECM are erected from molecules secreted by the developing cells. Cell growth, maturation, and circuit formation follows arrival of the neurites at their cortical destinations. The growing cells within each of the layers and areas develop their morphologically and electrophysiologically distinct axonal and dendritic structures, and they establish initial circuit-forming synaptic contacts. The initial contacts are then refined, and unwanted connections and cells are removed. At the end of this process more than 10^{10} neurons will have formed some 10^{15} precise synaptic connections.

Once the neurites reach their cortical destination and their migration ceases, they send out processes that become axons and dendrites that form the synapses and neural circuits. The growth cone (Figure 10.7) is a motile

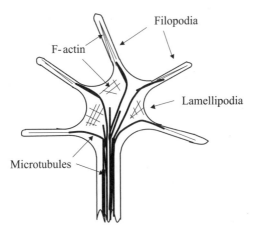

FIGURE 10.7. Structure of a growth cone: Microtubules are represented by thick lines and the actin cytoskeleton by thin lines.

sensory structure located at the tip of an advancing axonal or dendritic process. It contains dynamic extensions of its leading margin called *filopodia* and *lamellipodia* that can extend and retract, change position, and differentially adhere to surfaces. Paralleling the actions taken during the earlier cell migrations, the growth cone explores, interacts with, interprets, and responds to signaling molecules in its local microenvironment. It can navigate over distances of up to several centimeters corresponding to more than 1000 cell body diameters. This key structure contains a dense actin cytoskeleton and is enriched in adhesive and signaling proteins.

10.13 Diffusible, Anchored, and Membrane-Bound Glycoproteins in Neurite Outgrowth

The molecules that guide growth cones to their cortical targets include members of the same families as those involved in leukocyte movement. Integrins, cadherins, and IgCAMs are expressed in growth cones and all contribute to growth cone navigation. In addition to these cell adhesion molecules, several families of diffusible, anchored and membrane-bound signaling glycoproteins direct growth cone navigation by supplying attractive and repulsive signals. These additional molecules involved in growth cone guidance are evolutionarily ancient. They have been found and studied in the nematode worm *C. elegans*, in the fruit fly *D. melanogaster*, in the chick, and in mammals including humans. They are listed in Table 10.3.

The diffusible signaling molecules function as molecular markers. Growth cones maintain contact with surfaces while they navigate. Cells serving as signposts for the growth cones secrete diffusible markers. The markers may delineate forbidden regions where the growth cones are not

TABLE 10.3. Diffusible navigation markers and their receptors: The receptors are grouped according to the ligands—netrins, slits, semaphorins, and ephrins.

Ligand-receptor systems	Description
Netrins	
DCC family	Short or long range attraction; growth cone guidance
UNC-5 family	Short or long range repulsion; growth cone guidance
Slits	
Roundabout	Long range repulsion; growth cone guidance, axonal branching
Semaphorins	
Plexins	Axonal repulsion
Neuropilins	Neuronal sorting; axonal repulsion
Scatter factor receptors	Short range interactions; invasive growth; attraction
Ephrins	
Eph receptors	Contact-dependent repulsion; cell sorting, growth cone guidance

to enter, or mark intermediate targets toward which the growth cones are to turn, or provide a marker for the final destination. The way directional information is supplied is through *concentration gradients*. The basic idea, known as the *chemoaffinity hypothesis*, is that chemical markers are laid down on the surfaces over which the axons navigate. These markers form concentration gradients that are read by the growth cones to obtain the correct heading.

10.14 Growth Cone Navigation Mechanisms

A number of mechanisms operate in concert in growth cone navigation. Long range guidance by the secretion of diffusible molecules from signpost and target cells is supplemented by local (short range) positional cues and by contact-mediated cues. As indicated in Table 10.3 some of the guidance molecules attract growth cones while others repulse them. The designation *long range* refers to the large distance between the cells that secrete diffusible molecules such as netrins, slits, and semaphorins and the growth cones that sense these molecules when they are deposited in their local microenvironment. In many instances intermediate target cells secrete these molecules, and navigation is in short steps of several hundred microns. The intermediate target cells are referred to as *guidepost* or *signpost cells*, terms already mentioned. The immobilization of the diffusible molecules within the extracellular matrix or attached to accessible surfaces such as those of glial cells allows for detections of small variations in concentration by the growth cones. These molecules together with the cell adhesion molecules create highways for the growth cones to travel upon.

The term *short range* usually refers to diffusion that spreads markers no more than a few cell diameters from the source. The distance over which a marker will diffuse depends on the spatial distribution of receptors. If receptors for the marker are highly expressed by nearby cells the distance traveled will be reduced over that which results from a sparse distribution of receptors. The term *contact-dependent* is used to designate binding by membrane-associated receptors on the surface of the growth cones to membrane-associated ligands or counterreceptors expressed on the surface of adjacent cells, and to components of the extracellular matrix such as laminin. The distinction between a *ligand* and a *counterreceptor* is based on whether the signaling is one-way or two-way. If it is two-way then both signaling partners are receptors. If one of the signaling partners supplies a signal but does not transduce signals either directly or indirectly across its own plasma membrane, then that partner is a ligand.

The modern form of the chemoaffinity hypothesis has been greatly expanded to incorporate short and long range mechanisms—attraction, repulsion, and bifunctionality—and pathfinder axons, guidepost cells, and selective fasciculation. Here is how the last process is described: Axons in organisms such as *Drosophila* can grow on tracks laid down by earlier pathfinder axons. These later axons form axon bundles, or *fascicles*. The axons select a particular axon bundle, bypassing inappropriate bundles in the process, guided by molecular markers expressed on the surfaces. The overall mechanism is thus called *selective fasciculation*. As the axons approach their cellular targets they separate, or defasciculate, allowing the axons to individually form connections.

10.15 Molecular Marking by Concentration Gradients of Netrins and Slits

Concentration gradients of diffusible netrins and slits, and their receptors, serve as molecular markers. The idea advanced in the 1950s and 1960s that concentration gradients of diffusible chemical markers guide growth cones to their targets gained support with the discovery of netrins in the mid 1990s. Netrins and slits function as diffusible chemoattractants and repulsants. Their receptors are all multidomain mosaic proteins containing several immunoglobulin-like domains along with several fibronectin domains in their extracellular region along with a transmembrane segment (Figure 10.8). Netrins and slits are bifunctional. They can act either as chemoattractants or as chemorepellants. The specific response to a netrin or slit molecule is dictated by the receptor to which it binds, and in particular it depends upon the properties of the cytoplasmic domain of the receptor. When binding to a DCC receptor, netrins provide a positive signal, but when bound to an UNC-5 receptor they impart a repulsive cue. While the extracellular domain of a netrin receptor determines the binding specificity,

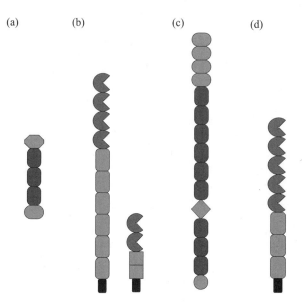

FIGURE 10.8. Netrins, slits, and their receptors: (a) Netrins consist of a laminin-like domain, 3 EGF domains, and a netrin C-terminus domain. (b) There are two netrin receptors. The first (left) is the DCC receptor contains 4 immunoglobulin (Ig) domains followed by 6 FNIII domains. The second receptor (right) is the UNC-5 receptor. It has 2 Ig domains plus 2 thrombospondin domains. (c) Slits consist of 4 leucine-rich repeats followed by 6 EGF domains, plus an agrin-laminin-perlecan-slit domain, plus 3 more EGF repeats and a cys knot. (d) The slit receptor Roundabout (Robo) contains 5 Ig domains followed by 3 FNIII repeats.

the cytoplasmic domain determines the cellular response elicited by the guidance cue. This aspect has been demonstrated in experiments where the cytoplasmic portion of a slit receptor was combined with the extracellular portion of a netrin receptor and vice versa: The functionality—attraction or repulsion—transferred with the cytoplasmic domain.

Both kinds of actions—attraction and repulsion—are required. An example of why this is so is provided by studies of *commissural axons*. These are axons that cross the midline of the developing central nervous system of bilaterally symmetric animals in order to contact neurons on the other side. Midline glial cells concurrently express netrins and slits. The netrins function as attractants and the slits as repellants. While navigating towards the midline the growth cones express the netrin DCC receptors more strongly than the slit Robo receptors, and the net effect is attractive. To avoid endlessly wandering back and forth across the midline due to netrin-mediated attraction, Robo expression is rapidly increased once the midline is crossed. Responsiveness to netrins is reduced, the net effect becomes repulsive, and the growth cone is sent on its way.

10.16 How Semaphorins, Scatter Factors, and Their Receptors Control Invasive Growth

Semaphorins and their plexin and neuropilin receptors provide guidance cues and work together with scatter factors and their receptors to control invasive growth. The semaphorins are a large family of secreted, GPI-anchored and transmembrane guidance proteins. Based on structural considerations they have been grouped into seven or eight classes. Classes 1 and 2 consist of invertebrate semaphorins, while the others classes contain vertebrate semaphorins. The defining characteristic of all semaphorins is the presence of a 500-amino acid residue long Sema domain in the extracellular region. Two families of receptors for semaphorins—plexins and neuropilins—have been found. The cytoplasmic domains of receptors belonging to these families are short and thus have limited signaling capabilities. The receptors trigger intracellular responses either through association with coreceptors, or by directly coupling to intracellular signaling proteins such as GTPases.

Scatter factor is also known as *hepatocyte growth factor* (HGF). It, like the ephrins to be discussed in the next section, is a polypeptide growth factor that binds to receptor tyrosine kinases, the main subject of the next chapter. Scatter factors are not only involved in neural growth cone guidance and growth, but also are central participants in invasive growth and branching morphogenesis. The term "branching morphogenesis" encompasses the growth, invasion, and proliferation of cells forming branched tubular structures that carry fluids in the vasculature, lungs, kidneys, and mammary glands. It is a type of invasive growth that encompasses organ development and regeneration, wound healing, axon guidance, and metastasis. The scatter factor receptor Met contains a Sema domain in its extracellular region, and next to that a short Met-related sequence (MRS), as do the semaphorins and plexins. Semophorin receptors and ligands interact with each other and with Met and other growth factor receptors and their ligands to transduce guidance and growth signals. For example, the semaphorin 4D receptor plexin B1 interacts with Met to control invasive growth.

10.17 Ephrins and Their Eph Receptors Mediate Contact-Dependent Repulsion

Eph receptors make up a large subfamily of tyrosine kinases that are involved in growth cone navigation, in directing migratory neurons during development, and in vascular cell assembly. These receptors and their ligands, the ephrins, are membrane-bound, and direct embryonic development by contact inhibition (repulsion). Eph receptors and ephrins are

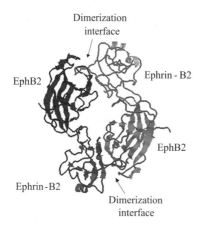

Dimerization
interface

EphB2

Ephrin - B2

EphB2

Ephrin - B2

Dimerization
interface

FIGURE 10.9. EphB receptor-ephrinB ligand tetramer: The figure was prepared using Protein Explorer with atomic coordinates deposited in the PDB under accession number 1kgy.

widely expressed by cells in the vertebrate embryo. There are at least 14 different Eph receptors and 8 different ephrin ligands. The ability of ephrins and their receptors to guide axons has been studied in the developing chick. Graded distributions of ephrins help guide axons from the chick retina to their visual cortex (tectum). They also assist in regulating axonal branchings and arborization in this retinotectal neural pathway.

Eph receptors fall into two groups—EphA and EphB. The extracellular portion of an EphB receptor contains an immunoglobulin-like domain in the N-terminal region followed by a cysteine-rich sequence followed by two fibronectin domains proximal to the plasma membrane. There is a single transmembrane chain and a cytoplasmic tyrosine phosphorylation domain. EphA receptors lack the transmembrane and cytoplasmic portions and instead are attached to the outer leaflet of the plasma membrane by a GPI anchor. The *ephrins*-fall into two groups, as well. The ephrin-A group contains GPI-anchored ephrins and the ephrin-B group consists of ephrins possessing a transmembrane segment and a cytoplasmic tail. In general, EphA receptors bind ephrinAs and EphB receptors bind ephrinBs.

A key step in activating the catalytic activity of the receptor is oligomerization. Like the scatter factor receptors just discussed, as well as other receptor tyrosine kinases, ligand binding brings together two or more receptor molecules (Figure 10.9). The formation of a dimer or a higher order oligomer brings the cytoplasmic domains into close physical proximity to one another and this triggers their autophosphorylation leading to recruitment of signaling molecules and the formation of a chain of signaling events ending at one or more control points. The chains of signaling events launched by binding of growth factors to receptor tyrosine kinases are a main focus of the next chapter.

References and Further Reading

Cell Migration

Laufenburger DA, and Horwitz AF [1996]. Cell migration: A physically integrated process. *Cell*, 84: 359–369.

Ridley AJ [2001]. Rho GTPases and cell migration. *J. Cell Sci.*, 114: 2713–2722.

Integrins

Boudreau NJ, and Jones PL [1999]. Extracellular matrix and integrin signaling: The shape of things to come. *Biochem. J.*, 339: 481–488.

Giancotti FG, and Ruoslahti E [1999]. Integrin signaling. *Science*, 285: 1028–1032.

Hogg, N, et al. [2003]. T-cell integrins: More than just sticking points. *J. Cell Sci.*, 116: 4695–4705.

Howe A, et al. [1998]. Integrin signaling and cell growth control. *Curr. Opin Cell Biol.*, 10: 220–231.

Hynes RO [2002]. Integrins: Bidirectional, allosteric signaling machines. *Cell*, 110: 673–687.

Shimaoka M, et al. [2003]. Structure of the α_L I domain and its complex with ICAM-1 reveals a shape shifting pathway for integrin regulation. *Cell*, 112: 99–111.

Takagi J, et al. [2002]. Global conformational rearrangements in integrin extracellular domains in outside-in and inside-out signaling. *Cell*, 110: 599–611.

Cadherins

Boggon TJ, et al. [2002]. C-cadherin ectodomain structure and implications for cell adhesion mechanisms. *Science*, 296: 1308–1313.

Gumbiner BM [2000]. Regulation of cadherin adhesive activity. *J. Cell Biol.*, 148: 399–403.

Yagi T, and Takeichi M [2000]. Cadherin superfamily genes: Functions, genomic organization, and neurologic diversity. *Genes Dev.*, 14: 1169–1180.

Yap AS, and Kovacs EM [2003]. Direct cadherin-activated cell signaling: A view from the plasma membrane. *J. Cell Sci.*, 160: 11–16.

Yap AS, Niessen CM, and Gumbiner BM [1998]. The juxtamembrane region of the cadherin cytoplasmic tail supports lateral clustering, adhesive strengthening, and interaction with p120[ctn]. *J. Cell Biol.*, 141: 779–789.

Immunoglobulin Family Cell Adhesion Molecules (IgCAMs)

Crossin KL, and Krushel LA [2000]. Cellular signaling by neural cell adhesion molecules of the immunoglobulin superfamily. *Developmental Dynamics*, 218: 260–279.

Fields RD, and Itoh K [1996]. Neural cell adhesion molecules in activity-dependent development and synaptic plasticity. *Trends Neurosci.*, 19: 473–480.

Klemke RL, et al. [1997]. Regulation of cell motility by mitogen-activated protein kinase *J. Cell Biol.*, 137: 481–492.

Rosales C, et al. [1995]. Signal transduction by cell adhesion receptors. *Biochim. Biophys. Acta*, 1242: 77–98.

Selectins

Kansas GS [1996]. Selectins and their ligands: Current concepts and controversies. *Blood*, 88: 3259–3287.

Vestweber D, and Blanks JE [1999]. Mechanisms that regulate the function of selectins and their ligands. *Physiol. Rev.*, 79: 181–213.

Adhesive Forces Probed by AFM and Optical Tweezers

Jiang G, et al. [2003]. Two-piconewton slip bond between fibronectin and the cytoskeleton depends on talin. *Nature*, 424: 334–337.

Kellermayer MS, et al. [1997]. Folding-unfolding transitions in single titin molecules characterized with laser tweezers. *Science*, 276: 1112–1116.

Oberhauser AF, et al. [1998]. The molecular elasticity of the extracellular matrix protein tenascin. *Nature*, 393: 181–185.

Rief M, et al. [1997]. Reversible unfolding of individual titin immunoglobulin domains by AFM. *Science*, 276: 1109–1112.

Firm Adhesion, Shear Stresses, and Rolling

Evans E [1998]. Energy landscapes of biomolecular adhesion and receptor anchoring at interfaces explored with dynamic force spectroscopy. *Faraday Discuss.*, 111: 1–16.

Genbacev OD, et al. [2003]. Trophoblast L-selectin-mediated adhesion at the maternal-fetal interface. *Science*, 299: 405–408.

Marshall BT, et al. [2003]. Direct observation of catch bonds involving cell-adhesion molecules. *Nature*, 423: 190–193.

Merkel R, et al. [1999]. Energy landscapes of receptor-ligand bonds explored with dynamic force microscope. *Nature*, 397: 50–53.

Smith MJ, Berg EL, and Lawrence MB [1999]. A direct comparison of selectin-mediated transient, adhesive events using high temporal resolution. *Biophys. J.*, 77: 3371–3383.

Axon Guidance and Growth

Araújo SJ, and Tear G [2003]. Axon guidance mechanisms and models: Lessons from invertebrates. *Nature Rev. Neurosci.*, 4: 910–922.

Goldberg JL [2003]. How does an axon grow? *Genes Dev.*, 17: 941–958.

Goodman CS, and Shatz CJ [1993]. Developmental mechanisms that generate precise patterns of neuronal connectivity. *Cell*, 72/*Neuron*, 10 (suppl.): 77–98.

Tessier-Lavigne M, and Goodman CS [1996]. The molecular biology of axon guidance. *Science*, 274: 1123–1133.

Netrins, Slits and Their Receptors

Bashaw GJ, and Goodman CS [1999]. Chimeric axon guidance receptors: The cytoplasmic domains of slit and netrin receptors specify attraction versus repulsion. *Cell*, 97: 917–926.

Brose K, et al. [1999]. Slit proteins bind Robo receptors and have an evolutionalily conserved role in repulsive axon guidance. *Cell*, 96: 795–806.

Goodhill GJ [2003]. A theoretical model of axon guidance by the Robo code. *Neural Comput.*, 15: 549–564.

Hong KS, et al. [1999]. A ligand-gated association between cytoplasmic domains of UNC5 and DCC family receptors converts netrin-induced cone attraction to repulsion. *Cell*, 97: 927–941.

Kidd T, Bland KS, and Goodman CS [1999]. Slit is the midline repellent for the Robo receptor in *Drosophila*. *Cell*, 96: 785–794.

Ming GL, et al. [2002]. Adaptation in the chemotactic guidance of nerve growth cones. *Nature*, 417: 411–418.

Semaphorins, Scatter Factors, Their Receptors, and Invasive Growth

Giordano S, et al. [2002]. The semaphorin 4D receptor controls invasive growth by coupling with Met. *Nature Cell Biol.*, 4: 720–724.

Goshima Y [2002]. Semaphorins as signals for cell repulsion and invasion. *J. Clin. Investig.*, 109: 993–998.

Maina F, and Klein R [1999]. Hepatocyte growth factor, a versatile signal for developing neurons. *Nature Neurosci.*, 2: 213–217.

Tamagnone L, and Comoglio PM [2000]. Signaling by semaphorin receptors: Cell guidance and beyond. *Trends Cell Biol.*, 10: 377–383.

Eph Receptors and Ephrins

Holder N, and Klein R [1999]. Eph receptors and ephrins: Effectors of morphogenesis. *Development*, 126: 2033–2044.

Kullander K, and Klein R [2002]. Mechanisms and functions of Eph and ephrin signaling. *Nature Rev. Mol. Cell Biol.*, 3: 475–486.

Mellitzer G, Xu QL, and Wilkinson DG [1999]. Eph receptors and ephrins restrict cell intermingling and communication. *Nature*, 400: 77–81.

Orioli D, and Klein R [1997]. The Eph receptor family: Axonal guidance by contact repulsion. *Trends Genet.*, 13: 354–359.

Wilkinson DG [2001]. Multiple roles of Eph receptors and ephrins in neural development. *Nature Rev. Neurosci.*, 2: 155–164.

Problems

10.1 Bond dissociation strengths are not constant quantities but instead are sensitive to the presence or absence of external forces. This dependence is captured by the expression

$$k_{\text{off}}(F) = k_{\text{off}}^0 \exp\left[-E(F)/k_{\text{B}}T\right]. \tag{10.1}$$

In this equation, the usual dependence of the rate on a constant barrier height E has been replaced by a force-dependent barrier $E(F)$. The quantity F represents the applied force on the bond formed during rolling. The applied force amplifies the off-(dissociation) rate for the receptor-ligand complex by reducing the height of the transition barrier. This force-driven reduction in barrier heights by an external force is depicted in the figure presented below. The amplification in dissociation rate due to the presence of an applied force is found by inserting the simplified expression for $E(F)$ into the off-rate equation. This gives

$$k_{off}(F) = k_{off}^0 \exp\left[Fx_{av}/k_BT\right], \tag{10.2}$$

where $k_{off}(0)$ is the unstressed rate; that is,

$$k_{off}(0) = k_{off}^0 \exp\left[-E/k_BT\right] \tag{10.3}$$

is the conventional value for the off-rate in the absence of an applied force. In the above, the angle-dependent expression has been replaced by the quantity x_{av} representing a thermally averaged distance between receptor and ligand interfaces over which the bond weakens but does not yet break The expression for the amplification in the off-rate is known as *Bell's equation*. According to Bell's equation the off-rate rises exponentially with increasing force. In situations where the barriers are sharp, the principal consequence is the linear lowering of the barrier heights as a function of distance, with little change in shape or location of the peaks and valleys. The influences of an applied force on the energy landscapes are illustrated in the figure for the cases where there are two barriers.

Dissociation rates are influenced not only by applied forces but also by how rapidly they are applied. A bond that will appear strong if the forces are rapidly ramped up to a maximum value will appear far weaker and more easily broken if the same external forces are applied slowly. One may define an *unbinding force* as that force required to just break a bond. In experiments performed in the laboratory, con-

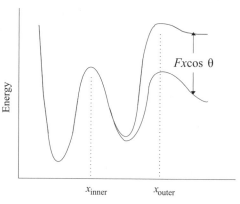

FIGURE 1 FOR PROBLEM 10.1. Lowering of the activation barrier by an applied force: The applied force F acts along a direction oriented at an angle θ with respect to the reaction coordinate x, reducing the energy E by an amount $F \cdot x \cdot \cos\theta$. Because of the dependence on the reaction coordinate x the outer barrier is driven down while the inner barrier is barely affected. If the forces are strong enough the outer barrier can be reduced below the height of the inner barrier, thereby revealing information about the latter's characteristics.

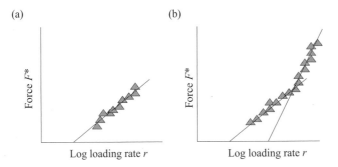

(a)   (b)

FIGURE 2 FOR PROBLEM 10.1. Receptor-ligand bond dissociation data: The most probable unbinding forces are plotted against the logarithm of the loading rate. (a) The data fall along a single line. (b) There is a sharp break in the data, producing two linear regimes.

stant loading forces are applied to the bonds; that is, forces F of the form $F = rt$, where r is the loading rate and t is the time. Under these conditions the *most probable unbinding force* F^* is related to the rate and to the characteristic length by

$$F^* = \frac{k_B T}{x} \ln \frac{r}{k_{off}(0) \dfrac{k_B T}{x}}. \tag{10.4}$$

When the most probable unbinding forces of receptor-ligand pairs are plotted against the logarithm of the loading rate the data typically fall along a straight line. Two commonly encountered situations are depicted below. In (a), the data are distributed about a single straight line. What quantity does the slope of this line represent, and how does it relate to an energy landscape? In (b), there is a sharp break in the straight line so that there are two slopes. Interpret these results within an energy landscape picture, i.e., what does the energy landscape look like? Recall that typical thermal energies are on the order of 0.6 kcal/mol. For a distance x of 0.5 Å what is the corresponding force, expressed in Newtons?

References on the Theory of Slip and Catch Bonds

Bell GI [1978]. Models for specific adhesion of cells to cells. *Science*, 200: 618–627.
Dembo M, et al. [1988]. The reaction-limited kinetics of membrane-to-surface adhesion and detachment. *Proc. R. Soc. Lond. B*, 234: 55–83.

11
Signaling in the Endocrine System

The endocrine system consists of a number of glands containing specialized secreting cells that release signaling molecules into the bloodstream, other bodily fluids, and extracellular spaces. Glands of the endocrine system include the adrenal gland, the hypothalamus, and the parathyroid, pineal, pituitary, and thyroid glands. Hormone-secreting cells are found in many locations in the body. The thymus produces a variety of hormones that regulate lymphocyte growth, maturation and homeostasis. The pancreas contains groups of hormone-secreting cells organized into pancreatic islets (the *Islets of Langerhans*) that secrete insulin, glucagon, and somatostatin. Hormone-secreting epithelial cells are strategically situated in the gastrointestinal tract, gonads, kidneys and placenta, where they direct the mitogenic activities of cells and tissues that wear down and require constant replenishment, and angiogenetic activities that produce new blood vessels.

Cells of the endocrine system produce several different kinds of hormones—peptide and polypeptide hormones, steroids synthesized from cholesterol, hormones derived from amino acids, and hormones made from fatty acids. Some hormones are lipophilic while others are water soluble. Lipophilic hormones such as steroids, retinoids, and thyroids are small, relatively long-lived hormones able to pass through the plasma membrane and enter the cell where they bind nuclear receptors. Water soluble hormones are fairly short lived and bind to receptors embedded in the plasma membrane of the target cells. Two kinds of transmembrane receptors bind nonlipophilic hormones: Polypeptide hormones that promote growth and wound-healing are bound by receptor tyrosine kinases; all others are bound by G protein-coupled receptors, which will be discussed in the next chapter.

Polypeptide hormones are small compact molecules, ranging in size from 6 to 45 kDa. Epidermal growth factor (EGF) and nerve growth factor (NGF) were the first polypeptide growth factors to be discovered. These and other growth factors bind to receptor tyrosine kinases expressed by the appropriate recipient cells. Receptor tyrosine kinases are single chain proteins that pass through the plasma membrane once. They, along with nonreceptor tyrosine kinases, are exclusively found in metazoans. Several different classes of proteins help transduce polypeptide hormone signals into the cell. These elements include, besides receptor tyrosine kinases and the aforementioned nonreceptor tyrosine kinases, a variety of highly modular adapters, and several families of GTPases. Each of these classes of signal transducers will be examined in this chapter.

11.1 Five Modes of Cell-to-Cell Signaling

In the last chapter, signaling between cells was characterized as being either contact-mediated or short range or long range. This catalog of signaling modes can be generalized to encompass cytokine signals in the immune system, long range hormone signals in the endocrine system, and neurotransmitters and neuromodulators in the nervous system. The expanded ensemble of signaling modes contains five entries—endocrine, paracrine, autocrine, juxtracrine, and synaptic. In endocrine signaling, the sending and receiving cells can be far apart. As noted in the introductory remarks, the signals are conveyed from sending glands to distant points in the body by bodily fluids.

Paracrine signaling refers to signaling in which molecules are secreted directly into the extracellular spaces by an originating cell, and these molecules travel no more than a few cell diameters to reach a target cell. The signaling molecules are often immobilized by elements of the extracellular matrix and bind to receptors on the surface of their cellular targets. This mode of signaling is utilized by hormone-releasing epithelial cells, by cytokine-releasing endothelial cells and fibroblasts, by chemotropic factor-releasing cells that direct growth cones and migrating cells, and by morphogen-releasing cells in organizing centers that direct cell fate during development (to be discussed in Chapter 13).

Some hormones that are secreted act back on the cells releasing them. In this *autocrine mode of signaling*, positive feedback serves to amplify the initial weak signals. This mode is encountered in the immune system (see, for example, signaling by IL-2 shown in Figure 9.1) and also in the endocrine system as part of growth factor signaling. It is made use of during development, and if not properly regulated leads to uncontrolled growth and proliferation in a variety of cancers.

Cell-to-cell (juxtracrine) signaling is frequently conveyed by direct contact between a receptor on one cell and a cell surface-bound ligand, or counterreceptor, on an adjacent cell. The nondiffusible signaling molecules participating in juxtacrine signaling may pass through the plasma membranes or may be tethered to the outer leaflet of the plasma membranes by GPI anchors. This mode of signaling includes situations where cell surface receptors bind ligands that are components of the extracellular matrix.

Synapses are junctions formed to promote sustained signaling between cells. Signaling through chemical synapses is the preeminent form of signaling between nerve cells. This *synaptic signaling* involves the diffusion of signaling molecules (neurotransmitters) across a synaptic cleft between the membranes of a pair of pre- and postsynaptic nerve cells. The membrane and submembraneous regions of synapses are specialized structures highly enriched in many different kinds of signaling molecules. Like synaptic signaling between neurons, immune system T cells use synaptic signaling when contacting antigen-presenting cells.

11.2 Role of Growth Factors in Angiogenesis

Several families of growth factors coordinate and control the formation of new blood vessels during angiogenesis. Whenever new tissue is made additional blood vessels must be created to supply oxygen and nutrients and remove waste products. The process of vascular development takes place in several stages. In the earlier stages, called *vasculogenesis*, precursor vascular endothelial cells migrate, differentiate, proliferate and assemble into an initial set of vascular connections of uniform size. In the later stages, termed *angiogenesis*, the initial latticework of tubules is refined through further differentiation, sprouting and branching to form a mature vascular system. The fully developed network contains large arteries that branch into progressively smaller blood vessels terminating in capillaries, and it contains a return system of progressively larger venous structures.

The endothelial and smooth muscle cells that form the lining and sheathing of blood vessels coordinate their activities by sending and receiving a variety of chemical signals. Some of the signaling molecules are vascular endothelial cell-specific; others such as platelet-derived growth factor (PDGF) and basic fibroblast growth factor (bFGF) are not. Members of the vascular endothelial growth factor (VEGF) family are prominent among the vascular specific growth factors. They are required for vasculogenesis and also play a role in angiogenesis. A second family of growth factors, the *angiopoietins*, binds to endothelial cell line-specific Tie receptors. They work together with endothelial-specific ephrins and with the VEGFs to coordinate and control angiogenesis. The VEGFs and angiopoietins operate through a paracrine mechanism to signal and recruit nearby cells. The ephrins remain attached to the plasma membrane of their originating cell and signal bidirectionally through a juxtacrine mechanism.

As shown in Figure 11.1, the VEGF, Tie, and Eph receptors involved in vascular development and repair are mosaic proteins. Their extracellular regions are composed of multiple domains arranged in a linear fashion. They differ from the receptors discussed in the last chapter by the presence of a tyrosine kinase domain in their cytoplasmic segment. The angiopoietins and ephrins promote the remodeling and branching, vascular maturation, and the attachment to surrounding support cells and extracellular matrix. For example, Ang1 promotes vascular maturation and stabilization while Ang2

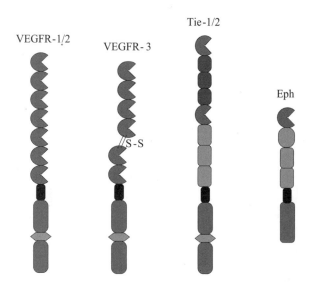

FIGURE 11.1. Angiogenesis receptors: The VEGFR-1/2 proteins contain 7 Ig-like domains followed by a transmembrane segment and a cytoplasmic kinase domain split by an insert. The VEGFR-3 chain differs slightly from VEGFR-1/2 in that the set of tandem Ig domains is shortened by one and contains a disulfide bridge. Tie receptors 1 and 2 contain an Ig-like domain followed by 3 EGF-like repeats and then another Ig domain followed by 3 FNIII repeats, a transmembrane segment, and the split kinase domain. Eph receptors contain an Ig domain, a cysteine-rich region, plus 2 FNIII repeats, a transmembrane segment, and a cytoplasmic kinase domain.

maintains the vascular system in a plastic state. The ephrins were discussed earlier in connection with growth cone navigation. The vascular-specific Ephrin-B2 ligand and EphB4 receptor function in a roughly analogous manner in angiogenesis. They are differentially expressed in endothelial cells. The Ephrin B2 ligand functions as an arterial marker, and the EphB4 receptor operates as a venous marker. These signaling proteins help delineate the boundary between blood vessels that become arteries and those that become veins.

11.3 Role of EGF Family in Wound Healing

Growth factors of the EGF family coordinate and control wound healing and the replacement of cells in tissues that undergo rapid turnover. Wound healing is a complex process. Even in the case of a simple skin cut it requires the participation of agents of the nervous system to generate a pain signal, and elements of the immune system to generate an inflammatory and immune response. New tissue must be grown to replace the damaged material. The old damaged vasculature must be taken down and new vasculature installed, and the underlying connective tissue must be remodeled and restored.

In a skin cut, gloss keratinocytes (skin cells) migrate into the region of injury, and upon arrival proliferate and mature. Signaling by members of the EGF family of ligands and their receptors is essential for the proper activities of keratinocytes during wound healing. Several EGF family ligands are involved in repairing a skin wound. Among these are TGFα, HB-EGF, AR, and betacellulin. These ligands are synthesized by keratinocytes, along with their EGF receptors and supply proliferation and migratory signals, through an autocrine mechanism. In response to these signals the keratinocytes increase their production of integrins. Binding by these integrins to the ECM mediates substratum adherence by the keratinocytes and triggers the expression of collagesase-1. This enzyme degrades the ECM, a step required for remodeling and repair of damaged tissue. Continued autocrine signals maintain the expression of collagenase-1 during migration and repair.

Besides the skin, wound repair occurs in the gastronintestinal tract, in muscle tissue following injury, and at sites of chronic inflammation, to name just a few common examples. Primary sources of epithelial growth factors include the submaxillary salivary gland, the duodenal Brunner's gland, epithelial cells in the mammary gland, and the kidneys. Growth factors are released into bodily fluids—saliva, serum, milk, and urine—and travel to sites of injury and growth. Epithelial cells at many places in the body are continuously renewed and growth factors are released locally near these sites as well. In these situations the growth factors operate in autocrine and paracrine manners rather than through an endocrine mechanism.

11.4 Neurotrophins Control Neuron Growth, Differentiation, and Survival

Programmed cell death is an important process in the developing nervous system. It is used to remove cells that are extraneous, cells that were transiently produced to help in development but are no longer needed, and cells that failed to wire up properly to other cells. Neurotrophins are trophic factors, chemicals that stimulate growth and development. Neurotrophins such as NGF regulate the decision to survive or die, promote axonal and dendritic outgrowth, branching and remodeling, and help control synapse formation and plasticity. Other examples of trophic factors are ciliary neurotrophic factor (CNTF), glial-derived neurotrophic factor (GDNF), PDGF, and bFGF.

The neurotrophin family of ligands and receptors contains four ligands and two receptors. The four ligands are NGF, brain derived neurotrophic factor (BDNF), neurotrophic factor 4 (NT4), and NT3. These ligands bind to two kinds of neurotrophin receptors, namely, $p75^{NTR}$ receptors and Trk receptors. As shown in Figure 11.2, the $p75^{NTR}$ receptor contains a cyto-

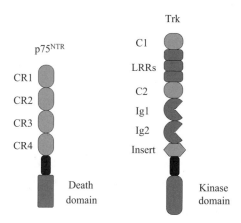

FIGURE 11.2. Neurotrophin receptors: The p75NTR receptor has four cysteine-rich repeats (CR1–CR4), a transmembrane segment, and a cytoplasic death domain. The Trk receptors have a cysteine segment (C1), 3 leucine-rich repeats (LRRs), a second cysteine segment (C2), 2 immunoglobulin-like domains (Ig1, Ig2), an insert trans-membrane segment, and a cytoplasmic kinase domain. All of the neurotrophins can bind p75^{NTR}. NGF binds the TrkA receptor; BDNF and NT4 bind the TrkB receptor, and NT3 binds the TrkC receptor.

plasmic death domain, while the Trk receptors possess a cytoplasmic tyro-sine kinase domain. The two receptors operate using different sets of adapter proteins to convey a variety of messages into the cell as indicated in Figure 11.3. The Trk receptors utilize Shc and Grb2 adapters, while p75 employs TRAFs and a set of four or more adapters (Figure 11.3b) to convey death and arrest messages. In general, p75^{NTR} receptors convey apoptotic messages and Trk receptors convey survival messages, but as illustrated in Figure 11.3 the full picture is far richer. These and other growth factor receptors work together along with each other and with integrins and cadherins to convey messages to most or all control points in the cell. These include regulatory sites for gene expression (to promote growth and dif-ferentiation), sites of focal contact between cells and ECM (to control movement, adhesion, cell shape, and outgrowth), and mitochondrial control sites for apoptosis (to regulate death versus survival decisions).

11.5 Role of Receptor Tyrosine Kinases in Signal Transduction

Receptor tyrosine kinases transduce signals from polypeptide growth factors into the cell. Except for receptors such as p75, polypeptide growth factors bind to members of the receptor tyrosine kinases (RTKs), single-pass transmembrane receptors possessing an intrinsic tyrosine kinase activity. These receptors contain three functional regions: an N-terminal

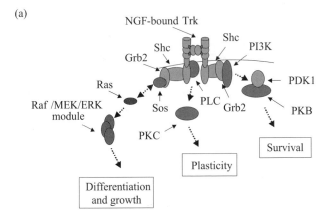

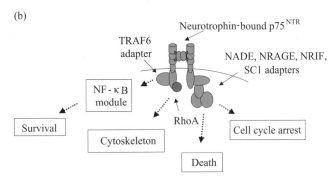

FIGURE 11.3. Ligand-bound Trk and p75 neurotrophin receptors, and associated adapters: (a) NGF-bound Trk receptor. Second messengers and associated signaling intermediates involved in activating the PKB and PKC have been omitted to keep the figure from becoming too cluttered. (b) Neurotrophin-bound p75 receptor. Abbreviations: p75^{NTR}-associated cell death executor (NADE), neurotrophin receptor-interacting melanoma-associated antigen (MAGE) homolog (NRAGE), neurotrophin receptor-interacting factor (NRIF), Schwann cell factor-1 (SC1).

extracellular region that contains one or more ligand-binding sites; a transmembrane segment; and a C-terminal cytosplasmic region that contains a catalytic domain and several phosphorylation sites.

The RTKs that bind polypeptide growth factors are highly modular linear arrays of domains. Among the domains present in the extracellular regions of these glycoproteins are Ig domains, fibronectin type III domains, cysteine-rich domains and the EGF ligand-binding domains. A juxtamembrane region is situated just inside the transmembrane helix region and, as mentioned above, there is a tyrosine kinase domain near the C-terminal region. Docking sites are opened when (auto)phosphorylation of tyrosine residues in the activation loop of the kinase domain turns on that domain's catalytic activity resulting in a subsequent phosphorylation of tyrosine residues. When this happens, proteins containing phosphotyrosine recognition modules are able to dock at these sites.

Like other single pass transmembrane receptors, ligand binding in the extracellular N-terminal region does not perturb the electrostatic environment enough to stabilize a conformation sufficiently different in its C-terminal cytoplasmic region to serve as a signal. Instead ligand binding promotes the formation of stable receptor dimers and oligomers. It is the bringing together of two or more cytoplasmic domains that initiates signaling inside the cell. In the case of RTKs, the close proximity of the two cytoplasmic kinase domains and accompanying phosplorylation sites enable cross or autophosphorylation of tyrosine residues.

The transmission of a signal into the cell interior by a receptor tyrosine kinase following ligand binding occurs in two stages. The first step is the autophosphorylation, or cross phosphorylation, of tyrosine residues in the activation loop of the catalytic domain. The bringing together of the cytoplasmic portions of the RTK triggers this step with the result that the catalytic activity of the kinase domain is turned on. The kinase domain then catalyzes the phosphorylation of one or more tyrosine residues in the cytoplasmic region outside of the catalytic domain. This second step makes available docking sites for signal proteins. Key to this second function is the presence of protein-protein recognition modules, compact domains that recognize short peptide sequences containing phosphorylated tyrosine, serine, and threonine residues.

Receptor dimers and oligomers can be formed in response to ligand binding in more than one way. Ligand-mediated dimerization triggers human growth hormone (hGH) signaling. As discussed in Chapter 9, a single ligand first binds to one receptor and then attaches to the second receptor to form a bridge that holds together a 1:2 ligand-receptor complex. A different strategy is observed in EGF-induced dimerization. In this case, the extracellular portions of the two receptors form the bridge resulting in a 2:2 complex. In more detail, EGF-EGFR binding triggers conformational changes that expose a loop on each receptor molecule. These loops bind to each other to form the bridge. As noted earlier, the ErbB2 receptor does not require a ligand for its activation. This happens because the bridging loop on this receptor is constitutively in the open for bridging conformation. The formation of a bridge by a pair of loops in contact with a ligand dimer is depicted in Figure 11.3.

11.6 Phosphoprotein Recognition Modules Utilized Widely in Signaling Pathways

Signaling proteins diffuse from one location to another, are activated and turned off by binding and posttranslational modifications, and are recruited, assembled, and disassembled into signaling complexes at cellular control points. The arrangements of these events into signaling pathways so that a sequence of signaling steps can occur in the correct order at the right time in the right place is made possible through the use of modular signaling domains. Some of these domains supply a sort of glue that enables one protein to bind another, and for that operation to be followed by yet another binding operation leading to the recruitment of proteins and formation of signaling complexes. Other domains are specifically designed to recognize the presence of posttranslational modifications such as the addition of phosphoryl groups. To see the utility of this kind of modular domain, consider the receptor tyrosine kinases just discussed. The result of ligand binding is the presence of one or more phosphorylated tyrosine residues near the cytoplasmic C-terminus of the receptor. If this event cannot be recognized and responded to, signaling ends.

Src homology-2 (SH2) and phosphotyrosine binding (PTB) domains recognize short peptide sequences containing phosphotyrosine residues. The SH2 domain is the prototypic recognition module. It consists of approximately 100 amino acid residues, and is found on proteins that assemble and disassemble in the vicinity of the plasma membrane in response to auto- and cross-phosphorylation of tyrosine residues in the cytoplasmic portion of transmembrane receptor molecules. It is the first discovered and largest family of modules that recognize phosphorylated tyrosines. SH2 domains recognize phosphorylated tyrosines and a specific flanking sequence of three to six amino acid residues located immediately C-terminal to pY. Phosphotyrosine binding (PTB) domains also recognize short peptide sequences containing phosphotyrosine, but, in contrast to SH2 domains, amino acid residues N-terminal and not C-terminal to pY belong to the recognition sequence. PTB domains recognize turn-loop structures and, in particular, hydrophobic amino acid residues five to eight residues located N-terminal to pY.

14-3-3 proteins are small, 28 to 30 kDa in size. There are at least seven isoforms of the mammalian 14-3-3 proteins; they are abundant and are expressed in all eukaryotic cells. Unlike most of the other domains discussed in this section, 14-3-3 domains are not embedded in larger proteins but exist as independent units that form homodimers. These proteins bind with high affinity to peptide sequences containing phosphoserine residues followed by a proline two positions towards the C-terminal. As dimers they are able to bind to signaling molecules such as Raf-1 containing tandem repeats of phosphoserine motifs. The 14-3-3 proteins localize their binding partners within the cytoplasmic compartment and keep them sequestered from the nucleus and membranes. These proteins operate in the growth and apoptosis pathways and in cell cycle control.

Forkhead-associated (FHA) domains are conserved sequences of 55 to 75 amino acid residues. Whereas SH2 and PTB domains recognize sequences containing phosphorylated tryosines, FHA domains can recognize peptide sequences containing phosphorylated threonines, phosphorylated serines, and phosphorylated tyrosines with a pronounced affinity for phosphothreonines. They are found in kinases, phosphatases, RNA-binding proteins, and especially in nuclear proteins such as transcription factors and proteins involved in detecting and repairing DNA damage.

11.7 Modules that Recognize Proline-Rich Sequences Utilized Widely in Signaling Pathways

Proline has several properties that favor its utilization in signaling pathways. First it exhibits a rather limited range of conformations since it alone of the amino acids has a side chain that closes back onto the backbone amino nitrogen atom. This restriction in conformation extends to the residue immediately preceding it and the result is a preference for a particular secondary structure called a *polyproline* (PP II) *helix*. This helix has an extended structure with three residues per turn, and the prolines form a continuous hydrophobic strip about the helix surface and also expose several hydrogen binding sites. All in all, the result is an amphipathic structure sometimes referred to as a "sticky arm." At the same time the binding is relatively weak and complexes so formed are easy to disassemble. Small changes in binding sequence or by phosphorylation produce large changes in the dissociation constant.

SH3, WW, EVH1, and GYF domains all recognize proline-rich sequences, but each recognizes a slightly different sequence or group of sequences. All recognize some variant of the sequence PxxP that forms a left handed PP II helix. The WW domain contains a pair of tryptophans (W) spaced some 20 to 22 amino acids apart and a block of two to four aromatic amino acids situated in between the two tryptophans, hence the name WW domain. There are several kinds of WW domains, each specific for a certain kind of proline-rich sequence. Group IV WW domains recognize sequences containing phosphorylated serine/threonine residues.

11.8 Protein–Protein Interaction Domains Utilized Widely in Signaling Pathways

Regulator-of-G protein-signaling (RGS) proteins contain a domain approximately 120 amino acids in length that acts as a G recognition module. It functions as a GAP to accelerate the GTPase activities of G_α subunits of heterotrimeric G proteins coupled to GPCRs. This module has been found in proteins containing other signaling modules such as PDZ domains. The next entry in Table 11.1 following RGS domain is the sterile alpha motif (SAM) domain. This domain forms homo- and hetero-oligomers with other domains and is encountered in a variety of signaling proteins. For example, it is found in the C-terminus of all Eph receptors.

TABLE 11.1. Protein interaction and phosphoprotein recognition modules: Phosphoprotein recognition motifs are denoted using p-Tyr, p-Ser, and p-Thr notation. Proline (P)-based protein interaction motifs are described using single letter codes.

Name of module	Abbreviation	Recognition motif(s)
Src homology-2	SH2	Phos-Tyr
Phosphotyrosine binding	PTB	Phos-Tyr, NPxY
Src homology-3	SH3	PxxP
WW	WW	PPxY; PPLP; P-R; Phos-Ser
Enabled/vasodilator-stimulated phosphoprotein homology-1	EVH1	FPPPP
Gly-Tyr-Phe	GYF	PPPPGHR
14-3-3	14-3-3	Phos-Ser
Forkhead-associated	FHA	Phos-Thr
Regulator-of-G protein signaling	RGS	G_α
Sterile α motif	SAM	SAM
Death domain	DD	DD
PSD-95, DLG, ZO-1	PDZ	PDZ, C-terminal motifs

The next-to-last entry in the table is the death domain (DD). Proteins containing this domain convey death (apoptosis) instructions. The p75^{NTR} receptor shown in Figure 11.1 contains a cytoplasmic death domain (DD). This domain enables cytoplasmic proteins bearing similar death domains to attach via a DD-to-DD linkage. The DD is not the only module of this sort. Instead there is an entire family of six (or seven)-helix bundle death domains that includes the DDs, death effector domains (DEDs), caspase recruitment domains (CARDs), and Pyrin domains (PYDs). Death domains will be discussed further in the chapter on apoptosis.

The final entry in the table is for the PDZ domain, named for the first three proteins found with this domain—the postsynaptic density protein of 95 kDa (PSD-95) found in postsynaptic terminals of neurons, the Discs Large (DLG) protein of *Drosophila*, and the zona occludens 1 (ZO-1) protein found in epithelial cells. These domains promote protein-protein interactions through PDZ-PDZ binding and also recognize specific sequences in the C-terminals of target proteins. Proteins containing PDZ domains are especially prominent in synaptic terminals where they combine with other domains to form scaffolding proteins. Members of the Shank family of proteins serve as examples of this theme. As shown in Figure 11.4, Shank proteins contain five domains. Their SH3, PDZ, and proline-rich regions either directly or indirectly bind to the three main kinds of glutamate receptors found in postsynaptic terminals, providing an anchorage for their assembly and clustering. This assembly is anchored to the actin cytoskeleton by intermediates that bind the ankrin repeats in the N-terminus and the proline-rich region in the middle. Finally, the SAM

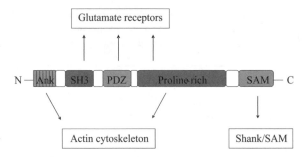

FIGURE 11.4. Domain organization of a Shank protein: From the N-terminus to the C-terminus the Shank protein has a set of ankrin repeats (Ank), an SH3 domain, a PDZ domain, proline-rich region, and a SAM domain.

TABLE 11.2. Families of nonreceptor tyrosine kinases.

Family	Family members
Abl	Abl, Ars
Csk	Csk, Ctk
FAK	CAKβ, FAK
Fes	Fps (Fes), Fer
Jak	JAK1-3, Tyk2
Src	Blk, Fgr, Fyn, Hck, Lck, Lyn, Src, Yes, Yrk
Syk	Syk, ZAP70
Tec	Bmx, Btk, Itk/Tsk, Tec, Txk/Rlk

domains near the C-terminus promote the linking of one Shank protein to another through SAM-SAM binding.

11.9 Non-RTKs Central in Metazoan Signaling Processes and Appear in Many Pathways

Metazoans make extensive use of tyrosine phosphorylation in their signal transduction pathways and in addition to the receptor tyrosine kinases there are several families of nonreceptor tyrosine kinases (NRTKs). These signaling protein tyrosines range in size from 50 to 150kDa. They are the archtypical examples of modular proteins. In addition to possessing catalytic domains, the NRTKs possess in various combinations phospholipid, phosphotyrosine, and proline-rich sequence recognition domains. Like an RTK, the signaling (catalytic) activity of an NRTK is triggered by tyrosine phosphorylation in the activation loop of the kinase domain. This can happen through autophosphorylation or through phosphorylation by another kinase.

The NRTKs can be grouped into eight families (Table 11.2) according to their catalytic domain sequences, domain composition, and posttranslational modifications. The Src family of NRTKs is the largest with nine members while several other families have as few as two members. Some NRTKs tether to the cytoplasmic face of the plasma membrane and are covalently modified to permit attachment. Most are cytoplasmic proteins, but Abl family members possess a nuclear localization signal (NLS) and are found both in the cytoplasm and in the nucleus.

11.10 Src Is a Representative NRTK

The Src family is representative of the NRTKs and the discussion will mostly focus on this family and on focal adhesion kinase (next section). The organization of the catalytic domain of Src follows the general pattern discussed above for the RTKs. The catalytic domain is bilobed with an active site cleft and an activation loop containing a critical tyrosine residue. Phosphorylation of this tyrosine activates the kinase domain. The overall structure of Src is as follows. The N-terminal region contains a myristylation site and sometimes a palmitoylation site. An SH3 domain, an SH2 domain, the catalytic domain, and finally the COOH terminal region containing a second critical tyrosine residue follow initial segment. Phosphorylation of the tyrosine (Tyr527) in the COOH tail by regulatory proteins such as Csk deactivates the protein kinase.

The SH2 and SH3 domains are located on the backside of the catalytic domain of Src and do not impede activation and catalysis (Figure 11.5). The SH2 domain of Src, as well as the SH2 domains of Fyn, Lck, and Fgr, select peptide sequences of the form pYEEI; but they also bind the sequence pYAEI of FAK and other similar sequences in other proteins. The SH3

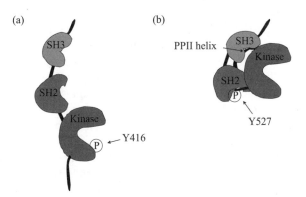

FIGURE 11.5. The Src protein in open and closed conformations: (a) The Src protein is in an open conformation in which a crucial tyrosine residue in the kinase domain is exposed and phosphporylated. (b) The site of tyrosine phosphorylation in the kinase domain is blocked and the binding surfaces of the SH2 and SH3 domains are sequestered.

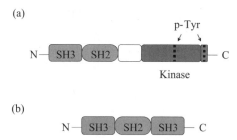

FIGURE 11.6. SH2- and SH3-bearing proteins: (a) Organization of c-Src nonreceptor tyrosine kinase. (b) Organization of the Grb2 adapter protein. Proteins such as Grb2 that lack domains that carry out enzymatic activities are referred to as *adapters*. If they are very large and can bind multiple proteins they are called *scaffolds*. Sites where the protein can be phosphorylated on tyrosine residues are labeled by the abbreviation p-Tyr.

domain of Src binds a number of PXXP-like sequences forming left handed PP II helices. Although these two domains are located on the backside of the catalytic domain they cooperate with one another to inhibit Src catalytic activity. In the simplest model of how this occurs Src has two states, an inactive closed conformation and an active open conformation (Figure 11.6). In its closed conformation the SH2 domain binds to pTyr527 in the COOH tail and the SH3 domain binds to a PP II helix in a portion of the linker between SH2 domain and the catalytic domain. These binding events are sufficient to shift the catalytic domain into a conformation where it cannot bind ATP and release ADP, and cannot bind its protein substrate. Thus phosphorylation at Tyr527 by Csk turns off the kinase activity of Src through cooperative autoinhibitory actions of Src's own SH2 and SH3 domains. The inhibitory interactions are relatively weak, and Src is activated when phosphotyrosine- and polyproline-containing sequences in other signal proteins favorably compete for Src SH2 and SH3 binding. Primary activators of Src include PDGF and other RTKs such as EGF, FGF, and NGF. These are not the only transmembrane receptors that signal through Src family members. Other signal pathways linked to these NRTKs include integrins, G protein-coupled receptors, and immune system receptors.

Proliferation is a highly regulated process. For proliferation to occur the appropriate adhesive and growth factor signals must be sent and received. Integrin receptors and receptor tyrosine kinases are localized in the plasma membrane close to one another and together work in signaling growth. The adhesive signals confirm that the cell remains in adhesive contact with the extracellular matrix, and thus growth is permitted. *Paxillin*, a 68-kDa protein associated with focal adhesions, contains multiple binding sites and serves as a platform for gathering adhesive signals relayed through integrin receptors and growth factor signals sent by receptor tyrosine kinases. Several NRTKs participate in the relay of messages from the transmembrane receptors to paxillin; prominent among these is the focal adhesion kinase (FAK).

11.11 Roles of Focal Adhesion Kinase Family of NRTKs

The FAK family of NRTKs promotes the assembly of signaling complexes at focal adhesions, and regulates motility and growth factor signaling. Focal adhesions are points of contact and adhesion between the cell and its supporting membranes. Focal adhesions are control points where growth and adhesion signals are integrated together and coordinated across multiple points of ECM-to-cell surface contact to govern the overall growth and movement of the cell. These control points not only regulate the assembly and disassembly of the focal adhesions but also convey signals that control cellular growth, proliferation, differentiation, and survival.

Extracellular matrix proteins, transmembrane proteins, and the actin cytoskeleton proteins participate in the adhesive contacts. Integrins and growth factor receptor co-localize at focal adhesions. The integrins bind to ECM proteins such as laminin and the growth factor receptors bind to growth factor ligands. In response to ligand binding by these receptors, a number of nonreceptor tyrosine kinases and adaptor/scaffold proteins are recruited to the plasma membrane. Among the nonreceptor tyrosine kinases recruited are Src, its negative regulator Csk, and focal adhesion kinase. These kinases along with paxillin, a key adapter, link the integrin and growth factor receptor-signaling to the actin cytoskeleton.

Paxillin is a fairly small protein; as mentioned previously it is only 68 kDa in mass, but it contains a large number of binding sites. Its structure is shown in Figure 11.7a. It possesses two tyrosine phosphorylation sites that are targeted by the nonreceptor tyrosine kinases such as Src, Csk and FAK, and bound by SH2 domain-bearing proteins subsequent to phosphorylation. A proline-rich region serving as an attachment site for SH3 domains is located in the same vicinity. These N-terminal sites provide a linkage to upstream integrins and growth factor receptors, and also downstream to proteins associated with the actin cytoskeleton through the five LD repeats. The C-terminal LIM domains anchor the paxillin protein at the plasma membrane.

The domain composition of FAK is presented in Figure 11.7b. It does not have any SH2 or SH3 domains but instead provides phosphorylation and anchoring sites for proteins with these domains. In place of the SH2 and SH3 domains FAK has two large domains of about 400 amino acids each, one on either side of the catalytic domain. FAK possesses six tyrosine phosphorylation sites. Two of these, Tyr397 and 407, lie just N-terminal to the kinase domain; two other, Tyr576 and 577, lie inside the kinase domain, and the last two, Tyr861 and 925, lie in the COOH terminal region N-terminal to the FAT. Autophosphorylation at Tyr397 exposes an SH2 docking site for Src.

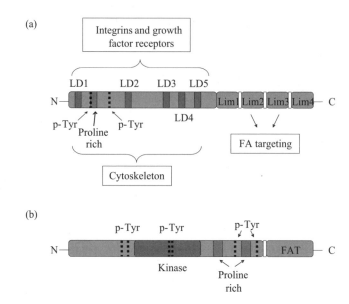

FIGURE 11.7. Focal adhesion proteins paxillin and FAK: (a) Paxillin—The four Lin-11, Isl-1, Mec-3 (Lim) protein-protein interaction domains in the C-terminus mediate targeting to focal adhesions (FAs). The leucine-rich sequences, or LD repeats, sequences of the form LDXLLXXL, where X is any amino acid residue, found in the N-terminus domain bind FAK and proteins such as vincullin associated with the cytoskeleton. The N-terminal domain contains a number of motifs (tyrosine phosphorylation sites and proline-rich regions) that bind to Sh2- and SH3-bearing proteins. (b) Focal adhesion kinase (FAK)—This protein has a focal adhesion targeting (FAT) domain in its C-terminus, and several tyrosine phosphorylation sites and proline-rich regions in its N-terminal domain.

Phosphorylation by Src at Tyr407, 576, and 577 maximally activates the kinase domain, and phosphorylation at the sixth site, Tyr925, provides a Grb2 docking site. The domain structure of Grb2 was presented in Figure 11.6b. It is an adapte protein that links FAK to the MAP kinase pathway. There are also two proline-rich regions in the COOH terminal domain that provide docking sites for adapte proteins bearing SH3 domains. Thus, like paxillin, FAK serves as integrator of adhesive and growth signals. The COOH-terminal region of FAK contains a focal adhesion targeting (FAT) sequence of about 160 amino acids that provides binding sites for paxillin and talin.

11.12 GTPases Are Essential Regulators of Cellular Functions

Since 1982, an ever-increasing number of GTPases have been found in eukaryotes. Most of these belong to the Ras superfamily. The Ras GTPases are small, 20–40 kDa monomeric proteins that bind guanine nucleotides,

TABLE 11.3. The Ras superfamily of GTPases.

Family	Function
Ras	Operates in the pathway that relays growth, proliferation, and differentiation signals to the nucleus
Rho	Relays coordinating signals to the actin cytoskeleton
Ran	Shuttles mRNAs and proteins in and out of the nucleus
Rab	Regulates the targeting and docking of cargo vesicles to membranes
Arf	Regulates the formation of cargo vesicles

either GDP or GTP. More than 100 Ras superfamily GTPases have been identified to date in eukaryotes. Each of these can be placed into one of five families—Ras, Rho, Ran, Rab, or Arf (Table 11.3). There are several other groups of proteins that operate as GTPases. Chief among these are the G_α subunits of the heterotrimeric G proteins that associate with G protein-coupled receptors and the EF-Tu elongation factors that help regulate protein synthesis.

Members of the Ras superfamily of GTPases carry out a variety of essential regulatory tasks. They regulate transcription, migration and focal adhesion, transport through the nuclear pore complex, assembly of the nuclear envelope and mitotic spindle, and vesicle budding and trafficking. Ras family members act as upstream switches in the pathway that conveys growth and differentiation signals to the nucleus. Rho family members Rho, Rac and Cdc42 coordinate growth and adhesion signaling. They regulate the reorganization of the actin cytoskeleton, and coordinate its structure with gene expression in response to extracellular signals. Ran GTPases regulate the inport and export of molecules from the cytoplasm to the nucleus. They regulate the assembly of the nuclear envelope and also that of the mitotic spindle. Rab proteins with over 30 family members are the largest family of the Ras superfamily of GTPases. They are regulators of vesicle trafficking, while Arf GTPases control vesicle budding.

11.13 Signaling by Ras GTPases from Plasma Membrane and Golgi

Ras is a major regulator of cell growth. Ras GTPases convey growth, proliferation, and differentiation signals to the nucleus from the plasma membrane and from the Golgi apparatus. Ras is implicated in a large number of human cancers, and in this context it will be examined further in Chapter 15. As shown in Figure 11.3 Ras operates as a master switch in the pathway connecting upstream receptors for growth factors with downstream MAP kinases that relay these signals to the transcription machinery in the nucleus. Once Ras becomes GTP-bound it activates the serine/threonine kinase Raf, which belongs to the Raf/MEK/ERK MAP kinase module.

Signaling through this module activates downstream transcription factors that stimulate growth and differentiation.

These downstream steps are preceded by upstream activation of receptor tyrosine kinases through ligand binding that creates binding sites for the Grb2 adapter protein. This adapter enables the RasGEF Son-of-sevenless (Sos) to bind and then accelerate the formation of GTP-bound Ras proteins. In the absence of GTP-bound Ras, the Raf kinase resides in the cytosol. GTP-bound Ras brings Raf to the plasma membrane. The Raf N-terminal domain binds Ras, thereby enabling the Raf C-terminal domain to bind and phosphorylate MEK.

Ras signaling is not restricted entirely to the plasma membrane. It can be activated at the Golgi through a calcium-dependent pathway involving a second pair of RasGEFs and RasGAPs. In place of the plasma membrane sequence of steps where the adapter Grb2 links receptor kinases to the RasGEF Sos, there is a pathway leading through Src. This nonreceptor tyrosine kinase activates phospholipase $C\gamma$ (PLCγ), which acts on PIP$_2$ to produce DAG and IP$_3$. The latter stimulates the release of Ca^{2+} from the intracellular stores. In response to release of DAG and Ca^{2+}, the RasGEF RasGRP1 translocates to the Golgi where it activates Ras, while at the same time the RasGAP calcium-promoted Ras inactivator (CAPRI) relocates to the plasma membrane where it deactivates any plasma membrane-associated Ras.

11.14 GTPases Cycle Between GTP- and GDP-Bound States

All GTPases function in a similar manner, passing through an assisted cycle of GDP/GTP-binding and release. Recall from Chapter 6 and Figure 6.4 that GTPases bind either GTP or GDP. They are converted from GDP-bound forms to GTP-bound forms by the catalytic actions of guanine nucleotide exchange factors (GEFs). GTPase-activating proteins (GAPs) catalyze their subsequent hydrolysis. In more detail, the GEFs "kick out" the GDP molecule. In most situations GTP is far more abundant than GDP and it readily binds the vacated pocket. In catalyzing the hydrolysis of the GTP, a molecule of a phosphoryl group is removed leaving a GDP molecule bound to the GTPase.

GTPases such as Ras function as molecular switches. These switches are turned "on" when GTP is bound and turned "off" when GDP is bound. In the "on" state the GTPases are continually active and become inactive only when turned off by a GAP. Ras is inactive when GDP-bound and it remains so until upstream signals conveyed by RasGEFs trigger the dissociation of GDP from Ras. Since the cellular concentration of GTP is far greater than that of GDP, the binding of GTP to Ras follows almost immediately. At this point Ras is activated and can bind to a downstream target, or effector,

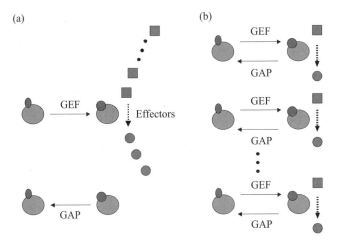

FIGURE 11.8. GTPase cycles: (a) Switch cycle in which GTP-binding turns on the GTPase, which can then interact with multiple effector molecules. The GTPase will stay on until turned off by hydrolysis. (b) Assembly-controller cycle in which GTP-binding initiates interactions with an effector molecule. If the assembly is not correct, the assembly step is aborted prior to hydrolysis, a process that is necessary for disengagement from the substrate.

molecule. It remains active (on) until it undergoes hydrolysis. During the time it remains on it can activate many Raf proteins, as illustrated in Figure 11.8a.

The GEF for plasma membrane associated Ras is Sos (Son of sevenless). It works in the following way to speed up the dissociation of GDP from Ras: The Ras molecule, consisting of 188 amino acids residues, contains two flexible regions on its surface. These surface regions can alternate between several conformational substates, and they are referred to as *Switch 1* and *Switch 2*. In its inactive state the Ras molecule binds GDP tightly with a dissociation rate of 10^{-5}/s. Sos-binding produces a tertiary Sos-Ras GDP complex that stabilizes Ras in a conformation in which Switch 1 has swung out to open the binding pocket. A portion of Sos stabilizes Switch 2 in an alternative conformation, and this change plus an altered electrostatic environment further weakens the binding of the phosphate group of GDP and its associated magnesium ion to Ras. The changes in the switch regions increase the dissociation rate by several orders of magnitude and allow GDP to exit the binding pocket in less than a second.

The Ras GAPs work in the following manner. Arginine and lysine residues have long side chains and are positively charged under physiological conditions. When bound to the Ras-GTP complex the Ras GAPs extend an "arginine finger" into the active site that neutralizes negative charges in its vicinity. A network of hydrogen bonds form, stabilizing the transition state and promoting the cleavage of the phosphodiester bond-linking gamma phosphate group to the GDP molecule. The result is an increase in the rate of hydrolysis from one GTP molecule every 35 minutes to 10^2 to 10^3 molecules per minute.

11.15 Role of Rho, Rac, and Cdc42, and Their Isoforms

Rho, Rac, and Cdc42, and their isoforms, coordinate the reorganization of the actin cytoskeleton in response to extracellular signals. Members of the Rho family of GTPases—Rho, Rac, and Cdc42—regulate cell polarity, cell morphogenesis and shape, and cell motility. Each of these GTPases operates in a different pathway. In fibroblasts, Rho promotes the formation stress fiber bundles, Rac proteins regulate actin polymerization resulting in the formation of lamellipodia, and Cdc42 proteins control the formation of filopodia. All three regulate the formation of focal adhesion complexes. These observations are not limited to fibroblasts but rather are believed to take place in all eukaryotic cells. In addition to their role in remodeling the actin cytoskeleton, Rho family members regulate gene transcription. Signaling through receptor tyrosine kinases, G protein-coupled receptors and cytokine receptors activate Rho GTPases. In many instances integrins and Rho family members work together to regulate the actin cytoskeleton.

An examination of the steps leading to bud formation in yeasts provides some insight into how cell polarity develops and is regulated by the Rho family GTPases. The Cdc42 GTPase, its GEF called Cdc24, its GAP, of which there are several, and a scaffold protein named Bem1 contribute to the start of budding. As is the case for all GTPases, Cdc42 goes through a cycle of GDP release catalyzed by its GEF immediately followed by GTP binding. This state is followed some time later by hydrolysis catalyzed by its GAP in which GTP is cleaved leaving GDP bound to the CDc42 protein. Unlike Ras, the Cdc42 protein does not act on multiple substrate proteins in its GTP-bound state. Rather it goes through repeated cycles of GDP-GTP-GDP binding and hydrolysis to regulate the polymerization of actin fibers. The difference between the Ras switchlike behavior and Cdc42 assembly controller-like behavior is depicted in Figure 11.8. In the Ras mode of cycling, the GTPase is on and can influence many substrate proteins before being switched off by hydrolysis. In the Cdc42 mode of operation (Figure 11.8b), hydrolysis is required. Substrate interactions are again initiated by GTP binding but the action is not completed until the hydrolysis part of the cycle is executed. The EF-Tu GTPase uses this method of repeatedly cycling on and off during the translation elongation process. This second mode allows for some checking and quality control over the process; the assembly step can be aborted prior to hydrolysis if errors are detected.

Cell polarization (and bud formation) is typically driven by environmental cues such as gradients of chemoattractants or by signals from neighboring cells. These cues tell the cell which way is "up" and which way is "down." In bud formation, the adapter protein Bem1 is recruited to the site of bud growth where it interacts with Ccd42 and its Cdc24 GEF. The GEF is recruited and stabilized at the plasma membrane by the Gβγ subunit activated by GPCR signaling. Both Cdc42 and Cdc24 bind to the scaffold and a positive feedback loop is set up in which additional Bem1 proteins are recruited. A crucial element in the choice of bud site selection is the presence of position landmarks in the cell that identify the location of the poles. If these landmarks are absent the yeast cell is still able to form buds by a process in which a bud site is randomly selected.

11.16 Ran Family Coordinates Traffic In and Out of the Nucleus

While small molecules can rapidly and freely diffuse in and out of the nucleus, macromolecules of size 40 to 50 kDa or greater cannot. Instead they are transported through selective and facilitated diffusion through nuclear pore complexes. Nuclear pore complexes (NPCs) are large units built up from more than 100 different proteins. The NPCs function as selective gates. Only the larger macromolecules containing the correct targeting or localization sequence are allowed through. Nuclear pore complexes have no motors to actively transport cargo. Instead, import and export occur by means of facilitated diffusion. This process does not consume energy and is nondirectional. Proteins, mRNAs, tRNA, ribosomal subunts, and other macromolecules are transported bound to transport molecules called *importins* and *exportins* that recognize the nuclear localization signals (NLSs) and nuclear export signals (NESs). These transport receptors together with adapter molecules, Ran GTPases, Ran GEFs, and Ran GAPs facilitate the diffusion of the macromolecules in and out of the nucleus.

Transport receptors belonging to the importin β family shuttle cargo in and out of the nucleus through the NPC. These receptors are encountered in two forms: as importins that shuttle cargo into the nucleus and as exportins that chaperone cargo the other way. The loading and release of cargo by the importin β receptors is regulated by the Ran GTPases. Ran, a 25-kDa protein, is an abundant gene product found in all eukaryotic cells. The key element in their ability to regulate cargo movement is the formation of a Ran concentration gradient across the nuclear envelope. This gradient in GTP-bound Ran is established and maintained by the Ran GEFs and Ran GAPs. The Ran GAPs are restricted to the cytoplasm and cannot enter the nucleus, while the Ran GEFs are localized exclusively in the nucleus. As a result the concentration of GTP-bound Ran is low in the cytoplasm and high in the nucleus.

The combined actions of importin β receptors and Ran GTPases are illustrated in Figure 11.9. As is the case for all GTPases its main actions occur during the GTP-bound part of the cycle, when it shuttles exportin + cargo and unbound importin molecules through the NPC from the nucleus to the

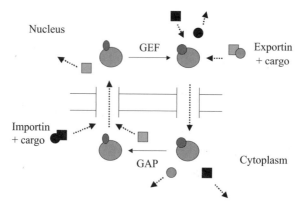

FIGURE 11.9. Ran GTPase-mediated import and export through nuclear pore complexes (NPCs): Importin (dark square) shuttles its cargo (dark circles) through the NPC into the nucleus. In the nucleus, GTP-bound Ran GTPase triggers the release of the cargo from the importin and shuttles the importin back out to the cytoplasm. The GTP-bound Ran GTPase shuttles exportin (light squares) plus its cargo (light circles) through the NPC from the nucleus to the cytoplasm where other proteins (not shown) help dissociate the complex. The unbound exportin then diffuses back to the nucleus.

cytoplasm. It differs from the Ras switching and Cdc42 cytoskeleton assembly in so far as its GEF and GAP actions take place in different compartments. Hydrolysis is not required for movement through the pore and release of the cargo, but this aspect changes when Ran mediates assembly of the nuclear envelope. In those assembly operations hydrolysis is necessary.

11.17 Rab and ARF Families Mediate the Transport of Cargo

Recall that in eukaryotes secreted soluble molecules and plasma membrane lipids and proteins are transported from compartment to compartment by transport vesicles. Newly synthesized proteins and lipids move from the ER to the Golgi apparatus and from there to the plasma membrane. The vesicles that transport these molecules are produced by budding from the membrane of the donor compartment, and upon arrival to the acceptor compartment fuse with the membrane of the acceptor compartment thereby transferring the cargo. The process of delivering cargo from the internal compartments to the plasma membrane is known as *exocytosis*. Cargo moves in two directions, outward and inward. Many membrane components are rapidly turned over. Vesicles transport macromolecules from

FIGURE 11.10. Rab GTPase cycle: The Rab GTPase cycle synchronizes vesicle transport and fusion between the membranes labeled A and B. In the cycle, a GDP-bound Ran protein is shuttled by its GDI to membrane A where GDP is released and GTP is bound in its place. While in its active form, a cargo vesicle pinches off from the membrane and is conveyed to membrane B where it docks and then fuses. Hydrolysis occurs and, assisted by the GDI, the GDP-bound Ran protein is released from the membrane and returned to the cytosol.

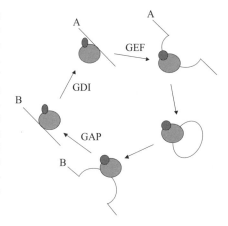

the plasma membrane either to the lysosomes, where they are degraded, or to endosomes, where they are recycled back to the plasma membrane. The process of transporting plasma membrane molecules to lysosomes for degradation is known as *endocytosis*, and the outward and inward movement of plasma membrane molecules is said to take place in the secretory and endocytic pathways.

The Rab family with more than 30 known members is the largest group of Ras superfamily GTPases. These proteins coordinate the docking and fusion of cargo vesicles operating in the secretory and endocytic pathways. Reciprocal pairs of proteins called v-SNAREs and t-SNAREs mediate the fusion of cargo vesicles. The binding of the SNAREs is preceded by tethering/protection protein-binding steps that ensure that only the appropriate fusion events takes place. The Rab proteins are localized to the cytoplasmic face of organelles and vesicles and participate in the preliminary binding operations.

The Rab protein cycle of GDP and GTP binding and release is synchronized with the movement and fusion of cargo vesicles. The combined set of steps is illustrated in Figure 11.10. In this figure the GEFs and GAPs are joined by a third set of accessory regulators—GDP dissociation inhibitors, or GDIs. These molecules bind and maintain pools of inactive Rab proteins in the cytosol and serve as recycling chaperones. They first help to release the GDPases from the membranes and then shuttle them back to their site of origin in the cytosol in preparation for their next cycle of use.

ADP-robosylation factors (Arfs) make up the fifth family of Ras GTPases. These 20-kDa proteins help regulate the formation of cargo vesicles through budding from donor membranes. In order for budding to take place, *coat proteins* must be assembled over the membrane surface. These binding agents, coat protein I (COPI), coat protein II (COPII), and clathrin, mediate the mechanical forces that pull a membrane into a bud, and they help capture membrane receptors and cargo. The Arf proteins recruit the coat proteins to the donor membrane.

References and Further Reading

Angiogenesis

Ferrara N, Gerber HP, and LeCouter J [2001]. The biology of VEGF and its receptors. *Nature Med.*, 9: 669–676.

Gale NW, and Yancopoulos GD [1999]. Growth factors acting via endothelial cell-specific receptor tyrosine kinases: VEGFs, angiopoietins and ephrins in vascular development. *Genes Dev.*, 13: 1055–1066.

Jones N, et al. [2003]. Tie receptors: New modulators of angiogenic and lymphangiogenic responses. *Nature Rev. Mol. Cell Biol.* 2: 257–267.

Neufeld G, et al. [1999]. Vascular endothelial growth factor (VEGF) and its receptors. *FASEB J.*, 13: 9–22.

Yancopoulos GD, et al. [2000]. Vascular-specific growth factors and blood vessel formation. *Nature*, 407: 242–248.

Neurotrophins

Chao MV [2003]. Neurotrophins and their receptors: A convergence point for many signaling pathways. *Nature Rev. Neurosci.*, 4: 299–309.

Lee R, et al. [2001]. Regulation of cell survival by secreted proneurotrophins. *Science*, 294: 1945–1948.

Zheng YZ, et al. [2000]. Cell surface Trk receptors mediate NFG-induced survival while internalized receptors regulate NFG-induced differentiation. *J. Neurosci.*, 20: 5671–5678.

Ligand-Induced, Receptor-Mediated Dimerization

Schlessinger J [2002]. Ligand-induced, receptor-mediated dimerization and activation of EGF receptor. *Cell*, 110: 669–672.

Phosphoprotein Recognition

Berg D, Holzmann C, and Reiss O [2003]. 14-3-3 proteins in the nervous system. *Nature Rev. Neurosci.*, 4: 752–762.

Durocher D, and Jackson SP [2002]. The FHA domain. *FEBS Lett.*, 513: 58–66.

Forman-Kay JD, and Pawson T [1999]. Diversity in protein recognition by PTB domains. *Curr. Opin. Struct. Biol.*, 9: 690–695.

Pawson T, Gish GD, and Nash P [2001]. SH2 domain, interaction modules and cellular wiring. *Trends Cell Biol.*, 11: 504–511.

Tzivion G, and Avruch J [2002]. 14-3-3 proteins: Active cofactors in cellular regulation by serine/threonine phosphorylation. *J. Biol. Chem.*, 277: 3061–3064.

Yaffe MB [2002]. Phosphotyrosine-binding domains in signal transduction. *Nature Rev. Mol. Cell Biol.*, 3: 177–186.

Recognition of Proline-Rich Motifs

Ball LJ, et al. [2002]. EVH1 domains: Structure, function and interactions. *FEBS Lett.*, 513: 45–52.

Freund C, et al. [1999]. The GYF domain is a novel structural fold that is involved in lymphoid signaling through praline-rich sequences. *Nature Struct. Biol.*, 6: 656–660.

Kay BK, Williamson MP, and Sudol M [2000]. The importance of being proline: The interaction of proline-rich motifs in signaling proteins with their cognate domains. *FASEB J.*, 14: 231–241.

Macias MJ, Wiesner S, and Sudol M [2002]. WW and SH3 domains, two different scaffolds to recognize proline-rich ligands. *FEBS Lett.*, 513: 30–37.

Mayer BJ [2001]. SH3 domains: Complexity in moderation. *J. Cell Sci.*, 114: 1253–1263.

Zarrinpar A, and Lim WA [2000]. Converging on proline: The mechanism of WW domain peptide recognition. *Nature Struct. Biol.*, 7: 611–613.

Protein–Protein Interaction Domains

Aravind L, Dixit MV, and Koonin EV [1999]. The domains of death: evolution of the apoptosis machinery. *Trends Biochem. Sci.*, 24: 47–53.

Harris BZ, and Lim WA [2001]. Mechanism and role of PDZ domains in signaling complex assembly. *J. Cell Sci.*, 114: 3219–3231.

Hepler JR [1999]. Emerging roles for RGS proteins in cell signaling. *Trends Pharmacol. Sci.*, 20: 376–382.

Hofmann K [1999]. The modular nature of apoptotic signaling proteins. *Cell. Mol. Life Sci.*, 55: 1113–1128.

Hung AY, and Sheng M [2002]. PDZ domains: Structural modules for protein complex assembly. *J. Biol. Chem.*, 277: 5699–5702.

Schultz J, et al. [1997]. SAM as a protein interaction domain involved in developmental regulation. *Protein Sci.*, 6: 249–253.

Zhong H, and Neubig RR [2001]. Regulator of G protein signaling proteins: Novel multifunctional drug targets. *J. Pharmacol. Exp. Ther.*, 297: 837–845.

Src Nonreceptor Tyrosine Kinase

Brown MT, and Cooper JA [1996]. Regulation, substrates and function of Src. *Biochim. Biophys. Acta*, 1287: 121–149.

Frame MC, et al. [2002]. v-Src's hold over actin and cell adhesions. *Nature Rev. Mol. Cell Biol.*, 3: 233–245.

Focal Adhesions

Ilic D, Damsky CH, and Yamamoto T [1997]. Focal adhesion kinase: At the crossroads of signal transduction. *J. Cell Sci.*, 110: 401–407.

Schlaepfer DD, Hauck CR, and Sieg DJ [1999]. Signaling through focal adhesion kinase. *Prog. Biophys. Mol. Biol.*, 71: 435–478.

Turner CE [2000]. Paxillin and focal adhesion signaling. *Nature Cell Biol.*, 2: E231–E236.

Renshaw MW, Ren XD, and Schwartz MA [1997]. Growth factor activation of MAP kinase requires cell adhesion. *EMBO J.*, 16: 5592–5599.

Ras Family of GTPases

Campbell SL, et al. [1998]. Increasing complexity of Ras signaling. *Oncogene*, 17: 1395–1413.

Hancock JF [2003]. Ras proteins: Different signals from different locations. *Nature Rev. Mol. Cell Biol.*, 4: 373–384.

Kolch W [2000]. Meaningful relationships: The regulation of the Ras/Raf/MEK/ERK pathway by protein interactions. *Biochem. J.*, 351: 289–305.

Rho Family of GTPases

Gladfelter AS, et al. [2002]. Septin ring assembly involves cycles of GTP loading and hydrolysis by Cdc42p. *J. Cell Biol.*, 156: 315–326.

Hall A [1998]. Rho GTPases and the actin cytoskeleton. *Science*, 279: 509–514.

Irazoqui JE, Gladfelter AS, and Lew DJ [2003]. Scaffold mediated symmetry breaking by Cdc42p. *Nature Cell Biol.*, 5: 1062–1070.

Ren XD, Kiosses WB, and Schwartz MA [1999]. Regulation of the small GTP-binding protein Rho by cell adhesion and the cytoskeleton. *EMBO J.*, 18: 578–585.

Schwartz MA, and Shattil SJ [2000]. Signaling networks linking integrins and Rho family GTPases. *Trends Biochem. Sci.*, 25: 388–391.

Ran Family of GTPases

Görlich D, Seawald MJ, and Ribbeck K [2003]. Characterization of Ran-driven cargo transport and the RanGTPase systems by kinetic measurements and computer simulation. *EMBO J.*, 22: 1088–1100.

Macara IG [2001]. Transport in and out of the nucleus. *Microbiol. Mol. Biol. Rev.*, 65: 570–594.

Melchior F, and Gerace L [1998]. Two-way trafficking with Ran. *Trends Cell Biol.*, 8: 175–179.

Weis K [1998]. Importins and exportins: How to get in and out of the nucleus. *Trends Biochem. Sci.*, 23: 185–189.

Weis K [2003]. Regulating access to the genome: Nucleocytoplasmic transport throughout the cell cycle. *Cell*, 112: 441–451.

Rab and Arf Families of GTPases

Moss J, and Vaughn M [1995]. Structure and function of Arf proteins: Activators of cholera toxin and critical components of intracellular vesicle transport. *J. Biol. Chem.*, 270: 12327–12330.

Zerial M, and McBride H [2001]. Rab proteins as membrane organizers. *Nature Rev. Mol. Cell Biol.*, 2: 107–117.

Recycling, and Signaling in the Endocytic Pathway

González-Gaitán M [2003]. Signal dispersal and transduction through the endocytic pathway. *Nature Rev. Mol. Cell Biol.*, 4: 213–224.

Sorkin A, and von Zastrow M [2002]. Signal transduction and endocytosis: Close encounters of many kinds. *Nature Rev. Mol. Cell Biol.*, 3: 600–614.

Problems

11.1 (a) *Binding interactions.* One of the standard questions addressed in investigations of signal transduction asks what interactions are possible in a particular molecular setting. A number of proteins

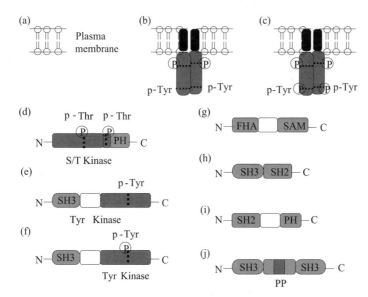

FIGURE FOR PROBLEM 11.1. Phosphorylated threonine and tyrosine residues are identified by a "P" enclosed in a circle. The symbol "PP" refers to proline-rich regions. Threonine and tyrosine phosphorylation sites with the correct flanking sequences for recognition by their kinases are indicated by the symbols p-Thr and p-Tyr. Note that phosphorylation at the sites located in the center of the catalytic domains of the NRTKs is required for activation.

and membranes are shown in the accompanying figure. Identify the possible interactions in a mixture containing multiple copies of each element (a) through (j).

(b) *Build your own.* Starting at the plasma membrane construct one or more signaling pathways using the elements (a) through (j).

11.2 Nerve cells are highly polarized structures, and their terminals are located far from the cell body where the nucleus is located. Signals sent from axon terminals to the cell body must travel distances of up to a meter. These signals are referred to as *retrograde signals* since they move in a direction opposite to action potentials, which travel down the axon to the nerve terminal. How might a neuron generate a rapid (i.e., faster than one conveyed by passive diffusion) retrograde signal, observing that receptor tyrosone kinases such as the Trks (and also G protein-coupled receptors) are internalized following ligand binding?

12
Signaling in the Endocrine and Nervous Systems Through GPCRs

The grouping formed by the G protein-coupled receptors (GPCRs) is the largest of its kind in the body. Recent estimates of the number of GPCRs in humans fall in the range of 800 or so. This number is divided roughly in half between *receptors* that bind sensory signals originating from outside the body (exogenous ligands) and *signals* produced internally by other cells (endogenous ligands). GPCRs transduce a remarkably diverse spectrum of messages into the cell. Among the signals transduced are light (phototransduction), extracellular calcium ions, tastants (gustatory), odorants (smell), pheromones (mating signals), warnings (pain), immunological (chemokines), endocrine (hormones), and neural (neuromodulators and neurotransmitters).

Members of the GPCR superfamily are sometimes referred to as the *7TM receptors* because their chains pass back and forth through the plasma membrane seven times. The GPCRs transmit messages into the cell by activating heterotrimeric G proteins and by sending signals through growth pathways. In the absence of GPCR ligand binding, heterotrimeric G proteins remain in close association with the GPCRs. Ligand binding leads to G protein dissociation and signaling through second messengers that target protein kinases/phosphatases and ion channels. In the first part of the chapter, the general characteristics of GPCR signaling will be examined, and second messengers initially discussed in Chapter 8 will be looked at in more detail.

The GPCRs are the targets of many therapeutic drugs due to the receptors' prominent involvement in the endocrine and nervous systems. It is estimated that some 40 to 60% of all therapeutic drugs target GPCRs. Some drugs act as agonists and others as antagonists. An *agonist* induces the same response in the receptor as that triggered by the natural ligand. An *antagonist* binds to the receptor but the receptor does not transmit a signal in response to the binding event. By binding the receptor the antagonist blocks access of the natural ligand to the receptor and thus prevents transmission of a signal. Drugs that bind in an antagonistic fashion are known as *blockers*. Prominent examples are *beta blockers*, which antagonize the beta-adrenergic receptor, and *antihistamines*, which inhibit the histamine H1 receptor. The second part of the chapter will provide an overview of signaling using hormones, neurotransmitters, and neuromodulators (endogenous ligands) and how light and other exogenous ligands are transduced into cellular responses.

12.1 GPCRs Classification Criteria

G protein-coupled receptors share little sequence homology with one another yet all are organized in a remarkably similar fashion. All are single chain polypeptides with a topology as depicted in Figure 12.1. The transmembrane segments are composed of hydrophobic amino acid residues arranged into an alpha helix. The helices are connected to one another by three extracellular loops E1–E3 and three cytoplasmic loops C1–C3. Each GPCR has an extracellular N-terminal region and a cytoplasmic C-terminal region. The extracellular and cytoplasmic regions can be quite large. The extracellular loop E2 and the N-terminus segment are especially well situated for ligand binding. These structures vary in size from one class of GPCR to another. The intracellular loops plus the cytoplasmic ends of the transmembrane helices participate in protein binding and activation. The cytoplasmic tail and third intracellular loop (C3), in particular, provide multiple sites for protein docking and interactions.

GPCRs can be grouped into families according to the most prominent characteristics of the extracellular and intracellular domains, loop properties, and the presence of structurally important disulfide bridges. GPCRs possess a number of highly conserved sequences that influence the structure and binding properties of the GPCRs. These, too, enter into the classification process. There are three main families of GPCRs and a growing

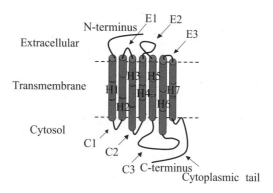

FIGURE 12.1. Stereotypic representation of a rhodopsin family (Class A) G protein-coupled receptor: Shown are the seven membrane-spanning alpha helices H1–H7, three extracellular loops E1–E3, and three cytoplasmic loops C1–C3.

TABLE 12.1. Three main classes of G protein-coupled receptors: Receptor subtypes are listed to the right of the ligand. G protein-coupled receptors are classified into families according to their structure and the presence or absence in them of a number of highly conserved sequences.

Class A: Rhodopsin family	Class B: Secretin family	Class C: Glutamate family
Adrenergic α_1, α_2, β_1, β_2	Calcitonin CTR	Calcium CaR
Angiotensin AT_{1A}, AT_{1B}, AT_2	Glucagon GR	GABA $GABA_BR1$ and $GABA_BR2$
Dopamine D_1 to D_5	Latrotoxin CL_1, CL_2, CL_3	Glutamate mGluR1 to mGluR8
Histamine H_1, H_2	PTH/PTHrP PTH/PTHrPR	
Melanocortin $MC_1, \ldots, MC_5$	Secretin SCTR	
Melatonin ML_{1A}, ML_{1B}	VIP/PACAP VIP_1, VIP_2	
Muscarinic acetylcholine m1, . . . , m5		
Neurokinen NK_1, NK_2, NK_3		
Opioid delta (δ), kappa (κ), mu (μ)		
Prostanoid $EP_1, \ldots, EP_4$, DP, FP, IP, TP		
Purine A_1, A_{2A}, A_{2B}, A_3, $P2Y_1, \ldots, P2Y_6$		
Serotonin 5-HT receptor subtypes		
Somatostatin $SST_1, \ldots, SST_5$		
Vasopressin V_{1A}, V_{1B}, V_2		

list of smaller groups with a few members each. Some of the more prominent G protein-coupled receptors belonging to the three major families are listed Table 12.1. Most of these receptors form small subfamilies. Each receptor has several subtypes and each subtype may be expressed in one of a number of alternative spliced forms.

Class A contains most of the known GPCRs. It is frequently referred to as the *rhodopsin family*, named for the rhodopsin molecule responsible for transducing photons in rod cells of the retina. Receptors for a diverse spectrum of ligands including most of the amide and peptide hormones and neuromodulators belong to this family. Class A receptors bind a number of glycoprotein hormones such as LH, TSH, and MSH using a large extracellular domain. The glycoprotein-binding receptors differ from the GPCRs for the smaller ligands. The latter receptors rely on extracellular loops or utilize a binding pocket formed by the H3 to H6 helices.

Class B GPCRs, the *secretin family*, include calcitonin, a 32-amino acid peptide hormone released by "C" cells in the thyroid gland that regulate calcium concentrations in the blood. Secretin, a hormone released by "S" cells in the small intestine, stimulates the pancreas to secrete fluids that regulate acidity during digestion. Parathyroid hormone (PTH) and parathyroid hormone related protein (PTHrP) regulate bone and mineral homeostasis. Many ligands for Class B GPCRs such as glucagon, secretin, and VIP/PACAP (vasoactive intestinal peptide/pituitary adenylate cyclase-activating polypeptide) are fairly large. Receptors in this family utilize an extracellular domain that is far larger than the one depicted in Figure 12.3, plus nearby extracellular loops, for ligand binding.

Class C receptors include the metabotropic glutamate and GABA$_B$ receptors mentioned earlier. This collection of GPCRs is also known as the *glutamate family*. Receptors that signal the presence of extracellular calcium (CaRs) also belong to this group as do a number of mammalian pheromone receptors. An extensive extracellular domain and a large intracellular domain characterize this group. The extracellular domain of these proteins contains two lobes that clamp down on the ligand. This extracellular domain is homologous to bacterial periplasmic binding proteins (PBPs). These are proteins found in the periplasmic space between the outer and inner membranes in gram-negative bacteria. It is speculated that at some time in the past genes encoding PBPs fused with those specifying integral membrane proteins resulting in the formation of Class C GPCRs.

12.2 Study of Rhodopsin GPCR with Cryoelectron Microscopy and X-Ray Crystallography

Cryoelectron microscopy and X-ray crystallography studies of rhodopsin have provided detailed information on how GPCR helix bundles are organized and how the GPCR activates its heterotrimeric G proteins. As shown in Figure 12.2 the seven alpha helices in rhodopsin are arranged sequen-

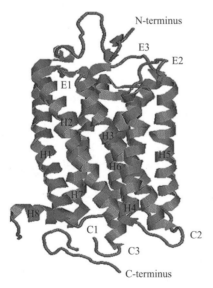

FIGURE 12.2. Structure of rhodopsin determined by means of X-ray crystallography: The rhodopsin molecule contains seven plasma membrane spanning helices (H1 through H7) plus a small C-terminus loop H8. The helices are connected to each other by extracellular loops E1–E3 and cytosolic loops C1–C3. The figure was prepared using the Protein Explorer and atomic coordinates deposited in the Brookhaven PDB under accession code 1F88.

tially in a clockwise manner when viewed from the intracellular side. Helices 1, 2, 3, and 6 are tilted. That is, their axes are inclined by about 25 degrees relative to an axis drawn perpendicular to the surface. Helices 4 and 7 are nearly perpendicular to the membrane bilayer. Helix 6 is bent; one part is nearly perpendicular to the plane of the membrane and the other part is inclined by about 25 degrees (helix 5 is also bent, but to a lesser extent). In the electron density plot of rhodopsin, the four tilted helices form a central band. Helix 4 lies on one side of the band and helices 6 and 7 lie on the other side.

The rhodopsin molecule operates in a switchlike manner to activate the G protein. Two of the helices, H3 and H6, project into the cytoplasm further than the others. Ligand binding, or photoisomerization in the case of rhodopsin, stabilizes the receptor in an alternative conformation in which there is a shift in the relative positions of helices H3 and H6. When activated by ligand binding the GPCR acts as a guanine nuclcotide exchange factor, or GEF, for its G protein. In its active shifted conformation, the rhodopsin GPCR is able to catalyze the release of GDP from the alpha subunit of the G protein leading to its binding GTP. The GDP-GTP exchange triggers the dissociation of $G_{\alpha\beta\gamma}$ into G_α and $G_{\beta\gamma}$ subunits and the subunits' migration towards their effectors. The switch is reset by the dissociation of the ligand from the GPCR.

12.3 Subunits of Heterotrimeric G Proteins

As the name implies, heterotrimeric G proteins are assembled from three distinct subunits. These subunits are designated as G_α, G_β, and G_γ. There are 20 different G_α subunits, 6 known G_β subunits and 12 distinct G_γ subunits. The G_α subunits function as GTPases. Like the Ras GTPase discussed in the last chapter the signaling activity of a G_α subunit is turned off when it is GDP-bound and switched on when it is GTP-bound. The GPCR provides the activation signal and also serves as the GEF, catalyzing the dissociation of bound GDP from G_α. A family of proteins called *regulators of G protein signaling* (RGS proteins) function as GAPs for the G_α subunits. They catalyze the hydrolysis of GTP by the G_α subunits, thereby rapidly switching off G_α signaling.

There are four kinds of G protein alpha subunits. For most GPCRs one type of GPCR couples to and activates only one of the four kinds of alpha subunits. However, in some cases, the coupling is richer and the GPCR can switch from one kind of subunit to another. Several portions of the GPCR contribute to the G protein-specific binding. Regions at the ends of helices H3, H5, and H6, loops C2 and especially C3, and the short helix (H8) located in the C-terminal region just after helix H7 are all involved in coupling and activating the various members of the heterotrimeric G protein family.

Once activated the G_α and $G_{\beta\gamma}$ subunits are free to diffuse laterally along the cytoplasmic surface of the plasma membrane, and bind nearby signaling targets, or effectors. The cycle of activation and signaling is completed with hydrolysis and reassociation of the G_α and $G_{\beta\gamma}$ subunits. Upon binding G_α, the $G_{\beta\gamma}$ induces substantial conformational changes and increases the affinity of G_α for GDP. When bound to G_α, the $G_{\beta\gamma}$ subunit cannot bind its effectors and signal because the binding sites on $G_{\beta\gamma}$ for its effectors overlap that for G_α.

12.4 The Four Families of G_α Subunits

Four families of G_α subunits are presented in Table 12.2. As shown in the table, G_s *subunits* bind to and stimulate adenylyl cyclases, while G_i *subunits* inhibit these effectors. G_q *subunits* stimulate phospholipase C, and the effectors of G_{12} *subunits* are as yet unidentified. Each G_α family contains several variants. Some variants and groups of variants are found in certain tissues while other are more broadly distributed. For example, G_0 subunits are brain-specific and the G_i family is sometimes designated as $G_{i/0}$ to reflect the inclusion of G_0 subunits in this family. Figure 12.3 illustrates how the G_s alpha subunit binds to the C_1 and C_2 domains of adenylyl cyclase.

$G_{\beta\gamma}$ subunits target many of the same second messenger generators as the G_α subunits. Like G_α subunits, some $G_{\beta\gamma}$ subunits stimulate adenylyl

TABLE 12.2. Mammalian G_α subunits: G proteins consist of a G_α subunit that is a GTPase, and G_β and G_γ subunits that remain attached and function as a unified $G_{\beta\gamma}$ subunit. The four different types of G protein alpha subunits are presented in column 1. Different subtypes are listed in column 2 and their effectors (substrates) are given in column 3. Tissue-specific G proteins are noted in column 4. The actions of the subunits on their effectors are given by the plus and minus signs. Plus signs denote stimulation and minus signs indicate inhibition. Abbreviations: adenylyl cyclase (AC), cyclic guanosine monophosphate (cGMP), phosphodiesterase (PDE), phospholipase A_2 (PLA$_2$), and phospholipase C (PLC).

Family	Gene variants	Effectors	Association
G_s	$\alpha_{s(S)}$, $\alpha_{s(L)}$	+AC	
	α_{olf}	+AC	Olfactory
G_i	α_{i1}, α_{i2}, α_{i3}	−AC	
	α_{0a}, α_{0b}	+PLC, +PLA$_2$	Brain
	α_{t1}, α_{t2}	+cGMP, PDE	Retina
	α_g	+PLC	Gustatory
	α_z	−AC	
G_q	α_q, α_{11}, α_{14}, α_{15}, α_{16}	+PLC	
G_{12}	α_{12}, α_{13}		

cyclases while others inhibit them. Still other $G_{\beta\gamma}$ subunits stimulate phosopholipase C. Thus, G_{α} and $G_{\beta\gamma}$ subunits jointly determine the overall action of a G protein on second messenger generators.

12.5 Adenylyl Cyclases and Phosphodiesterases Key to Second Messenger Signaling

There are nine types of adenylyl cyclases. These isoforms are designated as types I to IX. All of these isoforms can be stimulated by G_s subunits. However, such is not the case for the G_i subunits. Adenylyl cyclase isoforms I, V, and VI are the only ones inhibited by G_i subunits; the others are not affected. The isoforms also differ in their responses to $G_{\beta\gamma}$ subunits, and to the intracellular Ca^{2+} concentration $[Ca^{2+}]_i$. These response properties are summarized in Table 12.3. As can be seen in Table 12.3, five of the AC isoforms can be regulated by calcium. Thus, the two major second messenger systems are tightly coupled to one another. This coupling can produce a variety of effects including synchronized oscillations in the intracellular cAMP and Ca^{2+} concentrations.

The magnitude and duration of cyclic nucleotide second messenger signaling is regulated by nucleotide phosphodiesterases (PDEs). Recall that cAMP has a phosphate group attached to both the 3′ carbon and 5′ carbon of the sugar ribose. PDEs are enzymes that catalyze the hydrolytic cleavage of 3′ phosphodiester bonds in cAMP resulting in its degradation to inert 5′AMP. They also carry out the same operation in cGMP to yield inert

TABLE 12.3. Response properties of adenylyl cyclases to different regulators: The nine isoforms of adenylyl cyclase (AC) are listed in column 1. Columns 2–5 show how the four different kinds of regulators influence them. Stimulatory effects are denoted by plus (+) signs and inhibitory influences by minus (−) signs.

AC Isoform	G_s	G_i	$G_{\beta\gamma}$	Ca^{2+}
I	+	−	−	+
II	+		+	
III	+			−
IV	+		+	
V	+	−		−
VI	+	−		−
VII	+		+	
VIII	+			+
IX	+			

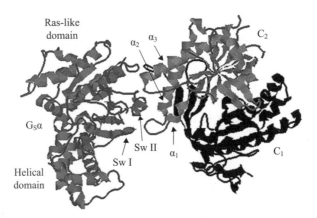

FIGURE 12.3. Complex of the G_s alpha subunit bound to the C_1 catalytic domain of adenylyl cyclase V and the C_2 domain of adenylyl cyclase II: Shown is a view of the complex looking up from inside the cell towards the plasma membrane. The main features of the interface between the G alpha subunit and the adenylyl cyclase catalytic units consist of Sw I and Sw II of the Ras-like domain of G alpha that protrude into a cleft formed by the alpha 1 to 3 helices of the C2 domain. The figure was prepared using Protein Explorer with atomic coordinates deposited in the Brookhaven Protein Data Bank under accession code 1AZS.

5'GMP, thereby terminating second messenger signaling. They modulate these signals with regard to their amplitude and duration, and through rapid degradation restrict their spread to other compartments in the cell.

12.6 Desensitization Strategy of G Proteins to Maintain Responsiveness to Environment

In order for a cell to maintain its responsiveness to future changes in environmental conditions as relayed through GPCR signaling, it must terminate current GPCR responsiveness to persistent ligands in a timely fashion. The process whereby a GPCR, or any other signaling entity, loses responsiveness to binding by its ligand is called *desensitization*. The turning off of the GPCR is accomplished in a sequential manner by G protein-coupled receptor kinases (GRKs) and arrestins.

The domain structure of GRK2 and β-arrestin 2 are presented in Figure 12.4. Recall from Chapter 6 and Table 6.1 that G protein-coupled receptor kinases are members of the AGC family of serine/threonine kinases. As shown in Figure 12.4a, the GRK2 member of this family has a central kinase domain plus two flanking regulatory domains. The N-terminal domain contains an RGS homology (RH) domain that characterizes the family. In addi-

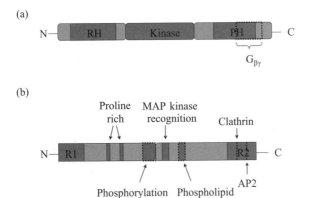

FIGURE 12.4. GRK2 and β-arrestin 2 regulators of GPCR signaling: (a) G protein-coupled receptor kinase 2 (GRK2) consists of three domains. The RGS homology (RH) and PH domains occupy the major portions of the N-terminal and C-terminal domains, respectively. (b) β-arrestin 2 has two domains, an N-terminal domain and a C-terminal domain. A pair of regulatory segments, R1 and R2, resides at the ends of the N- and C-terminals. The R2 domain contains clathrin- and AP2-binding motifs. Other binding motifs are distributed through the N- and C-terminal domains.

tion to these two domains, GRK2 (but not all GRKs) contains a C-terminal Plectstrin homology (PH) domain. PH domains bind phospholipids and also proteins. The GRK2 C-terminal half of this domain together with several residues lying just outside this domain binds $G_{\beta\gamma}$. The three-dimensional crystal structure of a $G_{\beta\gamma}$ subunit bound to GRK2 is presented in Figure 12.5.

The structure of β-arrestin 2 is presented in Figure 12.4b. β-arrestins are adapter proteins devoid of any catalytic ability. In place of a catalytic domain, members of this family possess a number of protein-protein binding domains and motifs that enable them to function as adapters and scaffolds. The β-arrestins contain an N-terminal domain and a C-terminal domain. Their proline-rich region in the N-terminal domain and MAP kinase recognition domain in the C-terminal region provide the protein with the capability of signaling to Src and MAP kinases. In addition, β-arrestins have a phosphorylation recognition domain that mediates binding to the cytoplasmic tail of the GPCR following its phosphorylation by the GRKs. They also have clathrin and AP2 motifs in the C-terminal region that mediate their interactions with the molecular machinery responsible for endocytosis.

The recruitment and subsequent termination of signaling by the arrestins is mediated by a negative feedback loop that automatically prevents excessively sustained signaling. Ligand binding activates the GPCRs, leading to activation of G proteins. The $G_{\beta\gamma}$ subunits bind (Figure 12.5) and activate

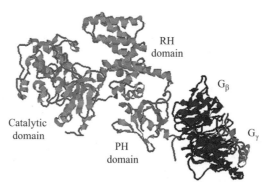

FIGURE 12.5. G protein-coupled receptor kinase (GRK) bound to the $G_{\beta\gamma}$ subunit of a heterotrimeric G protein: The structure of the complex determined by means of X-ray crystallography. The three main domains of GRK—the N terminal RGS homology (RH) domain, kinase domain and C-terminal PH domain—are labeled in the figure along with the G_{β} (dark gray) and G_{γ} subunits (light gray) of the G protein. The figure was prepared using Protein Explorer with atomic coordinates deposited in the PDB under accession number 1OMW.

the G protein-coupled receptor kinase, which phosphorylates the receptor. In response the β-arrestins are recruited to and bind the phosphorylated residues using their phosphorylation domain. The GPCR is then no longer able to active the G proteins and signaling ceases through this route. Thus, signaling through this receptor is automatically shut down after a certain time has elapsed.

Densensitization of the GPCRs is also promoted by second messenger-activated protein kinases. Protein kinase A activation mediated by G_s subunits and protein kinase C by G_q subunits both contribute to receptor desensitization. The third intracellular loop and the cytoplasmic tail of the GPCRs are primary sites of interaction with cytoplasmic proteins. The presence of consensus phosphorylation sites in these regions enables phosphorylation by these kinases. The phosphorylation by PKA of a GPCR to turn off signaling establishes a negative feedback loop that acts through G_s subunits to stimulate cAMP production resulting in PKA activation that shuts off further action by G_s subunits.

12.7 GPCRs Are Internalized, and Then Recycled or Degraded

The GPCRs go through a life cycle of activation, signaling, deactivation, and internalization resulting in either degradation or recycling. As illustrated in Figure 12.6, the first steps in this process are the same as discussed in the previous section involving $G_{\beta\gamma}$ subunits, namely, GRK2 binding and

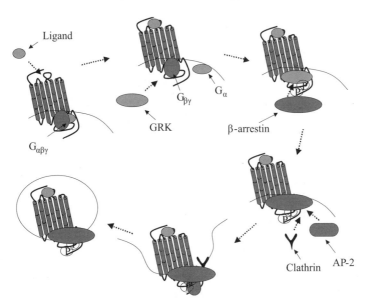

FIGURE 12.6. Receptor internalization: Activation and G protein signaling is followed by desensitization and signaling that is independent of G proteins. Ligand binding to a GPCR activates the G protein and signaling via second messengers. The $G_{\beta\gamma}$ subunit recruits a G protein-coupled receptor kinase (GRK) to the GPCR, which is then phosphorylated by the GRK. This action creates a docking site for β-arrestin, which upon binding to the GPCR blocks further G protein-mediated signaling. Proteins involved in endocytosis such as clathrin and AP-2 attach to the β-arrestin, which now acts as a scaffold for assembly for factors required for vesicle formation and internalization. In the internalization, the ligand-receptor complexes are dissociated and, depending on the type of GPCR, the receptors are either degraded or recycled rapidly or slowly.

phosphorylation, and β-arrestin recruitment. The subsequent recruitment of clathrin and AP-2 enables the packaging of the loaded GPCRs into endocytic vesicles that pinch off from the plasma membrane. The cargo consisting of ligand-bound receptors, β-arrestins, and perhaps additional signaling proteins such as Src are then delivered to endosomes for recycling to the cell surface or for shipment to lysosomes for degradation. This kind of life cycle is not restricted to GPCRs. Receptor tyrosine kinases such as the Trks discussed in the last chapter are also packaged into endodomal vesicles for internalization and recycling.

12.8 Hormone-Sending and Receiving Glands

The hypothalamus, anterior pituitary gland, adrenal gland, and the endocrine pancreas send and receive hormones. While polypeptide hormones signal through receptor tyrosine kinases, as was discussed in Chapter 11, other nonlipophilic hormones secreted by cells in the hypothalamus, anterior pituitary, adrenal gland, and endocrine pancreas signal through GPCRs.

Hypothalamus. The hypothalamus controls the release of hormones by the anterior pituitary. The hypothalamus produces two hormones that *inhibit* hormonal release by the pituitary, and five hormones that *stimulate* the release of pituitary hormones. One of these five, dopamine, is a *catecholamine*, a molecule composed of an aromatic (catechol) part, i.e., a benzene ring plus two hydroxyl groups, and an amine part. The other hormones are peptides that vary in length from 3 amino acids (TRH) to 44 amino acid residues (GHRH). These neurohormones all bind to GPCRs on cells of the anterior pituitary.

Anterior pituitary. The anterior pituitary produces hormones that control hormone release elsewhere in the body. Five types of cells secrete anterior pituitary hormones. *Somatotrophic cells* secrete growth hormone and *lactotrophic cells* produce prolactin. Growth hormone and prolactin are bound by single chain cytokine receptors, and not by GPCRs. Human growth hormone (hGH) receptor recognition was discussed in Chapter 8. Growth hormones circulate through the body and influence all cell types. Prolactin influences mammary gland development. *Corticotrophic cells* secrete corticotropin and melanocyte-stimulating hormone (MSH). Corticotropin influences the adrenal gland while MSH targets melanocytes, melanin-containing skin cells. ACTH and MSH are both derived from the precursor molecule, pro-opiomelanocortin (POMC). Another molecule derived from POMC is the neuromodulator beta-endorphin. *Gonadotrophic cells* secrete follicle-stimulating hormones (FSH) and lutenizing hormones (LH), and *thyrotrophic cells* produce thyroid-stimulating hormone (TSH). TSH, FSH, and LH are large glycoproteins. The addition of carbohydrate groups to these hormones extends their lifetime by protecting them from degradation. Gonadotropins stimulate a variety of responses in men and women including sperm development, androgen release, estrogen synthesis, and ovulation. These pituitary hormones with the exception of GH and PRL all bind to GPCRs expressed by cells of the above mentioned tissues and organs. The hormone sources and acronyms are summarized in Table 12.4.

Adrenal gland. Chromaffin cells in the adrenal gland produce two catecholamines—adrenaline (epinephrin) and noradrenaline (norepinephrin). These hormones, secreted in response to extreme stress, trigger an increased blood flow and other physiological "fight or flight" reactions. They are derived from tyrosine and bind to adrenoreceptors, adrenergic GPCRs distributed in smooth muscle of the vascular system and gastrointestinal tract,

TABLE 12.4. Hormones secreted by the hypothalamus, adrenal, anterior pituitary, and endocrine pancreas glands: Listed are the names and either common abbreviations or alternative names of the hormones.

Hypothalamus	Anterior pituitary
Corticotropin-releasing hormone (CRH)	Adrenocorticotropic hormone (ACTH)
Gonadotropin-releasing hormone (GRH)	Follicle-stimulating hormone (FSH)
Growth hormone-inhibiting hormone (GHIH)	Growth hormone (GH)
Growth hormone-releasing hormone (GHRH)	Lutenizing hormone (LH)
Prolactin-inhibiting hormone (PIH)	Melanocyte-stimulating hormone (MSH)
Prolactin-releasing hormone (PRH)	Prolactin (PRL)
Thyrotropin-releasing hormone (TRH)	Thyroid-stimulating hormone (TSH)

Adrenal gland	Endocrine pancreas
Adrenaline (Epinephrin)	Glucagon
Noradrenaline (Norepinephrin)	Insulin
Steroid hormones	Somatostatin

adipose tissue, heart muscle, skeletal muscle, brain, and lungs. The adrenal medulla is surrounded by the adrenal cortex, a second distinct structure in the adrenal gland. The adrenal cortex releases a large number of steroid hormones the most important being aldosterone and cortisol that help maintain salt balance and water homeostasis in the body.

As mentioned previously, steroid hormones do not bind to receptors on the cell surface but instead pass through the plasma membrane and bind to intracellular (nuclear) receptors. Peptide hormones stimulate the synthesis and release of these hormones. Angiotensin II and III produced in the kidneys stimulate aldosterone production, ACTH stimulates cortisol production, and the glycoprotein hormone LH stimulates the release of the gonadotropic steroid hormones progesterone and testosterone.

Endocrine pancreas. The *insulin* molecule, like ACTH and its siblings, is generated from a precursor molecule, or prohormone. The finished insulin molecule, consisting of two chains for a total length of 51 amino acids, is produced by cleavage from a single chain precursor, proinsulin. Once secreted by beta cells of the Islets of Langerhans in the pancreas, insulin binds to receptor tyrosine kinases on target fat, muscle, and red blood cells. *Glucogon*, a 29-amino acid protein, is produced by alpha cells of the islets and by cells distributed throughout the gastrointestinal tract. It binds to a GPCR. A third hormone, *somatostatin* is produced by cells in the islets as well as by cells in the hypothalamus. All three of these hormones, insulin, glucagons, and somatostatin, influence the flow of nutrients in the body, and maintain glucose homeostasis by regulating the storage of glucose and its transport in and out tissues. Insulin release is stimulated by high blood sugar levels and triggers the uptake of glucose by its target cells. Glucagon works in the opposite direction. Glucagon receptors are found in the liver.

Glucagon is secreted by the pancreas in response to low blood sugar levels and signals the liver to make and secrete glucose.

12.9 Functions of Signaling Molecules

Signaling molecules are often used in more than one way. Some function as hormones and also as neuromodulators, substances that modify the electrical properties (excitability) of neurons. Other substances function both as neuromodulators and as neurotransmitters. The distinction between functioning as a hormone or as a neurotransmitter or as a neuromodulator is based on how the signaling molecule is being used rather than on its chemical composition.

Neurotransmitters are secreted in a highly directional manner. They are released from the pre-synaptic terminal of a neuron and diffuse across the synaptic cleft to the postsynaptic terminal of a neighboring neuron. There are two kinds of receptors in the nervous system for neurotransmitters, ionotropic and metabotropic. Ionotropic receptors are referred to as *ion channels*. They respond to a binding event by transiently opening a channel, an aqueous pore through the plasma membrane for the passage of ions— usually Na^+, K^+, Ca^{2+} or Cl^-, depending on the type of neurotransmitter and receptor—in and out of the cell. This type of signaling is rapid and direct. Signaling through *metabotropic receptors*, another term for GPCRs, is slower and less direct.

Some of the most prominent neurotransmitters can signal through both ionotropic and metabotropic receptors. Glutamate and γ-aminobutyric acid (GABA) are two of the most often encountered neurotransmitters in the brain. There are ionotropic glutamate receptors and metabotropic glutamate receptors. Similarly there are ionotropic GABA receptors and metabotropic GABA receptors; the former are referred to as $GABA_A$ receptors and the latter as $GABA_B$ receptors. Another prominent neurotransmitter is acetylcholine. This neurotransmitter can bind to ionotropic (nicotinic) receptors and to G protein-coupled (muscarinic) receptors.

The four most common neurotransmitters found in the brain are listed in Table 12.5. The terms *excitatory* and *inhibitory* refer to the effects the neuro-

TABLE 12.5. Excitatory and inhibitory neurotransmitters used in the central nervous system: Common abbreviations for the neurotransmitters are shown in parentheses.

Excitatory	Inhibitory
Acetylcholine (ACh)	γ-Aminobutyric Acid (GABA)
Glutamate (Glu)	Glycine (Gly)

transmitters have on membrane excitability. Inhibitory neurotransmitters bind to anionic ion channels and decrease excitability, while excitatory neurotransmitters bind to cationic ion channels and increase excitability. The subject of electrical excitability will be explored in Chapter 19.

Neuromodulators are secreted in a broader manner than neurotransmitters, and this kind of release allows them to influence a large number of cells. They bind to GPCRs and modify the excitability of the target cells by regulating the activities of their ion channels mostly through inhibitory mechanisms. Neuromodulators acting through GPCRs provide a means whereby information from cells influenced by a population of excitable cells can be fed back to the originating population thereby regulating their electrical activities.

There are several routes of G protein regulation of ion channels. There is a direct one where G_α and $G_{\beta\gamma}$ subunits directly bind and regulate ion channels, and a number of indirect ones that operate through the second messenger-binding and second messenger-activated protein kinases. The direct route provides an efficient means for the regulation of electrical activity in excitable cells by chemical messages since GPCRs, G proteins, and ion channels are all membrane associated. A striking example of this form of regulation is the gating of cardiac potassium channels by $G_{\beta\gamma}$ subunits. The specific K^+ channels regulated in this manner are known as *G protein-linked inward rectified K^+ channels* (GIRKs). Binding of acetylcholine to the muscarinic (acetylcholine) GPCR leads to the activation of the GIRKs by $G_{\beta\gamma}$ subunits in cardiac pacemaker cells producing a slowing of the heart rate. This process is not restricted to the heart; GIRKs are also found in the brain and pancreas. A classic example of indirect regulation is phototransduction by rhodopsin and its retinal ligand.

The most common form of neuromodulation is through the indirect, second messenger-mediated regulation of ion channels. In this process, G_s, G_i, and G_q alpha subunits stimulate or suppress second messengers and second messenger-dependent protein kinases that phosphorylate Ca^{2+}, Na^+, and K^+ channels, thereby leading to changes in the ion channels' structural and kinetic properties. Because of the amplification inherent in second messenger systems the neuromodulatory signals are able to modulate the activities of many ion channels at the same time. This indirect method adjusts the overall firing properties of the excitable cells, and in many cases alters the way the neural circuit operates.

12.10 Neuromodulators Influence Emotions, Cognition, Pain, and Feeling Well

Neuromodulators produce changes in the electrical excitability of neurons and neural circuits that can last for hours and days. Some neuromodulators are peptides; others are nucleotides or amines. When acting in the brain

TABLE 12.6. Different categories of neuromodulators: The neuromodulators are classified according to their chemical derivation.

Amines	Peptides
Adrenaline	Angiotensins
Dopamine	Neurokinins
Histamine	Oxytocin
Noradrenaline	Substance P
Serotonin	Vasopressin

Nucleotides	Endogenous opioids
Adenosine	Dynorphins
ADP, ATP	Endorphins
GDP, GTP	Enkephalins

these neuromodulators influence cognition and emotions. Listed in Table 12.6 are a number of the most prominent neuromodulators that act in the brain, such as adenosine, serotonin, and dopamine. Serotonin is distributed at many places in the brain. It is synthesized from the amino acid L-tryptophan, and it is often referred to as 5-hydroxytryptamine (5-HT). Another amine—histamine—was discussed earlier as a mediator of inflammatory responses in the immune system. When secreted by mast cells it can bind to receptors expressed on peripheral nerve endings and serve as a neurmodulator.

Endorphins and tachykinins are examples of peptide neuromodulators. Endorphins—endogenous morphine—along with many other opiates influence perceptions of pain and pleasure. These substances bind to the opioid family of GPCRs expressed on the surface of neurons in the brain and spinal cord. Neuromodulators binding to opioid receptors have been placed under a separate heading in Table 12.6 reflecting their binding properties and actions. Endorphins have analgesic properties, and are also involved in maintaining water balance and other endocrine functions. Members of the endorphinlike family of opiates include beta-endorphin, enkephalins, and dynorphins. Beta-endorphins are synthesized in the pituitary as mentioned earlier, and enkephalins are produced in the adrenal medulla by chromaffin cells. These neuromodulators along with dynorphins are broadly distributed in several specific areas of the brain and spinal cord where they bind to their cognate opioid receptors. The tachykinins are another family of neuromodulators involved in the perception of pain. Mammalian members of this family, the neurokinins, include substance P (SP), neurokinin A (NKA) and neurokinin B (NKB).

Changes in the amount of neuromodulators in sensitive parts of the body produce sensations leading to alterations in behavior. The angiotensins are a family of regulators of blood pressure, blood volume, and cardiac vascular function. Production of angiotensins is stimulated by angiotensin-converting enzyme (ACE) and renin, a glycoprotein hormone produced by the kidney and other places in the body including the brain. An infusion of angiotensin II, the primary active angiotensin, into the brain of mammalian test subjects triggers the sensation of thirst and a craving for sodium (salt). Laboratory animals stop what they are doing and immediately begin drinking, and they exhibit an increased appetite for sodium when so infused.

12.11 Ill Effects of Improper Dopamine Levels

As mentioned earlier, G-protein coupled receptors and the signaling molecules that activate them are involved in a host of neurological disorders and are principal target of therapeutic drugs. *Dopamergic neurons* are those neurons that synthesize and secrete dopamine. Improper dopamine levels contribute to Parkinson's disease, ADHD, schizophrenia, and other disorders. There are three main groups of dopamine secreting cells: one group located in an area of the midbrain called the *substantia nigra* (SN), another in the ventral tegmental area (VTA), and a third in the hypothalamic nuclei. These cells, along with several other smaller groups of dopamine-secreting cells, are referred to as the *dopamine system*. This system of cells is involved in drug addiction, Parkinson's disease, Tourette syndrome, schizophrenia, and attention-deficit hyperactivity disorder (ADHD). In Parkinson's disease, cells in the SN that secrete dopamine die leading to a deficiency in dopamine in the brain. In schizophrenia, antagonists, drugs that inhibit dopamine signaling, are used to suppress an overactive dopamine system. The most common feature of drug addiction is an elevation in dopamine signaling.

Cells in these three areas send out dopamergic signals to a large number of brain areas. The cells receiving these signals express two kinds of dopamine receptors. Receptors of the D_1 type (D_1 and D_5) act through G_s subunits to activate adenylyl cyclase. Receptors of the D_2 type (D_2, D_3, and D_4) exert their influences through G_i subunits to suppress adenylyl cyclase and activate potassium channels.

Children and adults with ADHD exhibit behavior that is inattentive, impulsive, and hyperactive. In many of these patients there is reduction in size of the basal ganglia and frontal lobe responsible for controlling these behavioral responses. It appears that the level of dopamine signaling is too low, either because of the presence of too many dopamine transporters that remove dopamine from the extracellular milieu, or because of inadequate amounts of dopamine released from presynaptic terminals. These deficits are countered through the application of drugs such as methylphenidate [Ritalin®] resulting in increased levels of extracellular dopamine in the brain.

12.12 Inadequate Serotonin Levels Underlie Mood Disorders

The serotonergic system has, as its serotonin-producing core, cells in the midbrain and pons close to where the brain and spinal cord join. The serotonin-producing cells are organized into clusters called *Raphe nuclei*. The neurons in the nuclei are large and send out their processes to almost every locale in the brain. Serotonin receptors can be grouped into seven classes, designated as 5-HT_1 through 5-HT_7. All but the 5-HT_3 receptors are GPCRs. The 5-HT_3 receptors are ligand-gated ion channels (discussed in Chapter 19). The best-studied GPCR classes are the 5-HT_1 and 5-HT_2 receptors. Members of the 5-HT_1 grouping act through G_i and G_0 types of G-proteins alpha subunits. They inhibit adenylyl cyclase activity and stimulate the opening of hyperpolarization-producing potassium channels. Serotonin 5-HT_2 receptors work through G_q subunits to increase IP_3 and intracellular calcium levels, and depolarize neurons by closing potassium channels.

Inadequate serotonin levels are responsible in a variety of mood disorders. Serotonin levels are increased in several ways to treat depression and panic attacks as well as other anxiety disorders. Drugs that increase serotonin levels by inhibiting the reuptake of serotonin are especially popular. Prozac® (fluoxitine) is a selective serotonin reuptake inhibitor (SSRI), as are other popular drugs such as Paxil® and Zoloft®. Serotonin and noradrenaline are monoamines. Another class of drugs that elevate serotonin levels is the monoamine oxidase inhibitors, or MAOIs. Monoamine oxidase is an enzyme that degrades serotonin and noradrenaline. The MAOIs work by inhibiting the actions of monoamine oxidase. Yet another group of serotonin-increasing drugs are the tricyclic antidepressants (TCAs). These too prevent serotonin reuptake.

12.13 GPCRs' Role in the Somatosensory System Responsible for Sense of Touch and Nociception

G protein-coupled receptors play a central role in the sensing and transmission of messages of a warning nature in the somatosensory system. The *somatosensory system* handles four categories of touch information: proprioception, simple touch (pressure, texture, and vibration sensations), temperature (heat sensations), and pain. The term "proprioception" refers to body and body part position, orientation, and location. This faculty utilizes mechanoreceptors that sense forces in muscles, tendons, and joints. Simple touches such as pressure are sensed by special receptors in the skin. Sensations of a warning nature—of temperature and pain—are mediated by free peripheral nerve endings, called *nociceptors*, which extend through-

TABLE 12.7. Hormonal signaling during injury and pain.

Local hormone
Adenosine
ADP, ATP
Bradykinin
Histamine
Prostanoids
Serotonin
Substance P

out the body and use a variety of G protein-coupled receptors and ion channels to sense and transduce signals.

Nociceptors are sensitive to extracellular chemical, mechanical, and thermal environments. In response to inappropriate conditions, they transmit signals through spinal neurons to brain receptors resulting in the perception of pain. The "pain pathway" is the general term given to the signaling route from the peripheral neurons located in places such as the skin to the spinal cord to the brain. Nociceptors are distributed throughout the body, but are generally absent in the brain, deep tissues, and visceral organs such as the liver, spleen, and lungs. The cell bodies of the free nerves are located in dorsal root ganglia (DRG). Signals travel from the DRG to the dorsal horn of the spinal cord and from there to the brain. There are two kinds of free nerve fibers, Aδ and C. Aδ fibers are thin and myelinated, and rapidly conduct sharp pain signals. C fibers are unmyelinated and slowly conduct aching, itching, and burning signals. The C fibers contain receptors for chemical signals sent by injured cells and by cells of the immune system. (Note: Myelinated fibers are fibers, or axons, that are sheathed in myelin, a fatty protein material that insulates the axons and promotes fast conduction of nerve pulses.)

A variety of chemical signals associated with injury and inflammation are exchanged between the nervous and immune systems. Injured cells release some of these warning chemicals, and cells of the immune systems such as mast cells secrete others. These messages (Table 12.7) act as local hormones that bind to receptors on peripheral nerve cells. Some of these signaling molecules and their receptors were discussed earlier. Those not yet examined include prostanoids, bradykinin, and purines such as adenosine. These are discussed next.

12.14 Substances that Regulate Pain and Fever Responses

Nonsteroidal anti-inflammatory drugs (NSAIDs) have been relied on for reducing inflammatory responses including pain and fever for 3500 years. Early medicinal extracts from willow bark led to the first commercial

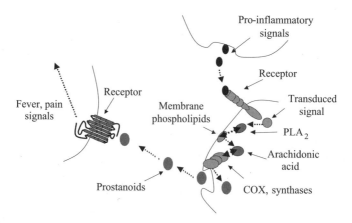

FIGURE 12.7. Generation of fever and pain signals: Pro-inflammatory signals activate phospholipase A2, which hydrolyzes membrane phospholipids (here depicted
as phospholipids in the plasma membrane) resulting in arachidonic acid. A sequence
of enzymes, beginning with COX, produces prostanoids from the arachidonic acid.
The prostanoids are short lived with half lives on the order of a minute or so and
are immediately secreted from the cell. Prostanoid receptors on peripheral nerve
endings bind the prostanoids, and the neurons convey the pain and fever signals to
the brain.

product aspirin one hundred years ago and to a host of more recent
NSAIDs such as ibuprofen and naproxen. All of the NSAIDs work the
same way—they inhibit the catalytic actions of a set of enzymes variously
called *endoperoxide H synthases* (PGHSs) or *cyclo-oxygenases* (COXs).
Two COX isoforms are targeted by anti-inflammatory NSAIDs. The first of
these, COX-1, is constitutively active and is found primarily in the stomach
and kidneys. The second, COX-2, is induced in many different cell types in
response to inflammatory signals conveyed by cytokines, growth factors, and
hormones. Inhibition of COX-1 can be harmful and the aim in drug design
is to preferentially target the COX-2 isoform.

Cells release *prostanoids* (prostaglandins) when injured by chemical,
thermal, or mechanical agents. Prostanoids are fatty acid derivatives of
membrane lipids. As depicted in Figure 12.7, the first step in their production is the hydrolysis of membrane lipids by phospholipase A_2 (PLA_2)
resulting in the release of arachidonic acid (AA). The AA intermediate is
then converted to prostanoids through the sequential actions of the COXs
and several other enzymes. The catalytic activities of PLA_2 are stepped
up in response to the cytokine, growth factor, and hormonal signals. In
response the cells produce and then secrete prostanoids. During the inflammatory response G protein-coupled prostanoid receptors expressed on
peripheral neurons bind prostaglandin. Signals sent on to thermoregulators

in the brain trigger an increase in body temperature. Aspirin and the other NSAIDs act to lower temperature and reduce pain by inhibiting the COXs, thereby preventing formation of and signaling by the prostanoids. The prostanoid signals are conveyed in a paracrine fashion to their cellular targets. For this reason prostanoids and other hormones acting in a paracrine manner are termed "local hormones."

Bradykinin is a nine-amino acid residue peptide rapidly produced after tissue injury occurs. It is a central element in the inflammatory response. It is involved in regulating blood flow (vascular dilation), increasing vascular permeability, smooth muscle contractions, stimulating the release of prostaglandins and other inflammatory mediators, and pain signaling. Bradykinins bind to two kinds of receptors, B_1 and B_2. B_2 receptors are constitutive in neurons and smooth muscle cells; B_1 receptors, in contrast, are rapidly upregulated when there is tissue injury. B_1 and B_2 receptors are found in free nerve endings and directly mediate pain signaling.

Adenosine, ATP, and *ADP* serve as neurotransmitters and neuromodulators in the central and peripheral nervous systems. When released into extracellular spaces these molecules bind to purine receptors. Purine receptors are divided into two families, P1 and P2, each containing a number of subtypes (Table 12.1). Adenosine attaches to P1 receptors while ATP, ADP, and the pyrimidines UTP and UDP bind to P2 receptors. Adenosine and the other nucleotides carry out a large number of signaling tasks in the body. In heart muscle, adenosine decreases heart rate and lowers blood pressure. Adenosine is associated with sleep; during sleep deprivation adenosine levels build up in the brain. When bound to adenosine receptors in the brain, adenosine decreases neural activity leading to sleep. Caffeine is structurally similar to adenosine, and is an adenosine antagonist acting on adenosine A_1 and A_{2A} receptors. Adenosine and ATP are released into the extracellular spaces during injury and inflammation. Mast cells, neutrophils, and free nerve endings express P1 and P2 receptors, and adenosine helps mediate communication between the immune and nervous systems, and contributes to the pain response.

12.15 Composition of Rhodopsin Photoreceptor

The rhodopsin photoreceptor is composed of an opsin GPCR and its ligand, 11-*cis*-retinal. G protein-coupled receptors function as sensors and transducers of information about the external and internal environments. GPCRs are involved in sight, taste, smell, touch, proprioception, and pain. Touch, proprioception, and pain were discussed in the previous sections. In the remaining sections of this chapter, the focus will be on how signals conveyed by exogenous ligands such as light are transduced into cellular responses. The starting point will be how light, or electromagnetic energy,

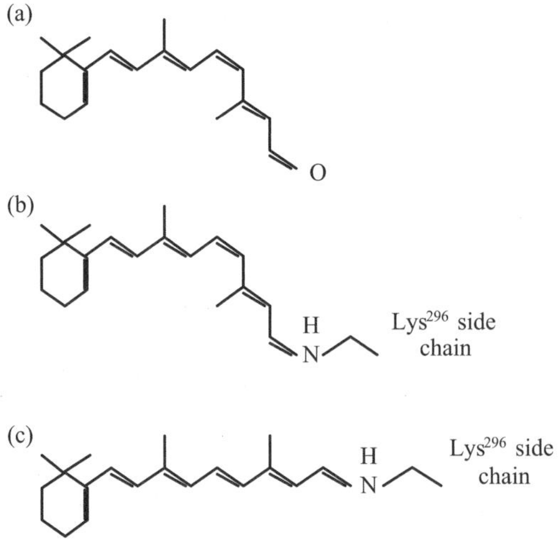

FIGURE 12.8. Retinal photoisomerization: A Schiff base is an organic compound formed by the double bonding of a nitrogen atom of an amino group to a carbon atom. Schiff bases are formed in rhodopsin by the bonding of nitrogen of the NH_3^+ group on a lysine residue to retinal containing a CHO group with the accompanying release of a water molecule. Photoisomerization is the process whereby light impinges on and is absorbed by the double bonded structure resulting in the shifting and twisting of the bonds so that a number of atoms assume a different spatial orientation. (a) The retinal molecule. (b) Retinal in its 11-*cis* form covalently attached to a Lys[296] side chain by a Schiff base. (c) The *all-trans* form of the covalently attached retinal molecule derived from the 11-*cis* form by photoisomerization.

striking a retinal molecule in converted into helix movements, a mechanical form of energy.

The retinal molecule, unlike most other ligands, is covalently linked to opsin and serves as a light-responsive chromophore. The retinal ligand is buried in a pocket that lies deep in the opsin protein. When a photon strikes the retinal chromophore it triggers a conformational change (retinal isomerization) from an 11-*cis* to an all-*trans* configuration, as shown in Figure 12.8. The retinal molecule is attached to the side chain of Lys[296] situated in the H7 helix. The movement associated with the conformational change is appreciable—when stabilizing the alternative all-*trans* configuration there is a 4.5-Å movement that is reflected in the positions of the cytoplasmic portions of the transmembrane helices. In its altered conformation rhodopsin activates the G-protein and initiates signaling in the vision pathway. The retinal is hydrolyzed and dissociated from the opsin after about a minute, enabling a new cycle of 11-*cis*-retinal attachment and activation.

12.16 How G Proteins Regulate Ion Channels

G proteins activated in response to light absorption regulate ion channels. Light absorption stimulates the GEF activity of the GPCR, enabling it to catalyze GDP dissociation from the G_α. The next set of steps used by vertebrates differs from those employed by invertebrates, but both are fairly typical of GPCR signaling. In vertebrates, the G_α subunits act on phosphodiesterases (PDEs), but in invertebrates such as *Drosophila* they target phospholipase Cβ. The ultimate effect of both kinds of second messenger signaling is to trigger changes in ion channel activities. Cyclic GMP is less frequently encountered as a second messenger than cAMP. One place where it has a prominent role is in phototransduction. In vertebrates, rod cells possess cGMP-gated ion channels. In the absence of stimulation by photons these channels are open resulting in membrane depolarization and transmitter release from the rod cells. G_t subunits, called *transducins*, are found in the retina where they couple to the phototransducer rhodopsin. When light strikes rhodopsin the transducins are activated. These subunits activate cGMP phosphodiesterases that hydrolyze cytosolic cGMP to GMP, thereby reducing its concentration. The cGMP-gated ion channels close, leading to reductions in intracellular calcium levels and shifts in the membrane potential towards more negative values (Figure 12.9b).

In *Drosophila*, phospholipase C acts on PIP_2 to generate IP_3 and DAG leading to the opening of members of the transient receptor potential (Trp) family of cation channels. A scaffolding protein called *inactivation no afterpotential* (InaD) helps organize the signaling events that follow G_α separation. This scaffolding protein comprises five PDZ domains. Recall from the last chapter that PDZ domains mediate several different kinds of protein-protein interactions. In *Drosoplila*, InaD binds to the Trps and their regulators, serving as a platform for assembly of signaling complexes (Figure 12.9c). Similar proteins are found in vertebrate nerve terminals where they, too, help organize signaling complexes formed about ion channels.

12.17 GPCRs Transduce Signals Conveyed by Odorants

Olfactory neurons situated in the nasal cavity express receptors for an enormously wide range of chemical signals. The chemical compounds that can be sensed, or odorants, number in the thousands. They include aromatic and alipathic alcohols, aldehydes, esters, ethers, and ketones; aromatic hydrocarbons; and alipathic acids, alkanes, and amines. Tiny changes in structure can be sensed and converted into different odor precepts. The olfactory receptors that function as chemical sensors in these neurons belong to a large family of evolutionarily ancient Group A GPCRs. Many of these receptors have little sequence homology to one another, especially in

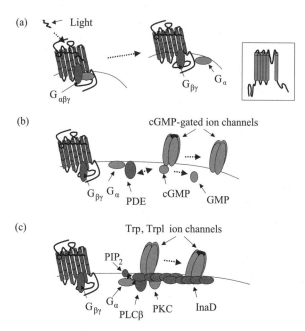

FIGURE 12.9. Phototransduction in vertebrates and invertebrates: (a) Light absorbed by the retinal chromophore results in photoisomerization and GPCR conformational changes leading to activation of heterotrimeric G proteins. (b) Vertebrate signal transduction in which transducin stimulates phosphodiesterase (PDE) activity leading to hydrolysis of cGMP and the closing of cGMP-gated ion channels. The insert depicts the 6-pass transmembrane topology of the subunits of both the cGMP-gated and Trp family ion channels. (c) Invertebrate signal transduction that targets phospholipase Cβ, leading to activation of phospholipid and calcium second messengers, and assembly of signaling complexes organized by InaD scaffolding proteins.

transmembrane regions H3 through H6, a fact consistent with the role of these regions in forming the ligand-binding pocket.

Olfactory signal transduction is depicted in Figure 12.10. Cyclic AMP is now used as a second messenger in place of cGMP. Binding of an odorant activates the G protein signaling to adenylyl cyclase type III, which catalyzes the production of cAMP from ATP. The cAMP molecules bind to the cyclic nucleotide-gated (CNG) ion channels resulting in their opening. The initial signal is then amplified. Calcium ions entering the cell through the CNG channels bind and activate chloride channels so that not only does positive change enter the cell but negative change leaves as well.

The olfactory system uses a combinatorial coding scheme to distinguish between different odorants. A given olfactory neuron expresses a single type of odor receptor (OR) gene on its surface. Each OR can recognize multiple orodant ligands and each kind of odor stimulates a number of

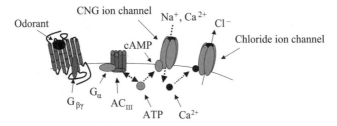

FIGURE 12.10. Olfactory signal transduction: Binding of an odorant activates the G protein signaling to adenylyl cyclase type III that catalyzes the production of cAMP from ATP. The cAMP molecules bind to the cyclic nucleotide-gated (CNG) ion channels resulting in their opening. Calcium ions entering the cell through the CNG channels bind and activate chloride channels. Increased production of calcium and cAMP second messengers activate protein kinases such as protein kinase A (not shown) that phosphorylate the GPCR, leading to its desensitization.

different odorant receptors. Neurons expressing a specific OR gene are dispersed throughout one of four regions in the nasal cavity. The range of ligands recognized by different ORs contains overlaps, and a particular odor is coded by a combination, or pattern, of ORs situated on different cells. The odorant signals are sent from the nose to the olfactory bulb and from there to the piriform cortex. The organization and operation of the olfactory system is discussed in greater detail in Chapter 20.

There are two other families of olfactory receptors besides the odorant receptors. These are expressed in a second olfactory structure, the vomeronasal organ. The two families are known as the V1R family with about 35 members and the V2R family with approximately 150 members. These receptors may function as receptors for mammalian pheremones. Interestingly, V1Rs are Class A GPCRs and V2Rs are Class C GPCRs, paralleling the situation for taste where two families of receptors are found, one (T1Rs) belonging to Class A and a second larger one (T2Rs) to Class C.

12.18 GPCRs and Ion Channels Respond to Tastants

There are five taste modalities—salty, sweet, sour, bitter, and umami. Two of these, salty and sour (acidic), are sensed through interactions of salts and acids with specialized ion channels. These ion channels allow for the direct entry of H^+, K^+, and Na^+ ions into cells localized in taste buds in the tongue. The influx of these ions triggers neurotransmitter release leading to the excitation of other sensory cells resulting in the perceptions of salty and sour. Sweet, bitter, and umami are more complex. These taste modalities are sensed through G protein-coupled receptors.

Umami is the sensation produced by food additive monosodium glutamate (MSG). Glutamate is found in many protein-rich foods such as meat, milk products, and seafood, and is an important nutrient. Umami is sensed by a GPCR that is derived from mGluR4, a metabotropic glutamate receptor belonging to Class C GPCRs. The taste receptor differs from the neurotransmitter-detecting form in that it is missing 50% of the extracellular domain. This modification converts the receptor from a high affinity glutamate detector to a low affinity form suited for sensing amino acid and sweet tastants. There are three receptors in this family of Class A GPCRs. They are designated as T1R1, T1R2, and T1R3. They form heterodimers with one another that transduce sweet (T1R2/T1R3) and umami (T1R1/T1R3) tastants.

Bitter is an exceptionally important modality since it can signal the presence of alkaloids and other potentially harmful toxins. A separate family of GPCRs transduces bitter signals. This family, consisting of 30 or more Class C receptors, is referred to as the T2Rs. The T2Rs are coexpressed with G protein alpha subunits known as *gustducins* (G_g). Gustducin is closely related to transducin (G_t) and is part of the pathway that conveys bitter signals within the cell. The distribution of receptors for taste differs from that for olfaction. The goal in olfaction is not only to recognize a wide range of odors but to discriminate among them as well. As noted above, each neuron expresses one type of olfaction receptor. Bitter tastes serve as warnings of potentially dangerous substances, and it is not necessary for the body to discriminate among the different sensations of bitter. Many types of bitter receptors are coexpressed in each cell thus maximizing sensitivity at the expense of specificity.

References and Further Reading

G Protein-Coupled Receptors

Baldwin JM, Schertler GFX, and Unger VM [1997]. An alpha carbon template for the transmembrane helices in the rhodopsin family of G protein-coupled receptors. *J. Mol. Biol.*, 272: 144–164.

Bockaert J, and Pin JP [1999]. Molecular tinkering of G protein-coupled receptors: An evolutionary success. *EMBO J.*, 18: 1723–1729.

Gether U [2000]. Uncovering molecular mechanisms involved in activation of G protein-coupled receptors. *Endocr. Rev.*, 21: 90–113.

Gether U, and Kobilka BK [1998]. G protein-coupled receptors II: Mechanism of agonist activation. *J. Biol. Chem.*, 273: 17979–17982.

Ji TH, Grossmann M, and Ji I [1998]. G protein-coupled receptors I: Diversity of receptor-ligand interactions. *J. Biol. Chem.*, 273: 17299–17302.

Palczewski K, et al. [2000]. Crystal structure of rhodopsin: A G protein-coupled receptor. *Science*, 289: 739–745.

GPCR Regulation

Bockaert J, et al. [2003]. The 'magic tail' of G protein-coupled receptors: An anchorage for functional protein networks. *FEBS Lett.*, 546: 65–72.

Grimes ML, and Miettinen HM [2003]. Receptor tyrosine kinase and G protein-coupled receptor signaling and sorting within endosomes. *J. Neurochem.*, 84: 905–918.

Hall RA, and Lefkowitz RJ [2002]. Regulation of G protein-coupled receptor signaling by scaffold proteins. *Circ. Res.*, 91: 672–680.

Hall RA, Premont RT, and Lefkowitz RJ [1999]. Heptahelical receptor signaling: Beyond the G protein paradigm. *J. Cell Biol.*, 145: 927–932.

Koenig JA, and Edwardson JM [1997]. Endocytosis and recycling of G protein-coupled receptors. *Trends Pharmacol. Sci.*, 18: 276–287.

Lefkowitz RJ [1998]. G protein-coupled receptors III: New roles for receptor kinases and arrestins in receptor signaling and desensitization. *J. Biol. Chem.*, 273: 18677–18680.

Lodowski DT, et al. [2003]. Keeping G proteins at bay: A complex between G protein-coupled receptor kinase 2 and Gβγ. *Science* 300: 1256–1262.

Luttrell LM, and Lefkowitz RJ [2002]. The role of β-arrestins in the termination and transduction of G protein-coupled receptor signals. *J. Cell Sci.*, 115: 455–465.

G Proteins and Their Effectors

Hamm HE [1998]. The many faces of G protein signaling. *J. Biol. Chem.*, 273: 669–672.

Wedegaertner PB, Wilson PT, and Bourne HR [1995]. Lipid modifications of trimeric G proteins. *J. Biol. Chem.*, 270: 503–506.

Adenylyl Cyclases and Nucleotide Phosphodiesterases

Beavo JA [1995]. Cyclic nucleotide phosphodiesterases: Functional implications of multiple isoforms. *Physiol. Rev.*, 75: 725–748.

Cooper DMF, Mons N, and Karpen JW [1995]. Adenylyl cyclases and the interaction between calcium and cAMP signaling. *Nature*, 374: 421–424.

Houslay MD, and Milligan G [1997]. Tailoring cAMP-signaling responses through isoform multiplicity. *Trends Biochem. Sci.*, 22: 217–224.

Taussig R, and Gilman AG [1995]. Mammalian membrane-bound adenylyl cyclases. *J. Biol. Chem.*, 270: 1–4.

The Somatosensory System and Nociception

Marceau F, Hess JF, and Bachvarov DR [1998]. The B1 receptors for kinins. *Pharmacol. Rev.*, 50: 357–386.

Negishi M, Sugimoto Y, and Ichikawa A [1995]. Molecular mechanisms of diverse actions of prostanoid receptors. *Biochem. Biophys. Acta*, 1259: 109–120.

Smith WL, Garavito RM, and DeWitt DL [1996]. Prostaglandin endoperoxide H synthases (cyclooxygenases)-1 and -2. *J. Biol. Chem.*, 271: 33157–33160.

Phototransduction

Baylor D [1996]. How photons start vision. *Proc. Natl. Acad. Sci. USA*, 93: 560–565.

Clapham DE, Runnels LW, and Strübing C [2001]. The Trp ion channel family. *Nature Rev. Neurosci.*, 2: 387–396.

Hardie RC, and Raghu P [2001]. Visual transduction in Drosophila. *Nature*, 413: 186–193.

Kramer RH, and Molokanova E [2001]. Modulation of cyclic-nucleotide-gated channels and regulation of vertebrate phototransduction. *J. Exp. Biol.*, 204: 2921–2931.

Odorants and Pheromones

Buck LB [2000]. The molecular architecture of odor and pheromone sensing in mammals. *Cell*, 100: 611–618.

Clyne PJ, et al. [1999]. A novel family of divergent seven-transmembrane proteins: Candidate ororant receptors in Drosophila. *Neuron*, 22: 327–338.

Firestein S [2001]. How the olfactory system makes sense of scents. *Nature*, 413: 211–218.

Malnic B, et al. [1999]. Combinatorial receptor codes for odors. *Cell*, 96: 713–723.

Zhao HQ, et al. [1998]. Functional expression of a mammalian odorant receptor. *Science*, 279: 237–242.

Tastants

Adler E, et al. [2000]. A novel family of mammalian taste receptors. *Cell*, 100: 693–702.

Chandrashekar J, et al. [2000]. T2Rs function as bitter taste receptors. *Cell*, 100: 703–711.

Chaudhari N, Landin AM, and Roper SD [2000]. A metabotropic glutamate receptor variant functions as a taste receptor. *Nature Neurosci.*, 3: 113–119.

Lindemann B [2001]. Receptors and transduction in taste. *Nature*, 413: 219–225.

Margolskee RF [2002]. Molecular mechanisms of bitter and sweet taste transduction. *J. Biol. Chem.*, 277: 1–4.

Nelson G, et al. [2001]. Mammalian sweet taste receptors. *Cell*, 106: 381–390.

Problems

12.1 The β-arrestins can function as switches that first shut down signaling through the G proteins and then turn on signaling through growth pathways. When switching to non-G protein modes of signaling, the arrestin molecule and the GPCR serve as a platform for the assembly of Src and other signaling proteins. The essential features that mediate these associations are the presence of peptide sequences that bind to SH2, SH3, and PDZ domains, either in the third intracellular loop or in the cytoplasmic COOH-terminal tail, and the Src non-receptor tyrosine kinase that can bind to the proline rich motifs of the β-arrestins by means of its SH3 domain. Construct a growth pathway leading to the nucleus that begins at the GPCR and is routed through Src that binds to the β-arrestins.

12.2 One of the G protein alpha subunits, G_s, stimulates adenylyl cyclases to produce cAMP, which activates protein kinase A. In β_2 adrenergic GPCR (β_2AR) signaling, protein kinase A phosphorylates the GPCR, and this negative feedback loop not only desensitizes the receptor but also switches its G protein preference from G_s to G_i. What are two, more conventional targets of protein kinase A signaling? Draw the routes starting from ligand-receptor binding. Using information presented in Tables 12.2 and 12.3, diagram some the routes of activation of protein kinase B and protein kinase C by the G proteins.

13
Cell Fate and Polarity

There are approximately 10^{14} cells in the human body. Some of these are heart cells; others are liver, kidney, nerve, or muscle cells. During development sets of identical progenitor cells undergo different cells fates. They diverge at the correct time to form different cell types along the appropriate body axes and boundaries. Mutual signaling between these cells drives many of these decisions so that each cell knows what kind of cell to become. These signals are transmitted from cell to cell, transduced across the plasma membrane, and sent on to the nucleus where they activate specific sets of developmental genes. Some genes are turned on and others are turned off in response to these signals.

A small number of signaling pathways guide embryonic development. In the first part of this chapter, four pathways of particular importance with respect to development—Notch, transforming growth factor-β (TGF-β), Wnt, and hedgehog—will be examined. These pathways regulate the programs of gene expression so that at the correct time in the right place cells with the same propensity for a particular cell fate give rise to daughters exhibiting differences in morphology and the mix of proteins being expressed.

A variety of stratagems are used to achieve the developmental goals. One of these is to lay down gradients of signaling proteins either on cell surfaces or in extracellular spaces that help determine cell fate when they activate receptors on a cell. These signaling proteins are known as *morphogens*. Another stratagem is to utilize hierarchical sequences of gene expression so that over time different progeny will become different kinds of cells. The focus in the first part of the chapter will be on how signals are sent from the cell surface to the nucleus through the four pathways. In the second part of the chapter, the goal will be to see how morphogen gradients and hierarchical patterns of gene expression guide cell fate decisions.

The Notch, TGF-β, Wnt, and Hedgehog signaling pathways are named either for the transmembrane receptor or for the molecules that serve as ligands for the receptors. These four pathways are highly conserved in multicellular organisms. They have been studied extensively in the fly (*Drosophila*), worms (*C. elegans*), and vertebrates. A variety of rather colorful names have been given to signaling proteins belonging to these pathways. The names are usually derived from the types of developmental defects seen in *Drosophila* when the genes encoding the proteins suffer a mutation, usually of the loss-of-function type. For example, when the Hedgehog gene is mutated, a spiky process called denticles is seen, and hence the name hedgehog was given to the protein. In the Notch pathway, partial loss-of-function defects in the Notch receptor produced notches in the wing. Defects in the Groucho protein, a downstream-acting element that participates in several pathways, result in the production of bristles around the eye. These aberrant structures resemble the eyebrows of the well-known comedian and hence the name Groucho was given to the gene and its protein product.

13.1 Notch Signaling Mediates Cell Fate Decision

The Notch pathway mediates numerous cell fate decisions. The pathway has a central role in determining which cells become neurons and which do not during the early stages of development. This same pathway is utilized in a variety of cellular contexts to generate a broad spectrum of cell fate decisions. Depending on cellular context Notch pathway mediates patterning, terminal differentiation, mitosis, and apoptosis fates.

There are three core components of the Notch signaling pathway: ligands, receptors, and effectors functioning as transcription factors. Like the other central signaling pathways, these components are highly conserved across phyla, and corresponding members of the pathway for vertebrates, fly, and worm are listed in Table 13.1. Notch signals through a juxtacrine mechanism in which a Notch receptor expressed on the surface of one cell binds to a Delta/Serrate/Lin (DSL) ligand (Table 13.1) expressed on the surface of an adjacent cell. Notch molecules are 300-kDa transmembrane receptors. They possess large extracellular domain containing 29–36 tandem epidermal growth factor (EGF) repeats, and three cysteine-rich Lin/Notch repeats (LNRs). The extracellular EGF and LNR repeats mediate DSL ligand binding and Notch activation. The intracellular domain contains an NLS, 6 Cdc10/ankyrin repeats and a PEST motif, the latter a region rich in prolinc (P), glatamic acid (E), serine (S), and threonine (T) residues.

TABLE 13.1. The Notch signaling pathway: Abbreviations—Suppressor of Hairless [Su(H)].

Ligands	Vertebrates	*Drosophila*	*C. elegans*
Ligands	Delta1, 2, Jagged1, 2	Delta, Serrate	LAG-2, Apx-1
Receptors	Notch1–4	Notch, LIN-12	GLP-1
Transcription factors	CBF-1	Su(H)	LAG-1

FIGURE 13.1. Structure of the Notch protein: Shown are the 180kDa extracellular chain and the 120kDa transmembrane/intracellular chain. Three cleavage sites are indicated in the figure; these are labeled by the corresponding proteolytic enzymes.

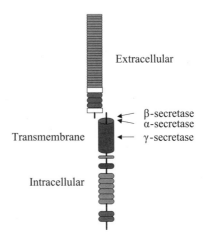

Extracellular

Transmembrane

β-secretase
α-secretase
γ-secretase

Intracellular

Notch is synthesized as a 300-kDa precursor molecule. This primary transcript is cleaved in two in the trans-Golgi network. The two fragments remain associated with one another during translocation and insertion in the plasma membrane resulting in the formation of a heterodimer. The heterodimer consists of an N-terminal 180-kDa molecule containing the extracellular EGFs and LNRs, and a smaller C-terminal 120-kDa molecule possessing a short extracellular segment, the transmembrane sequence and the cytosolic region (Figure 13.1).

The 120-kDa C-terminal chain is cleaved at several locations to create the Notch intracellular domain (NICD), a fragment that is released from the membrane and can move to the nucleus where it forms a complex with several other proteins. Su(H) binds the NICD and together the two proteins enter the nucleus (Figure 13.2). They act as transcription factors to active genes belonging to the enhancer of a split cluster, which acts to suppress neural development. In more detail, Notch and Su(H), more generally, CSL (CBF1, Su(H), Lag-1), work together to stimulate the transcription of genes belonging to the enhancer of split E(spl) cluster. The E(spl) gene products, in turn, inhibit transcription of a cluster of proneural genes referred to as the *achaete-scute complex*. Since these genes are not transcribed, cells transducing the Notch signals in response to Delta ligand binding are inhibited from adopting a neural cell fate. A positive feedback loop operates to help drive unambiguous decisions and the overall process is referred to as *lateral inhibition*.

13.2 How Cell Fate Decisions Are Mediated

Lateral inhibition and positive feedback mediates cell fate decisions. In the absence of Notch signaling most cells in an equivalence group will adopt the same primary fate. Notch signaling restricts the number of cells travel-

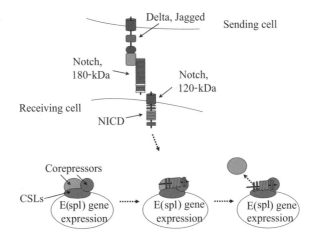

FIGURE 13.2. Signaling through the Notch pathway: The sending cell expresses the Notch ligand Delta or Jagged, while the receiving cell expresses the Notch receptor. The extracellular chain of Notch binds the ligand and in response the transmembrane segment is cleaved at several places to form the NICD. The NICD, together with Su(H), translocates to the nucleus where they interact with several other proteins to activate gene transcription. As depicted in the figure, they may promote gene transcription by displacing corepressors.

ing along this fate pathway by promoting either a different pathway or by maintaining the cells in an uncommitted state. The general mechanism for restricting cell fate decisions is called *lateral inhibition* because a cell adopting a particular cell fate inhibits its neighbors from adopting the same fate.

A key factor that promotes cell fate decisions is the use of a positive feedback loop where small stochastic differences in expression levels are amplified and drive the decision. Lateral inhibition works in the following manner (Figure 13.3). Cells express both Notch and Delta on their surface. Notch receptors on one cell bind to Delta ligands on the other. When a cell's Notch receptor binds a Delta ligand on an opposing cell it develops a diminished capacity for expressing Delta ligands on its own surface. Cells expressing Notch more strongly will receive a stronger inhibitory signal and inhibit its neighbor less strongly. This generates a feedback loop that amplifies initially small stochastic differences in receptor expression.

13.3 Proteolytic Processing of Key Signaling Elements

Proteolytic processing plays a key role in many signaling pathways. Notch is cleaved after ligand binding, and the cytosolic fragment translocates to the nucleus where it functions as a transcription factor. Thus, proteolytic

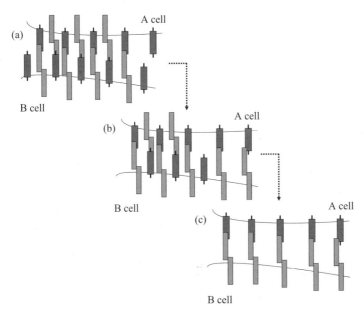

FIGURE 13.3. Lateral inhibition through cell-to-cell signaling and positive feedback: (a) Because of stochastic fluctuations there are more Notch receptors on the B cell than on the A cell and, other things being equal (i.e., the density of Delta ligands on the two cells being the same), signaling into the B cell is stronger than in the A cell. (b) In the B cell, Notch signal transduction upregulates Notch receptor expression and reduces Delta ligand expression. On the A cell the strength of Notch signaling fails to compensate for the decay of Notch receptors over time. Thus on the B cell Delta ligand expression goes down while on the A cell Notch receptor expression declines. (c) This tendency is amplified over time leading to an A cell expressing Delta ligands and a B cell expressing the Notch receptors.

processing enables a single gene product to function first as a receptor and then as a transcription factor. As noted earlier proteolytic processing of the 300-kDa precursor takes place in a trans-Golgi network resulting in formation of the two-chain receptor.

Proteolytic processing plays key role in signaling through the Hedgehog and Wnt pathways, too. As will be seen shortly in Sections 13.6 and 13.7 these processes are, in turn, regulated by protein phosphorylation. In these pathways, these two forms of posttranslational modification work together to activate elements of the signaling pathways located within the cell, in the cytosol, in response to ligand binding at the plasma membrane.

Several families of enzymes participate in cleaving the transmembrane form of the Notch receptor leading to release of the NICD. One of the families of proteolytic enzymes is the *Presenilins*. These enzymes have been the subjects of intensive study in the past few years due to their similar involve-

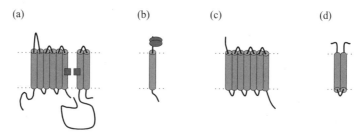

FIGURE 13.4. Components of the γ-secretase: (a) Presenilin—The small squares denote the location of a pair of aspartate residues crucial for the catalytic activities of presenilin; (b) Nicastrin—Ovals located on extracellular part of the chain denote essential glycosylation sites. (c) Aph-1. (d) Pen-2.

ment in cleaving the amyloid precursor protein (APP) leading to the release of the Aβ amyloid protein, an important step in the onset of Alzheimer's disease. Like the Notch process depicted in Figure 13.1 the APP undergoes several cleavages. These are mediated by enzymatic complexes referred to as α-secretases, β-secretases, and γ-secretases. The Presenilins are part of the γ-secretase complex, which has three other members—Nicastrin, Aph-1 and Pen-2. As illustrated in Figure 13.4 the Presenilins are 8-pass transmembrane (TM) proteins proteolytically cleaved into two chains. A pair of aspartate residues positioned in opposition to one another is crucial for Presenilin's γ-secretase actions. Notch and APP pass through the gap formed by the TM segments containing the aspartates, and are cleaved. The Presenilins carry out their catalytic activities together with the single-pass Nicastrin protein, the 7-pass Aph-1 protein, and the 2-pass Pen-2 protein, which help to assemble and stabilize the complex.

The APP is cleaved twice, one in its extracellular domain just outside the membrane and the other near the middle of the intermembrane region. In the first step, the APP proteins are cleaved at one of two alternative extracellular locations termed the α- and β-sites. The γ-secretases are responsible for subsequent cleavages within the transmembrane segment. In the case of cleavage at β-sites, but not α-sites, the products are 40- and 42-amino acid residue forms of the Aβ amyloid protein. The 42-residue form, favored by the mutated Presenilins, is especially prone to form clumps due to exposure of hydrophobic patches, as was discussed in Chapter 5. The α-secretases and β-secretases responsible for cleaving the APPs at the α- and β-sites are members of two other families of proteases. One or more members of the "a disintegrin and metalloprotease" (ADAM) family are the α-secretases, while aspartyl proteases belonging to the membrane-associated aspartyl protease (memapsin) family serve as the β-secretases.

The ADAM enzymes are members of a large group of multidomain enzymes that possess not only metalloproteinase domains, but also integrin-binding regions and a cytoplasmic domain that bind a variety of signaling proteins. Members of the ADAM family of proteases degrade collagens and other extracellular matrix proteins. This activity is particularly important during embryonic development. During that time the ECM has to be continuously remodeled to accommodate new growth and emerging organs and appendages. The ADAM proteins also promote the creation of soluble signaling proteins from membrane-bound forms in another developmentally important process called *ectodomain shedding*. In this process, membrane-bound proteins are cleaved at sites just outside the plasma membrane, thereby converting ligands and receptors that signal in a short range, juxtacrine manner into soluble signaling proteins that can signal in a longer-range, paracrine manner. One of the ADAM family members, a protein called *tumor necrosis factor-α converting enzyme* (TACE) can cleave the TNF-α ligand, as its name suggests, and functions as the α-secretase for Notch and the APP perhaps along with another protein called ADAM10 in mammals and Kuzbanian in *Drosophila*. Lastly, degradation of ECM molecules is necessary for cell migration. The upregulation of metalloproteases (metalloproteinases) accompanies the downregulation of cell adhesion molecules, increased angiogenesis, and increased motility in the progression of cancer to metastasis.

13.4 Three Components of TGF-β Signaling

The TGF-β signaling pathway has three main components: ligands, receptors and Smad signal transducers. The transforming growth factor-β pathway, like the Notch pathway, has many different effects on cell fate depending on the stage of development. Among the processes triggered through this pathway are patterning, wound healing, bone formation, ECM production, and homeostasis. There are three core components in a TGF-β pathway—a ligand, a cell surface receptor complex, and a set of cytoplasmic signal transducers called *Smads*. The TGF-β pathway is named for the TGF-β subfamily of diffusible polypeptide ligands. Other prominent ligands involved in signaling though the TGF-β pathway are the bone morphogenetic proteins (BMPs), growth and differentiation factors (GDFs), members of the Activin subfamily, and Nodal.

TGF-βs, BMPs, GDFs, and related ligands signal through receptors belonging to the TGF-β receptor family. These receptors are transmembrane glycoproteins possessing cytoplasmic domains that phosphorylate serine and threonine residues of target proteins. Thus they belong to the family of receptor serine/threonine kinases. A sequence of events involving one or more members of two distinct TGF-β receptor groups is required in order to signal across the plasma membrane. The two receptors groups are designated Type I and Type II receptors, abbreviated as TβRI and TβRII. Members of each group possess an extracellular ligand-binding domain and a cytoplasmic kinase domain. Type I receptors have an additional cytoplas-

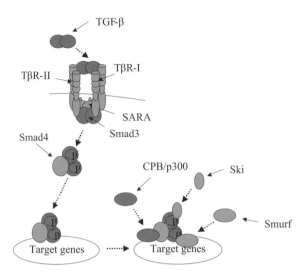

FIGURE 13.5. Signaling through the TGF-β pathway: Binding of a ligand (TGF-β) dimer to a TβRII-TβRI receptor complex leads to phosphorylation of serine/threonine residues in the GS region of the TβRIs resulting in the activation of the kinase domain. The activated receptors phosphorylate the Smad3 proteins leading to the latter's dissocation from the SARA anchors. They form a heterotrimeric complex with Smad4, and in that form translocate to the nucleus where they form a scaffold for assembly of transcriptional activators and cofactors such as CPB/p300 and Ski to induce the transcription of target genes.

mic domain referred to as a *GS domain* that functions as a key regulatory region.

Two different mechanisms of ligand-receptor binding are utilized by TGF-β receptors. BMP ligands have a high affinity for Type I receptors and a low one for Type IIs. Once BMP ligands bind the Type Is, these receptors have a high affinity for Type II receptors. The assembly setups are, as follows. First, a complex is formed containing a BMP ligand dimer bound to a Type I receptor dimer. This assembly then binds to two Type II receptors to make a receptor foursome, as shown in Figure 13.5. TGF-β and Activin utilize a different assembly method. These ligands prefer to bind to Type II receptors. Their first step is the binding of ligands to Type II receptor molecules. These binding events serve to recruit Type I receptor molecules to the assemblage, resulting again in the formation of a complex consisting of a ligand dimer bound to a receptor foursome. As a consequence of either series of binding events the cytoplasmic domains of the Type II and Type I proteins are brought into close proximity and stabilized.

The next step in transducing the signal into the cell is transphosphorylation of serine and threonine residues in the GS domains of the Type I receptors, assisted by the Type II receptors. This step creates docking sites for a number of different adapter proteins. One of these, the Smad anchor for receptor activation (SARA), couples Smad2/3 to the activated receptor complex. The SARA protein possesses a lipid-binding FYVE domain that enables it to tether to the plasma membrane. Two other domains permit it to bind simultaneously to the receptor and to the Smad protein. Once the Smad proteins have been recruited to the receptor complex they are phosphorylated, triggering their release into the cytosol (Figure 13.5).

13.5 Smad Proteins Convey TGF-β Signals into the Nucleus

The Smad proteins convey messages from the cell surface to the nucleus. The name "Smad" is a contraction of the names of the first two Smad-type proteins to be discovered—the *C. elegans* Sma protein and the *Drosophila* mothers against dpp, or Mad, protein. There are at least eight known Smads, designated Smad1 through Smad8, and they fall into three categories (Table 13.2). Five of the Smads are receptor-activated and they are called R-Smads. Smad4 is required for signaling through all pathways and is called a common Smad, or co-Smad. The other two Smads, Smad6 and Smad7 are inhibitory. They turn off signaling and form an anti-Smad grouping. Smad 6 inhibits signaling through the BMP branch while Smad7 inhibits signaling through all of the branches of the TGF-β signaling pathway.

Smad proteins contain a pair of globular signaling domains connected by a linker region. The N-terminal Mad Homology-1 (MH1) and C-terminal

TABLE 13.2. The TGF-β signaling pathway: As indicated in the table, the TGF-β pathway branches at the Type I receptors (Type I Rs) level into an Activin-like branch and a BMP-like branch. Abbreviations—Activin receptor-like kinase (ALK); bone morphogenetic protein (BMP); transforming growth factor (TGF); receptor (R); inhibitory (I).

	Activin	TGF-β	TGF-β	BMP
Ligands	Activins	TGF-βs	TGF-βs	BMPs
Type II Rs	ActII/IIB	TβRII	TβRII	BMPRII, ActRII/IIB, MISRII
Type I Rs	ALK4	ALK5	ALK1	ALK3, ALK6, ALK2
R-Smad	Smad2, Smad3	Smad2, Smad3	Smad1, Smad5, Smad8	Smad1, Smad5, Smad8
Co-Smad	Smad4	Smad4	Smad4	Smad4
I-Smad	Smad7	Smad7	Smad6, Smad7	Smad6, Smad7

Mad Homology-2 (MH2) domains mediate protein-DNA (MH1) and protein-protein (MH2) interactions. In addition, the R-Smad proteins contain a phosphorylation site near their C-terminal. Once activated by TβRI, the R-Smads translocate to the nucleus. On the way to the nucleus the R-Smads associate with the co-Smad (Smad4) as illustrated in Figure 13.5 for the case of Smad3.

The specific cell fate decision arrived at through the TGF-β pathway depends on context. The steps outlined above are general and are neither cell-specific or developmental stage-dependent. Partner and accessory signaling molecules that combine with the Smads to form the active transcriptional complexes supply the contextural information. Coactivators, corepressors and other partners provide cell type and developmental stage inputs, and they fix parameters such as the duration of transcription. A few of these cofactors are presented in Figure 13.5. The p300 and CBP proteins are coactivators. A number of corepressors associate with the Smad proteins. Examples of these additional factors are the Sloan-Kettering Institute proto-oncogene (Ski), included in the figure, and the Ski-related novel gene N (SnoN) and the TG3-interacting factor (TGIF), not included.

Smads are regulated by ubiquitination. Recall that ubiquitination prepares substrate proteins for proteolytic destruction by the proteosome. It involves sequential operations by E1 ubiquitin activating enzyme, E2 ubiquitin conjugating enzyme, and E3 ubiquitin ligase enzyme. As is the case for all four of the pathways discussed in this chapter, selective, regulated proteolysis is an important mechanism for controlling what signals are sent on to the nucleus. In the TGF-β signaling pathway, proteolytic regulation of the Smads takes place both in the cytosol and in the nucleus. Members of the Smurf family of E3 ligases are recruited to the Smad scaffolds. The WW domains of the Smurfs bind PPXY motifs in the linker region connecting the MH1 and MH2 domains of the Smads. Members of the UbcH5 family of E2s are present, as well. These operations are represented in Figure 13.5 by the presence of a generic Smurf protein.

13.6 Multiple Wnt Signaling Pathways Guide Embryonic Development

The Wnt pathway is named for the Wnt family of secreted glycoproteins, which function as morphogens during development in vertebrates and invertebrates. Wnt ligands are typically 350 to 400 amino acid residues in size and are characterized by a highly conserved cysteine-rich domain (CRD), a pattern of 23 cysteine residues. A large number of Wnt proteins have been identified. There are at least 25 vertebrate Wnt genes, 7 *Drosophila* genes, and 5 *C. elegans* genes. The most prominent of the *Drosophila* Wnt family members is called *Wingless* (Wg), and is the homolog to the Wnt-1 protein found in humans. This ligand is responsible

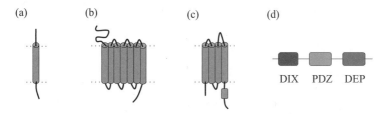

FIGURE 13.6. Wnt signal receptors and transducers: (a) LRP coreceptor. (b) Friz-zled receptor. (c) Strabismus co-receptor. The rectangle attached to the cytoplasmic C-terminal tail of the receptor denotes a PDZ binding domain. (d) Dishevelled adapter. Abbreviations—Dishevelled, Egl-10, and Pleckstrin (DEP); Dishevelled and axin (DIX). The DIX domain mediates attachment to the cytoskeleton, and the DEP domain promotes binding to the plasma membrane.

for a well-studied anterior/posterior decision that defines tissue polarity. (Note: Wnt is a contraction of wingless (Wg), and the murine, or mouse, homolog Int.)

Wnt ligands bind to *Frizzled receptors.* These receptors are named for the *Drosophila* frizzled gene involved in the tissue polarity decisions. Frizzled proteins have a large extracellular CRD responsible for ligand binding, a 7-pass (serpentine) transmembrane region and a short cytoplasmic C-terminal region. Several coreceptors operate in conjunction with the Frizzled receptors to transduce signals into the cell following ligand binding. The chain topology of Frizzled and two of its coreceptors, LRP and Strabismus, are shown in Figure 13.6. As shown in the figure, LRP is a single-pass recep-tor while Strabismus passes back and forth through the plasma membrane four times. The first cytoplasmic step following ligand binding in some Wnt pathways is the recruitment of an adapter protein, Dishevelled, to the plasma membrane. Coreceptors such as Strabismus exert their influences by interacting with this adapter protein.

The Wnt signaling pathway has several branches. One of these is referred to as the *canonical* or *β-catenin pathway.* Another is termed the *planar cell polarity* (PCP) pathway. As depicted in Figure 13.7, this pair of pathways splits when signals reach the cytoplasmic adapter protein Dishevelled. The decision of whether to activate the β-catenin or the planar polarity pathway is determined by interactions between the coreceptors and this adapter. The Strabismus coreceptor interacts with Dishevelled through their PDZ domains. It shifts the routing of the signals away from the β-catenin pathway and towards the PCP pathway by isolating the adapter from the Frizzed receptor, while the LRP coreceptor assists in signaling through the canon-ical pathway.

The canonical Wnt pathway leads through β-catenin. This protein is a transcription activator, but in the absence of Wnt signaling is prevented from performing this function. It forms a complex in the cytoplasm with

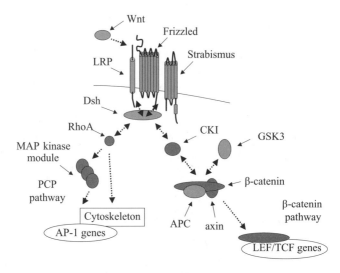

FIGURE 13.7. Signaling through the Wnt β-catenin and PCP pathways: Ligand binding activates either the canonical β-catenin pathway or the planar cell polarity pathway, depending on the actions at the Dishevelled switching point. In the canonical pathway, Dishevelled acts through the CKI kinase to influence the decision whether LEF/TCF gene transcription is repressed or activated. Alternatively, Dishevelled acts through the small GTPase protein RhoA and downstream MAP kinases such as JNK to promote cytoskeleton polarization and gene expression.

axin, the adenomatous polyposis coli (APC) protein, and two serine/threonine kinases—GSK3 and CKI. This module works in the following way. The kinases phosphorylate the other members of the complex. Phosphorylation of axin stabilizes it; phosphorylation of APC enhances its interactions with β-catenin. β-catenin cannot signal because phosphorylation tags it for proteolytic destruction. As a consequence of the phosphorylations, β-catenin does not translocate to the nucleus and stimulate transcription. Dishevelled acting through CKI destabilizes the complex, and leads to dephosphorylation of β-catenin. The number of mobile and untagged β-catenin molecules increases, and they are able to translocate to the nucleus and stimulate transcription of TCF/LEF genes (Figure 13.7).

The presence of two serine/threonine kinases in the complex leads to a two-step phosphorylation/degradation process. CKI acts first to phosphorylate substrate residues within the complex. This action "primes" the system for subsequent phosphorylation by GSK3, which is then followed by the recruitment of proteolytic elements to β-catenin. In more detail, CKI phosphorylates β-catenin at Ser45. This action enables GSK3 to phosphorylate β-catenin at three neighboring sites (Ser33/Ser37/Thr41). In the absence of the priming CKI phosphorylation, GSK3 cannot phosphorylate β-catenin at the aforementioned sites.

13.7 Role of Noncanonical Wnt Pathway

The noncanonical Wnt pathway guides the development of planar cell polarity and convergent extension. *Planar cell polarity* (PCP) is the term used to describe coordinated patterns of polarization of cells lying within the planar epithelium or sheet. The two best-known examples of PCP are in the *Drosophila* wing and compound eye. In the wing, hairs produced by epithelial cells all point in the same direction, towards the distal tip. In the compound eye, oriented hexagonal arrays of photoreceptor cell clusters (ommatidia) are formed. In vertebrates such as frogs and fish, cells in a tissue shift about and change shape so that their distribution becomes narrower along one axis and longer about an axis perpendicular to the first one. These developmental rearrangements are referred to as *convergent extension*. They, too, are guided by signaling through the PCP pathway.

Signals sent through the planar cell polarity pathway influence the organization of the cytoskeleton and gene expression. Whereas signals were routed by Dishevelled to the CKI and GSK3 kinases and to the β-catenin module in the β-catenin pathway, they are routed by Dishevelled to the RhoA GTPase and then to JNK family of MAP kinases in the PCP pathway resulting in activation of c-Jun and AP-1 dependent transcription (Figure 13.7). A third route, which may actually be a branch of the PCP pathway since it appears to involve Dishevelled and Strabismus, utilizes G proteins, increased Ca^{2+} second messenger production, and activation of calcium-dependent serine/threonine kinases and phosphatases such as CaMKII, PKC, and calcineurin.

One of the crucial steps in generating planar cell polarity is symmetry breaking—the creation of asymmetries in the cell population using components that are initially distributed uniformly about each cell. Several proteins working in conjunction with Strabismus and Dishevelled accomplish this task. One of these proteins is an adapter protein called Prickle. This protein binds to Strabismus and sequesters Dishevelled near the coreceptor and away from Frizzled. A feedback loop amplifies initial random and small differences in Prickle concentration to produce sharp asymmetries between cells that signal to the nucleus through the PCP pathway and those for which this pathway is shut down.

13.8 Hedgehog Signaling Role During Development

The fourth signaling pathway to be discussed in this chapter is the signaling pathway named for the *Hedgehog (Hh) family* of diffusible cell-to-cell signaling molecules. Members of the Hh family include Hedgehog in *Drosophila* and at least seven vertebrate Hedgehogs, the most prominent of which are *Sonic hedgehog* (Shh), *Desert hedgehog* (Dhh), and *Indian hedgehog* (Ihh). This family regulates a wide spectrum of developmental

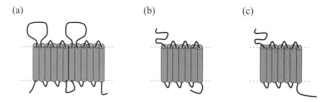

FIGURE 13.8. Hedgehog receptors Patched and Smoothened: (a) Patched receptor. (b) Smoothened receptor in a closed conformation. (c) Smoothened receptor in an open configuration.

events. For example, the Hh pathway is responsible for body, wing, eye, genitals, and leg patterning in *Drosophila* while the Shh pathway regulates eye, portions of the brain, hair, lung, gut, bladder, and urethra patterning in mammals.

The *Patched* receptor is a 12-pass transmembrane protein (Figure 13.8). It functions together with a partner receptor called *Smoothened*. The Smoothened protein is a 7-pass transmembrane receptor similar to the Wnt receptor Frizzled. Patched is responsible for ligand binding and Smoothened is responsible for transducing the signal across the plasma membrane. Two different conformations of the Smoothened receptor are depicted in Figure 13.8. In the first, (b), the cytoplasmic tail cannot bind to the downstream signaling partners, but, in the second, (c), binding sites along the cytoplasmic tail become available. In the absence of ligand binding, Patched acts catalytically to maintain Smoothened in its closed conformation, but ceases to do so when bound to Hedgehog.

The cytoplasmic tail (CT) of Smoothened in its open conformation interacts with two cytoplasmic proteins—Costal-2 (Cos2) and Fused (Fu). The first of these, Cos2, is a kinesin-like protein that binds to microtubules while the other, Fu, is a serine/threonine kinase. The Cos2 protein functions as a scaffold for assembly of several signaling elements. These include, besides Smoothened, CT and Fused, *Suppressor of Fused* (Su(Fu)) and a downstream-acting transcription factor called *Cubitus interruptus* (Ci). Two additional serine/threonine kinases—PKA and GSK3—complete the specification of this Hedgehog signaling node (Figure 13.9).

13.9 Gli Receives Hh Signals

The primary recipients of Hh signals are Gli transcription factors. *Glis* (so named because of their involvement in malignant gliomas) are zinc finger, DNA-binding proteins. They have five zinc fingers, three of which grip the DNA molecule, a CBP binding domain, and numerous phosphorylation sites (Figure 13.10). The best studied of these proteins is the *Drosophila* Cubitus interruptus protein. As just described this protein forms a complex

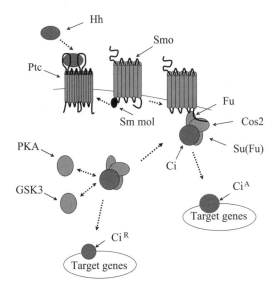

FIGURE 13.9. Signaling through the Hedgehog pathway: In the absence of ligand binding, Patched (Ptc) represses Smoothened (Smo), depicted in the figure as acting through a small molecular (Sm mol) intermediary. In response to ligand binding, the repression is relieved and Smo undergoes a conformational change from a closed configuration to an open one that can bind Cos2 and Fu. This binding action prevents phosphorylation of Ci by PKA and GSK3. The complex dissociates and the full length 155 kDa Ci^A translocates to the nucleus where it promotes transcription. In the absence of ligand binding, the Ci protein is cleaved following phosphorylation, and the 75 kDa Ci^R protein translocates to the nucleus where it represses transcription.

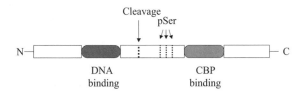

FIGURE 13.10. Organization of the *Drosophila* Cubitus interruptus (Ci) protein: Five tandem zinc fingers situated in the N-terminal half of the protein are responsible for binding DNA. A CBP binding domain located in the C-terminal portion of Ci binds a CBP cofactor. Cleavage of the protein produces the Ci^R protein containing the N-terminal portion and acting as a transcriptional repressor. Three of the many S/T phosphorylation sites (pSer) are shown. The sites indicated are the priming sites phosphorylated by protein kinase A (PKA). Sites located both N-terminal and C-terminal to these sites are phosphorylated by GSK3.

with Cos2, Fu, and Su(Fu) that is bound to microtubules. In the absence of Hh signaling, another member of the signaling pathway, protein kinase A (PKA) phosphorylates Ci. This phosphorylation event tags Ci for proteolysis. The result of this process is the formation of a 75-kDa Ci^R protein (Figure 13.9), which translocates to the nucleus where it functions as a repressor of the transcription of several genes. When Hh is present, phosphorylation of Ci by PKA is blocked, the complex dissociates, and the full-length 155-kDa protein is left intact and free to move into the nucleus. This intact Ci^A protein functions as an activator of transcription of a number of genes including patched, dpp, and wg.

13.10 Stages of Embryonic Development Use Morphogens

The fertilized egg of multicellular animals goes through a sequence of developmental stages. The egg first goes through a cleavage stage where it divides mitotically several times to form a ball of smaller cells, or *blastomeres*. In the next phase, the cells move to the outside forming an epithelial sheet that encloses a fluid-filled chamber. In the third stage, *gastrulation*, the single-layered blastula develops into a gastrula consisting of three layers of cells—the endoderm, mesoderm, and ectoderm, roughly corresponding to gut, connective tissue and muscles, and epidermis (respectively) of the adult organism.

In more detail, the lungs, components of the digestive system such as stomach and liver, and associated glands and structures, develop from the endoderm. In the developing mesoderm, left and right sides of the body are delineated by a *notocord* that defines the central axis of the body. Sections of mesodermal cells progressively bud off on both sides of the notocord to form *somites*, which then develop further into the individual vertebra and muscle groups. The vascular system, including the heart, bone, and cartilage, develop from the mesoderm, while the nervous system and associated sensory organs develop from the ectoderm in a developmental stage called *neurulation*.

The term *morphogen* was introduced at the beginning of the chapter. Morphogens are signaling proteins that are expressed either on cell surfaces or secreted into the extracellular spaces in the form of concentration gradients. These gradients are subsequently "read" by cells to determine their developmental fate. Cells adopt different cell fates according to their position in the gradient relative to the signal source.

Cells in organizing centers, at boundaries between different layers or regions, and within regions, in the embryo secrete morphogens during development. Organizing centers are localized groupings of one or more kinds of cells that secrete morphogens in order to impart patterns of cell fates to fields of progenitor cells. The morphogens are secreted not only at specific locations but also at specific times during embryonic development. These morphogens and associated signaling proteins are expressed sequentially; that is, through a hierarchical pattern of gene expression with each family of gene products preparing the way for the next family of gene products.

13.11 Gene Family Hierarchy of Cell Fate Determinants in *Drosophila*

In *Drosophila*, five sets of gene products have major roles in determining cell fate. The gene families form a hierarchy with one set of gene products preparing the way for the next set of genes and their protein products. Each set contributes to the emergence of the body plan, producing a succession of progressively finer partitions of the embryonic body into segments and compartments that eventually become adult body parts such as head, thorax, abdomen, wings, and legs. Although the details differ from phylum to phylum these families are highly conserved among multicellular organisms up to an including vertebrates. These families, their functions, and members are presented in Table 13.3.

The earliest set of gene products is the *maternal effect genes*. As their name indicates, they are supplied maternally. Some of these gene products function as morphogens, while others assist in morphogen localization. The morphogen gradients are generated internally within the single cell, the egg, rather than externally by many cells, and contribute to the establishment of cell polarity and asymmetric cell division. Maternal effect genes prepare the cell for expression of the gap genes. These are distributed in broad bands

TABLE 13.3. Hierarchy of *Drosophila* patterning genes.

Gene family	Function	Members
Maternal effect	Establish anterior-posterior and dorsal-ventral body axes; regulate gap and pair rule gene expression pattern	A/P axis: Bicoid, caudal, hunchback, nanos, oskar, stauffen; D/V axis: cactus, dorsal, pelle, spätzle, toll, tube
Gap	Partition body into three broad regions; regulate pair rule gene expression	Zygotic caudal, giant, zygotic hunchback, huckebein, knirps, krüppel, tailless
Pair rule	Partition body into bands; regulate segmentation gene expression	Even-skipped, fushi tarazu, hairy, odd-paired, odd-skipped, paired, runt, sloppy paired
Segment polarity	Establish anterior-posterior axis within each compartment	Armadillo, cubitus interruptus, engrailed, frizzled, fused, hedgehog, patched, wingless
Hox cluster	Establish body part identity	Abd-A, Abd-B, Antp, Dfd, Lab, Pb, Scr, Ubx

to anterior, middle, and posterior regions of the embryo. If a *gap gene* is missing the corresponding portion of the body does not develop, producing a gap, and hence the name. These genes prepare the way for the expression of *pair rule genes*. The pair rule genes further partition the broad segments produced by the gap genes into seven embryonic segments, or parasegments. If one of these genes is missing every other segment in the developing larva is absent, and hence the name "pair rule." The next set of genes, *segment polarity genes*, specify the polarity within each parasegement; that is, they define the anterior-posterior axes of each of the parasegments. Finally, the *Hox genes* help delineate the various adult body parts.

13.12 Egg Development in *D. Melanogaster*

The establishment of asymmetry and polarity in *D. melanogaster* begins prior to fertilization, during the development of the egg. In the egg development (oogenesis) stage, an initial germ cell divides four times, producing a single egg cell and 15 nurse cells. The nurse cells are connected to the egg by cytoplasmic bridges that permit mRNAs to flow into the egg cell from the nurse cells. Somatic follicle cells surround and nurture the resulting assemblage. Whereas the point of entry of the sperm into the egg determines the polarity of the *C. elegans* zygote, the orientation of the *Drosophila* oocyte is determined by signals exchanged between it and the most posterior follicle cell. The Gurken protein plays a key role in this signaling pathway, which involves two-way signaling between the egg cell and its surrounding cells in which directional information is encoded through a polarized microtubule cytoskeleton.

Once an initial orientation is specified through communication with the neighboring cells, distributions of mRNAs and proteins are set up in regions of the egg and zygote delineated by anterior/posterior and dorsal/ventral axes. Some mRNAs and proteins are localized either to the anterior or posterior regions of the cell, while others are localized in a dorsal/ventral manner. The proteins being localized belong to the signal transduction pathways and include most importantly a considerable number of transcription factors. When the cells divide, the programs of gene expression in the daughters differ from one another because of the asymmetric distributions of transcription factors, and the daughters have different cell fates.

Two of the proteins involved in delineating the anterior and posterior regions are Bicoid and Oskar. Bicoid localizes to the anterior pole and Oskar protein collects near the posterior pole (Figure 13.11). These proteins, along with several others, guide the formation of graded distributions of mRNA for Hunchback and Caudal in the anterior and posterior regions, respectively. These last-named proteins are intracellular morphogens and their graded distributions help determine cell fate. Genes expressed in the

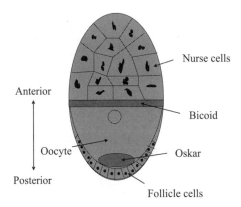

FIGURE 13.11. Schematic depiction of a *Drosophila* oocyte: Shown in the figure are the polynuclear nurse cells, oocyte, and posterior array of follicular cells. Bicoid mRNAs are laid down in a strip in the anterior end and diffuse towards the posterior pole to form a concentration gradient. Oskar mRNAs and proteins are laid down at the posterior pole and diffuse out to form a countergradient.

oocyte, acting as they do in the absence of paternal gene products, are referred to as "maternal effect genes." The bicoid and oskar genes are maternal effect genes acting to establish cell polarity. Bicoid and Nanos, another maternal effect gene product, act as morphogens. Bicoid is laid down to form a gradient peaked at the anterior pole, while Nanos is laid down opposite way to Bicoid, having its highest concentration in the posterior pole.

Maternal effect gene products belonging to the Toll pathway are involved in delineating the dorsal/ventral distributions. Recall that the Toll pathway is an evolutionary ancient pathway that mediates innate immune responses. *Drosophila* counterparts to the mammalian Toll gene products are involved in establishing the early dorsal/ventral patterning. Spätzle is the *Drosophila* Toll ligand; Cactus is the *Drosophila* counterpart to the IκB inhibitor, and Dorsal the counterpart to the NF-κB transcription factor. After fertilization occurs, the Toll signal cascade is initiated resulting in activation of Dorsal in a graded fashion due to ventral restriction of Spatzle signaling. Tube and Pelle are intermediaries in the Toll signaling pathway.

13.13 Gap Genes Help Partition the Body into Bands

Gap genes are expressed in broad regions of the developing *Drosophila* embryo. Maternal effect gene products initiate the expression of these genes. Working in concert the Hunchback, Knirps, Giant, and Krüppel genes products subdivide the embryo into regions along the anterior posterior axis. As shown in Figure 13.12, Krüppel is expressed in a wide band in the

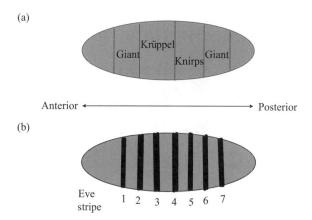

FIGURE 13.12. Patterns of distributions of developmental genes in the *Drosophila* embryo: (a) Distribution of gap gene products Krüppel, Giant, and Knirps in broad bands. (b) Distribution of pair rule gene product Even-skipped (Eve) in the *Drosophila* embryo. Shown is the 7-striped zebra pattern of eve gene expression. Stripes are numbered from the anterior end to the posterior end.

center of the embryo. Knirps is expressed the in a broad posterior region adjacent to the central Krüppel region, and Giant genes are expressed on the two outer sides of the centrally situated Krüppel and Knirps regions. A second, small Knirps region is situated near the anterior pole (not shown in the figure).

Tailless and huckebein are terminal gap genes. They are expressed at both the anterior pole and posterior pole to complete the A/P partitioning of the embryo. Hunchback inhibits expression of posterior gap genes knirps and giant in the anterior part of the embryo. If hunchback is not expressed the head will be missing; if Knirps is missing the abdomen will be missing, and if Krüppel is missing the thorax and abdomen will not appear. Thus, the proteins encoded by these genes are cell fate determinants that divide the growing embryo into head, abdomen, and terminal bands.

13.14 Pair-Rule Genes Partition the Body into Segments

Like the gap genes, the pair rule genes are transcription factors. Three of them—even-skipped, hairy and runt—are regarded as *primary pair rule genes*. The other pair rule genes are referred to as *secondary pair rule genes*; they are expressed later and are governed by the expression patterns of the primary pair rule genes. The pair rule genes specify segment boundaries; they are laid down in zebra-like patterns giving rise to the 7-segment striped body of the developing larva. As is the case for the other primary pair rule genes, the regulatory region for even-skipped (eve) contains a set of special binding sites, called *enhancer elements*, for transcription factors, one for each stripe. These enhancers correspond in a unique way to the gradient information in the region of the stripes.

Gap and maternal effect genes both regulate the pair rule gene expression patterns. The gene regulatory region (enhancer) for eve in stripe 2 illustrates the general theme. Bicoid and Hunchback bind to sites in the enhancer and function as activators of gene expression. Giant and Krüppel also bind to sites in the enhancer and operate as repressors of gene expression. Since Giant is expressed at the anterior end and Krüppel in the middle, stripe 2 is inhibited in those regions but is encouraged by Bicoid and Hunchback in a stripe in the anterior region located in between the two repressed regions. Similarly, enhancers of the other stripes reflect the gradients of the corresponding sets of maternal effect and gap genes in the regions where the other stripes are to be sited. The net result is the emergence of seven stripes of expression for eve, and similarly, seven stripes for hairy and seven stripes for runt. The gene products appear in regions that partially overlap one another and so they operate in combinatorial manner to delineate the parasegments.

13.15 Segment Polarity Genes Guide Parasegment Development

The development of the *Drosophila* larva continues with the expression of segment polarity genes. The expression of gap and pair rule genes occurs during the precellular blastomere stage of embryo development. The blastomere stage ends with cellularization where movements of the membrane about the nuclei take place, separating the nuclei and forming distinct cells. The segment polarity genes are expressed at the start of gastrulation. Whereas the gap and pair rule gene products function as transcription factors, the segment polarity gene products include both transcription factors and signaling proteins that mediate communication between the newly formed cells and coordinate their programs of gene expression.

Genes that encode several receptor-ligand combinations appear in the list of segment polarity genes. Among the entries in Table 13.3 are patched and hedgehog, and frizzled and the gene encoding the Wnt ligand Wingless. Downstream signal transducers such as Cubitus interruptus are expressed; as well. As this stage begins Engrailed, a transcription factor, is expressed along with two secreted signaling proteins—Wingless and Hedgehog. Pair rule gene products such as Eve and Ftz establish patterns of Wg, En, and Hh gene expression, and cell-to-cell signaling sustains them. The result of this activity is the formation of 14 stripes of Engrailed/Hh gene expression and 14 stripes of Wingless gene expression.

The boundary between cells expressing Engrailed/Hedgehog and those expressing Wingless is the parasegment boundary, and the regions between pairs of boundaries are the parasegments. Cells residing in the anterior half of each parasegment express Wg and those situated in the posterior compartments express En/Hh. The cell fates in the two compartments differ: cells in the anterior compartment become "A" (anterior) cells while those in the posterior compartment become "P" (posterior) cells.

13.16 Hox Genes Guide Patterning in Axially Symmetric Animals

Bilaterally symmetric animals such as humans express a family of highly conserved regulatory genes, called *Hox genes*. These genes play an important role in specifying which cells in the mesoderm become which body parts. In vertebrates, these genes guide the emergence of the axial skeleton (for example, the breastbone, ribcage, and spine), voluntary muscles, and dermis of the back from the somites mentioned earlier in the chapter, that is, from the repeated, identical segments of cells belonging to the mesoderm.

These genes have been studied extensively in the fly and other insects, the nematode, chick, and mouse. In all organisms studied, the genes appear as a cluster, one gene after the other. In vertebrates there are four clusters of genes, each cluster containing from 9 to 11 genes. In insects and the nematode there is a single gene cluster. The members of the *Drosophila* Hox cluster are listed in Table 13.3. Three of the Hox genes—Dfd, Lab, and Pb—delineate the head. Two gene products—Antp and Scr—specify the thorax, and the remaining three—Abd-A, Abd-B and Ubx—specify the abdomen.

In many metazoans, a small number of regulatory genes control the development of organs and morphological structures. These genes, called *selector genes*, have been intensively studied in *Drosophila*, where they control the formation of body parts and bilateral partitioning. These regulatory genes, of which Hox genes are prominent members, function as transcription factors. Selector genes do not work alone. Rather, they work together with other regulatory components in a combinatorial fashion to create specific patterns of gene expression, leading to the development and arrangement of body parts. The segment polarity gene, engrailed, discussed previously, is a good example of a selector gene. It works in concert with other segment polarity genes to determine which cells become A cells and which ones become P cells.

Hox genes are activated in a spatially sequential manner from anterior (head) to posterior (tail) end. Cells progressively bud off from the main axis to form the somites. The differentiation of the cells within the somites to form vertebrae and muscles takes the form of a wave of activity whose leading edge moves down the A/P axis, with presomitic mesoderm in front of the wave and somites behind it. The movement of this wave must be carefully timed. The timing activity keeps the movement of the wave in phase with the time needed for development of the somites.

References and Further Reading

Notch Signaling

Artavanis-Tsakonas S, Rand MD, and Lake RJ [1999]. Notch signaling: Cell fate control and signal integration in development. *Science*, 284: 770–776.

Lai EC [2002]. Keeping a good pathway down: Transcriptional repression of Notch pathway target genes by CSL proteins. *EMBO Rep.*, 3: 840–845.

Milner LA, and Bigas A [1999]. Notch as a mediator of cell fate determination in hematopoiesis: Evidence and speculation. *Blood*, 93: 2431–2448.

Schroeter EH, Kisslinger JA, and Kopan R [1998]. Notch-1 signalling requires ligand-induced proteolytic release of intracellular domain. *Nature*, 393: 382–386.

Struhl G, and Adachi A [1998]. Nuclear access and action of Notch in vivo. *Cell*, 93: 649–660.

Presenilins and Matrix Metalloproteinases

De Strooper B [2003]. Aph-1, Pen-2, and Nicastrin with Presenilin generate an active γ-secretase complex. *Neuron*, 38: 9–12.

De Strooper B, and Annaert W [2000]. Proteolytic processing and cell biological function of the amyloid precursor protein. *J. Cell Sci.*, 113: 1857–1870.

Fortini ME [2001]. Notch and Presenilin: A proteolytic mechanism emerges. *Curr Opin. Cell Biol.*, 13: 627–634.

Schlondorff J, and Blobel CP [1999]. Metalloprotease-disintegrins: Modular proteins capable of promoting cell-cell interactions and triggering signals by protein-ectodomain shedding. *J. Cell Sci.*, 112: 3603–3617.

Vu TH, and Werb Z [2000]. Matrix metalloproteinases: Effectors of development and normal physiology. *Genes Dev.*, 14: 2123–2133.

TGF-β Signaling (Receptor Serine/Threonine Kinase Pathway)

Moustakas A, Souchelnytskyi S, and Heldin CH [2001]. Smad regulation in TGF-β signal transduction. *J. Cell Sci.*, 114: 4359–4369.

Qin BY, et al. [2002]. Smad3 allostery links TGF-β receptor kinase activation to transcriptional control. *Genes Dev.*, 16: 1950–1963.

Shi YG, and Massagué J [2003]. Mechanism of TGF-β signaling from cell membrane to the nucleus. *Cell*, 113: 685–700.

Ten Dijke P, et al. [2002]. Regulation of cell proliferation by Smad proteins. *J. Cell Physiol.*, 191: 1–16.

Ten Dijke P, Miyazono K, and Heldin CH [2000]. Signaling inputs converge on nuclear effectors in TGF-β signaling. *Trends Biochem. Sci.*, 25: 64–70.

Wnt Signaling Pathway

Cardigan KM, and Nusse R [1997]. Wnt signaling: A common theme in animal development. *Genes Dev.*, 11: 3286–3305.

Dale TC [1998]. Signal transduction by the Wnt family of ligands. *Biochem. J.*, 329: 209–223.

Kalderon D [2002]. Similarities between the Hedgehog and Wnt signaling pathways. *Trends Cell Biol.*, 12: 523–531.

Ma D, et al. [2003]. Fidelity in planar cell polarity signaling. *Nature*, 421: 543–547.

Mlodzik M [1999]. Planar polarity in the *Drosophila* eye: A multifaceted view of signaling specificity and cross-talk. *EMBO J.*, 18: 6873–6879.

Moon RT, Brown JD, and Torres M [1997]. Wnts modulate cell fate and behavior during vertebrate development. *Trends Genet.*, 13: 157–162.

Wallingford JB, Fraser SE, and Harland RM [2002]. Convergent extension: The molecular control of polarized cell movement during embryonic development. *Dev. Cell*, 2: 695–706.

Hedgehog Signaling

Collins RT, and Cohen SM [2003]. The secret life of Smoothened. *Dev. Cell*, 5: 823–824.

Hammerschmidt M, Brook A, and McMahon AP [1997]. The world according to Hedgehog. *Trends Genet.*, 13: 14–21.

Ingham PW [1998]. Transducing Hedgehog: The story so far. *EMBO J.*, 17: 3505–3511.

Ingham PW, and McMahon AP [2001]. Hedgehog signaling in animal development: Paradigms and principles. *Genes Dev.*, 15: 3059–3087.

Taipale J, et al. [2002]. Patched acts catalytically to suppress the activity of Smoothened. *Nature*, 418: 892–897.

Morphogens

Lawrence PA, and Struhl G [1996]. Morphogens, compartments, and pattern: Lessons from *Drosophila*? *Cell*, 85: 951–961.

Teleman AA, Strigini M, and Cohen SM [2001]. Shaping morphogen gradients. *Cell*, 105: 559–562.

Vertebrate Organizers and Body Plan

Lemaire P, and Kodjabachian L [1996]. The vertebrate organizer: Structure and molecules. *Trends Genet.*, 12: 525–531.

Schier AF, and Shen MM [2000]. Nodal signaling in vertebrate development. *Nature*, 403: 385–389.

Sokol SY [1999]. Wnt signaling and dorsal-ventral axis specification in vertebrates. *Curr. Opin. Genet. Dev.*, 9: 405–410.

Feedback Loops in Development

Eldar A, et al. [2002]. Robustness of the Bmp morphogen gradient in *Drosophila* embryonic patterning. *Nature*, 419: 304–308.

Freeman M [2000]. Feedback control of intercellular signalling in development. *Nature*, 408: 313–319.

Selector Genes, Gene Regulatory Networks, and the Segmentation Clock

Curtiss J, Halder G, and Mlodzik [2002]. Selector and signaling molecules cooperate in organ patterning. *Nature Cell Biol.*, 4: E48–E51.

Davidson EH, et al. [2002]. A genomic regulatory network for development. *Science*, 295: 1669–1678.

Guss KA, et al. [2001]. Control of a genetic regulatory network by a selector gene. *Science*, 292: 1164–1167.

Hirata H, et al. [2002]. Oscillatory expression of the bHLH factor Hes1 regulated by a negative feedback loop. *Science*, 298: 840–843.

Saga Y, and Takeda H [2001]. The making of the somite: Molecular events in vertebrate segmentation. *Nature Rev. Genet.*, 2: 835–845.

Problems

13.1 The core components and basic features of the four primary developmental pathways have now been introduced. The underlying picture is a fairly simple one in which ligand binding at the plasma membrane triggers a sequence of signaling events culminating in the activation of gene transcription in the nucleus. The signaling pathways possess a number of characteristics that enable them to guide the development of complex metazoans possessing a large variety of different cell types organized into tissues and organs. The first of these characteristics is that signaling is context dependent. The term "context" refers to the mix of proteins being expressed. How a cell responds to a signal depends on what other signaling events are taking place at the same time and in previous times that have altered the mix of proteins being expressed. How might context dependence come into play at the plasma membrane top influence signaling?

13.2 Control points are situated at the plasma membrane, in the cytoplasm, and in the nucleus. Examples of cytoplasmic control points are the beta catenin complex formed in the Wnt signaling pathway and the Ci signaling complex formed in the Hedgehog signaling pathway. These along with MAP kinase and NF-κB modules function as major signaling nodes. Give some examples of how cellular context may influence what happens at these nodes in the signaling pathway.

13.3 Two kinds of posttranslational modifications—phosphorylation and proteolysis are especially prominent in the four developmental pathways. These modifications operate individually and jointly in these pathways. List the various ways these modifications influence signaling.

13.4 Despite the presence of different proteins, the developmental pathways have many features in common. In response to ligand binding, a series of signaling events takes place resulting in the translocation from the cytoplasm to the nucleus of a transcriptional factor. In these pathways downstream signaling elements can often function both as activators of transcription or as repressors, with the choice depending on cellular context. How does context come into play in the nucleus? How was this dual capability used in the pathways discussed in this chapter?

14
Cancer

Cancers arise from malfunctions in the cell control layer that lead to the unregulated proliferation of cells. The underlying causes of cancers are mutations and other alterations in DNA, and attendant inappropriate expression levels, of genes encoding proteins that either promote growth or restrain it, or direct the apoptosis machinery, or are responsible for DNA damage repair and signaling, and chromatin remodeling. The mutations may be heritable, that is, they may be present in the germ cells, or they may be produced in somatic cells.

DNA can be damaged in several ways. Hydrolytic processes and oxidative byproducts of normal cellular metabolism can damage DNA. Ionizing radiation from cosmic rays and from natural-occurring radioactive materials in the soil, water, and air such as uranium, thorium, and radon can damage DNA. In addition, ultraviolet (UV) radiation from the sun can damage DNA. The main step leading to DNA damage by environmental and endogenous stimuli is the generation of oxidative free radicals near the DNA. Free radicals are molecular species with unpaired electrons, making them highly reactive. The most damaging of these reactive oxidative species (ROS) is the hydroxyl radical. When produced in the vicinity of a DNA molecule the hydroxyl radicals and other ROS attack sugars and bases producing single strand breaks, base losses, and modified bases. In addition to this ROS mechanism, ionizing radiation such as X-rays and gamma irradiation can directly generate double strand breaks in DNA. When a cell that has been damaged divides without having the damage repaired the changes are carried on to its daughter cells and become a mutation.

In metastasis, malignant cancer cells break away from where they are immobilized, enter the circulatory system, and invade other organs. They form colonies, or secondary tumors, at the new locations and cause damage to their neighbors by appropriating their sources of nutrition. Cancers derived from epithelial cells are the most common type of cancer. In order for the cancerous epithelial cells to break away from their initial location they must detach from other epithelial cells and from the ECM. To do so they must disrupt the cell-to-cell adhesive contact established by E-

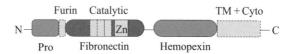

FIGURE 14.1. Matrix metalloproteinase structure: Most MMPs contain an N-terminal signal peptide (not shown) that targets the enzyme for secretion, followed by a prodomain. The catalytic domain contains a zinc-binding domain and all MMPs except MMP7 and MMP26 contain a Hemopexin domain C-terminal to the catalytic domain and separated from it by a linker region. Dashed boxes represent domains present in some MMPs but not others. Included in this set are Furin-like domains, Type II Fibronectin repeats, and transmembrane + cytoplasmic segments.

cadherins, and cell-to-ECM contacts maintained by the integrins. Once they have detached from one another and from the ECM they have to pass through the basement membrane to reach and enter the circulatory system. This is accomplished using matrix metalloproteinases to degrade matrix proteins. The cancer cells make and secrete sufficient quantities of these proteolytic enzymes to weaken the membrane and allow passage of the cells through it into the blood vessels.

Matrix metalloproteinases (MMPs) are a family of over two dozen secreted proteolytic enzymes that (i) cleave components of the ECM, and (ii) cleave signaling proteins associated with the ECM and cell surfaces resulting in their activation and solubilization. These enzymes are normally synthesized in small quantities but production is increased in response to cytokines and stress, and accompanies the transformation of normal cells to cancerous ones.

As shown in Figure 14.1, MMPs contain a *prodomain*, which is cleaved to activate the enzyme, and fibronectin, hemopexin, and collagen V (not shown) domains that promote substrate and inhibitor binding. The furin domain provides an alternative cleavage site. Members of a family of four proteins called *tissue inhibitors of metalloproteinases*, or TIMPs, bind the hemopexin domain of MMPs, forming MMP-TIMP complexes that regulate the activities of the MMPs. A zinc-binding site is present in the catalytic domain. Zinc binding (mediated by a trio of histidine residues) is necessary, and this requirement gives rise to the name "metalloproteinase."

14.1 Several Critical Mutations Generate a Transformed Cell

The transformation of a normal cell into a cancerous one is a multistep process. Mutations accumulate over time and over cell generations, and the susceptibility to cancer increases rapidly with age. As the mutations accumulate a variety of genetic modifications are produced. These include alterations in chromosome number (aneuploidy) involving the loss or gain of entire chromosomes, chromosome translocations that can generate a new gene by fusing two different genes, and gene duplications (gene amplification). One or more of these dramatic changes in chromosome organization are present in most human tumors.

The genetic targets of mutations that promote cancerous growth can be grouped into four functional classes. The first of these are *genes that encode proteins that function in the growth signaling pathways*. These gene products may convey progrowth signals or serve as brakes on growth. Mutated forms of the receptor tyrosine kinases, GTPases, and nonreceptor tyrosine kinases discussed in Chapters 10 and 11 are present in many different kinds of cancer. The developmental pathways discussed in the last chapter are not simply turned off at the end of embryogenesis, but rather remain active in one form or another during adult life. Mutated and/or overexpressed elements of these pathways are encountered in cancers as well.

The second set of mutations is to *genes that encode proteins that regulate cell suicide*. These target proteins either trigger or inhibit the cellular apoptosis program activated when aberrant conditions are detected. The third group of crucial mutations is to *genes that encode cellular caretakers*. These proteins carry out DNA repair and maintain chromosome integrity. The fourth and last set of crucial gene products consists of the central regulators of growth, repair, and death. These gene products are referred to as *controller proteins*. They are responsible for ensuring an orderly progression through the cell cycle, halting or advancing it when necessary and signaling to the apoptosis machinery when that outcome in required. Listed in Table 14.1 are a number of prominent members of each of these classes of proteins.

Oncoproteins are proteins that have been structurally altered by mutations in the genes that encode them. These proteins operate in the signal

TABLE 14.1. Protein with mutated and altered forms associated with cancer: Abbreviations—Oncoprotein (o); tumor suppressor (ts); DNA caretaker (c).

Protein	Function/pathway	Class	Cancer role
Ras	Growth	o	Many cancers
Src	Growth	o	Sarcomas
Abl	Growth	o	Leukemias
APC	Growth	ts	Colorectal cancers
β-catenin	Growth	o	Colorectal cancers
Myc	Growth	o	Many cancers
Bax	Apoptosis	ts	Many cancers
Bcl-2	Apoptosis	o	Many cancers
hMLH1	Repair	c	Colon cancers
hMSH2	Repair	c	Colon cancers
ATM	Repair	c	Ataxia telangiectasia
NBS	Repair	c	Nijmegen breakage syndrome
BRCA1,2	Repair	c	Breast/ovarian cancers
p53	Controller	ts	Most cancers
pRb	Controller	ts	Most cancers

transduction, integration, and regulatory pathways involved in cellular growth, multiplication, differentiation, and death. As a result of the structural alterations these proteins do not function normally, but instead are changed in a manner that stimulates unregulated cell growth and proliferation thus promoting the development of cancer. Tumor suppressors are similar to oncoproteins except that they normally act as brakes on growth. When they suffer critical mutations these brakes on growth are removed.

14.2 Ras Switch Sticks to "On" Under Certain Mutations

The first group of entries in Table 14.1 consists of proteins that relay growth signals from the plasma membrane to target sites in the cell interior. Ras and Src are prominent members of this group. Src is implicated in about 80% of human colon cancers. Ras oncoproteins are even more widespread in their cancer occurrences. They are present in 30 to 40% of human cancers. Not all mutations are equally important. Recall that a codon is a sequence of three nucleotide bases that encodes an amino acid. In Ras mutations, codons 12, 13, and 61 serve as hot spots for oncogene activity. The mutations occurring at these sites are point mutations that change one of the base pairs in the codons encoding glycine (12 and 13) and glutamine (61) into those encoding a different amino acid.

Ras is a crucial relay operating in the pathway that relays growth signals to downstream targets. It functions as a binary switch. Like all GTPases it is turned on by a GEF and turned off by a GAP. In the absence of growth signals, the switch is in its off position, but turns on in response to the appropriate signals. The mutations leading to cancer result in the switch being stuck in the on position, unable to turn off, and continually sending growth signals into the cell.

X-ray crystallography of Ras in complexes with its GEF (Figure 14.2) and GAP (Figure 14.3) provide insights into how the binary switch operates and how certain mutations leave Ras stuck in its on position. As can be seen in Figure 15.1a the Sos protein is organized into two domains, an N-terminal domain that is largely structural in nature, and a C-terminal that catalyzes the release of GDP. The catalytic domain forms a bowl about Ras. The Ras protein has two switch regions, called Sw 1 and Sw 2. These regions create a cavity within which GTP and GDP along with Mg^{2+} bind and release.

The Ras switch operates in the following manner. In the absence of growth signals, Ras is in its off position with GDP bound firmly in the pocket. When growth signals are present Sos is recruited to the plasma membrane and binds to the Ras-GDP complex. The αH helix of Sos engages the switches and flips Sw 1 to an open position in which it has rotated away from Sw 2. The GDP molecule dissociates from the complex, followed shortly thereafter by the dissociation of Sos. The GTP

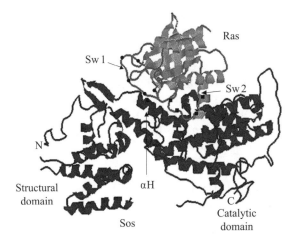

FIGURE 14.2. Structure of Ras in a complex with its Sos GEF as determined using X-ray crystallography: Ras is shown in light gray while Sos is depicted in black. Small filled circles and squares superimposed on the figure serve to outline Sw 1 and Sw 2, respectively. The figure was generated using Protein Explorer with atomic coordinates deposited in the Brookhaven Protein Data Bank (PDB) under accession code 1BKD.

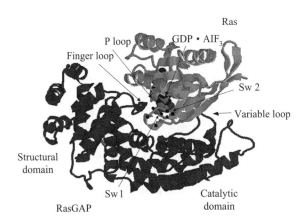

FIGURE 14.3. Structure of Ras in a complex with its GAP as determined using X-ray crystallography: Ras is shown in light gray while the RasGAP is depicted in black. Small filled circles and squares superimposed on the figure serve to outline Sw 1 and Sw 2, respectively. The Gly12 residue is located at the tip of the Ras P-loop while a second crucial residue, Glu61, located in the Ras Sw 2 region in close opposition to the arginine finger loop of the RasGAP. The figure was generated using Protein Explorer with atomic coordinates deposited in the Brookhaven PDB under accession code 1WQ1.

molecule is plentiful and binds in the pocket that was vacated by the GDP molecule.

Mutations of the glycine residue at position 12 in the Ras chain convert Ras into a form that is active all the time. That is, the RasGAP is unable to turn off Ras. The reason for this can be discerned from the crystal structure exhibited in Figure 14.3. The Gly12 residue is positioned at a crucial place at the very end of the P loop. Because of its small size, replacement of glycine by any other residue blocks the arginine in the finger from interacting with the ATP molecule bound in the cleft, and hydrolysis is consequently impeded.

14.3 Crucial Regulatory Sequence Missing in Oncogenic Forms of Src

Both Src and Abl are nonreceptor tyrosine kinases and were discussed in Chapter 11. The Src gene was first identified in *retroviruses*. These are viruses that use RNA as their genetic coding medium rather than DNA. When a retrovirus invades a cell the viral RNA is transformed into DNA and integrated into the host genome. Retroviruses sometimes acquire genes with oncogenic capabilities from an early host and deliver these into later hosts. These genes are referred to *viral oncogenes*. The Src oncogene was first identified in the chicken Rous sarcoma virus, and then a cellular counterpart to the viral oncogene was found in normal cells in the chicken and then in humans. The viral forms of Src and other viral oncogenes usually differ in some way from their cellular counterparts. For that reason the viral form of Src is denoted as v-Src while its cellular form is designated as c-Src, and a similar situation obtains for other oncogenes.

Recall from Chapter 11 that a critical residue Tyr527 located in its COOH tail controls the catalytic activity of c-Src. Phosphorylation of this residue by Csk deactivates c-Src, and the Tyr527 phosphorylation site is required for proper function of the kinase. In v-Src, the tail region containing Tyr527 is missing, and the truncated protein cannot be turned off. The result is that the v-Src protein is constitutively active sending uncontrolled cell growth and proliferation signals to the nucleus. The cellular form of Src can be mutated in several ways to generate oncogenic forms. Point mutations in the codon for Tyr527 converting this amino acid to phenylalanine can transform Src as can specific mutations that disrupt the ability of the SH2 and SH3 domains to cooperate with the COOH tail in inhibiting Src activation.

14.4 Overexpressed GFRs Spontaneously Dimerize in Many Cancers

Growth factors and growth factor receptors (GFRs) support the development of malignant tumors in several ways. One prominent contributor to the onset of malignancy is the vascular endothelial growth factor (VEGF).

In order for a solid tumor to grow and thrive it must have an adequate blood supply. In response to this need for vascular expansion, tumor angiogenesis takes place. The expression of messenger RNAs for VEGF ligands is enhanced in most human tumor cells. Increased VEGF mRNAs are present in rapidly growing glioblastoma multiform brain tumors, and in cancers of the lung, breast, gastrointestinal tract, female reproductive organs, thyroid gland, and urinary tract.

Growth factor receptor dimerization is a critical step in relaying signals conveyed by polypeptide growth factors into the cell interior. As discussed in Chapter 11, receptor tyrosine kinases are brought into close physical proximity through ligand binding, resulting in the formation of receptor dimers or oligomers. Autophosphorylation in the activation loop occurs next followed by recruitment of cytoplasmic signaling molecules. In the absence of a ligand the receptors do not dimerize and there is no signaling across the plasma membrane. In contrast to this normal situation, spontaneous dimerization occurs in many cancers. In these abnormal situations the receptors dimerize in the absence of ligand binding. Ligand-free dimerization can be produced in several different ways. Most often it is generated through overexpression of receptors arising as a consequence of gene amplification. It can also be generated through point mutations and exon deletions.

Epidermal growth factor receptors (EGFRs) undergo spontaneous dimerization in many cancers including those of the breast, lung and ovarian cancers and gliomas, and brain tumors of glial origin. Amplification of the EGFR (ErbB1) gene occurs in about 40% of gliomas, and the amplification of the ErbB2 gene takes place in about 30% of breast cancers. In many of the brain tumors, gene rearrangements accompany gene amplification and these alterations often involving truncations of portions of the molecule. The main effect of spontaneous dimerization is to activate a pathway that sends inappropriate growth/proliferation signals to the nucleus.

14.5 GFRs and Adhesion Molecules Cooperate to Promote Tumor Growth

Alterations in the mix of cell adhesion molecules being expressed accompany tumor progression. The altered expression patterns occur not only during metastasis but also during solid tumor growth. Most cancers develop from epithelial cells, and loss of E-cadherins is a common occurrence. Recall that E-cadherins help maintain tight adhesive contacts in populations of these cells. Loss of adhesive junctions and changes in cytoskeleton organization accompanies the transformation to malignancy. Among the changes in expression patterns are the upregulation of $\alpha_V\beta_3$ and $\alpha_6\beta_4$ integrins, and the switching from N-cadherins to E-cadherins and back when adhesive contacts are again needed.

Integrins along with cadherins and Ig superfamily cell adhesion molecules form complexes with growth factor receptors. Examples of cooperativity between adhesion and growth factor receptors are $\alpha_6\beta_1$ and $\alpha_6\beta_4$ and EGFRs, NCAM, and N-cadherins with FGFRs, and VE-cadherins with VEGF receptors. One result of this form of association is the ability of integrins and/or growth factor receptors to convey signals into the cell without having to engage their natural ligands. Clustering brings the receptors into close proximity with one another, and promotes phosphorylation and the recruitment of cytoplasmic signaling transducers, thereby alleviating the need for ligand engagement.

A second consequence of the growth factor receptor–adhesion molecule clustering is the strengthening of signals that would otherwise be too weak to elicit a cellular response if conveyed by one or the other alone. An example of this form of cooperativity is that which occurs between Met and $\alpha_6\beta_4$ integrins. Recall from Chapter 10 that Met is the receptor for HGF/SF, a set of diffusible ligands that are central participants in invasive growth and branching morphogenesis. SF does not stimulate growth, but rather triggers the dissociation, or scattering, of cells. By forming clusters the cytoplasmic segments of the Met receptors and integrins come into close contact; they are able to promote phosphorylation and provide multiple docking sites for adapters and other cytoplasmic signaling elements. In this second signaling role, the $\alpha_6\beta_4$ acts as an amplifier to increase the magnitude of the cellular response to a growth signal and promote invasive growth independent of binding to the ECM.

14.6 Role of Mutated Forms of Proteins in Cancer Development

The likelihood of getting a colorectal tumor exceeds 50% by age 70. Most of these tumors do not progress to a lethal stage, but nevertheless colorectal tumors are the second leading cause of cancer death in the United States. Mutated forms of several gene products contribute to the onset of colorectal tumors. Two of these, the APC protein and β-catenin will be discussed in this section, while mutations in another pair of gene products, the mismatch repair proteins hMLH1 and hMSH2, will be examined in a later section.

The adenomatous polyposis coli (APC) protein and β-catenin participate in the Wnt signaling pathway. The Wnt signaling system is thought of as a developmental pathway, and was discussed in the last chapter. APC is localized in the basolateral compartment of epithelial cells along with glycogen synthase kinase (GSK) that regulates its signaling activity. Recall that in the absence of a Wnt signal, GSK phosphorylates β-catenin thereby tagging that molecule for destruction. In the presence of Wnt signaling GSK is antagonized and does not tag β-catenin. The latter is stabilized as a cytoplasmic monomer and can then translocate to the nucleus. APC forms a complex with GSK and β-catenin. If APC is absent or is mutated, β-catenin is not properly regulated by GSK. Instead, the β-catenin levels are raised mimicking the effect of active Wnt signaling. Similarly, certain mutations to β-catenin render that molecule insensitive to APC/GSK regulation leading to the same result. Mutations in APC or β-catenin are encountered in many colorectal tumors.

One of the main endpoints of signaling through APC/GSK/β-catenin is the transcription machinery in the nucleus. Upon translocation to the nucleus, β-catenin binds to the transcription factor Tcf-4/LEF to form a dimeric complex. Thus, the program of gene expression is altered when β-catenin is not properly regulated. A second end point of signaling through APC is the cytoskeleton. During cell division APC interacts with proteins in the ends of microtubules that form the mitotic spindles. APC contributes to the linking of microtubules to *kinetochores*, attachment sites on the chromosomes, and it contributes to signaling that the correct attachments are made. Cells containing mutated forms of APC exhibit chromosome instability, that is, they have incorrect numbers of chromosomes.

The Wnt pathway is not the only developmental pathway whose components can contribute to cancer. Aberrant signaling in the Hedgehog and TGFβ pathways can promote cancer as well. Mutated forms of Patched and Smoothened, the two receptors that operate in the Hedgehog pathway, are encountered in basal cell carcinoma, the most common human cancer. Recall from the last chapter that Patched normally binds and sequesters Hh, and thus restricts its growth-promoting effects. It is therefore a tumor suppressor, while Smoothened, which transduces growth signals into the cell, acts as a growth stimulator. Gli transcription factors acting downstream from the aforementioned receptors in the Hedgehog pathway convey the growth signals to the nucleus. Raised levels of Gli activity are found in these cancers as a result of Gli mutations and aberrant receptor behavior.

As noted in the last chapter, Smads, acting as downstream signal transducers in the TGFβ pathway, can promote or inhibit growth. In their capacity as growth inhibitors they inhibit G1 cyclin-dependent kinases Cdk4 and Cdk2 that act during the G1 phase of the cell cycle. (These kinases will be discussed later in the chapter.) The Smads stimulate production of p15Ink4b, a protein that binds to and inhibits these cdks, and prevents assembly of Cyclin D-cdk4 and Cyclin E-Cdk2 complexes. Mutations in components of the TGFβ pathway—ligands, receptor, Smads (Smad2 or Smad4) and cofactors—are present in the majority of colon and pancreatic cancers.

14.7 Translocated and Fused Genes Are Present in Leukemias

Member of the Bcl-2 family of proteins are central regulators of apoptosis. When these proteins are not expressed at the proper levels, the option of triggering apoptosis to remove cells that are either damaged or infected or growing out of control is no longer possible. As will be discussed in the next chapter, there are several types of Bcl-2 proteins. Some promote apoptosis while others inhibit it. Bcl-2 is a prominent member of the antiapoptosis group. Aberrant bcl-2 genes are present in 90% of follicular lymphomas or B-cell lymphomas for which the Bcl-2 protein is named. In these cells, a translocation has occurred. The gene encoding Bcl-2 located on chromosome 18 has been translocated and fused with an immunoglobulin heavy chain (IgH) gene located on chromosome 14. The result of this translocation, designated symbolically as t(14:18), is that the bcl-2 gene is positioned next to the enhancer for the antibody gene. Antibody genes are vigorously expressed and as a result of the translocation the Bcl-2 gene is continually overexpressed. The decision between proliferation and apoptosis is shifted towards the former, and the B-cells do not die off, as they should when they age. Furthermore, the B cells are resistant to radiation therapy and chemotherapy, since these forms of treatment work at least in part by stimulating apoptosis.

Several other fusion products play prominent roles in leukemias. Burkitt's lymphoma is characterized by the translocation t(8:14) and subsequent fusion of the c-Myc gene with an IgH gene. Chronic myelogenous leukemia and acute lymphocytic leukemia involve the translocation of the abl gene from chromosone 9 to chromosome 22, where it fuses with the BCR gene. The altered form of chromosome 22 is known as the *Philadelphia chromosome*. All of these examples involve the joining of a gene with oncogenic capabilities (Bcl-2, c-Myc, and Abl) with a strongly expressed gene (BCR and IgH) leading to the continual overexpression of the oncogene.

14.8 Repair of DNA Damage

DNA repair systems are responsible for maintaining the integrity of the genome throughout the life of the cell. There are about 130 known genes that encode DNA repair proteins. These are organized into five DNA damage repair systems (Table 14.2). The base excision repair (BER) and nucleotide excision repair (NER) systems treat single strand damage. Another, the mismatch repair (MMR) system, corrects for mismatched base pairings generated during DNA replication and recombination. Finally, two interlocking systems, the homologous recombination (HR) and nonhomologous end joining (NHEJ) systems, handle double-strand breaks. Each of these systems contains an ensemble of enzymes and regulatory molecules that repair and rejoin DNA strands through sequences of chemomechanical operations.

As noted in Table 14.2, the base excision repair system removes bases damaged by UV radiation and X-rays, and by endogenous oxygen radicals and alkylating chemicals. The nucleotide excision repair system handles

TABLE 14.2. The four kinds of DNA repair.

Type of damage repair	System(s)	Description
Base excision repair	BER	Removes bases damaged by UV, X-rays, oxygen radicals, and alkylating agents
Nucleotide excision repair	NER	Removes DNA lesions brought on by environmental agents
Mismatch repair	MMR	Repairs damage occurring during DNA replication and meiotic recombination
Double-strand break repair	HR, NHEJ	Repairs double-strand breaks caused by ionizing radiation, oxidative stresses, and other environmental factors

bulky DNA lesions and damage arising from environmental agents such as polycyclic aromatic hydrocarbons compounds contained in cigarette smoke. Portions of chromosomes bearing genes that are actively undergoing transcription receive greater NER attention than DNA segments containing genes that are only rarely transcribed. Several disorders including skin cancers and Cockayne's syndrome are associated with mutations in genes encoding members of the NER system.

The mismatch repair system monitors and treats damage occurring during DNA replication and meiotic recombination. Among AGTC combinations corrects A-G and T-C mismatches, as well as improper sequence insertions and deletions. Mutations in this repair systems lead to increases in the mutation rate and to cancer development. Mutations in two members of the MMR system, hMLH1 and hMSH2, are prominently linked to colorectal and other cancers.

The machinery for repairing double strand breaks is utilized for several purposes in the cell. It is central to V(D)J recombination in lymphocytes. DNA undergoing replication is subject to breaks and the machinery for repairing DNA participates in remedying broken replication forks. This machinery is involved in genetic recombination, the process whereby segments of DNA are exchanged between chromosomes. It is involved in both mitotic recombination, involving sister chromatids, and meiotic recombination, involving homologous chromosomes. Homologous recombination and nonhomologous end joining are responsible for repairing DNA double-strand breaks. HR utilizes regions of DNA sequence homology from the sister chromatid to repair damaged DNA in an error-free manner. NHEJ does not utilize extensive regions of sequence homology to effect repairs and is not necessarily error-free.

Double-strand breaks (DSBs) can be generated endogenously by oxidative agents and environmentally by ionizing radiation. This form of damage, while not as common as the other forms of DNA injury, can be extremely harmful. DSBs can lead to chromosomal translocations, deletions, and fragmentation. Mutations in several members of the machinery that repair double-strand breaks are associated with a variety of different cancers. This machinery will now be looked at in more detail.

14.9 Double-Strand-Break Repair Machinery

Proteins belonging to the DSB repair machinery form a number of distinct repair complexes. The protein responsible for repairing and rejoining broken strands of DNA are organized into several complexes, each consisting of proteins that come into physical contact with one another to carry out their repair functions. Proteins comprising the modules have been grouped together in Table 14.3. One of the key findings in studies of the DSB repair machinery is that some of the proteins are centrally involved in a number of rare, inherited genetic disorders whose study reveals important details. These participants in DSB repair are named for the disorders in which they were discovered. Examples of this naming convention include Ataxia-telangiectasia mutated (ATM) and Nijmegen breakage syndrome (NBS) proteins. Other rare, cancer-associated diseases involving DSB proteins are Fanconi anemia and Bloom syndrome.

The first entry in the table is ATM/ATR, two proteins that function as sensors and as signaling elements that convey damage signals to the p53 cellular controller and to the regulatory units of the repair modules. The ATM/ATR proteins are kinases and convey their signals by catalyzing the transfer of phosphoryl groups to their substrates. Three protein complexes follow the ATM/ATR entry. Each complex is made up of proteins, some

TABLE 14.3. Double-strand-break sensing, repair, and signaling.

Component	Function	Description
ATM/ATR	Signaling	Phosphorylates p53, NBS1, and BRCA1 in response to DSBs
Rad50	HR, NHEJ	Forms an end-processing complex with Mre11 and NBS1
Mre11	HR, NHEJ	Endonuclease activity
NBS1	HR, NHEJ	Regulatory
Rad51	HR	Searches for DNA homology; forms a complex with Rad52, Rad 54 and BRCA1,2
Rad52	HR	Stimulates Rad51 activity
Rad54	HR	Chromatin remodeling
BRCA1,2	HR	Regulatory; chromatin remodeling
Ku70	NHEJ	Forms a heterodimer with Ku80; recruits and activates DNA-PKcs
Ku80	NHEJ	KU70 and KU80 bind to the free ends of the DSB
DNA-PKcs	NHEJ	Regulatory; signals p53
XRCC4	NHEJ	Forms a complex with DNA ligase IV
DNA ligase IV	NHEJ	Ends reattachment

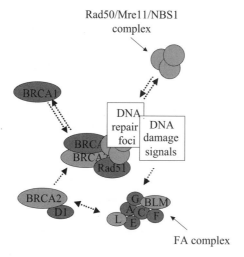

FIGURE 14.4. Onset of double-strand-break repair: The ATM proteins senses double-strand breaks and conveys damage signals to the NBS1 protein in the Rad50/Mre11/NBS1 complex and to the BRCA1 protein in the Rad51/Rad52/Rad54/BRCA1,2 complex. Damage signals are also sent to the Fanconi anemia (FA) complex, which includes the Bloom syndrome helicase BLM. In response, the FA ubiquitin ligase FANCL (L) activates the BRCA2 protein, which then joins the other DSB repair proteins at the repair site.

with regulatory functions and others with repair functions. The modules listed in Table 14.3 work synergistically and sequentially to make the repairs to the DNA. The early steps in operation of this repair system are depicted in Figure 14.4.

The first module, referred to as the Mre11 complex, is composed of the Rad50, Mre11, and NBS1 proteins. This module participates in both homologous recombination and nonhomologous end joining along with several other DNA strand manipulations involved in mitosis and meiosis. The key function of this complex is to form flexible bridges between ends of broken DNA strands. The Rad50 proteins contain hooks, which join opposing Rad50 proteins protruding from the ends of broken DNA strands. The hooks join pairs of Rad50 proteins in the middle of the break while the ends of the Rad50s remain attached to the DNA strands via the Mre11 proteins.

The next module is the Rad52 group. The Rad52 family of proteins is responsible for homologous recombination. The first step in HR is the coating of the single strand DNA (ssDNA) with molecules of replication protein A (RPA). In the next step, molecules of Rad51 and BRCA2 are loaded onto the strands so that Rad51 displaces RPA. Both RPA and Rad51 remove secondary structure from the DNA to facilitate the pairing of sister chromatids. As indicated in Table 14.3, Rad52 facilitates the replacement of RPA by Rad51. The Rad51 and BRCA2 proteins form nucleoprotein filaments—BRCA2 anchors the filaments to the DNA and Rad51 proteins form the body of the filaments. The pairing of sister chromatids follows growth of the filaments.

The nonhomologous end joining (NHEJ) proteins complete the list. The last two entries in the table, XRCC4 and DNA Ligase IV, are needed in the final steps of DSB repair. These proteins, operating synergistically with members of the Ku module, rejoin the two ends of the DNA to complete V(D)J recombination and NHEJ.

14.10 How Breast Cancer (BRCA) Proteins Interact with DNA

The Ataxia-telangiectasia mutated, Nijmegen breakage syndrome, and breast cancer proteins belong to two families of protein that operate in DNA damage detection, checkpoint signaling, and repair. One of these is a large family of proteins operating in DNA damage-signaling, and characterized by the presence of one or more BRCT (defined below) repeats. The second family is characterized by the presence of a COOH terminal phosphoinositide-3 kinase- (PI3K) like domain.

The BRCT proteins are characterized by the presence of a motif consisting of about 95 amino acid residues, sometimes repeated several times. This domain may be best thought of as an adapter module that enables members of this family of proteins to interact and participate with a variety of other proteins in DNA repair and recombination, transcription regulation, and cell cycle control. First discovered in the breast cancer 1 (BRCA1) protein, this motif is called a BRCA1 C-terminal (BRCT) domain. Its secondary structure consists of four beta strands plus two alpha helices.

BRCA1 (1863 amino acids) and BRCA2 (3418 amino acids) along with other members of the BRCT family are large multidomain proteins. BRCA1 contains a RING finger domain, a pair of nuclear localization signal sequences, and several BRCT repeats in its transcriptional activation domain. BRCA2 possesses several BRC repeats, an NLS and a transcriptional activation domain. Its BRCT region is followed by a COOH terminal region containing multiple single strand and double strand DNA-binding domains (Figure 14.5). Its tower domain is the site of several cancer-inducing mutations.

Inherited mutations in BRCA1 and BRCA2 genes predispose women to breast and ovarian cancer. Cancer cells having defective BRCA1 or BRCA2 proteins exhibit several kinds of chromosome instability. They have chromosome breaks and incorrect chromosome numbers and abnormal centrosomes. These instabilities are produced by defective transcription-coupled DSB repair, failures in checkpointing, and improper regulation

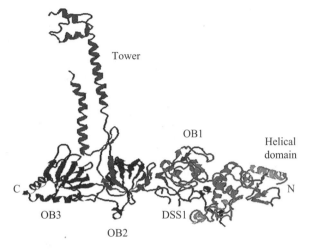

FIGURE 14.5. Crystal structure of the COOH terminal portion of the BRCA2 tumor suppressor protein: Shown are the main secondary structure elements' helices (corkscrew-shaped ribbons) and strands (flat arrow-shaped ribbons), and the overall organization of the COOH terminal portion of the molecule into five domains. The tower domain binds DNA and is the locus of a large number of cancer-inducing mutations. The oligonucleotide/oligosaccharide binding (OB) domains bind ssDNA. Two fragments of a DSS1 protein found in association with BRCA2, and needed for crystallization, are included in the figure. The figure was generated using Protein Explorer. The Brookhaven PDB accession number is 1MIU.

of centrosome duplication. The ensuring genetic instability then leads to further mutations and to tumor formation.

14.11 PI3K Superfamily Members that Recognize Double-Strand Breaks

ATM, ATR, and DNA-PKcs are members of the phosphatidylinositol-3 kinase (PI3K) superfamily that recognize DNA double-strand breaks, and are involved in their repair. Members of this grouping, which includes ATM, ATR, and DNA-PKcs, have a characteristic PI3K homologous domain in their COOH terminal region. The PI3K-like domain of these molecules does not appear to possess any lipid kinase activity. For that reason these proteins are called *PIK-related kinases*. They are large serine-threonine kinases; the ATM gene product is a 350-kDa protein, and DNA-dependent protein kinase (DNA-PK) is even larger. The DNA-PK molecule contains a Ku heterodimer consisting of 70-kDa and 80-kDa Ku subunits, plus a 460 kDa catalytic subunit, DNA-PKcs. The Ku subunits are named after the first two letters of a patient found to be suffering from an autoimmune disorder associated with the Ku proteins.

DNA-PKcs and ATM are structurally similar, and both molecules are capable of binding DNA. DNA-PK not only participates in repair of DSBs caused by ionizing radiation but also is involved in V(D)J recombination used to generate diversity in cells of the immune system. A mutation in the kinase domain produces severe combined immunodeficiency (Scid) in mice and defects lead to ionizing radiation hypersensitivity in humans. ATM and ATR are found on meiotic chromosomes and have a role in processing double-strand breaks generated during that process. These molecules serve as sensors of ionizing radiation-generated DSBs. They not only function as damage sensors but also as signaling molecules that relay damage signals to regulatory proteins that halt the cell cycle progression while repairs are made.

The ATM protein functions as a sensor of DNA double-strand breaks in the following manner. In the absence of DSBs, the ATM proteins exist as inactive dimers. Double strand breaks produced by, for example, ionizing radiation, induce alterations in chromatin structure. The alterations in chromatin structure trigger the dissociation of the ATM dimers resulting in autophosphorylation and activation. Once activated, ATM migrates over to p53, to NBS1 and BRCA1, and to cell cycle checkpoint proteins, and phosphorylates them at one or more sites.

14.12 Checkpoints Regulate Transition Events in a Network

The cycle of cell growth and division passes through four stages (Figure 14.6). The first of these stages (G_1) is a *cellular growth stage* that prepares the way either for entry into a cell division series of stages (S, G_2, and M) or to *growth arrest* (G_0), or to *apoptosis*. Cells that do not undergo growth arrest or apoptosis enter a *synthesis* (S) stage where the DNA is replicated in preparation for mitosis. This stage is followed by a further *mitosis preparatory stage* (G_2) where RNAs and proteins required for mitosis are synthesized. Finally, the cell enters into *mitosis* (M) where the cell divides

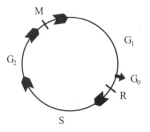

FIGURE 14.6. The cell cycle: Chevrons mark the divisions of the cell cycle into its four stages. The slash placed late in G_1 phase denotes the restriction (R) point and the second slash placed in mitosis represents the spindle checkpoints. The three nonmitosis stages are commonly referred to as *interphase*.

to produce two offspring. The three stages preceding mitosis are collectively referred to as *interphase*.

Checkpoints are signaling pathways that ensure that a process does not begin before a prior process is completed. They control the order and timing of transition events in a network. DNA damage checkpoints, for example, sense the presence of damage, producing signals that halt progression through the cell cycle while the damage is repaired, and stimulate the transcription of repair genes to deal with it. The cell cycle includes several checkpoints. The decision to undergo cell division or not occurs late in G_1 phase. This point is called the G_1/S *checkpoint*, or alternatively, START in yeast and the restriction (R) point in mammals. Growth conditions signified by receipt of growth (mitogenic) signals from outside the cell are evaluated, DNA is checked for damage, and a decision whether to proceed through the cell division cycle is made. A second halt for checkpointing occurs during S phase and the third takes place late in G_2 phase and is known as the G_2 *checkpoint*. Checkpointing takes place during mitosis, as well. The checkpoints halt mitosis if the spindle is damaged or if the chromosomes are not properly attached to the microtubules.

14.13 Cyclin-Dependent Kinases Form the Core of Cell-Cycle Control System

The timing and duration of events occurring during the cell cycle are tightly regulated by a control system containing at its core a family of serine/threonine kinases called *cyclin-dependent kinases* (cdks) and a set of regulatory subunits called *cyclins*. Cyclin-dependent kinases are present throughout the cell cycle but must bind to their regulatory cyclin to become active. There are several different kinds of cyclin-dependent kinases, each kind is associated with a particular cyclin subunit (Figure 14.7).

The concentrations of the different cyclins and cyclin-dependent kinases oscillate with the cell cycle, and different cdk-cyclin pairs are active different times in the cell cycle. The correspondence between cdks-cyclins and the phases of the cell cycle are presented in Table 14.4. The cyclin-dependent kinases carry out their regulatory roles by phosphorylating the retinoblastoma protein pRb, which along with p53 lie at the heart of the cell's control system.

14.14 pRb Regulates Cell Cycle in Response to Mitogenic Signals

In multicellular organisms, neighboring cells send mitogenic signals that coordinate differentiation during development and trigger growth and proliferation. These events must be coordinated in order for the cells to work

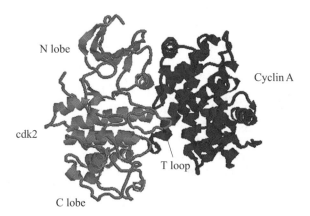

FIGURE 14.7. Crystal structure of the Cdk2—Cyclin A complex: Binding of Cyclin A to the kinase results in partial activation. Binding of the cyclin results in the extension of the catalytically important T loop. A further extension and full activation of the cdk is achieved through phosphorylation by a separate kinase. The figure was generated using Protein Explorer. The Brookhaven PDB accession number is 1FIN.

TABLE 14.4. Cyclins and cyclin-dependent kinases (cdks).

Cyclin	Protein kinase	Cell-cycle phase
Cyclins D1 to D3	Cdk4, Cdk6	Late G1 phase
Cyclin E	Cdk2	G1/S phase transition
Cyclin A	Cdk2	S-phase
Cyclin A	Cdk1 (Cdc2)	S-phase, G2-phase
Cyclin B	Cdk1 (Cdc2)	G2-phase, M-phase

together in a tissue or organ. The retinoblastoma protein, pRb, functions in a checkpoint role by inhibiting activation of an array of transcription factors, collectively designated as E2Fs, required for cell cycle progression. It keeps the gate shut at the G_1 checkpoint where most of the decisions are made whether to proceed to mitosis. The pRb gatekeeper integrates a variety of positive and negative mitogenic signals conveyed as phosphorylation events and if conditions are propitious it opens the gate and enables progression through to mitosis.

The retinoblastoma protein binds to members of the E2F family of transcription factors. During G_0 and G_1 stages of the cell cycle the E2Fs are maintained in an inactive state by their pRb binding. During this time, pRb is underphosphorylated (hypophosphorylated), and is able to form stable complexes with these transcription factors. Its phosphorylation state changes near the G_1/S boundary when the protein becomes hyperphosphorylated. Cyclins and their associated cyclin-dependent kinases play

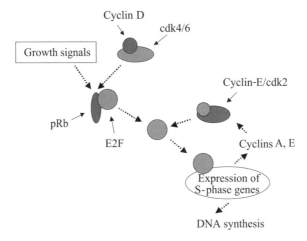

FIGURE 14.8. Growth signal stimulation of the transition from G_1 phase to S phase: Activated Cdk4s and Cdk6s along with kinases activated by growth signals hyper-phosphorylate the retinoblastoma protein (pRb). In response, the E2Fs dissociate from pRb and stimulate transcription of genes required for S-phase. Activation of cyclin E is associated with the passage to S phase from G_1 phase. The transcription of cyclin E along with cyclin A and DNA polymerase is stimulated by the E2Fs. These actions are terminated when Cyclin A-Cdk2 build up and turn off the E2Fs transcriptional activities.

a key role in inactivating pRb. Phosphorylation of pRb by these cell cycle regulators and by the growth factor-activated kinases frees the E2Fs from their pRb inhibition and they can then carry out their cell cycle-progression-driving transcriptional activities (Figure 14.8).

Two noncontiguous domains, referred to as the *A and B domains*, collectively form a binding locus known as an *A/B pocket* that binds many proteins to pRb (Figure 14.9). For that reason, pRb and other members of its family are known as *A/B pocket proteins*. This importance of this interaction manifests itself by the occurrence of cancer-causing mutations in residues located in the A/B pocket. The cyclin cell cycle regulators are among the key proteins that bind to the A/B pocket. The cyclins, most notably cyclins D and E, recruit cyclin-dependent kinases to the A/B pocket and to a docking site located in the C terminus.

14.15 p53 Halts Cell Cycle While DNA Repairs Are Made

The integration of the G_1/S DNA damage checkpoint pathway into the cell cycle occurs in the following way. Damage to DNA is sensed by the ATM/ATM and Rad3-related (ATR) kinases. When they detect damage to

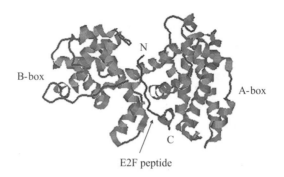

B-box

A-box

N

C

E2F peptide

FIGURE 14.9. Binding of an E2F peptide fragment to the pRb pocket: Shown in the figure is a 18-residue fragment derived from the transactivation domain of E2F (residues 410–427) in a complex with the A/B pocket domain of pRb. The figure was generated using Protein Explorer with atomic coordinates deposited in the Brookhaven PDB under accession number 1N4M.

DNA they phosphorylate a pair of kinases called checkpoint 1 (Chk1) and checkpoint 2 (Chk2). The Chk1 and Chk2 kinases phosphorylate a protein phosphatase known as cell division cycle 25 (Cdc25). The Cdc25 protein phosphatase is a cyclin-cdk activator. It is active when it is in a dephosphorylated state and is inactivated and degraded when it is phosphorylated. Phosphorylation of Cdc25 by the checkpoint kinases inactivates the phosphatase and tags it for proteolysis, leading to phosphorylation and inactivation (arrest) of the cyclin-cdks through actions of another protein kinase called Wee1.

When damage to the cellular DNA is sensed, ATM, ATR, and also DNA-PKcs convey that information to p53 by phosphorylating the molecule at sites specific to the type of damage sensed. In response to these signals, p53 acts as a transcription factor and upregulates a number of proteins. One of these, p21, halts the progression through the cell cycle until DNA repairs can be made. The p21 protein upregulated by p53 is a universal regulatory subunit of cyclin-cdks. When p21 is activated it binds to the Cyclin E/Cdk2 complex, resulting in Cdk2 inactivation. The Cdk2s cannot phosphorylate pRb, which then becomes hypophosphorylated and inhibits the E2Fs (Figure 14.10) thereby halting progression from G_1 to S phase. If the damage is regarded as unrepairable, p53 functions in a second different way. It triggers an apoptosis circuit that targets the cell for suicide.

14.16 p53 and pRb Controllers Central to Metazoan Cancer Prevention Program

The p53, pRb, and connecting signaling proteins and other key regulators form a circuit that regulates the progression through the cell cycle. Proper operation of this circuit is central to a cell's cancer prevention program. As

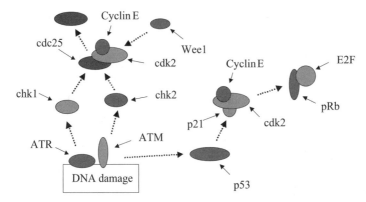

FIGURE 14.10. Cell cycle arrest triggered by DNA damage checkpoint signals: ATR and ATM phosphorylate Chk1 and Chk2, which then phosphorylate the protein phosphatase Cdc25 thereby inactivating it. Cdk2 is phosphorylated by the protein kinase Wee1 and its actions are arrested. ATM signals through p53 to activate p21, which arrests entry into S phase by inhibiting the transcriptional activities of the E2Fs.

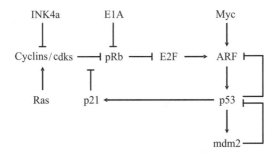

FIGURE 14.11. Controller circuit coordinating p53 and pRB activities: Mitogenic signals are relayed into the cell via Ras, Ink4a and c-Myc. Pointed arrows denote stimulatory influences and flat-headed arrows denote repressive influences.

shown in Table 14.1, mutations in p53 or pRb leading to malfunctions in the circuitry are found in most human cancers. A simplified representation of this circuitry highlighting some of the most crucial protein-protein interactions is depicted in Figure 14.11. INK4a, Ras and c-Myc relay mitogenic (growth and proliferation) signals. These signals converge upon ARF which sits just upstream of p53. As shown in the figure, when it is activated, ARF stimulates p53 activity. It does so by disrupting mdm2's inhibition of p53. Because of its key location, ARF mutations are encountered in a host of human cancers. Another prominent gene encountered in many cancers is the adenovirus early region 1A (E1A) gene. This gene product interacts with a variety of control proteins including pRb. When E1A binds pRb, it disrupts formation of pRb-E2F heterodimers and thus promotes establishment of cancer cell cycles resembling those produced by mutations in pRb itself.

The pRb and p53 work together. Early in the cell cycle, pRb inhibits the E2Fs, and Mdm2 keeps p53 at a low level of activity. When pRb becomes hyperphosphorylated its block on the E2F is relieved, ARF is stimulated and it, in turn, stimulates p53 so that as the cell enters S phase, p53's surveillance activities are stepped up. If p53 detects DNA damage it signals p21, which blocks the cyclin/cdk complexes from hyperphosphorylating pRb. The retinoblastoma protein then inhibits the E2Fs while the repairs to DNA are carried out.

The ATM/ATR and DNA-PKcs proteins activate p53 by stimulating proteolysis of the Mdm2 inhibitor. Under normal cellular conditions, the Mdm2 protein represses the checkpointing and apoptosis promoting activities of p53. The Mdm2 protein acts as a p53-specific, E3 ubiquitin ligase. It suppresses the transcriptional actions of p53 by tagging it for proteolytic destruction and keeping its expression levels low. One of the genes turned on by p53 is the Mdm2 gene. This activity leads to the formation of an autoregulatory loop in which p53 regulates Mdm2 at the level of transcription, and Mdm2 regulates p53 at the level of its activity. The Mdm2-p53 system is balanced through these mutual dependencies so that the appropriate signals can activate p53. These signals dislodge Mdm2 from p53, leading to the transcriptional activation of p53.

14.17 p53 Structure Supports Its Role as a Central Controller

As might be expected from the appellation "controller," the p53 and pRb proteins receive input signals from many proteins and in turn influence many other proteins. The first of these, p53, functions as a transcription factor while the second protein, pRb, acts through the E2F family of transcription factors. In the language of networks, p53 and pRb operate as highly connected nodes, and therefore it is not surprising that disabling these proteins through mutations has such strong negative consequences. The signals convergent upon p53 take the form of phosphorylations and acetylations of specific residues. Some of the best characterized of the sites of phosphorylation and acetylation are depicted in Figure 14.12.

The flexibility in the chain connecting the core unit to the DNA-binding domain allows the p53 molecule to contact the DNA in any of a number of ways. This property is illustrated in Figure 14.13 where three DNA-binding domains bind the DNA molecule, each in a slightly different way. The most frequent mutations in p53 found in cancer cells are all situated in the DNA-binding domain. Of these, five involved arginine residues and one a glycine residue. The leading mutation is to Arg248, which mediates the direct

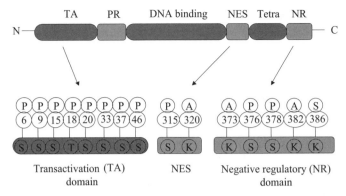

FIGURE 14.12. Organization of the p53 protein: Shown in the upper part of the figure is the domain structure of the p53 monomer. It contains an N-terminal transactivation (TA) domain, a proline-rich (PR) region, a DNA-binding domain, and in its C-terminal region a nuclear export sequence (NES), tetramerization (Tetra) domain and a negative regulatory (NR) domain. Expanded depictions of some of the key regulatory regions are shown in the lower part of the figure. The specific amino acid residues that are phosphorylated (P), acetylated (A) or sumoylated (S) are shown. Abbreviations: serine (S), threonine (T), lysine (K).

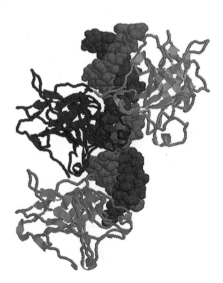

FIGURE 14.13. p53-DNA binding: The DNA-binding domains of the p53 protein are depicted as ribbons while the dsDNA molecule is shown in a space-filled model. The figure was generated using Protein Explorer with atomic coordinates deposited in the Brookhaven PDB under accession numbers 1TUP and 1TSR.

binding of the p53 molecule to the minor groove of DNA. Another often mutated residue, Arg273, makes contact with the backbone phosphate. The other residues help stabilize the interface between the p53 and DNA surfaces.

14.18 Telomerase Production in Cancer Cells

The ends of chromosomes are capped by *telomeres*, a series of TTAGGG repeats and associated proteins, and terminated at the very end by a lasso-like structure called a *T-loop*. Telomeres are shortened by about 100 base pairs every cell division. When the capping structure is degraded sufficiently the cell's DNA repair machinery is able to sense the ends of the DNA molecules and interprets the ends as double strand breaks. If the cell keeps dividing several negative outcomes become possible. These include degradation, chromosome recombination leading to loss of genetic information, aberrant rearrangements of chromosomes, and genomic instability. To avoid such dangerous situations the cell ceases to divide after several kilo base pairs of telomere are lost and instead enters a nondividing stage called *senescence*.

Several proteins associate with the telomeres. One of these is the (human) telomerase reverse transcriptase (hTERT), which catalyzes the addition of multiple TTAGGG repeats at the ends of chromosomes and protects them from the DSB repair machinery. Among the others are two proteins, named telomeric repeat binding factors 1 and 2 (TRF1 and TRF2), that bind the TTAGGG repeats. In addition, a number of double strand break repair proteins form complexes with the TRFs and the telomerase components (the hTERT catalytic subunit and a human telomerase RNA (hTR) that contains the template for adding telomeric repeats). This grouping includes members of the Rad50/Mre11/NBS1 double strand break repair complex, and Ku86 and DNA-PK$_{cs}$ involved in NHEJ. Members of these complexes are thought to convey prosenescence signals to the p53 circuitry when telomeres become shortened.

Signals sent through p53 and pRb ensure that the cell ceases to divide. In response to critical telomere shortening, production of regulators of p53 and pRb activity such as ARF, p21, and Ink4a is increased. The pRb protein becomes hypophosphorylated and binds E2F thereby suppressing proliferation. In more detail, ARF suppresses Mdm2 leading to the release of p53 from Mdm2 inhibition; p53 stimulates p21, which contributes to hypophosphorylation of pRb. The Ink4a protein contributes to pRb hypophosphorylation by suppressing the Ckd4/6 inhibition of pRb.

Cancer cells avoid senescence by increasing the production of the telomerase chromosome-capping enzyme. This activity greatly expands the number of cell divisions possible, effectively immortalizing the cells. The cells avoid passing into senescence, and instead continue to divide and progress towards more lethal states. Mutations in the p53 and pRb that disable these proteins can contribute to the immortalization by disrupting the conveyance of prosenescence signals.

References and Further Reading

MMPs

McCawley LJ, and Matrisian LM [2001]. Matrix metalloproteinases: They're not just for matrix anymore! *Curr. Opin. Cell Biol.*, 13: 534–540.

Nagase H, and Woessner JF, Jr [1999]. Matrix metalloproteinases. *J. Biol. Chem.*, 274: 21491–21494.

Stamenkovic I [2003]. Extracellular matrix remodeling: The role of matrix metalloproteinases. *J. Pathol.*, 200: 448–464.

Growth Factor Signaling

Birchmeier C, et al. [2003]. MET, metastasis, motility and more. *Nature Rev. Mol. Cell Biol.*, 4: 915–925.

Blume-Jensen P, and Hunter T [2001]. Oncogenic kinase signaling. *Nature*, 411: 355–365.

Downward J [2003]. Targeting Ras signaling pathways in cancer therapy. *Nature Rev. Cancer*, 3: 11–22.

Growth Factor Receptor, and Adhesion Molecule Cooperativity

Comoglio PM, Boccaccio C, and Trusolino L [2003]. Interactions between growth factor receptors and adhesion molecules: Breaking the rules. *Curr. Opin. Cell Biol.*, 15: 565–571.

Conacci-Sorrell M, Zhurinsky J, and Ben-Ze'ev A [2002]. The cadherin-catenin adhesion system in signaling and cancer. *J. Clin. Invest.*, 109: 987–991.

Hood JD, and Cheresh DA [2002]. Role of integrins in cell invasion and migration. *Nature Rev. Cancer*, 2: 91–100.

Developmental Pathways

Fearnhead NS, Britton MP, and Bodmer WF [2001]. The ABC of APC. *Human Mol. Genet.*, 10: 721–733.

Giles RH, van Es JH, and Clevers H [2003]. Caught up in a Wnt storm: Wnt signaling in cancer. *Biochim. Biophys. Acta*, 1653: 1–24.

Massagué J, Blain SW, and Lo RS [2000]. TGFβ signaling in growth control, cancer and heritable disorders. *Cell*, 103: 295–309.

Peifer M, and Polakis P [2000]. Wnt signaling in oncogenesis and embryogenesis—A look outside the nucleus. *Science*, 287: 1606–1609.

Taipale J, and Beachy PA [2001]. The Hedgehog and Wnt signalling pathways in cancer. *Nature*, 411: 349–354.

DNA Repair Mechanisms

Critchlow SE, and Jackson SP [1998]. DNA end-joining: From yeast to man. *Trends Biochem. Sci.*, 23: 394–398.

De Laat WL, Jaspers NGL, and Hoeijmakers JHJ [1999]. Molecular mechanisms of nucleotide excision repair. *Genes Dev.*, 13: 768–785.

Hoeijmakers JHJ [2001]. Genome maintenance mechanisms for preventing cancer. *Nature*, 411: 366–374.

Kanaar R, Hoeijmakers JHJ, and van Gent DC [1998]. Molecular mechanisms of DNA double-strand break repair. *Trends Cell Biol.*, 8: 483–489.

Lindahl T, and Wood RD [1999]. Quality control by DNA repair. *Science*, 286: 1897–1904.

Yang W [2000]. Structure and function of mismatch repair proteins. *Mutation Res.— DNA Repair*, 460: 245–256.

Double-Strand-Break Repair

D'Amours D, and Jackson SP [2002]. The Mre11 complex: At the crossroads of DNA repair and checkpoint signaling. *Nature Rev. Mol. Cell Biol.*, 3: 317–327.

D'Andrea AD, and Grompe M [2003]. The Fanconi anaemia/BRCA pathway. *Nature Rev. Cancer*, 3: 23–34.

Hopfner KP, et al. [2002]. The Rad50 zinc-hook is a structure joining Mre11 complexes in DNA recombination and repair. *Nature*, 418: 562–566.

Leuther KK, et al. [1999]. Structure of DNA-dependent protein kinase: Implications for its regulation by DNA. *EMBO J.*, 18: 1114–1123.

Song BW, and Sung P [2000]. Functional interactions among yeast Rad51 recombinase, Rad52 mediator, and replication protein A in DNA strand exchange. *J. Biol. Chem.*, 275: 15895–15904.

Yang HJ, et al. [2002]. BRCA2 function in DNA binding and recombination from a BRCA2-DSS1-ssDNA structure. *Science*, 297: 1837–1848.

The ATM Protein

Bakkenist CJ, and Kasten MB [2003]. DNA damage activates ATM through intermolecular autophosphorylation and dimer dissociation, *Nature*, 421: 499–506.

Shiloh Y [2003]. ATM and related protein kinases: Safeguarding genome integrity. *Nature Rev. Cancer*, 3: 155–168.

The Cell Cycle, E2Fs, and the Retinoblastoma Protein

Dyson N [1998]. The regulation of E2F by pRb-family proteins. *Genes Dev.*, 12: 2245–2262.

Harbour JW, et al. [1999]. cdk phosphorylation triggers sequential intramolecular interactions that progressively block Rb functions as cells move through G1. *Cell*, 98: 859–869.

Malumbres M, and Barbacid M [2001]. To cycle or not to cycle: A critical decision in cancer. *Nature Rev. Cancer*, 1: 222–231.

Muller H, and Helin K [2000]. The E2F transcription factors: Key regulators of cell proliferation. *Biochim. Biophys. Acta.*, 1470: M1–M12.

Muller H, et al. [2001]. E2Fs regulate the expression of genes involved in differentiation, development, proliferation, and apoptosis. *Genes Dev.*, 15: 267–285.

Sherr CJ [2000]. The Pezcoller Lecture: Cancer cell cycles revisited. *Cancer Res.*, 60: 3689–3695.

Weinberg RA [1995]. The retinoblastoma protein and cell-cycle control. *Cell*, 81: 323–330.

The p53 Protein

Espinosa JM, and Emerson BM [2001]. Transcriptional regulation by p53 through intrinsic DNA/chromatin binding and site-directed cofactor recruitment. *Mol. Cell*, 8: 57–69.

Sherr CJ [2001]. The INK4a/ARF network in tumor suppression. *Nature Rev. Mol. Cell Biol.*, 2: 731–737.

Vogelstein B, Lane D, and Levine AJ [2000]. Surfing the p53 network. *Nature*, 408: 307–310.

Vousden KH, and Lu X [2002]. Live or let die: The cell's response to p53. *Nature Rev. Cancer*, 2: 594–604.

Telomere Maintenance

Blackburn EH [2001]. Switching and signaling at the telomere. *Cell*, 106: 661–673.

Chan SWL, and Blackburn EH [2002]. New ways to make ends meet: Telomerase, DNA damage proteins and heterochromatin. *Oncogene*, 21: 553–563.

De Lange T [2002]. Protection of mammalian telomeres. *Oncogene*, 21: 532–540.

Itahana K, Dimri G, and Campisi J [2001]. Regulation of cellular senescence by p53. *Eur. J. Biochem.*, 268: 2784–2791.

Maser RS, and DePinho RA [2002]. Connecting chromosomes, crisis and cancer. *Science*, 297: 565–569.

Problems

14.1 Make a list of the ways a cancer cell differs from a normal one. Identify the signaling proteins that promote the altered response and how these proteins are altered to promote cancer for each entry in the list. For example, mutations in the Ras proteins that leave the GTPase stuck in the on position contribute to aberrant growth signaling that cannot be turned off.

14.2 Cancer is a multistep process. Arrange the entries in the list generated in Problem 14.1 in a temporal order with early acting changes at the top and late acting changes at the bottom. Which kinds of alterations promote metastasis? Which ones help generate genome instabilities and chromosome abnormalities?

15
Apoptosis

Kerr, Wyllie, and Currie coined the term *apoptosis* in an article that appeared in the *British Journal of Cancer* in 1972. The term was taken from the Greek and has the meaning of "falling off" as in the dropping of petals from flowers, and leaves from trees, which occur in a programmed manner every autumn. Apoptosis, or programmed cell death, is widespread during animal development where it is used to sculpt and refine tissues and organs. Apoptosis makes possible the regression of tadpole tails, permitting the emergence of an adult tailless form. It is used to remove larval organs in insects, and to sculpt digits—fingers and toes—from undelineated limb buds in mammals.

Apoptosis is widely encountered during development of the nervous system. Some neurons are only transiently formed, and once their task is completed they are removed. In many parts of the nervous system, cells are initially overproduced. This overproduction ensures that there is adequate input to target neurons during the initial phase of the nervous system connection. Excess cells are removed by numerically matching pre- and post-synaptic structures. The initial set of neural connections is refined, and cells that are inappropriately wired are removed. In some neural populations, most of the cells are removed in order to produce a set of precise neural connections.

Apoptosis is an essential ingredient in the operation of the immune system. Cells of the immune system such as B cells die off as they age and when they are no longer needed. Virally infected cells, and damaged cells that cannot be repaired, are targeted for apoptosis to prevent their harming undamaged cells in the tissue or organ where they reside. The same strategy is used by the immune system to rid the body of cancer cells.

Apoptosis is different from necrosis. The plasma membrane does not rupture in apoptosis as it does in necrosis, but instead cellular components are degraded and packaged, and then digested. During apoptosis cells undergo an orderly sequence of morphological changes. These changes include cell shrinkage, chromatin condensation, DNA fragmentation, and membrane blebbing, in which membrane-wrapped pieces of cell boil off of the cell surface as apoptotic bodies containing fragments of DNA and other macromolecules.

Malfunctions in the cellular machinery that controls apoptosis are encountered in many disease situations. Excessive apoptotic cell death occurs in Alzheimer's disease, Parkinson's disease, Huntington's disease, and ALS (Lou Gehrig's disease). Too little cell death is a hallmark of cancer. In B-cell leukemia, for example, key regulators of the decision circuitry that determines whether a cell survives or dies are overexpressed. Apoptosis is suppressed and because population control is lost leukemia develops.

15.1 Caspases and Bcl-2 Proteins Are Key Mediators of Apoptosis

Apoptosis is largely carried out by *caspases*, proteolytic enzymes that catalyze the cleavage of specific molecules and groups of molecules in response to activating signals. Caspases target critical repair, splicing, and replication components, they cut up membranes and cytoskeleton regulators, and they destroy cellular DNA. They also stimulate the expression of markers on the cell surface that tag the cell for orderly destruction and engulfment by neighboring cells. This orderly disassembly of a cell occurs in a way that prevents damage due to leakage.

Bcl-2 proteins are a second group of proteins intimately involved in apoptosis. They function as sensors and regulators of the apoptosis program. They were first identified in B-cell lymphomas and, since then, mutated forms have been found in a variety of cancers. These proteins are characterized by the presence of one or more Bcl-2 homology (BH) domains, the ability of some to form pores in internal membranes, and their propensity to either promote or inhibit the release of apoptotic signals and agents from the internal membranes.

Apoptosis can be initiated by death signals sent into the cell from other cells and by stress signals generated within the cell. Signals sent by other cells instructing a cell to undergo apoptosis are received by death receptors belonging to the tumor necrosis factor (TNF) family. When a death ligand binds the death receptors, a death inducing signaling complex is formed that initiates the apoptosis process. Death signals are also triggered by cellular stresses detected internally in organelles such as the endoplasmic reticulum, Golgi, nucleus, and mitochondria. Apoptosis signaling is sent in response to conditions such as irrevocable DNA damage in the nucleus, unfolded protein stresses in the ER, and oxidative stresses in the mitochondria. Two loci, one within the mitochondria and the other just outside that organelle, serve as the main control points for the launching of apoptotic responses.

15.2 Caspases Are Proteolytic Enzymes Synthesized as Inactive Zymogens

The activity of enzymes that chop up and digest molecules is tightly controlled in the cell. One common strategy for controlling proteolytic enzymes, or *proteases*, is to synthesize them in an inactive form that requires further processing for their activation. One common kind of inactive form is a *zymogen*, a proenzyme containing a prodomain that must be removed in order to create an active form of the enzyme. Zymogens are converted to catalytically active forms by their proteolytic cleavage into two or three pieces followed by assembly of the catalytic subunits into complexes. Examples of proteases synthesized as zymogens include digestive enzymes that reside in the stomach (pepsin) and pancreas (trypsin), and also include blood-clotting enzymes (thrombin).

Caspases are cysteine aspartate-specific proteases. They break peptide bonds after Asp residues, i.e., at Asp-X sites, and possess a highly conserved cysteine residue in their catalytic site. Caspases are synthesized as zymogens. They contain a prodomain in the amino terminal region that regulates the proenzyme, followed by a large domain, approximately 20 kDa in size, and then a small domain, about 10 kDa in size. The proenzyme is activated by proteolytic cleavage at two Asp-X sites, one situated at the end of the prodomain and the other separating the large and small domains (Figure 15.1).

The tetramer is the functional (active) form of the caspase. It is constructed in several stages. In the first stage, two zymogens associate to form a zymogen homodimer. Adjacent large and small subunits (left hand pair and right hand pair) are part of the same polypeptide chain with the small units placed on the inside and the large subunits on the outside. In the next stage, each of the polypeptide chains making up the caspase zymogen dimer is cleaved at the Asp-X sites. These cuts induce conformational changes in the subunits that are part of the caspase activation process because they bring the caspases closer to conformation supporting catalysis. The similarity in conformations can be seen in Figure 15.2 where zymogen and caspase homodimers are compared side-by-side. Differences in the crucial loops L1

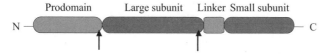

FIGURE 15.1. Caspase domain structure: The N-terminal prodomain is followed by a large subunit, a linker, and a small subunit. The location of the two Asp-X cleavage sites is indicated in the figure by arrows. Following cleavage, two large and two small subunits associate to form a caspase tetramer with the two small subunits inside and two large subunits on the outside.

(a)   (b)

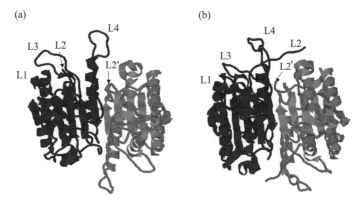

FIGURE 15.2. Structure of caspase-7 homodimers as determined by means of X-ray diffraction measurements: (a) Procaspase-7 zymogen homodimer, and (b) Caspase-7 homodimer. The figures were prepared using Protein Explorer using atomic coordinates deposited in the Brookhaven Protein Data Bank under accession codes 1K88 (a) and 1K86 (b).

to L4 that determine the catalytic cleft are clearly visible in the left and right hand panels of the figure. The caspases are still not fully activated in the more open caspase form, but a final set of changes occurring during substrate binding renders the caspases fully active.

15.3 Caspases Are Initiators and Executioners of Apoptosis Programs

More than a dozen different kinds of caspases have been identified in mammals. Those that have been found in humans have been placed in one of three groups in Table 15.1. Caspases belonging to Group I are associated with inflammatory responses. These caspases were first named for interleukin-1β converting enzyme (ICE), and then renamed caspases 1, 4, and 5. These enzymes process pro-inflammatory cytokines. The remaining two groups of caspases are specifically associated with apoptosis, either as initiators that convey signals through their proteolytic actions or as effectors that degrade cellular components. Group II caspases are effectors. They proteolytically degrade a variety of cellular components and are thus the executioners of the apoptosis program.

Group III caspases are apoptosis initiators. They act upstream of the effectors and activate them in response to proapoptotic signals and events. The four caspases appearing as Group III caspases all have large prodomains in which there is either a DED (death effector domain) or a CARD (capsase recruitment domain) protein–protein interaction domain.

TABLE 15.1. Mammalian caspases and their roles in the cell: Group III initiator caspases act upstream of the Group II effector caspases. The executioners have small prodomains and require assistance of the initiators for their activation. Four element consensus sequences that the caspases recognize and cleave are presented in column 4. The most conserved residue in the consensus sequence is the Asp (D) residue proximal to the cleavage site while the residue in the fourth position mostly determines the substrate specificity.

Group	Caspase	Consensus sequence	Prodomain
I: ICE	1	(WL)EHD	Large, CARD
	4	(WL)EHD	Large, CARD
	5	(WL)EHD	Large
II: Effectors	3	DExD	Small
	6	(ILV)ExD	Small
	7	DExD	Small
III: Initiators	2	DExD	Large, CARD
	8	(ILV)ExD	Large, DED
	9	(ILV)ExD	Large, CARD
	10	(ILV)ExD	Large, DED

These regulatory sequences target the procaspases either to adapters bound to death receptors at the cell surface or to adapters positioned near mitochondria. At these locations the initiator caspases are positioned to respond to proapoptotic signals. Caspases 8 and 10 contain a pair of DEDs. Caspase 2 and 9 contain CARDs. The effector (Group II) caspases do not have a large prodomain in their N-terminus but instead possess a small N-terminal peptide. One other initiator caspase has been found—Caspase 12, a murine (mouse) caspase that is localized to the endoplasmic reticulum.

15.4 There Are Three Kinds of Bcl-2 Proteins

Bcl-2 proteins are central regulators of caspase activity and of the cell decision concerning whether or not to undergo apoptosis. Bcl-2 proteins can be grouped into three subfamilies according to their domain structure and their pro- or antiapoptotic activities (Table 15.2). The defining characteristic of the Bcl-2 proteins is the presence of one or more BH domains. The Bcl-2 and Bax subfamilies contain multiple BH domains while the Bad subfamily only possesses a BH3 domain.

There are four kinds of BH domains, designated as BH1 through BH4. A typical Bcl-2 subfamily member contains at least three of the BH domains, namely, BH1, BH2, and BH3. In addition, it has a hydrophobic tail that anchors the protein to the outer membrane of mitochondria, the endoplasmic reticular membrane, and the outer nuclear envelope. Members of the Bcl-2 subfamily inhibit apoptosis by restricting membrane permeabil-

TABLE 15.2. Bcl-2 family of apoptosis regulators: Listed are the numbers of amino acid residues in the proteins.

Bcl-2 subfamily: Antiapoptotic	size (aa)	Bax subfamily: Proapoptotic	size (aa)	Bad subfamily: Proapoptotic	size (aa)
A1	172	Bak	211	Bad	197
Bcl-2	239	Bax	192	Bid	195
Bcl-xL	233	Bcl-xS	170	Bik	160
Bcl-w	193	Bok	213	Bim	196
Boo	191			Blk	150
Mcl-1	350			Bmf	186
				Hrk	91
				Noxa	103
				Puma	193

ity through interactions with mitochondrial membrane components and by binding to and sequestering members of the proapoptotic Bax group.

The proapoptotic Bax subfamily consists of proteins that possess a BH3 domain, a hydrophobic transmembrane tail, and at least one other BH domain. Some have a BH4 domain; others have BH1 and BH2 domains and perhaps a BH4 domain. Their 3-D structure is similar to pore-forming bacterial toxins. The Bax proteins interact with the proteins embedded in the outer mitochondrial membrane to increase membrane permeability, and they can form pores by themselves in membranes when they oligomerize.

The pro- and antiapoptotic, multi-BH domain proteins have electrostatic and structural properties that enable them to not only operate in the cytoplasm but also insert into the membrane to make pores. Their polypeptide chains are organized into sets of eight α-helices. There are three layers of two α-helices and a pair of short capping helices at one end of the chain. The organizaztion of the protein and the correspondence between helices and BH domains is presented in Figure 15.3. Several structural features support pore forming. The structure is fairly flexible and so can rearrange itself with little energy penalty. Two of the helices—α5 and α6—are able to span the membrane. There are several disordered, flexible regions, and there are three cavities. Charge-wise, the bottom of the protein is lined with basic (positively charged) residues that complement the acidic (negatively charged) membrane surface, and there is a pronounced hydrophobic cleft surrounded by basic residues. The picture that emerges from examinations of these structures is that of α5 and α6 along with the corresponding α5 and α6 helices from dimerization partners forming a pore, with α2 to α4 forming a binding groove, and the C-terminus forming an anchor.

The Bad subfamily is referred to as the *BH3-only family*. Some members of this group have a transmembrane anchor while others do not. When activated by proapoptotic stimuli, these proteins translocate to the mitochondria and stimulate apoptotic responses. BH3-only proteins function as cellular sentinels. In unstressed cells Bid, Bim, and Bmf are immobilized in

(a)

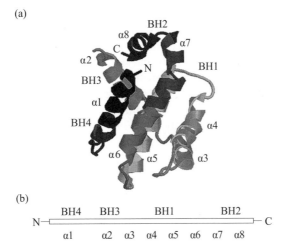

(b)

FIGURE 15.3. Structure of the antiapoptotic protein Bcl-x$_L$: Shown in part (a) of the figure is a ribbon diagram of the portions of the protein whose structure could be determined through X-ray crystallography (i.e., highly disordered regions are not included in the model). Gray-scale shadings highlight the four BH regions. The correspondence between BH regions and the eight α-helices is presented in part (b) of the figure. The figure was generated using Protein Explorer from Brookhaven Protein Data Bank entry 1AF3.

the cytoplasm. Bid is a sensor of death signals sent into the cell through the death receptors, and is localized in the vicinity of the death receptors. Bim is sequestered at microtubule associated myosin V motors where it awaits activation by cytokine and other stress signals. Bmf is also immobilized at myosin V motors where it responds to loss of cell attachment (*anoikis*) signals. Two other BH3-only proteins, Bik and Blk, function in the endoplasmic reticulum as sensors of cellular stress. The remaining BH3-only proteins are regulated at the transcriptional level. Noxa and Puma are transcribed in a p53-dependent manner and may be regarded as DNA damage sensors, while Hrk and Bim are upregulated in response to growth factor deprivation and cytokine withdrawal.

15.5 How Caspases Are Activated

Caspases are activated by external suicide instructions and internal stress signals. Cells receive death instructions from other cells. These messages are conveyed by cell-to-cell messengers called *death ligands*. The messages are received by death receptors embedded in the plasma membrane. The death messages are transduced into the cell interior through a multiprotein signaling complex formed by the activated death receptors. This complex is called the *death-inducing signaling complex*, or DISC. The DISC is the control point for external signal activation of initiator Caspases 8 and 10 and signal to the BH3-only sensor protein Bid.

The counterpart to DISC for internal stress signals is a signaling complex called the *apoptosome*. This multiprotein signaling complex is formed just outside the mitochondria in response to internal stress signals. Initiator caspase 9 is activated at this control point in response to the stress-induced release of proapoptotic factors from the mitochondria. The release of the mitochondrial factors is triggered by activity at the mitochondrial control point called the *permeability transition pore complex*, or PTPC, where multidomain Bcl-2 proteins are localized and BH3-only sensor proteins converge when activated.

Several families of positive and negative regulators control the caspase machinery. The regulatory proteins ensure that apoptosis is not triggered inappropriately in response to random perturbations and aberrant signals. External and internal signals are integrated together, and both contribute to the live or die decision. If strong signals are sent into the cell instructing it to undergo apoptosis, and strong stress signals are also present within the cell, the decision is fairly simple—caspases will be activated and the cell will die. If, as is normally the case, neither external nor internal death signals are present the cell will live. All other situations are more complex. Cellular context comes into play through adjustments in the expression levels of the positive and negative regulators that determine the set, balance, or commitment point for apoptosis. The remainder of the chapter is devoted to exploring how the apoptosis control system works.

15.6 Cell-to-Cell Signals Stimulate Formation of the DISC

The death receptors that transduce the death messages into the cell belong to the tumor necrosis factor (TNF) superfamily. The TNF superfamily in humans includes 19 Type II ligands, single-pass transmembrane proteins with cytoplasmic N-terminals, and at least 29 receptors, mostly Type I (extracellular N-terminals). Recall from Chapter 9 that these receptors are widely expressed in the immune system where they respond to variety of growth, proliferation, and death signals. Their extracellular region is characterized by the presence of from 2 to 5 repeats of a cysteine-rich motif containing a number of disulfide bridges. Their cytoplasmic portions contain docking sites for several different kinds of adapters that mediate the recruitment of key signaling elements to the receptors.

The death receptor family includes the TNF-α, Fas/Apo-1, and TNF-related apoptosis-inducing ligand (TRAIL) receptors (Table 15.3). The receptors and ligands operate as homotrimeric proteins. Signaling begins when a trio of ligands binds to a trio of receptors. This event triggers the

TABLE 15.3. Members of the TNF family of receptors containing death domains in their cytoplasmic region: Abbreviations—Death receptor (DR); TNF-related apoptosis inducing ligand (TRAIL).

Death receptor	Alternative name(s)
Fas	Apo-1, CD95
TNF R1	CD120a
DR3	Apo-3
TRAIL R1	Apo-2, DR4
TRAIL R2	DR5
DR6	

recruitment of the adapters to the cytoplasmic portion of the receptors leading to the assembly of a DISC. In forming a DISC, the TNF-associated death domain (DD) proteins are first recruited to the cytoplasmic portion of the TNF receptors and bind by means of their death domains. Proteins with death effector domains (DEDs) bind next, and other binding events follow. In this manner, a DISC is formed with FADD, TRAF2, TRADD adapters serving as a starting point for three pathways—Caspase 8, NF-κB, and MAP kinase, respectively.

15.7 Death Signals Are Conveyed by the Caspase 8 Pathway

The Caspase 8 pathway (Figure 15.4) begins when the Fas-associated death domain (FADD) protein is recruited to the nascent DISC. Zymogens with large prodomains such as Caspase 2, Caspase 8, and Caspase 10 are brought into close proximity with one another at the FADDs and as a result can form zymogen dimers and act on themselves to remove their prodomains. Several caspase molecules can enter, become activated, and leave, one after the other. Once activated these enzymes make contact with and activate the effector caspases such as Caspase-3.

One of the Caspase-3 substrates is the caspase-activated deoxyribonuclease (CAD) protein and its inhibitor ICAD. The CAD and ICAD proteins are the catalytic and regulatory subunits of a protein referred to as the *DNA fragmentation factor*, or DFF. The ICAD subunit remains bound to the CAD subunit in the absence of Caspase 3 activities and inhibits its enzymatic actions. Caspase 3 cleaves the ICAD thereby freeing the CAD DNase and allowing it to move into the nucleus where it cleaves chromatin.

The BH3-only protein, Bid, is sited at the DISC. It functions as a sensor and as part of the circuitry that integrates externally and internally generated signals. As indicated in Figure 15.4, Caspase 8 cleaves the 22-kDa Bid sensor protein to create the 15-kDa tBid. Once formed, the tBids translo-

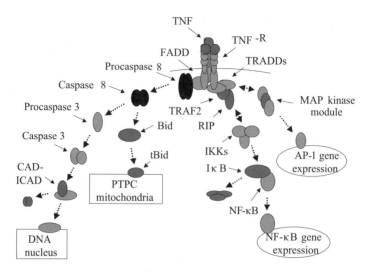

FIGURE 15.4. Signaling through the DISC located at the plasma membrane: Depicted are the main positive-regulating signaling elements. Homotrimeric TNF ligands bind homotrimeric TNF receptors. In response, several adapter molecules are recruited to the cytoplasmic portion of the receptors. The FADD adapter mediates recruitment and activation of the Caspase 8 pathway. The TRAF2 and RIP proteins mediate signaling and activation of the IKKs, which disinhibit NF-κB from the IKKs. The TRADDs also signal to the JNK MAP kinase cascade.

cate to the mitochondrial PTPC where they promote the activities of proapoptotic Bcl-2 proteins. If the overall mix of BH proteins at the PTPC favors apoptosis, proapoptotic factors are released from the mitochondria leading to stimulation of the apoptosome and the sequential activation of Caspase 3 and then Caspase 6, which further stimulates Caspase 8 activities in situations where the externally driven stimulation is weak.

15.8 How Pro- and Antiapoptotic Signals Are Relayed

Pro- and antiapoptotic signals are relayed to the nucleus by NF-κB proteins and MAP kinases. The two other pathways activated by death ligand relay signals via NF-κB proteins and MAP kinase modules as is usual for other members of the TNF superfamily. These pathways were discussed earlier in Chapter 9. Downstream signaling proteins establish contact with receptor interacting proteins (RIPs) and tumor necrosis factor receptor associated factor (TRAFs) that are recruited into the DISC. These pathways promote the expression of both pro- and antiapoptotic genes. The NF-κB module usually, but not always, acts to promote survival by raising the threshold for apoptosis. The IKKs are the key point of convergence of a variety of regulatory signals triggered by cellular stresses. The IKKs are activated when recruited to and phosphorylated at the DISC. They, in turn, phosphorylate the IκBs, resulting in the activation of NF-κB. In their prosurvival mode, the NF-κB dimers translocate to the nucleus where they stimulate transcription of negative regulators of not only DISC signaling but also of mitochondrial proapoptotic signaling elements. As a consequence, the balance between pro- and antiapoptotic factors is shifted in favor of the antiapoptotic ones and apoptosis is prevented.

Recall from Chapter 9 that MAP signaling pathways convey stress (JNK and p38) and growth (ERK) signals from the plasma membrane to the nucleus where they influence transcription of a different sets of target genes. As shown in Figures 9.4 and 15.4, the JNK pathway begins in the DISC, where the TRADDs recruit and activate MEKK1, the first of the kinases in the MAP kinase cascade. The last kinase in the cascade is JNK. Once activated this kinase translocates to the nucleus where it phosphorylates members of the AP-1 family of transcription factors. A similar set of signaling steps occurs in the p38 pathway.

Transcription factors such as AP-1 family members and NF-κB reflect cellular conditions and prior signaling events in their transcriptional activities. Depending on the specific mix of coactivators and corepressors present, subunit composition, and the set of residues that have been phosphorylated (and acetylated), these transcription factors will either promote or inhibit apoptosis. For instance, the c-Jun transcription factor, an AP-1 family member activated at the end of the MAP kinase cascade, usually functions as a transcription activator, but can also function as transcription repressors when associated with corepressors. In NF-κB signaling, flexibility of response is provided by variations in subunit composition. Depending on Rel subunit composition, NF-κB will either promote apoptosis by expressing TRAIL receptors or inhibit apoptosis by expressing antiapoptotic survival factors.

15.9 Bcl-2 Proteins Regulate Mitochondrial Membrane Permeability

Mitochondria occupy a central place in internal stress-induced apoptosis. As noted earlier in the chapter, the PTPC located in the mitochondria serves as a key control point for internal stress responses. The PTPC, or alternatively, the permeability transition pore (PTP), is formed at points of contact between the inner and outer mitochondrial membranes. These complexes are a conduit for the passage of agents such as Cytochrome c and Smac/DIABLO that trigger apoptosome assembly and activation of Caspase 9 (Figure 15.5). The PTPC encompasses the crucial inner membrane (IM) and outer membrane (OM) proteins along with key constituents

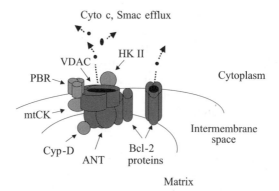

FIGURE 15.5. Central components of the mitochondrial PTPC: Depicted in the figure are the central elements—the ANT located in the inner membrane and the VDAC situated in the outer membrane. They form a pore through which Cyto c, Smac/DIABLO, and a number of different effector molecules can diffuse through and into the cytoplasm. Also shown are a variety of key Bcl-2 family members. A proapoptotic Bcl-2 protein, for example, Bax, is depicted bound to the ANT/VDAC, having displaced an antiapoptotic Bcl-2 protein, while several other Bax/Baks form a pore.

of the intermembrane space and the matrix. It is composed of voltage-dependent anion channels (VDACs) located in the outer mitochondrial membrane and the adenosine nucleotide translocator (ANT) situated in the inner membrane. The PTPC also includes the peripheral benzodiazepine receptors, creatine kinases, hexokinase II, and cyclophilin D.

The mitochondrial PTPC is the main target of Bcl-2 signaling and is the main site of their pro- and antiapoptotic actions. If a molecule can cross a membrane by means of simple passive diffusion, driven only by a concentration gradient, then that membrane is permeable to that molecule. In apoptosis, the Bcl-2 proteins regulate the permeability of the mitochondrial membrane to the apoptosis-promoting molecules. Bcl-2 proteins act as sensors of stress signals and as regulators of mitochondrial membrane permeability. They regulate membrane permeability by binding to and altering the pores formed by the ANT and VDAC, and by oligomerizing and forming pores by themselves.

BH3-only proteins stimulate the release from mitochondria of Cytochrome c and other apoptosis-promoting factors. The BH3-only protein tBid, for example, promotes the permeability-increasing activities of Bax and Bak proteins, stimulates the remodeling of the mitochondrial cristae, and triggers the release of Cytochrome c. Members of the anti-apoptotic branch of the family inhibit apoptosis by sequestering BH3-only proteins thereby preventing their stimulation of Bax and Bad. There are two kinds of BH3 protein actions. Bid-like BH3 proteins facilitate the oligomerization of Bax and Bak, while Bad-like BH3 proteins bind Bcl-2s and displace Bid-like peptides from them.

15.10 Mitochondria Release Cytochrome c in Response to Oxidative Stresses

Oxidative phosphorylation is carried out in mitochondria. One of the key agents in this process is Cytochrome c, an evolutionary ancient electron carrier. It is a small protein, consisting of 104 amino acid residues and a covalently attached heme group. Cytochrome c is part of the machinery that transfers electrons through several protein complexes located in the inner mitochondrial membrane. It carries electrons from the cytochrome reductase complex to the cytochrome oxidase complex. The electron transport activity leads to the pumping of protons from the matrix side to the cytoplasmic side of the inner mitochondrial membrane. The biochemical process, oxidative phosphorylation, generates ATP by utilizing the energy released during the electron transfer from NADH or $FADH_2$ to O_2.

Recall that atoms in stable molecules are tied together by bonds consisting of pairs of electrons, one with spin up and the other with spin down. When bonds are broken the molecules may end up with one or more unpaired electrons. Molecules possessing unpaired electrons are referred to as *free radicals*. The presence of an unpaired electron renders the molecules highly reactive. Because of the presence of an electron the molecule has a propensity to "steal" electrons from other molecules, in many cases breaking bonds to acquire the electrons. Chain reactions can be produced in the cell, and free radicals are highly dangerous. Most commonly encountered free radicals in the cell involve oxygen and these are called *reactive oxygen species* (ROS).

Mitochondria are the major cellular source of reactive oxygen species. In mitochondria, a small fraction, perhaps 2 to 5%, of the molecular oxygen being reduced by the respiratory electron transport chain is converted to superoxide (O_2^-) and then to hydrogen peroxide (H_2O_2) or to the hydroxyl radical (OH^-). The ROS cause cellular (oxidative) stresses because they can oxidize DNA, proteins, and lipids. The presence of ROS influences cellular physiology in many ways and triggers protective reactions involving the upregulation of antioxidants and detoxifying enzymes. Antioxidants are free radical scavengers. They supply electrons to the free radicals, allowing them to form bonds and converting them nonreactive forms.

Cytochrome c is loosely bound to the inner mitochondrial membrane by cardiophilin and other anionic lipids. When ROS levels rise it disrupts the binding of Cytochrome c thereby making it easier to release. Once Cytochrome c is released it stimulates assembly of the apoptosome leading to activation of Caspase 9 and Caspase 3. Thus, oxidative stress conditions arising when ROS activity levels in mitochondria become excessive and are no longer controlled by antioxidant scavengers are signaled by Cytochrome c release from the mitochondria. The Cytochrome c molecules act as local signaling molecules that convey stress messages from the mitochondria to the apoptosome.

15.11 Mitochondria Release Apoptosis-Promoting Agents

Several different kinds of molecules are sent out through the PTPC, as indicated in Table 15.4. Cytochrome c and deoxyadenosine triphosphate (dATP) convey signals that are required for assembly of the apoptosome and the activation of Caspase 9. Two other regulators of the apoptosome, Smac/DIABLO and Omi/HtrA2 are also released from the mitochondria. A number of apoptotic effectors are released in addition to the activators and regulators of apoptosome and Caspase 9 functions. These include Apoptosis inducing factor (AIF) and Endonuclease G, which target nuclear chromatin and degrades it.

Chromatin condensation and fragmentation of nuclear DNA is an important part of apoptosis. A number of proteins fragment DNA and target chromatin in response to proapoptotic signals. One of these, the CAD DNase, was discussed in Section 15.7. Two apoptotic enzymes are released by mitochondria that target chromatin and fragment DNA. One of the DNA chopping proteins, or *DNases*, is Endonuclease G. This enzyme is released from mitochondria in response to stimulation by proapoptotic proteins. Once it is released into the cytosol it translocates to the nucleus where it fragments DNA into nucleosomal-sized fragments. In the first nuclear steps of apoptosis, DNA repair is halted and chromatin condensation occurs, which turns off DNA transcription. Another apoptosis promoting agent sent out from the mitochondria is Apoptosis-inducing factor. This protein is released from mitochondria at an early stage in the apoptosis process. It is responsible for peripheral chromatin condensation and for large-scale chromatin fragmentation.

TABLE 15.4. Agents released by mitochondria: Abbreviations—Second mitochondrial activator of caspases (Smac); direct IAP binding protein with low pI (DIABLO).

Agent	Action
Cytochrome c	Required for assembly of the apoptosome
dATP	Required for assembly of the apoptosome
Smac/DIABLO	Proapoptotic regulator
Omi/HtrA2	Proapoptotic regulator
AIF	Early acting nuclear factor
Endonuclease G	Later acting nuclear factor

15.12 Role of Apoptosome in (Mitochondrial Pathway to) Apoptosis

The apoptosome is the main control point in the mitochondrial pathway to apoptosis. When released from the mitochondria, Cytochrome c interacts with a 130-kDa adapter protein called Apaf-1 located in the cytoplasm just outside the mitochondria. This adapter contains three domains (Figure 15.6a). It has an N-terminal caspase recruitment domain (CARD), a central domain that binds dADP/ATP, and a C-terminal domain containing a series of WD-40 repeats. When Cytochrome c binds to the WD-40 repeats to override the autoinhibition, and deoxyadenosine triphosphate (dATP) hydrolysis occurs, Apaf-1 is able to form oligomers leading to the formation of a 1-MDa apoptosome.

The apoptosome is a sevenfold symmetric platform. It is wheel-shaped with seven spokes radiating out from a central hub (Figure 15.6b), and functions to activate Procaspase 9 when these zymogens are incorporated into it. When fully saturated, seven Procaspase 9 molecules are bound to the seven Apaf-1 CARD domains and seven Cytochrome c molecules are bound to the Y domains. The formation of an apoptosome containing activated Caspase 9 molecules triggers the apoptosis process by interacting with and activating Caspase 3.

The apoptosome is a major control point where proapoptotic agents released by the mitochondria converge and where Caspase 9 and Caspase 3 interact and activate one another. At the apoptosome, Caspase 9 processes and activates Caspase 3. A feedback loop in which processed Caspase 3 cleaves and activates Caspase 9 amplifies the production of activated caspases. Several other proteins are present, and form complexes with the caspases. Among these are caspase inhibitors and caspase counterinhibitors.

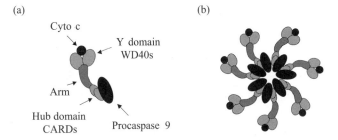

FIGURE 15.6. Apaf-1 and the apoptosome: (a) Apaf-1 molecule bound to Cytochrome c and Procaspase 9—Shown are a pair of WD-40 repeats in the C-terminal Y domain that bind Procaspase 9, and a pair of CARD domains in the N-terminal that bind Cyto c. A flexible arm containing a CED4 homology motif that binds dATP/ADP connects the N- and C-terminal regions to one another. (b) Fully assembled apoptosome illustrating how the individual Apaf-1 molecules associate into a sevenfold symmetric platform.

15.13 Inhibitors of Apoptosis Proteins Regulate Caspase Activity

Inhibitor of apoptosis proteins (IAPs) bind to and inhibit the activities of caspases once they have been converted by proteolysis from zymogens to caspase monomers and homodimers. The presence of these regulators (and their counterregulators) establishes a second tier of control over the catalytic activities of caspases. The defining feature of the IAPs is the presence of one or more BIR (baculoviral IAP repeat) domains (Figure 15.7). These are found in the eight mammalian IAPs discovered to date. A second prominent feature of the IAPs is the presence of a RING domain in the C-terminal of some but not all IAPs.

Caspases must form homodimers to become catalytically active. The homodimerization partner supplies a sequence crucial for catalysis, a sequence that that monomeric form of the caspases lacks. Crystal structure studies of Caspase 9 in complex with the BIR3 domain of an X-linked AIP (XIAP) protein show how the IAP proteins inhibit the catalytic activities of the caspases. The BIR3 domain binds to the homodimerization domain of monomeric caspase, thereby preventing formation of caspase homodimers. This mechanism differs from that used by the IAPs to inhibit effector caspases such as Caspase 3 and Caspase 7. Rather than preventing formation of the homodimers, XIAP uses a sequence just forward of its BIR2 domain to block the active site of the effector caspases.

Inhibitor of apoptosis proteins possess a ubiquitin ligase activity. As shown in Figure 15.7, IAPs such as XIAP and c-IAP1 and c-IAP2 possess a RING finger domain in their C-terminus. This motif is often encountered in E3 ubiquitin ligases. In the case of IAPs bearing this motif, there seem

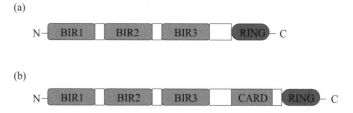

FIGURE 15.7. Domain organization of the inhibitor of apoptosis proteins (IAP): (a) Structure of XIAP. (b) Structure of c-IAP1 and c-IAP2. ILP-2 and Livin resemble XIAP except that they only have one BIR domain (BIR3 in the case of ILP-2 and BIR2 in the case of Livin). The other three mammalian AIP proteins—NAIP, Survivin, and Apollon/Bruce—have from one to three BIR domains but lack a RING finger domain. Abbreviations: X-linked IAP (XIAP); IAP-like protein (ILP); neuronal inhibitory apoptosis protein (NAIP); baculoviral IAP repeat (BIR); caspase recruitment domain (CARD).

to be two choices of substrates. In response to proapoptotic signals, the IAPs trigger their own ubiquitination. Alternatively, in the absence of proapoptotic signals, the IAPs act in an antiapoptoptic capacity by promoting the ubiquitination of activated Caspases 3 and 7.

15.14 Smac/DIABLO and Omi/HtrA2 Regulate IAPs

Smac/DIABLO is an IAP counterregulator. It is released from mitochondria in response to proapoptosis signals, and once released disrupts the ability of IAPs to inhibit the caspases. The IAP family member XIAP is one of its main targets. In the absence of Smac/DIABLO, XIAP binds to the small subunit of Caspase 9. It does not bind to Caspase 9 in its procaspase form, but only to the cleaved and assembled Caspase 9 monomer. Smac/DIABLO disrupts the ability of XIAP to inhibit Caspase 9 by binding to its BIR3 domain. A sequence of four residues in the N-terminal recognizes and binds a surface groove in the BIR3 domain.

The protein Omi/HtrA2 is another IAP counterregulator released by mitochondria in response to proapoptotic signaling. Like Smac/DIABLO it promotes apoptosis by antagonizing the IAPs ability to bind to and inhibit activation of Caspases 3, 7, and 9. Upon release from the mitochondria Omi/HtrA2 forms complexes with XIAP and disrupts the latter's caspase-inhibiting functions. Although Omi/HtrA2 binds to the IAP in a manner resembling that of Smac/DIABLO, its manner of action is different. Unlike Smac/DIABLO, Omi/HtrA2 is a serine protease and may act in a proteolytic manner to degrade and thus limit the inhibitory activities of the IAPs.

15.15 Feedback Loops Coordinate Actions at Various Control Points

The apoptosome and the mitochondrial PTPC are major control points for apoptosis. These control points communicate with each other and with the DISC in order to arrive at live or die decisions. The communication between PTPC and apoptosome is summarized in Figure 15.8. In the figure, a stress-generating perturbation causes the loss of inner mitochondrial transmembrane potential Ψ_m, produces ROS, and/or elevates the free calcium concentration $[Ca^{2+}]_m$ in the mitochondrial matrix. In response, the mitochondrial membranes become more permeable to Cytochrome c and other proapoptotic agents. These agents leave the mitochondria. Some (cytochrome c and dATP) trigger formation of the apoptosome and others (Smac/DIABLO and Omi/HtrA2) activate the caspases by binding and inhibiting the IAPs. Caspase 3 activated at the apoptosome diffuses to and enters the mitochondria. It then cleaves a number of components of the mitochondrial electron transport chain. These feedback operations trigger the

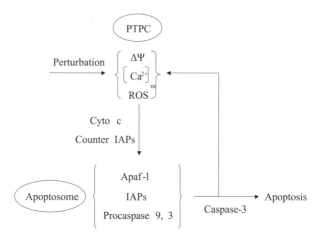

FIGURE 15.8. Mitochondrial PTPC—Apoptosome circuit: Perturbations of the mitochondrial permeability transition pore complex (PTPC) stimulates the release of apoptosis-promoting factors such as Cytochrome c and counter inhibitor of apoptosis proteins (IAPs). These act at the apoptosome to activate Caspases 3 and 9.

stepped-up production of ROS and efflux of Cytochrome c, which then acts at the apoptosome to amplify the amount of activated Caspase 9 and Caspase 3.

Another feedback loop connects events at the apoptosome with those occurring at the DISC. The connecting link is Caspase 6. Caspase 3 cleaves and activates Caspase 6, which then migrates to the DISC where it stimulates Caspase 8, thereby amplifying the strength of the apoptotic signal at the cell surface as was discussed in Section 15.7. The BH3-only protein Bid completes the circuit by connecting actions taking place at the DISC with those occurring at the PTPC.

Apoptosis is controlled by a plethora of positive feedback loops that ensure that once the appropriate thresholds are passed there will be a firm commitment to apoptosis. The presence of thresholds ensures that random excursions and perturbations do not unnecessarily commit the cell to apoptosis when it ought not to. An equally important ensemble of negative feedback loops generates the threshold dependences.

15.16 Cells Can Produce Several Different Kinds of Calcium Signals

Calcium is an important signaling intermediary, or second messenger, and was introduced as such in Chapter 8. Calcium can also function as a first messenger. Calcium is well suited to function as a signaling molecule. It is able to accommodate from 4 to 12 oxygen atoms in their primary coordination sphere. This property allows the calcium ions to contact multiple partners and trigger large conformational changes when binding a protein. In addition, calcium does not diffuse very far because of the presence of a large number of calcium buffers in the cytosol.

The normal intracellular calcium ion concentration is about 100 nM. This level is several orders of magnitude lower than the calcium ion concentrations in the extracellular spaces, which is roughly 2 mM. To maintain the intracellular calcium concentrations at the resting levels, the Na$^+$/Ca^{2+} ion exchanger and pumps such as the plasma membrane calcium ATPase (PMCA), remove calcium from the cell. In addition, the sarco-endoplasmic reticulum calcium ATPase (SERCA) embedded in the SR/ER ships calcium from the cytosol into the intracellular stores. Because of the presence of cellular machinery keeping calcium levels low, transient local increases in calcium concentration can be produced easily and serve as a signal.

Lipid second messengers relay signals from the plasma membrane to the intracellular stores. These molecules trigger the release of calcium when they bind inositol (1,4,5) triphosphate receptors (InsP$_3$Rs). A second kind of calcium release channel, the ryanodine receptor (RyR), releases calcium from the intracellular stores, as well. Several different kinds of calcium signals can be produced. One kind of calcium signal, the calcium "wave," is a global signal that propagates over large distances across the cytosol. The propagation of these waves over large intracellular distances is made possible by positive feedback in which calcium released from stores in one locale diffuses to and triggers the release of calcium from nearby stores that sets off further releases, thereby generating a wave of calcium. Calcium can be released from stores in a more localized manner. The local releases of calcium from intracellular stores are called "puffs" when produced by InsP$_3$Rs and "sparks" when facilitated by RyRs.

15.17 Excessive [Ca^{2+}] in Mitochondria Can Trigger Apoptosis

Calcium signals are sent to the mitochondria under normal conditions and also under abnormal conditions. Calcium signals are sent to the mitochondria in response to increased signaling at the plasma membrane in order to spur increases in metabolic activity to support the elevated signaling load. This coupling of metabolism and calcium is made possible by the presence in the mitochondrial matrix of a number of calcium sensitive metabolic enzymes. Increased calcium signaling to the mitochondria also occurs under abnormal conditions of cytosolic calcium overload (i.e., disruptions in calcium homeostasis) and ER stresses. In these situations, the changes in mitochondrial physiology drive the cell towards apoptosis.

Calcium ions are cycled between the SR/ER and mitochondria. The membranes of these organelles are in close proximity to one another and highly local and directional signaling similar to that occurring at chemical synapses takes place. One of the early events taking place in stress-generated apoptosis is an increased permeability of the PTPC to calcium entry. This event leads to increases in permeability to cytochrome c, resulting in its increased efflux from the mitochondria.

Combinations of proapoptotic conditions such as oxidative stresses and calcium signaling between the ER and mitochondria can initiate apoptotic responses under condition where one or the other condition by itself cannot. The cooperation between the two is mediated by a positive feedback loop in which ROS generated in the mitochondria promote increased Ca^{2+} release from the ER. The calcium accumulates in the mitochondria and triggers further increases in ROS production. This process generates the progressive depolarization of the inner mitochondrial membrane and increases membrane permeation leading to cell death.

The PTPC is not the only control point where Bcl-2 family proteins exert their regulatory influences. They also act at the endoplasmic reticulum where they influence the release of Cytochrome c occurs through their modulation of calcium signaling. They help maintain homeostatic control over movement of calcium into and between the two organelles (the ER and the mitochondria), and mobilize calcium release. Proapoptotic Bax and Bak proteins localize to both the endoplasmic reticulum and mitochondria where they control calcium trafficking between the two organelles. In the ER, Bax and Bak help initiate the apoptosis process by contributing to Caspase 12 activation, and in the mitochondria they help activate Caspase 7.

15.18 p53 Promotes Cell Death in Response to Irreparable DNA Damage

In response to DNA damage signals, the p53 protein will either halt the cell cycle to allow for repairs or promote apoptosis if the damage is irreparable. The signaling proteins that convey messages to p53 do so through phosphorylations and acetylations of specific residues. The specific mix of phosphorylations and acetylations determines in large measure the response that will be made by p53. For example, phosphorylation of p53 on Ser46 stimulates transcription of apoptosis-promoting genes, but does not stimulate transcription of cell cycle genes. A second factor that guides the p53 response is the mix of cofactors present at the promoter sites. As is the case for other transcription factors, the cofactors help determine which genes get transcribed and which ones do not. In the case of p53 two other p53 family members—p63 and p73—work together with p53 to transcribe proapoptosis genes.

The p53 protein can shift the balance of pro- and antiapoptosis factors in at least two distinct ways. First, in its role as a transcription factor, the p53 proteins can step up the transcription of proapoptotic genes. As

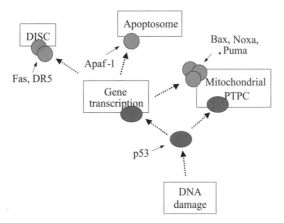

FIGURE 15.9. Regulation of apoptosis by p53: DNA damage signals are conveyed to p53 in the form of posttranslational modifications of specific residues such as phosphorylation on Ser46. In response, p53 (along with p63 and p73, not shown) stimulates transcription of apoptosis-promoting proteins acting at the DISC, PTPC, and apoptosome. It also diffuses to the mitochondria where it interacts with antiapoptotic Bcl-2 family members to increase membrane permeability.

depicted in Figure 15.9, p53 can increase the numbers of propapoptotic proteins. These include death ligands and receptors at the DISC, the Apaf-1 scaffold proteins at the PTPC, and proapoptotic multidomain Bax, and BH3-only proteins, at the apoptosome. Second, p53 can translocate to the mitochondria where it can interact with the antiapoptotic Bcl-x_L and Bcl-2 proteins to increase the efflux of Cytochrome c from the mitochondria. Recall from the last chapter that some of the most lethal mutations to p53 occur in its DNA binding domain. This domain contains the binding site for attachment to the Bcl-2 proteins so these mutations not only destroy p53's ability to bind DNA in the nucleus but also negate its ability to interact with the Bcl-x_L proteins at the mitochondria.

15.19 Anti-Cancer Drugs Target the Cell's Apoptosis Machinery

The machinery involved in regulating apoptosis is a major target of anti-cancer therapies. The goal of these therapeutic strategies is to induce apoptosis in the malignant tissue while leaving healthy tissue alone. Components of the key control loci—DISC, apoptosome, and PTPC—are the primary targets of many of these strategies. One set of strategies revolve about causing apoptosis-inducing damage to DNA and other cellular components, while another set of approaches focuses on causing oxidative damage in the mitochondria in a way that stimulates the release of proapoptotic factors.

The majority of chemotherapeutic approaches to killing tumors is based on the idea of selectively inducing apoptosis in the diseased cells. Many anti-cancer drugs target the mitochondria with the aim of producing a decrease in membrane potential leading to the efflux of pro-apoptotic molecules and the activation of the apoptosome/Caspase9 pathway. There are a number of ways that a decrease in mitochondrial membrane potential may be induced. One way is to supply ligands for either the ANT or the VDAC. Alternatively, drugs may be used that trigger increases in ROS leading to an increase in membrane permeability.

All of the major components of the apoptosis machinery are tightly con-trolled. Negative feedback loops ensure that small perturbations and random releases of Cytochrome c or Smac/DIABLO from the mitochon-dria do not set off apoptosis. Restorative processes in the form of anti-oxidant scavenger systems prevent ROS generated within the mitochondria from causing excessive damage. Calcium homeostasis within the cytosol and organelles such as the ER and mitochondria are under negative feedback control. Finally, the inner mitochondrial transmembrane potential Ψ_m is a quantity with an important role in apoptosis and endowed with restorative properties. In order to be effective, therapeutic approaches that produce stresses and perturb the mitochondria must overcome the plethora of pro-tective measures used by the cell to deal with such stresses. These issues compound the difficulties in dealing with machinery that doesn't work quite right because of mutations in genes that encode its key elements.

References and Further Reading

General References

Igney FH, and Krammer PH [2002]. Death and anti-death: Tumor resistance to apoptosis. *Nature Rev. Cancer*, 2: 277–288.

Johnstone RW, Ruefli AA, and Lowe SW [2002]. Apoptosis: A link between cancer genetics and chemotherapy. *Cell*, 108: 153–164.

Mattson MP [2000]. Apoptosis in neurodegenerative disorders. *Nature Rev. Mol. Cell Biol.*, 1: 120–129.

Meier P, Finch A, and Evan G [2000]. Apoptosis in development. *Nature*, 407: 796–801.

Vila M, and Przedborski S [2003]. Targeting programmed cell death in neuro-degenerative diseases. *Nature Rev. Neurosci.*, 4: 1–11.

Caspases

Boatright KM, et al. [2003]. A unified model for apical caspase activation. *Mol. Cell*, 11: 529–541.

Chai JJ, et al. [2001]. Crystal structure of a procaspase-7 zymogen: Mechanisms of activation and substrate binding. *Cell*, 107: 399–407.

Chang DW, et al. [2003]. Interdimer processing mechanism of procaspase-8 activa-tion. *EMBO J.*, 22: 4132–4142.

Shi YG [2002]. Mechanisms of caspase activation and inhibition during apoptosis. *Mol. Cell*, 9: 459–479.

Stennicke HR, and Salvesen GS [2000]. Caspases—Controlling intracellular signals by protease zymogen activation. *Biochim. Biophys. Acta*, 1477: 299–306.

Bcl-2 Proteins

Cheng EHYA, et al. [2001]. Bcl-2, Bclx$_L$ sequester BH3 domain only molecules preventing BAX- and BAK-mediated mitochondrial apoptosis. *Mol. Cell*, 8: 705–711.

Cory S, and Adams JM [2002]. The Bcl-2 family: Regulators of the cellular life-or-death switch. *Nature Rev. Cancer*, 2: 647–656.

Gross A, McDonnell JM, and Korsmeyer SJ [1999]. Bcl-2 family members and the mitochondria in apoptosis. *Genes Dev.*, 13: 1899–1911.

Letai A, et al. [2002]. Distinct BH3 only domain either sensitize or activate mitochondrial apoptosis, serving as prototypes cancer therapeutics. *Cancer Cell*, 2: 183–192.

Moreau C, et al. [2003]. Minimal BH3 peptides promote cell death by antagonizing anti-apoptotic proteins. *J. Biol. Chem.*, 278: 19426–19435.

Scorrano L, et al. [2002]. A distinct pathway remodels mitochondrial cristae and mobilizes cytochrome c during apoptosis. *Dev. Cell*, 2: 55–67.

DISC Signaling

Enari M, et al. [1998]. A caspase-activated DNase that degrades DNA during apoptosis, and its inhibitor ICAD. *Nature*, 391: 43–50.

Nagata S [1997]. Apoptosis by death factor. *Cell*, 88: 355–365.

Scaffidi C, et al. [1998]. Two CD95(APO-1/Fas) signaling pathways. *EMBO J.*, 17: 1675–1687.

Weber CH, and Vincenz C [2001]. The death domain superfamily: A tale of two interfaces? *Trends Biochem. Sci.*, 26: 475–481.

Regulation of the Apoptotic Signaling Pathways by NF-κB

Karin M, and Lin A [2002]. NF-κB at the crossroads of life and death. *Nature Immunology*, 3: 221–227.

Mayo CY, et al. [1998]. NF-κB antiapoptosis: Induction of TRAF1 and TRAF2 and c-IAP1 and c-IAP2 to suppress caspase 8 activation. *Science*, 281: 1680–1683.

Permeability Transition Pore Complex

Crompton M [1999]. The mitochondrial permeability transition pore and its role in cell death. *Biochem. J.*, 341, pt. 2: 233–249.

Ott M, et al. [2002]. Cytochrome c release from mitochondria proceeds by a two-step process. *Proc. Natl. Acad. Sci. USA*, 99: 1259–1263.

Susin SA, Zamzami N, and Kroemer G [1998]. Mitochondria as regulators of apoptosis: Doubt no more. *Biochim. Biophys. Acta.*, 1366: 151–165.

Apoptosome Assembly

Acehan D, et al. [2002]. Three-dimensional structure of the apoptosome: Implications for assembly, pro-caspase-9 binding, and activation. *Mol. Cell*, 9: 423–432.

Bratton SB, et al. [2001]. Recruitment, activation and retention of caspase-9 and -3 by Apaf-1 apoptosome and associated XIAP complexes. *EMBO J.*, 20: 998–1009.

Inhibitor of Apoptosis Proteins

Huang HK, et al. [2000]. The inhibitor of apoptosis, cIAP2, functions as a ubiquitin-protein ligase and promotes in vitro monoubiquitination of caspases 3 and 7. *J. Biol. Chem.*, 275: 26661–26664.

Salvesen GS, and Duckett CS [2002]. IAP proteins: Blocking the road to death's door. *Nature Rev. Mol. Cell Biol.*, 3: 401–410.

Suzuki Y, Nakabayashi Y, and Takahashi R [2001]. Ubiquitin-protein ligase activity of X-linked inhibitor of apoptosis protein promotes proteosomal degradation and caspase-3 and enhances its anti-apoptotic effect in Fas-induced cell death. *Proc. Natl. Acad. Sci. USA*, 98: 8662–8667.

Yang Y, et al. [2000]. Ubiquitin protein ligase activity of IAPs and their degradation in proteosomes in response to apoptotic stimuli. *Science*, 288: 874–877.

IAP Counterregulators Smac/DIABLO and Omi/HtrA2

Holley CL, et al. [2002]. Reaper eliminates IAP proteins through stimulated IAP degradation and generalized translational inhibition. *Nature Cell Biol.*, 4: 439–444.

Nicholson DW [2002]. Baiting death inhibitors. *Nature*, 410: 33–34.

Srinivasula SM, et al. [2001]. A conserved XIAP-interaction motif in caspase-9 and Smac/DIABLO regulates caspase acytivity and apoptosis. *Nature*, 410: 112–116.

Yoo SJ, et al. [2002]. Hid, Ror and Grim negatively regulate DIAP1 levels through distinct mechanisms. *Nature Cell Biol.*, 4: 416–424.

Mitochondrial Physiology

Duchen MR [2000]. Mitochondria and calcium: From cell signaling to cell death. *J. Physiol.*, 529: 57–68.

Jacobson J, and Duchen MR [2002]. Mitochondrial oxidative stress and cell death in astrocytes—Requirement for stored Ca^{2+} and sustained opening of the permeability transition pore. *J. Cell Sci.*, 115: 1175–1188.

Kowaltowski AJ, Castilho RF, and Vercesi AE [2001]. Mitochondrial permeability transition and oxidative stress. *FEBS Lett.*, 495: 12–15.

Marchetti P [1997]. Redox regulation of apoptosis: Impact of thiol oxidation status on mitochondrial function. *Eur. J. Immunol.*, 27: 289–296.

Ricci JE, Gottlieb RA, and Green DR [2003]. Caspase-mediated loss of mitochondrial function and generation of reactive oxygen species during apoptosis. *J. Cell Biol.*, 160: 65–75.

ER and Calcium Signals to the Mitochondria

Boehning D, et al. [2003]. Cytochrome c binds to inositol (1,4,5) triphosphate receptors, amplifying calcium-dependent apoptosis. *Nature Cell Biol.*, 5: 1051–1061.

Li C, et al. [2002]. Bcl-x_L affects Ca^{2+} homeostasis by altering expression of inositol 1,4,5-triphosphate receptors. *Proc. Natl. Acad. Sci. USA*, 99: 9830–9835.

Nutt LK. et al. [2002]. Bax-mediated Ca^{2+} mobilization promotes cytochrome c release during apoptosis. *J. Biol. Chem.*, 277: 20301–20308.

Orrenius S, Zhivotovsky B, and Nicotera P [2003]. Regulation of cell death: The calcium-apoptosis link. *Nature Rev. Mol. Cell Biol.*, 4: 552–565.

Scorrano L, et al. [2003]. BAX and BAK regulation of endoplasmic reticulum Ca^{2+}: A control point for apoptosis. *Science*, 300: 135–139.

p53 Regulation

Flores ER [2002]. p63 and p73 are required for p53-dependent apoptosis in response to DNA damage. *Nature*, 416: 560–564.

Mihara M, et al. [2003]. p53 has a direct apoptogenic role at the mitochondria. *Mol. Cell*, 11: 577–590.

Vousden KH, and Lu X [2002]. Live or let die: The cell's response to p53. *Nature Rev. Cancer*, 2: 594–604.

Cancer Therapy

Herr I, and Debatin KM [2001]. Cellular stress response and apoptosis in cancer therapy. *Blood*, 98: 2603–2614.

Johnstone RW, Ruefli AA, and Lowe SW [2002]. Apoptosis: A link between cancer genetics and chemotherapy. *Cell*, 108: 153–164.

Kaufmann SH, and Earnshaw WC [2000]. Induction of apoptosis by cancer chemotherapy. *Exp. Cell Res.*, 256: 42–49.

Nicholson DW [2000]. From bench to clinic with apoptosis-based therapeutic agents. *Nature*, 407: 810–816.

Problem

15.1 Some cell types are more susceptible to apoptosis than others. What classes of cells are especially responsive to apoptosis signals? What kinds of cells send out proapoptosis messages as part of their normal functions? Briefly describe some of the ways a cell has of regulating its own sensitivity to apoptosis.

16
Gene Regulation in Eukaryotes

The finding that metazoan genomes are as small as they are was remarked upon in Chapter 1. In the introductory chapter, several aspects of eukaryotic cell organization that make this possible were noted. One of the most important consequences of the changes in going from prokaryotic to eukaryotic cell organization was creation of many different ways of regulating gene and protein expression, and for fine-tuning and specializing the functions being performed. As a result of changes in the way the cells are organized and gene expression and protein function are regulated, large increases in complexity become possible without requiring large numbers of additional genes. This central aspect will be examined in detail in this chapter, which is devoted to the regulation of gene expression and protein synthesis. Transcription will be discussed first, followed by alternative splicing and translation.

Transcription in eukaryotes, even in the unicellular yeast, is more complex than in prokaryotes because of the presence of chromatin. Recall from Chapter 2 that eukaryotic DNA is highly condensed. Most of the chromatin is transcriptionally silent; that is, the DNA is tightly wound about the histone core and promoter sites are not accessible. The molecular machinery responsible for transcription not only contacts and manipulates the DNA molecule but also interacts with and alters the chromatin structure. The molecular machines involved in modifying the chromatin structure and transcribing genes are large. They can contain 50 or more subunits and have molecular masses of several megadaltons. An example of how large these machines may become is supplied by an examination of yeast cells. A set of complexes that might be found in a typical yeast cell is presented in Table 16.1.

The first kind of complex is the *basal (core) transcription machinery*, which includes RNA polymerase II and a set of general transcription factors. It binds at the transcription start site and catalyzes the polymerization of RNA chains from DNA templates. This complex is commonly referred to as the preinitiation complex (PIC). The second kind of protein complex is exemplified in yeast by a set of about 20 proteins collectively

TABLE 16.1. Protein complexes involved in transcription in yeast cells.

Machine	Number of subunits	Molecular mass (MDa)	Function
Basal transcription	>40	2	Transcription driver (PIC)
Mediator	20	1	Signal integrator
SWI/SNF complex	11	2	Chromatin remodeler
SAGA	15	2	Histone modifier

called *the Mediator*. Complexes similar to the Mediator have been found in all metazoans examined. They contain suppressor of RNA polymerase B (another name for RNA polymerase II) proteins (Srb proteins) and Mediator proteins (Med proteins), and serve as intermediaries between the transcription factors and the basal transcription machinery. The last two complexes modify chromatin by chemomechanical means. The first (SWI/SNF) breaks and reforms histone-DNA bonds, while the second (SAGA) alters the chemical affinity of histones for the DNA thereby either promoting or inhibiting transcription through acetylation of the histone tails. If all the contributions from the four complexes are combined the result is a machine with about 90 subunits and a molecular mass of 7 MDa. (A few subunits are shared by more than one complex but this changes the additive result only slightly.) These machines and the underlying mechanisms are highly conserved from yeast to man.

16.1 Organization of the Gene Regulatory Region

The molecular machines introduced in the last section carry out their functions at promoters. This is the name given to regions of DNA that contain binding sites for the transcription control elements and transcription start sites for RNA polymerase II (RNAP II). The *core promoter* is a region of approximately 40 bp that directs the start of transcription by the preinitiation complex. The core promoter contains several short DNA sequences that are recognized by DNA-binding proteins belonging to the PIC. One of these is the TATA box, an A/T rich, 8 bp sequence located 25 to 30 bp upstream of the start site for transcription. A second sequence known as initiator (Inr) is positioned at the start site. Other important sequences contained within the core promoter include the TFIIB recognition element (BRE) and the downstream promoter element (DPE).

The core promoter correctly positions and orients RNAP II at the start site. The TBP subunit of the PIC recognizes the TATA box and the TFIIB subunit recognizes the BRE. The architecture of the core promoter in yeast and metazoans is depicted in Figure 16.1. Not all promoters contain TATA

(a)

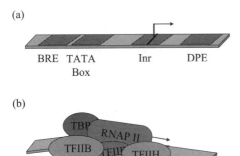

(b)

FIGURE 16.1. Eukaryotic core promoter: Promoters are major endpoints of signaling pathways and have multiple sites for binding proteins. (a) The core promoter contains several binding sites for components of the basal transcription machinery to bind. The TATA box is located 25 to 30 bp upstream from the transcription start site, Inr, centered about the +1 position. The TFIIB recognition element (BRE) is located just above the TATA box while the downstream promoter element (DPE) is located about 30 bp from Inr. (b) The first event to occur is the binding of the TATA box binding protein (TBP) to the TATA box, followed by recruitment of TFIIB to the TBP and the BRE, followed by recruitment of RNA polymerase II and other elements of the PIC.

boxes. Both the TATA box and the Inr can direct transcription initiation by RNAP II. TATA box sequences are present in promoters for many abundantly transcribed genes. Housekeeping genes and genes encoding growth factors and transcription factors often have TATA-less promoters. They utilize the Inr for correct positioning of RNAP II at the start site. Some promoters, lacking both the TATA box and the Inr, rely on the DPE for positioning.

16.2 How Promoters Regulate Genes

One of the most important aspects of gene expression is the presence of multiple binding sites for regulatory proteins in the promoter. The core promoter was discussed in the last section. There are two other components— the *proximal control region* and the *distal control region*. These regions contain binding sites that are recognized by regulatory proteins in a sequence-specific manner. Several binding sites are usually found in the proximal region located just upstream of the core promoter. These binding sites are referred to as *response elements*. Proteins that stimulate transcription when they bind at these sites are called *activators* while those that impede transcription are called *repressors*. A third kind of protein, which

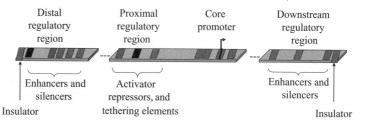

Figure 16.2. Eukaryotic promoter consisting of the core promoter plus proximal, distal, and downstream regulatory sequences called responsive elements (REs): The REs are represented in the figure as dark patches separated from the core promoter and from one another by spacers. A pair of insulators, one at each end, bounds the gene regulatory region.

provides sites for attachment of coactivators and corepressors and mediates long range enhancer-promoter interactions, is called a *tethering element* (Figure 16.2).

Distal sites can be located either upstream of the core promoter or downstream, in between coding regions and even inside *introns*. These are the sites where the long-range interactions between regulatory proteins and the basal transcription machinery take place. The positively acting proteins that bind the DNA are called *enhancers* and those that inhibit transcription are called *silencers*. The distal regulatory regions are generally larger than the proximal ones and supply binding sites for up to a dozen or so proteins.

Besides proximal and distal sites, a third kind of site—one that marks a boundary—is called an *insulator*. Insulators are DNA sequences that mark boundaries between independent sections of DNA. They protect the genes they delineate from influences outside the protected regions, serving as blockers of enhancer functions or as barriers against elements that inactivate chromatin. When situated between an enhancer and a promoter the insulators are able to block action of that enhancer on the promoter. They also block the actions of gene silencers that act on the chromatin, thereby marking the ends of chromatin domains (Figure 16.2).

Promoters are major endpoints of signaling pathways. Transcription factors (TFs), the signaling elements that relay signals into the nucleus, bind to responsive elements in promoters in a sequence-specific manner. The TFs contain DNA-binding motifs or domains. These motifs enable the proteins to bind to specific DNA sequences just as protein recognition modules enable one protein to bind to another in an amino acid sequence-specific manner. A typical promoter contains several responsive elements. This architecture allows a number of transcription factors to bind and jointly regulate transcription.

Each gene has its own promoter site consisting of the core promoter and a number of responsive elements for transcription factors. A signaling pathway that activates downstream transcription factors can regulate the expression of multiple genes. The activation process is a differential one because of differences in promoter architecture from promoter site to promoter site and in the mix of proteins that are present at each. The sites can differ from one another in the mix of responsive elements, in the presence or absence of other transcription factors and coactivators (Mediator and the like), the state of the chromatin, and disposition of chromatin modifiers. This type of regulation, where multiple proteins jointly determine the regulatory outcome, is called *combinatorial control*.

16.3 TFs Bind DNA Through Their DNA-Binding Domains

Transcription factors enter the nucleus, bind to DNA in a sequence-specific manner, and regulate gene expression. Transcription factors, like other elements of the control layer, are highly modular in design. Sequences that code distinct functions are organized into modules that can be arranged in different ways to generate a variety of functional proteins. The separable units in a typical transcription factor are

- DNA-binding domains,
- transcriptional activation domains,
- dimerization domains,
- protein-protein interaction domains, and
- nuclear localization sequences.

In the above, the protein-protein interaction domains mediate their interactions with upstream elements of the signaling pathways. The nuclear localization sequences include nuclear import sequences and/or nuclear export sequences. The DNA-binding domains are modules designed to recognize specific nucleotide sequences. Recall that the edge of each base pair is exposed on the outside of the DNA helix. The DNA-binding domains recognize distinctive patterns of hydrogen bond donors and acceptors, as well as hydrophobic patches along the edges located in the major and minor grooves. As is the case of protein-protein binding interfaces, electrostatic surface complementarity is accompanied by geometric complementarity, that is, by geometric matching of the shape of the DNA helix and protein surface. Protein-DNA binding interfaces were discussed in general terms in Chapter 4. It was noted that some DNA-binding proteins bind the major groove; others bind the minor groove; and some simultaneously contact both. Besides hydrogen bonds and hydrophobic patches, salt bridges and van der Waals contacts contribute to the establishment of good surface complementarity.

There are a number of prominent protein-DNA binding motifs. One of these is the helix-turn-helix (HTH) motif. Others include the helix-loop-helix (HLH), zinc finger, and basic region-leucine zipper (bZIP) domains.

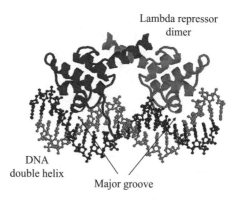

Lambda repressor
dimer

DNA
double helix

Major groove

FIGURE 16.3. Lambda repressor in a complex with DNA determined by means of X-ray crystallography: A pair of lambda repressors makes contact with the DNA double helix in its major groove. Protein helices are depicted as ribbons and DNA molecules as balls-and-sticks. The figure was prepared using Protein Explorer with atomic coordinates deposited in the Brookhaven Protein Data Bank under accession number 1LMB.

The HTH domain consists of two short alpha helices connected by a sequence of four amino acids. The first helix executes two or three turns and the second helix has four turns. The two helices are oriented 120 degrees to each other. The second helix contacts the DNA in the major groove, and the first helix assists in the positioning. HTH motifs are common in bacterial DNA-binding proteins, and are found embedded in eukaryotic homeodomains. These domains are often found in proteins needed for *Drosophila* development, and are widely distributed among eukaryotes from yeast to man. An example of the helix-turn-helix motif is the lambda repressor protein shown in Figure 16.3 (The lambda repressor will be discussed in Chapter 18).

The HLH motif is found in many of the transcription factors involved in eukaryotic development and regulation of adult physiological function. This motif consists of an alpha helix followed by a loop that connects the first helix to a longer second alpha helix. The loop is highly flexible and this property enables the two helices to fold and pack against one another. This ability promotes the formation of dimers. When forming a dimer a portion of the longer helix contacts the DNA, and the shorter helix contacts the corresponding shorter helix in the HLH motif of its dimerization partner. The overall result is the formation of a four-helix bundle that grips the DNA molecule.

Leucine zipper domains are present in many proteins that form dimers. These proteins consist of a pair of alpha helices. A portion of one of the helices is enriched in hydrophobic leucine residues—there is a leucine residue in every seventh amino acid position. The leucines from one protein

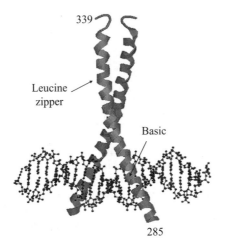

FIGURE 16.4. CREB bZIP dimer bound to DNA determined using X-ray crystal-lography: The basic region (residues 285 to 307 straddle and grip the DNA helix while leucine zipper residues 308 through 336 form a dimerization interface. The basic residues contact the major groove of the CRE region in the promoter. Protein helices are depicted as ribbons and DNA molecules as balls-and-sticks. The figure was prepared using Protein Explorer with atomic coordinates deposited in the Protein Date Bank under accession number 1DH3.

are interdigitized with the leucines from the partner molecule to form a short coiled-coil dimerization region, hence the name "leucine zipper." These proteins also contain a region rich in basic residues that interfaces with the DNA. The basic region begins at a location seven residues N-terminal to the first leucine in the leucine zipper domain. The overall domain is referred to as a *basic region leucine zipper* (bZIP) domain; the resulting bZIP dimer tightly grips the DNA molecule in the major groove as illustrated in Figure 16.4 for the CREB transcription factor (CREB transcription factors will be discussed in Chapter 21). A number of transcription factors possess a basic region plus HLH and leucine zipper domains. These DNA-binding proteins are referred to as bHLH-LZ proteins. The bHLH-LZ and bZIP proteins form a variety of homodimers and het-erodimers with one another.

The final category of DNA-binding proteins to be discussed is the *zinc fingers*. These proteins utilize one or two zinc atoms to help form compact domains of about 30 amino acid residues. The zinc finger motif is often repeated, and each small module is organized about a central zinc atom. Adjacent zinc finger modules form an overall DNA-binding domain whose zinc fingers grip the DNA molecule. This motif is widely distributed among eukaryotic transcription factors. The specifc composition of these modules is variable but all contain zinc atoms and independently fold into small compact structures as depicted in Figure 16.5.

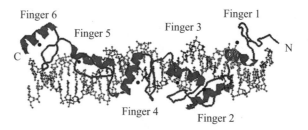

FIGURE 16.5. Zinc fingers of the TFIIA protein in contact with DNA determined using X-ray crystallography: The DNA-binding domain of the transcription factor TFIIA has six zinc fingers each organized by a central zinc atom (indicated in the figure by small filled circles). Protein helices are depicted as ribbons and DNA molecules as balls-and-sticks. The figure was prepared using Protein Explorer with atomic coordinates deposited in the Protein Data Bank under accession number 1TF6.

16.4 Transcriptional Activation Domains Initiate Transcription

The tight packaging of the DNA into nucleosomes inhibits transcription. Transcription factors stimulate transcription by: recruiting to transcription sites components of the transcription machinery; recruiting cofactors such as the Mediator; and by attracting to these sites molecules and complexes that remodel and modify chromatin structure.

Transcription factors possessing DNA-binding and dimerization domains alone cannot trigger transcription. Instead, an activation domain must be present. Transcription factors utilize several kinds of activation domains. These domains are enriched in specific amino acids and are classified according to their composition rather than by any particular secondary or supersecondary structures. Some are enriched in acidic amino acid residues and are called *acid blobs*. Others are enriched in glutamine or proline residues.

The activation domains of transcription factors are protein-binding regions. One of the ways the activation regions stimulate transcription is by recruiting components of the general transcriptional machinery. The transcription activators bind to sites along the DNA several hundred base units upstream of the transcription start site using their DNA-binding domains. Components of the transcription machinery are then recruited to and bind the activation domains, thus seeding the further assembly of transcription components and, so, stimulating onset of transcription.

Chromatin remodeling complexes are attracted to the promoter site by the activation domains of the transcription factors. Referring back to the yeast modules listed in Table 16.1, The SWI/SNF module is recruited by the acidic activation domain present in many transcription factors. The SAGA

TABLE 16.2. Nuclear receptors and their ligands: Common abbreviations for the receptors are given within the parentheses.

Steroid receptors	Nonsteroid receptors
Androgens (AR)	Ecdysone (EcR)
Estrogens (ER)	All-*trans* retinoids (RAR)
Cortisol (GR)	9-*cis* retinoids (RXR)
Aldosterone (MR)	Thyroid hormone (TR)
Progestins (PR)	Vitamin D_3 (VDR)

histone acetyltransferase module can also be recruited in this manner. The recruitment of the chromatin remodeling machinery is not dependent on the presence of either the TBP (TATA binding protein) or RNA polymerase II at the promoter.

16.5 Nuclear Hormone Receptors Are Transcription Factors

The genomes of complex metazoans contain large numbers of transcription factors. An early estimate is that there are over 3000 transcription factors encoded in the human genome. Nuclear hormone receptors, with over 150 members, form the largest family of transcription factors. These receptors can be grouped into two large families, designated as Class I (steroid) and Class II (nonsteroid). Representative members of these two families are listed in Table 16.2.

Steroid hormones and other small lipophilic molecules such as vitamin D_3 are able to pass through the plasma membrane and enter the cell interior where they bind to nuclear receptors. Following ligand binding, the steroid receptors dimerize, translocate into the nucleus from the cytoplasm, and bind their cognate responsive elements. Class II receptors differ from those of Class I in that they can bind to responsive elements in the absence of ligand binding. Many Class II receptors readily form heterodimers and bind to a variety of coactivators.

16.6 Composition and Structure of the Basal Transcription Machinery

The basal transcription machinery consists of RNA polymerase II and a set of general transcription factors (GTFs). These components, listed in Table 16.3, make up a minimal set of transcription initiation factors that support recognition and low (basal) levels of transcription from most promoter

TABLE 16.3. Basal transcription machinery: Listed are the components, the number of subunits in each component, their mass, and their role in initiating transcription (discussed further in the main text).

Component	Number of subunits	Molecular weight (kDa)	Function
TFIID-TBP	1	38	TATA binding protein
TFIID-TAF$_{II}$s	12+	15–250	Positive/negitive regulators
TFIIB	1	35	TFIIF/Poly II binding
TFIIF	2	30, 74	Shuttles Poly II
RNA Poly II	12	10–220	Catalytic
TFIIE	2	34, 57	Stim. TFIIH melting
TFIIH	9	35–89	Stim. melting/escape

sites. The minimal set of general transcription factors assembled at the promoter includes in order of recruitment TFIID, TFIIB, TFIIF, TFIIE, and TFIIH. Another GTF, TFIIA, is sometimes recruited to the promoter, and this (disposable) unit assists in TBP binding and stabilization. The assembly of the minimal set at the core promoter is depicted schematically in Figure 16.1(b).

TFIID is believed to be the first component of the transcription initiation complex to be recruited to the promoter site. It is the major sequence-specific GTF, and is a common target of TF activation domains. TFIID contains the TATA box-binding protein (TBP), and a collection of TBP-associated factors (TAFs). The TBP subunit is a general transcription factor. It recognizes the TATA sequence in its minor groove and is present in all RNA polymerization operations. The TAFs are promoter-selective and different mixes of these factors are present in different promoter sites. Recruitement of TFIID triggers the recruitment of TFIIA and then TFIIB followed by TFIIF bound to RNA polymerase II, and after that TFIIE, and then TFIIH. These additional GTFs that follow establish protein-protein contacts and link to the promoter site through the TBP. The TBP bends the DNA and forms a saddle with its inner surface in contact with the DNA molecule covering an 8-bp region of DNA.

16.7 RNAP II Is Core Module of the Transcription Machinery

The RNAP II molecule is composed of 12 subunits designated as Rpb1 through Rpb12. The yeast RNAP II molecule contains 4565 amino acid residues and has a mass of 514 kDa. The sizes of the subunits vary considerably. The largest two, Rpb1 (192 kDa) and Rpb2 (139 kDa), account for the bulk of the mass of the polymerase. A top view of the entire assembly is presented in Figure 16.6.

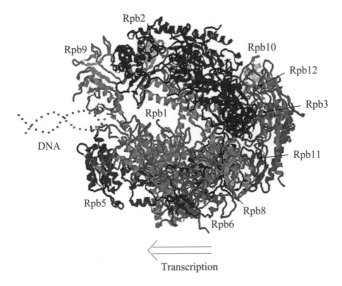

FIGURE 16.6. Top view of RNAP II determined by means of X-ray diffraction measurements at 3.1 Å and 2.8 Å resolution: Ten of the subunits are shown in this figure. Superimposed on this structure is a representation of a DNA molecule entering the RNAP II, and an arrow indicating the direction of transcription. The figure was prepared using Protein Explorer with atomic coordinates deposited in the Protein Data Bank under accession number 1I50.

As shown in Figure 16.6, the DNA enters the assembly through a cleft formed by Rpb1 and Rpb2. The DNA is gripped by protein "jaws" but cannot pass straight through the molecule because of a protein "wall" and makes a right angle bend facilitating the addition of nucleotide triphosphates (NTPs). The NTPs enter through a "funnel" and pass through a pore to reach the active site for polymerization. Exit of the nascent messenger RNA chain is guided by several other structural elements, most notably by rudder, lid, and clamp elements. Most of these functional elements belong to the large Rpb1 and Rpb2 subunits. These subunits and the assignments of functional roles to portions of each are depicted in Figure 16.7.

16.8 Regulation by Chromatin-Modifying Enzymes

Chromatin-modifying enzymes regulate the accessibility of the DNA to transcription factors and the basal transcription machinery. Recall from Chapter 2 that histone tails undergo several different kinds of posttranslational modifications. The main targets of the modifications are amino groups situated at the end of lysine side chains (see Figure 2.13). Regulatory

(a)   (b)

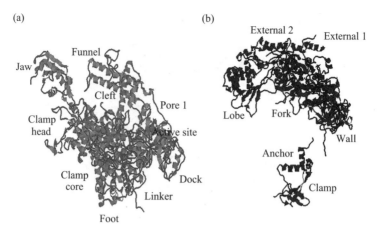

FIGURE 16.7. Functional roles of the large subunits of RNA polymerase II: (a) Rpb1 subunit and (b) Rpb2 subunit. The figure was prepared using Protein Explorer with atomic coordinates deposited in the Protein Data Bank under accession number 1I50.

enzymes functioning as histone acetyltransferases (HATs) catalyze the addition of acetyl groups, while histone deacetylases (HDACs) do the opposite, catalyzing the removal of these groups. Acetyl groups are not the only ones added and removed from histone tails. Phosphoryl, methyl, and SUMO groups are added and removed as well, as was discussed in Chapter 2.

Transcriptionally inactive heterochromatin is tightly compacted. Sites where transcription initiation and regulation take place are inaccessible to the proteins responsible for these actions. In contrast, transcriptionally active euchromatin has a far more open shape, where transcription factors and the basal transcription machinery bind are accessible to the responsible proteins. The covalent addition of acetyl groups to the side chains reduces the net positive charge on the histone tails thereby weakening their attraction to the DNA strands. The addition of these groups counteracts the natural tendency for the chromatin fibers to fold into compact nucleosomal units making transcription difficult to impossible. Upon acetylation, the promoter sites become far more accessible to the transcription machinery.

The enzymes that mediate the modifications on the H3 and H4 tails are quite specific. This aspect, together with the presence of so many potential modification sites, has led to the suggestion that there is a histone code that determined whether a particular gene or cluster of genes is silenced or not. An example of how this might work is supplied by data regarding the H3 lysine 9 (K9) site. As indicated in Figure 2.13 this is a site that can be either acetylated or methylated. The selection process is a competitive one. Acetylation at K9 blocks methylation at that site, and vice versa. The silencing

(a)   (b)

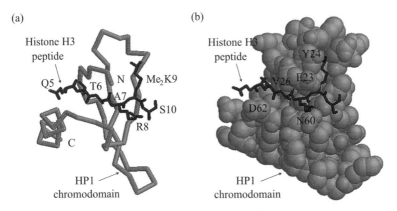

FIGURE 16.8. A peptide fragment of histone H3 doubly methylated on Lys9 (shown in black) complexed with the chromodomain of HP1 (shown in gray): Residues Gln5, Thr6, Ala7, and Arg8 of the H3 peptide form beta sheet interactions with residues Glu23, Tyr24, Val26, Asn60 and Asp62 of the chromodomain. The peptide residues are labeled in (a) while the chromodomain residues are labeled in (b). In (a) the chromodomain backbone is highlighted, while in (b) a space-filled representation is presented. The atomic structures were determined using X-ray crystallography. The figure was prepared using Protein Explorer with atomic coordinates deposited in the Protein Data Bank under accession number 1KNA.

process for that site works in the following way. An HDAC acts on K9 to remove an acetyl group. Next, a histone methyltransferase-specific for that site catalyzes the covalent addition of a methyl group. Then, heterochromatin (adapte) protein 1 (HP1) binds to the methylated K9 site and blocks transcription. The binding of HP1 to the methylated histone tail is shown in Figure 16.8.

16.9 Multiprotein Complex Use of Energy of ATP Hydrolysis

There are two basic kinds of chromatin modifiers. There are modifiers of chemical affinity between histones and DNA, and there are hydrolyzing enzymes that break and reform bonds between histones and DNA. The first group consists of enzymes such as the HATs and HDACs that act on the histone tails, covalently modifying them through the addition or subtraction of acetyl, methyl, or phosphoryl groups as just discussed. The second group encompasses several large multisubunit complexes that use the energy of ATP hydrolysis to disrupt interactions between histones and the DNA. There are three classes of ATP-dependent multisubunit complexes. Each contains at least one ATPase plus additional subunits. Some of the

TABLE 16.4. Three families of ATP-dependent SWI chromatin remodeling complexes: Abbreviations—ATP-dependent chromatin assembly and remodeling factor (ACF); chromatin accessibility complex (CHRAC); chromatin organization modifier (Chromo); Imitation SWI (ISWI); nucleosome remodeling factor (NURF); remodels the structure of chromatin (RSC); sucrose nonfermenting (SNF); switch (SWI); Yeast (y); *Drosophila* (d); human (h).

Family	Complexes	Recognition domain
SWI/SNF	ySWI/SNF, hSWI/SNF, yRSC	Bromodomain
ISWI	dCHRAC, dNURF, dACF	SANT domain
Mi-2	Mi-2	Chromodomain

complexes identified to date in various organisms are summarized in Table 16.4.

16.10 Protein Complexes Act as Interfaces Between TFs and RNAP II

Chromatin modifying enzymes do not interact with the basal transcription machinery, but instead make it possible for that machinery to find and attach to the promoters. Other complexes then take over and further support transcription. These interfacing complexes link activators and enhancers to the basal machinery attached to the core promoter. These complexes function as interfaces integrating and conveying to RNA polymerase II and its suite of general transcription factors instructions embodied in the transcription factors.

The best studied of these interfacing complexes is the Mediator complex found in yeast. This complex consists of about 20 different proteins arranged into three modules. Figure 16.9 illustrates how this complex can serve as an interface between the core machinery and the transcription factors bound to activator and enhancer sites. As can be seen in this figure the mediator complex forms a bridge between the regulatory proteins and basal transcription machinery at the core promoter. This bridging action is especially important for proteins bound at distal enhancer sites brought into close proximity to the core promoter through bending and looping of the DNA.

The yeast mediator complex has a number of metazoan counterparts. These complexes each contain homologs of a number of the yeast subunits along with additional members specifically needed to deal with the metazoan regulatory pathways. Some of the metazoan complexes are quite large with more than a dozen members, while others are smaller composed of subsets of the larger complexes. As is the case for the Mediator, the physical dimensions are large enough to act as bridges. For example, the planar dimensions of the *activation-recruited coactivator-large* (ARC-L) complex

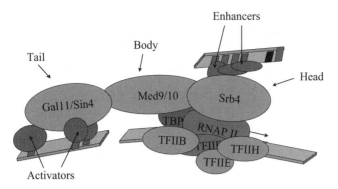

FIGURE 16.9. Yeast Mediator complex: Shown are the three Mediator modules—
The Srb4 "head" domain, the Med9/10 "body" domain, and the Gal11/Sin4 "tail"
domain, each named for prominent members of the module.

are 420 Å by 185 Å with distinct head, body, and tail regions formed by specific groups of subunits. The smaller *coactivator required for Sp1 activation*
(CRSP) complex, which consists of a subset of the ARC-L subunits, is
still 360 Å by 145 Å. Another mammalian Mediator complex, called
SMCC/TRAP (SRB- and MED-containing cofactor complex/thyroid
hormone receptor-associated protein), has about 25 members. These large
mediator complexes attach to the RNAP II CTD and form holoenzymes
with a total mass of about 1.5 MDa.

16.11 Alternative Splicing to Generate Multiple Proteins

The remarkable efficiency of genetic coding is exemplified by the relatively
modest increases in genome size in going from the yeast to the worm and
the fly, and to vertebrates and man. A central feature in making this possible is the modular organization of the genome and the proteins so
encoded. In higher eukaryotes, genes consist of short coding sequences, or
exons, separated from one another by longer noncoding or intervening
sequences, or *introns*. The exons are typically 50 to 400 nucleotides in length,
while introns can be as long as 200,000 nucleotides. Whereas most yeast pre-
mRNAs do not contain introns, and those that do usually contain only a
single one, in higher eukaryotes a single gene may contain as many as 50
introns. The introns are included in pre-mRNAs produced by transcription
from the DNA templates. In splicing, introns situated in between pairs of
flanking exons are removed. These intervening sequences are removed
from *pre-messenger* RNAs (pre-mRNAs) to form mature mRNAs containing a selected set of exons that is used for protein synthesis.

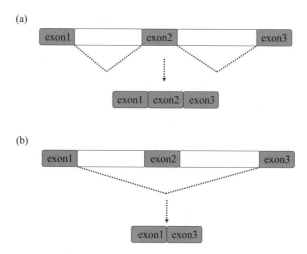

FIGURE 16.10. Alternative splicing of pre-messenger RNAs: Alternative splicing is an efficient method of generating multiple proteins from a single primary transcript in a controlled way. (a) The dotted lines indicate the intervening DNA sequences that are spliced out to produce a mature messenger RNA molecule containing three exons. (b) A different mix of exons is selected; the middle exon is skipped; that is, it is spliced out too, resulting in a mature mRNA molecule consisting of exon1 and exon3.

In alternative splicing, different sets of pre-mRNA sites are selected for splicing. This process can generate several different isoforms, or splice variants, of a single gene. By varying the mix of splice sites a variety of messenger RNAs (mRNAs) can be generated differing from one another by the presence of certain exons and the absence of others that have been skipped. This selection process is illustrated schematically in Figure 16.10. Alternative splicing generates a variety of tissue and compartment-specific proteins from a single gene; it is commonplace in the expression of genes that encode control layer proteins, especially those that function in the nervous and immune systems.

16.12 Pre-Messenger RNA Molecules Contain Splice Sites

The pre-mRNA molecule contains exons, introns, and regulatory sequences that provide binding sites for the splicing machinery and regulatory proteins. Three sites are depicted in Figure 16.11. The first is the 5′ site characterized by the presence of a binding sequence containing a guanine-uracil (GU) pair within a longer GURAGU-like sequence, where R is a purine.

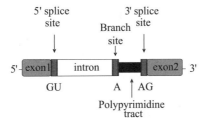

FIGURE 16.11. Organization of the pre-messenger RNA molecule: A representative pre-messenger RNA molecule is depicted consisting of two exons, an intron, and three accompanying regulatory sites. The introns are usually much longer than the exons, but are shown shortened for illustrative purposes. Nucleotides characteristic of the regulatory sequences are shown at the bottom of the diagram.

The second binding site is the branch site. This site is characterized by the nucleotide sequence YNYURAY, where Y is a pyrimidine. The letter "A" in Figure 16.11 denotes this site. The next portion of the RNA molecule is the pyrimidine tract, which as its name suggests is a string of pyrimidine nucleotides. The last regulatory unit is the 3′ splice site characterized by either a CAG sequence or a UAG sequence. Splicing is mediated by proteins that bind to particular stretches of DNA just as transcription is mediated by proteins that bind to promoter sites.

16.13 Small Nuclear RNAs (snRNAs)

A family of small nuclear RNA (snRNA) molecules found in the cell nucleus forms the catalytic core of the *spliceosome*, the name given to the machinery responsible for splicing out introns from the pre-mRNAs. These snRNAs are enriched in uridine and are referred to as the U1 snRNA, U2 snRNA, and so on. The U4 and U6 snRNAs are bound to one another through base pairing and operate as a single unit. The snRNAs function in tight association with from 60 to 100 proteins to form ribonucleoprotein particle (RNP) complexes. These RNP complexes are often referred to as small nuclear ribonucleoprotein particles (snRNPs) or "snurps." The pre-mRNAs produced during transcription are also complexed with proteins. The combined structures are referred to as *heterogeneous nuclear RNP* (hnRNP) particles, and the proteins are called *hnRNP proteins*.

The spliceosome is assembled in several stages. Each stage is identified by the presence of a particular set, or complex, of snRNPs as depicted in Figure 16.12. The first step in spliceosome assembly is the recruitment of the U1 snRNP to the 5′ splice site. Several auxiliary proteins assemble at the branch point and polypyrimidine (Py) tract at this time. The first stage where the U1 snRNP binds to the 5′ splice site sequence is known as the

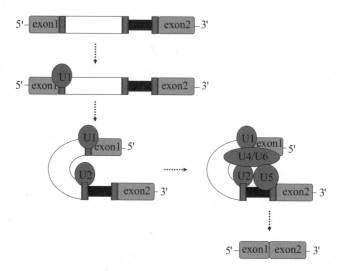

FIGURE 16.12. Splicing of a pre-mRNA by the spliceosome: Splicing takes place through the formation of splicing complexes. In the first step, U1 binds to the 5′ splice site forming the early (E) complex. In the next step, U2 binds the branch point forming the A complex. This action is followed by recruitment of U4/U6 and U5 to form the B complex. U4 and U1 leave and the others complete the splicing process.

commitment stage (complex E). The next stage, the assembly of the U2 snRNP at the branch point, assisted by the auxiliary proteins, is referred to as the *presplicing stage* (complex A). In the third stage, a preassembled U4/U6 U5 complex is recruited to the U2 snRNP and these components are joined by the U1 snRNP. All components of the spliceosome are now assembled into an *early splicing machine* (complex B) and, as shown in Figure 16.12, the 5′ and 3′ ends of the intron have been brought together to form a loop.

Once the first phase of the splicing process is complete, the *catalytic phase*—early splicing and late splicing (complex C)—begins. In the catalytic phase the spliceosome removes the intron and joins the flanking exons. The catalytic process involves several steps and is accompanied by extensive conformational changes and rearrangements of the spliceosome. The U2, U5, and U6 snRNPs form the catalytic core of the spliceosome. These components remain attached to the removed intron, while U1 and U4 leave the spliceosome earlier. At the end of the catalytic stage the introns and Py tracts have been removed leaving the selected set of exons joined together in a mature mRNA molecule.

Transcription and splicing are coupled processes. Splicing begins while the pre-mRNA is still being formed, during the transcription elongation process, rather than after completion. As noted above, long introns separate rather short exons from one another. The coupling of transcription with splicing helps bring together the exons in the correct manner over the large distances and enables components of the transcription machinery to recruit proteins required for splicing.

16.14 How Exon Splices Are Determined

In addition to the principal splice sites shown in Figure 16.11, there are a number of auxiliary splice sites. These auxiliary sites may be located in either exons or in introns and provide places where splicing regulators may bind. Auxiliary sites located within exons are called *exonic splice enhancer* (ESE) and *exonic splice silencer* (ESS) sites, depending on which regulatory outcome is supported. Similarly, intronic sites are termed *intronic splice enhancer* (ISE) and *intronic splice silencer* (ISS) sites. Two kinds of proteins—hnRNPs and SR proteins—bind to these auxiliary regulatory RNA sequences.

SR proteins are a family of essential splicing factors ranging in size from 20 to 75 kDa. Members of the family are characterized by the presence of an arginine-serine-rich (RS) domain of variable length at their carboxyl terminal and one or more RNA recognition motifs (RRMs) in their amino terminal. These proteins are sequence-specific RNA binding proteins that have multiple roles in splicing. They play essential roles in the assembly of the spliceosome at splice sites and in the selection of alternative splice sites.

Figure 16.13 depicts the assembly of a representative set of splicing factors along a section of a pre-mRNA molecule. Two auxiliary splicing factors (U2AF65 and U2AF35) associated with the U2 snRNP have been recruited to the Py tract and 3′ splice site by SR proteins bound at nearby ESEs. These proteins, and other auxiliary factors, recruit or inhibit components of the spliceosome from assembling at the pre-mRNA and recognizing nearby splice sites. Proteins that bind enhancer or silencer sites can be either SR proteins or hnRNPs.

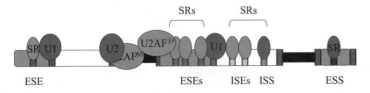

FIGURE 16.13. SR (serine-arginine-rich) proteins recruit U1 and U2 snRNPs to splice and branch sites: Auxiliary splicing sites and regulatory proteins that bind them determine which exons are spliced together. Abbreviations—Exonic splice enhancer (ESE) site; exonic splice silencer (ESS) site; intronic splice enhancer (ISE) site; intronic splice silencer (ISS) site.

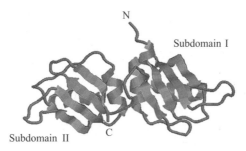

FIGURE 16.14. Crystal structure of the N-terminal region (UP1) of the hnRNP A1: Each subdomain contains an RRM and the overall structure of the module is that of an extended RNA binding surface. The crucial amino acid sequences are located on the two middle beta strands of each subdomain. The figure was prepared using Protein Explorer with atomic coordinates deposited in the Brookhaven Protein Data Bank under accession number 1UP1.

Like other regulatory proteins the hnRNP proteins are modular in design. They contain one or more RNA recognition motifs (RRMs) that facilitate their participation in the complexes. An example of an RNA binding domain for an hnRNP called A1 is presented in Figure 16.14. The RNA binding module consists of a pair of subdomains, I and II. Each subdomain consists of a four-stranded antiparallel beta-pleated sheet and two alpha helices off to one side. The overall structure is that of an extended RNA binding surface.

The relative concentration of splicing proteins, and the mix of regulatory sites, jointly determine which combination of exons is assembled into the mature messenger RNA molecule. Different cell types in a multicellular organism will express a different mix of proteins that bind to the auxiliary regulatory sites, and a given cell may alter the mix of splicing regulators over time to produce proteins with slightly altered properties.

16.15 Translation Initiation Factors Regulate Start of Translation

The machinery involved in translation initiation serves as an important end point for stress signaling. Before discussing the mechanisms underlying the control of translation it is worthwhile to examine the organization of mRNAs and the machinery that carries out translation. Messenger RNA molecules contain special structures at their 5′ and 3′ termini. A cap structure is added to its 5′ end. The cap consists of a 7-methylguanylate (a CH_3 group appended to the 7 position of the G base) plus a triphosphate group that links the 7-methylguanylate to the sugars at the 5′ end of the RNA molecule. The cap is referred to as an m^7GpppN structure, where N is any nucleotide. A string of perhaps as many as 200 adenylates is appended to the 3' terminus of the mRNA. This addition is referred to as a polyadenylate tail or poly (A) tail. The cap and poly (A) additions are separated from the mRNA coding region by flanking noncoding regions and start and stop codons (Figure 16.15).

A set of proteins called *eukaryotic initiation factors*, or eIFs, is responsible for assembling the ribosomal 40S and 60S subunits at the mRNA molecule. These proteins are listed in Table 16.5. Translation initiation begins at the 5' end of the mRNA. An initiator transfer RNA (tRNA) is a tRNA molecule loaded with a methionine (Met), the amino acid encoded by the AUG start codon. For translation to begin, the Met-tRNA initiator and the small and large ribosomal subunits must be recruited and positioned at the AUG start site of the mRNA. This occurs in a staged manner as shown in Figure 16.16. In the first main step the loaded tRNA molecule and small ribosomal subunit associate. The Met-tRNA molecule is guided to its correct position by the initiation factor eIF2 (eukaryotic initiation factor 2) that forms a complex with the Met-tRNA and a GTP molecule. This terti-

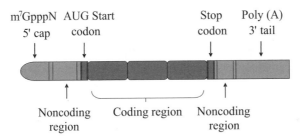

FIGURE 16.15. Organization of the mature mRNA molecule: The coding region is flanked by a pair of noncoding regions, a 5' end cap, and a 3' end polyadenylate tail.

TABLE 16.5. Vertebrate initiation factors.

Initiator factor	Molecular mass (kDa)	Function
eIF1A	17	Promotes Met-tRNA, 40S, mRNA binding
eIF2	126	GTP binding protein, guides Met-tRNA onto 40S
eIF2B	277	GEF for eIF2
eIF3	330	Binds to 40S, prevents 60S binding
eIF4F	291	Complex of eIF4E, eIF4G, and eIF4A; eIF4E binds to the cap; eIF4G interacts with eIF3 to augment binding of the preinitiation complex; eIF4A is an RNA-dependent ATPase and an RNA helicase
eIF4B	69	Facilitates ribosome-mRNA binding
eIF5	45	Stimulates hydrolysis of GTP associated with eIF2

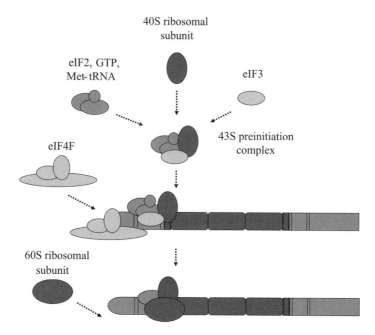

FIGURE 16.16. Assembly of the ribosomal 40S and 60S subunits at the mRNA molecule: At the end of the process the 40S subunit, 60S subunit, and Met-tRNA are positioned at the AUG start codon. This positioning is guided by the eIFs with the eIF4E subunit gripping the cap and the eIF4G subunit serving as a scaffold.

ary complex associates with the small (40S) ribosomal subunit to form a 43S preinitiation complex.

Next, the preinitiation complex attaches to the mRNA near its 5′ end. An initiation factor complex, eIF4F, binds to the m⁷GpppN cap and prepares the way for the attachment of the preinitiation complex to the mRNA. Upon attachment the preinitiation complex scans in the 5′ to 3′ direction for the start AUG codon and halts upon arrival. Another initiation factor, eIF5, stimulates the hydrolysis by eIF2 of GTP to GDP, resulting in the dissociation of the initiating factors from the assemblage leading to the attachment of the large (60S) ribosomal subunit, and translation elongation can begin.

16.16 eIF2 Interfaces Upstream Regulatory Signals and the Ribosomal Machinery

The translation initiation factors have a crucial role in guiding the 40S and 60S ribosomal subunits to the start codon. As a result, if the translation initiation factors are prevented from carrying out their tasks, translation will cease. Prevention can be accomplished by phosphorylating certain serine residues on eIF2 and eIF2B, thereby placing translation under the firm control of upstream signals conveyed by protein kinases.

The eIF2 translation initiation factor is a GTPase. As noted in Table 16.5, eIF2B functions as a GEF for eIF2, and eIF5 operates as its GAP. The eIF2 initiation factor contains several subunits. Its alpha subunit can be phosphorylated on Ser51 by a number of upstream kinases. One of the kinases that phosphorylates the subunit is protein kinase R (PKR). As discussed earlier in Chapter 9, interferon signaling acts in part through this kinase to halt protein synthesis when a virus infection is detected (through the presence of dsRNAs). Among the other kinases that target eIF2 are GCN2, a kinase that phosphorylates eIF2 in response to poor nutrient conditions, and PERK, a kinase that phosphorylates eIF2 in response to ER stresses.

The eIF2 initiation factor is deactivated by phosphorylation and activated by dephosphorylation. When eIF2 is phosphorylated it binds eIF2B tightly, thereby sequestering both and preventing their translational activities. The translation initiation factor eIF2B catalyzes the conversion of GDP back to GTP, a step that is required for continued cycling and reformation of Met-tRNA-GTP-eIF2 complexes. It, too, can be phosphorylated, by GSK3 among others, to prevent its activities. In response to growth and survival signals, Akt is activated and when this kinase phosphorylates GSK3, eIF2B is freed of the inhibitory activities of GSK3.

16.17 Critical Control Points for Protein Synthesis

The eIF2 and eIF2B proteins are not the only targets of upstream regulatory kinases and phosphatases. The cap-binding, translation initiation factor eIF4E is phosphorylated at Ser209 in response to growth factor, hormonal and mitogenic signaling, and it is dephosphorylated in response to heat shock. Phosphorylation enhances the ability of eIF4E to bind to eIF4G and/or the cap and thus promotes translation. The extent to which eIF4Es are phosphorylated matches the overall translation level at different stages of the cell cycle. These levels are low during G_0 phase, increase through G_1 and S phases, and then rapidly drop during M phase. Several MAP kinase pathways are involved in phosphorylating eIF4E. These include the Ras-Raf-Mek-Erk that relays growth/mitogenic signals, and the p38 MAP kinase that relays stress signals. Both of these pathways activate Mnk1, a kinase that integrates MAP kinase signals, and in response phosphorylates eIF4E on Ser209.

The cap-binding protein eIF4E is the target of negative regulation by a family of small 10 to 12 kDa-eIF4E binding proteins (4E-BPs). These regulators compete with eIF4G for binding to eIF4E. The ability of the 4E-BPs to compete for binding is dependent on their phosphorylation status. Phosphorylation decreases the binding affinity of the 4E-BPs for eIF4E and leads to dissociation. The best characterized member of the family is 4E-BP1

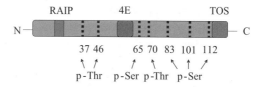

FIGURE 16.17. Structure of the human 4E-BP1 protein: The cap-binding protein eIF4E, along with its 4E-BPs, is a critical control point for protein synthesis. The seven identified phosphorylation sites are indicated along with the N-terminal RAIP (Arg-Ala-Ile-Pro) motif, the C-terminal TOS (Phe-Glu-Met-Asp-Ile) motif and the eIF4E binding site. The RAIP and TOS motifs are essential for proper phosphorylation by upstream kinases.

(Figure 16.17). The regulation of eIF4E by 4E-BP1 was discussed in Chapter 6. Recall from that earlier discussion that, under good growth/nutrient conditions, target of rapamycin (TOR) proteins phosphorylate and thus activate a protein called Tap42 that ensures that 4E-BP1 does not bind and inactivate eIF4E (Figure 6.10). Under poor growth conditions Tap42 is not activated by TOR and 4E-BP1 is free to inhibit the activity of the translation initiation factor. The 4E-BP proteins are phosphorylated on a number of serine/threonine residues by several kinases in a hierarchical fashion. For example, they are phosphorylated on Thr37 and Thr46 by Akt activated in the PI3K pathway, and once primed by Akt they are further phosphorylated by hTOR on Ser65, Thr70 and Ser83. These stratified phosphorylation events induce the dissociation of 4E-BP1 from eIF4E.

References and Further Reading

Promoter Architecture

Bell AC, West AG, and Felsenfeld G [2001]. Insulators and boundaries: Versatile regulatory elements in the eukaryotic genome. *Science*, 291: 447–450.

Blackwood EM, and Kadonaga JT [1998]. Going the distance: A current view of enhancer action. *Science*, 281: 60–63.

Calhoun VC, Stathopoulos A, and Levine M [2002]. Promoter-proximal tethering elements regulate enhancer-promoter specificity in the *Drosophila antennapedia* complex. *Proc. Natl. Acad. Sci. USA*, 99: 9243–9247.

Structure of RNAP II

Cramer P, Bushnell DA, and Kornberg RD [2001]. Structural basis of transcription: RNA polymerase II at 2.8 Ångstrom resolution. *Science*, 292: 1863–1876.

Gnatt AL, et al. [2001]. Structural basis of transcription: An RNA Polymerase II elongation complex at 3.3 Å resolution. *Science*, 292: 1876–1882.

Woychik NA, and Hampsey M [2002]. The RNA polymerase II machinery: Structure illuminates function. *Cell*, 108: 453–463.

Chromatin-Modifying Enzymes

Boyer LA, et al. [2000]. Functional delineation of three groups of the ATP-dependent family of chromatin remodeling enzymes. *J. Biol. Chem.*, 275: 18864–18870.

Eissenberg JC [2001]. Molecular biology of the chromo domain: An ancient chromatin module comes of age. *Gene*, 275: 19–29.

Horn PJ, and Peterson CL [2002]. Chromatin higher order folding: Wrapping up transcription. *Science*, 297: 1824–1827.

Narlikar GJ, Fan HY, and Kingston RE [2002]. Cooperation between complexes that regulate chromatin structure and transcription. *Cell*, 108: 475–487.

Vignali M [2000]. ATP-dependent chromatin-remodeling complexes. *Mol. Cell Biol.*, 20: 1899–1910.

Zeng L, and Zhou MM [2002]. Bromodomain: An acetyl-lysine binding domain. *FEBS Lett.*, 513: 124–128.

Mediator Complexes

Asturias FJ, et al. [1999]. Conserved structures of Mediator and RNA polymerase holoenzyme. *Science*, 283: 985–987.

Boube M, et al. [2002]. Evidence for a mediator of RNA polymerase II transcriptional regulation conserved from yeast to man. *Cell*, 110: 143–151.

Dotson MR, et al. [2000]. Structural organization of the yeast and mammalian mediator complexes. *Proc. Natl. Acad. Sci. USA*, 97: 14307–14310.

Lewis BA, and Reinberg D [2003]. The Mediator coactivator complex: Functional and physical roles in transcriptional regulation *J. Cell Sci.*, 116: 3667–3675.

Malik S, and Roeder RG [2000]. Transcriptional regulation through Mediator-like coactivators in yeast and metazoan cells. *Trends Biochem. Sci.*, 25: 277–283.

Taatjes DJ, et al. [2002]. Structure, function, and activator-induced conformations of the CRSP coactivator. *Science*, 295: 1058–1062.

Theory

Levine M, and Tjian R [2003]. Transcriptional regulation and animal diversity. *Nature*, 424: 147–151.

Orphanides G, and Reinberg D [2002]. A unified theory of gene expression. *Cell*, 108: 439–451.

Alternative Splicing

Cáceres JF, and Kornblihtt AR [2002]. Alternative splicing: Multiple control mechanisms and involvement in human disease. *Trends Genet.*, 18: 186–193.

Faustino NA, and Cooper TA [2003]. Pre-mRNA splicing and human disease. *Genes Dev.*, 17: 419–437.

Graveley BR [2000]. Sorting out the complexity of SR protein function. *RNA*, 6: 1197–1211.

Krecic AM, and Swanson MS [1999]. hnRNP complexes: Composition, structure, and function. *Curr. Opin. Cell Biology*, 11: 363–371.

Maniatis T, and Reed R [2002]. An extensive network of coupling among gene expression machines. *Nature*, 416: 499–506.

Smith CWJ, and Valcárcel J [2000]. Alternative pre-mRNA splicing: The logic of combinatorial control. *Trends Biochem Sci.*, 25: 381–388.

Staley JP, and Guthrie C [1998]. Mechanical devices of the spliceosome: Motors, clocks, springs, and things. *Cell*, 92: 315–326.

Regulation of Translation Initiation

Dever TE [2002]. Gene-specific regulation by general translation factors. *Cell*, 108: 545–586.

Gingras AC, Raught B, and Sonnenberg N [2001]. Regulation of translation initiation by FRAP/mTOR. *Genes Dev.*, 15: 807–826.

Jacinto E, and Hall MN [2003]. TOR signaling in bugs, brain and brawn. *Nature Rev. Mol. Cell Biol.*, 4: 117–126.

Kosak M [1999]. Initiation of translation in prokaryotes and eukaryotes. *Gene*, 234: 187–208.

Pain VM [1996]. Initiation of protein synthesis in eukaryotic cells. *Eur. J. Biochem.*, 236: 747–771.

Proud CG [2002]. Regulation of mammalian translation factors by nutrients. *Eur. J. Biochem.*, 269: 5338–5349.

Wang XM, et al. [2003]. The C terminus of initiation factor 4E-binding protein 1 contains multiple regulatory features that influence its function and phosphorylation. *Mol. Cell. Biol.*, 23: 1546–1557.

Problems

16.1 Transcription factors and cofactors can promote or inhibit transcription in several ways. List some of these ways. For each way give the location/site of action along the DNA, the specific interaction that mediates the response, and the effect of the binding interaction on transcription.

16.2 The translation initiation factor eIF2 is a GTPase. Several different modes of action of GTPases were presented in Chapter 13. For example, Ras acted one way as a switch while Cdc42 cycles continuously, and thus operates in a different manner from Ras. Which of the several GTPase modes of operation does eIF2 follow?

16.3 One of the early events in apoptosis is the shutting down of the translation machinery. Give some examples of how the apoptosis regulatory machinery discussed in the last chapter might signal and regulate protein synthesis through interactions of apoptosis signaling/regulatory proteins with the translation initiation factors.

17
Cell Regulation in Bacteria

Bacteria are not only highly efficient organisms, but they are also the unseen majority wherever they exist. Bacteria are found in large numbers in the open oceans (1.2×10^{29}), in soils (2.6×10^{29}), in oceanic subsurfaces (3.5×10^{30}), and in terrestrial subsurfaces ($\sim 1 \times 10^{30}$). Each human body hosts some 10^{13} bacteria. One of the locales of high concentration is the gut. From 500 to 1000 different species of bacteria populate the human intestinal tract. The greatest concentration of bacteria is in the colon where there are about 10^{11} bacteria per gram of bulk material. This number is far greater than the amount of bacteria living on the surface of the skin. From 1000 to 10,000 bacteria live on each square centimeter of skin. These numbers yield a total skin population of about 300,000,000 bacteria.

Most of the bacteria that reside in the gastrointestinal (GI) tract are anaerobic, harmless, and even beneficial. One of the species of resident bacteria is the Gram-negative bacterium *Bacteroides thetaiontaomicron*. This bacterium acquires and hydrolyzes dietary polysaccharides that would otherwise be indigestible. *Bacteroides thetaiontaomicron* possesses a large number of genes that encode signaling proteins and gene regulators allowing the bacteria to communicate with and regulate their local environments. A symbiotic relationship is established between bacterial flora and their host environment. The bacteria stimulate vascularization and development of intestinal villi and stimulate epithelial cells to secrete bacterial nutrients. The bacteria generate simplified carbohydrates, amino acids, and vitamins.

Not all bacteria found in the GI tract are harmless. One of the occupants is the Gram-positive bacterium *Enterococcus faecalis*. This is an opportunistic pathogen that occupies a variety of ecological niches including soil, sewage, and water. Inside the GI tract, this bacterium causes uninary tract infections and is highly resistant to vancomycin, the preferred antibiotic for treating gram-positive bacterial infections. The propensity for this bacterium to harm its host is tied to genetic material acquired through horizontal gene transfer (HGT) from other bacteria. More than 25% of its genome consists of DNA acquired from other species.

17.1 Cell Regulation in Bacteria Occurs Primarily at Transcription Level

Messenger RNAs are rapidly turned over in bacteria. Their half-lives are far shorter than eukaryotic mRNAs. Whereas eukaryotic mRNAs have half-lives on the order of an hour, bacterial mRNA are degraded within a few minutes. Since the steady state concentration of any molecule is directly proportional to its half-life, the rapid degradation of selected molecules ensures that the bacterium can respond rapidly to changing environmental conditions. The rapid degradation of proteins is coupled to an equally rapid capability for synthesizing new proteins.

The machinery for regulating gene expression in bacteria is highly streamlined compared to that for eukaryotes. Bacterial DNA is not wrapped in histones, and chromatin-modifying enzymes are not required. Their genes do not possess introns, and a splicing stage is not needed. There are few noncoding regions, and the fraction of DNA that codes for proteins ranges from 87 to 95%. Transcription and translation are contiguous, and the two processes take place at the same time in the same place. In many species extra-chromosomal DNA is present. It is organized into one or more small circular plasmids containing collections of genes that perform specialized functions.

Bacteria such as *Streptomyces coelicolor*, *Escherichia coli*, *Bacillus subtilis* and *Caulobacter crescentus* possess genomes comparable in size to the unicellular yeast genome, and some, for example, *Calothrix*, have genomes approaching *Drosophila* in size. These bacterial species possess a sophisticated control layer that enables them to alter their metabolism, morphology, and lifestyle. Some can execute developmental programs that produce differentiated progeny and make possible colonial behavior. The first part of this chapter is devoted to an examination of gene regulation in bacteria, the counterpart to the discussions in the last chapter dealing with gene regulation in eukaryotes. The second part of this chapter deals with the more sophisticated forms of cell regulation in bacteria. Special attention is given to how some of these modes help cause disease in humans while other modes promote beneficial symbiotic relationships.

17.2 Transcription Is Initiated by RNAP Holoenzymes

Transcription in bacteria is carried out by RNA polymerase (RNAP) holoenzymes. Whereas eukaryotes possess three kinds of RNA polymerases, bacteria utilize a single kind. The bacterial RNAP molecule is composed of a large and rather nonspecific core RNAP unit and a small regulatory and targeting subunit, the *sigma factor* (Table 17.1).

The core unit is composed of four subunits—a pair of alpha subunits, and two larger subunits, β and β', responsible for the enzymatic activities of the

TABLE 17.1. Components of the *E. coli* RNA polymerase holoenzyme.

Subunit	Molecular weight (kDa)	Function
α	36.5	Targeting and assembly
β, β′	150.6, 155.2	Catalytic activities
σ	20 to 70	Promoter recognition

(a)

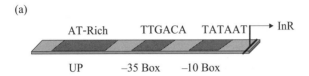

(b)

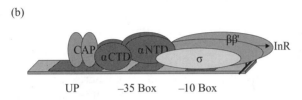

FIGURE 17.1. Organization of the bacterial promoter: (a) The primary DNA binding regions are shown together with the transcription start site, InR. (b) Arrangement of the subunits of RNA polymerase together with a catabolite activator protein (CAP) transcriptional activator dimer.

polymerase. The alpha subunit contains two domains connected by a flexible hinge. As shown in Figure 17.1b, each alpha subunit has an N-terminal domain (αNTD) that interfaces the subunit with the beta subunits, and a C-terminal domain (αCTD) that mediates interactions with the DNA promoter and with regulatory proteins. The core, written symbolically as $\alpha_2\beta\beta'$, has a total mass of about 380 kDa. There are several different kinds of sigma factors, and their masses vary from 20 to 70 kDa. The sigma factors are DNA sequence-specific binding proteins that recognize and bind the two specific promoter sites listed in Table 17.2. The bacterial transcription-competent RNAP holoenzyme is thus of the form $\alpha_2\beta\beta'\sigma$.

The association of a sigma factor and the core RNAP unit is transient. Shortly after initiation of transcription, the sigma factor dissociates from the core unit, which proceeds with transcript elongation. This dissociation makes possible the attachment of a different sigma factor subsequent to termination of the transcription event. If alternative sigma factors are present in sufficient numbers, they will out-compete the initial sigma factor for binding to the core unit. The core is thus reprogrammed to bind to an alternative promoter site and transcribe a different set of genes.

TABLE 17.2. Bacterial σ^{70} promoter architecture.

Regulatory region	Consensus sequence	Binding
–10 box	5′-TATAAT-3′	σ
–35 box	5′-TTGACA-3′	σ, TFs
UP element	AT-rich tracts	αCTD, TFs

17.3 Sigma Factors Bind to Regulatory Sequences in Promoters

DNA promoters that bind bacterial RNAP subunits and transcription factors contain a pair of core regulatory sequences separated by a spacer region. The regulatory sequence nearest the transcription (+1) start site is the –10 box, and it is centered 10 bp immediately upstream of that site. This region is separated by a spacer from the –35 box, the second core regulatory region centered 35 bp upstream. The amount of agreement between –10 and –35 box sequences and the consensus sequences has a regulatory role. Genes are strongly expressed at sites where the agreement is good and are weakly transcribed at sites where the agreement is not optimal. In many promoters, there is a third regulatory site located further upstream. Referred to as the *upstream* (UP) *sequence* its precise location varies from promoter to promoter (Figure 17.1a).

Different sigma factors recognize different promoter sequences. The *consensus sequences* listed in the second column of Table 17.2 are recognized only by primary sigma factor (i.e., by σ^{70}). Each of the groups of alternative sigma factors listed in Table 17.3 has its own characteristic –10 and –35 consensus sequences. There are two families of sigma factors, σ^{70} and σ^{54}. Members of the σ^{54} differ from those in the σ^{70} in several ways. One of the ways they differ is that instead of recognizing sequences –10 and –35 bp upstream of the start site they recognize sequences centered –12 and –24 bp upstream.

17.4 Bacteria Utilize Sigma Factors to Make Major Changes in Gene Expression

Bacteria such as *E. coli* and *B. subtilis* express several kinds of sigma factors. The primary sigma factors, σ^{70} in *E. coli* and σ^{A} in *B. subtilis*, regulate genes active all the time that perform housekeeping functions and support normal vegetative growth. This latter term denotes situations where nutrients are plentiful. Rapid cell growth and cell division typically produces exponential increases in the bacterial population. A second sigma factor is utilized in situations where exponential growth is not supported. The

TABLE 17.3. Sigma factors of *E. coli* and *B. subtilis*.

Sigma factor	Bacteria and designation	Control function
Primary	*E. coli* σ^{70}, *B. subtilis* σ^{A}	Housekeeping, vegetative growth
	E. coli σ^{S} (σ^{38})	Stationary phase
Heat shock	*E. coli* σ^{H} (σ^{32})	Stress response genes
	E. coli σ^{E} (σ^{24})	Stress response genes
	B. subtilis σ^{B} (σ^{37})	Stress response genes
Flagellar and chemotactic	*E. coli* σ^{F} (σ^{28}),	Genes for motility and
	B. subtilis σ^{D}	chemotaxis
Sporulation	*B. subtilis* σ^{H}	Postexponential, competence and early sporulation genes
	B. subtilis σ^{E}	Early mother cell genes
	B. subtilis σ^{F}	Early forespore genes
	B. subtilis σ^{G}	Late forespore genes
	B. subtilis σ^{K}	Late mother cell genes
Extracytoplasmic function (ECF)	*E. coli* σ^{FecI} (σ^{19}), *B. subtilis* σ^{V}, σ^{Z}	Genes for iron uptake, antibiotic production, virulence factors, outer membrane proteins
σ^{54} family	*E. coli* σ^{N} (σ^{54}), *B. subtilis* σ^{L}	Nitrogen fixation, genes for degradative enzymes

bacteria modify many of their internal activities in these cases using the stationary phase sigma factor σ^{S}. This sigma factor stimulates physiological changes needed in the presence of nutrient deprivation and toxic chemicals.

Many stressful conditions have to be dealt with by bacteria. Some conditions, like temperature extremes, osmotic stresses, and pH effects, are related to the chemical and physical environment. These stresses are dealt with by stress response sigma factors. High temperature is a representative example of a stressful condition. When high temperatures are encountered bacteria such as *E. coli* and *B. subtilis* use σ^{H} and σ^{B} to trigger the synthesis of heat shock proteins. These stress response proteins refold proteins that have become partially unfolded due to thermal effects, and if the proteins cannot be refolded tag them for destruction. Sporulation and extracytoplasmic function (ECF) sigma factors are two more families of sigma factors involved in controlling specific stress responses. The sporulation sigma factors control the development of spores in *B. subtilis* in response to starvation conditions. The ECF sigma factors control a diverse set of responses (for example, iron uptake, heavy metal resistance, antibiotic production, and induction of virulence factors) mostly involving the remodeling of the transport machinery located in bacterial membranes. Finally, some members of the σ^{54} family are involved in control of nitrogen metabolism while others are involved in regulating stress responses.

17.5 Mechanism of Bacterial Transcription Factors

Most bacterial transcription factors contain a helix-turn-helix (HTH) motif consisting of about 20 amino acid residues that binds DNA and is part of a larger DNA binding domain. The DNA binding domain often has additional sites that contact DNA and aid in recognition. The transcription factors bind to regulatory sequences in the promoters as dimers or as higher order oligomers. Besides contacting the DNA, transcription factors also contact one or more subunits of the RNAP holozenzme. Most often they contact σ and/or αCTD. However, some contact β, β', or αNTD. Promoters often contain sites for attachment of multiple transcription factors, and by this means integrate a variety of environmental signals into the transcription decision.

Bacterial TFs can be grouped into families according to characteristic DNA sequences present in family members and related physiological functions. Members of more than a dozen families of transcription factors are present in environmental bacteria such as *Escherichia coli*, *Bacillus subtilis*, *Vibrio cholerae*, and *Pseudomonas aeruginosa*. A short list of the most prevalent families present in these organisms is presented in Table 17.4. As discussed in Chapter 1, the size of the control layer contracts rapidly when the genome sizes are reduced. Minimalist bacteria such as *Mycoplasma genitalium* have only a few transcription factors, or none at all. Social, differentiating bacteria, which will be discussed later in this chapter, possess genomes that are even larger than those of the environmental bacteria. These organisms may well possess families of transcription factors needed for differentiation that are absent in Table 17.4.

Bacterial transcription factors are HTH DNA binding proteins that integrate signals, and fine tune and regulate transcription in response to changes

TABLE 17.4. Transcription factors, their actions, and typical cellular roles.

Family	Actions	Cellular role
AraC/XylS	Activator	Virulence, sugar metabolism
ArsR	Repressor	Heavy metal resistance
AsnC	Dual	Amino acid biosynthesis
Crp	Dual	Global responses
DeoR	Repressor	Sugar metabolism
GntR	Repressor	Carbon metabolism
LacI/GalR	Repressor	Carbon uptake
LuxR	Activator	Biosynthesis, glycerol metabolism
LysR	Dual	Amino acid biosynthesis
MarR	Repressor	Antibiotic resistance
MerR/SoxR	Dual	Heavy metal resistance
TetR/AcrR	Repressor	Antibiotic resistance

in their environment. First and foremost, the transcription factors regulate metabolic activities. Sugar metabolism, carbon uptake and metabolism, and amino acid biosynthesis are heavily represented in the list of cellular roles in Table 17.4. The second main function of the transcription factors is to regulate physiological responses to potentially harmful substances in the environment. Noteworthy in the table are transcription factors for arsenic resistance (ArsR), for mercury, copper, zinc, and cobalt resistance (MerR), and for tetracycline and other antibiotics resistance (TetR).

17.6 Many TFs Function as Response Regulators

Many transcription factors function as response regulators in two-component signal transduction systems. Two-component signal transduction systems are the main devices used by bacteria to sense their environments. The bacterial chemotactic system discussed in Chapter 6 is the prototype example of a two-component system. It is used to sense nutrients and noxious compounds in the local environment and relay these signals to a flagellar motor so that a bacterium can swim towards nutrients and away from harmful substances. More commonly, two-component systems are used to relay environmental information directly to sites on chromosomes and plasmids that regulate transcription.

Recall that two-component systems consist of a *sensor module* and a *response regulator module*. In two-component systems that regulate transcription the sensor module is a single transmembrane protein and the response regulator is a single cytoplasmic protein. The sensor, a histidine protein kinase, contains an N-terminal input unit and a C-terminal transmitter unit. The input unit responds to environmental stimuli by stimulating the autophosphorylation of the transmitter on a specific histidine residue. The high energy phosphoryl group is then transferred to the second component of the system, the response regulator. The response regulator contains an N-terminal regulatory domain and a C-terminal DNA binding domain. The N-terminal domain catalyzes the transfer of the phosphoryl group to a conserved aspartate residue in its own domain. The C-terminal output unit possesses an HTH DNA-binding domain that functions as a transcription factor. The output unit is activated in response to phosphorylation state-dependent conformational changes in the regulatory domain.

The second column of the Table 17.4 describes the regulatory function (action) of the transcription factors. Some TFs are *activators*: They are positive regulators that stimulate transcription at higher frequencies and/ or for greater durations. Other transcription factors are *repressors*: They are negative regulators that block or mechanically inhibit transcription. Still other transcription factors can act either as activators or as repressors: They are *dual regulators* and their specific actions will vary with cellular context.

Transcription activators bind to sites upstream of the −35 box. Some bind in the vicinity of the UP element while others bind to sites further upstream. Many of these transcription activators exert their influences by binding to αCTD. Because of its flexibility, αCTD can bind to activators over distances from −35 to −100 bp. Transcription factors that bind to distant upstream sites are often referred to as *transcription enhancers*. The DNA molecule itself has a certain intrinsic flexibility. Sequences such as A-T tracts are especially bendable. This flexibility plus bending brought on by protein binding enables activators to come into physical contact with subunits of the holoenzyme. *Transcription repressors* bind to sites close to the transcription start site. These proteins prevent the holoenzyme from initiating transcription either by sterically inhibiting access or by binding the holoenzyme so tightly that it cannot release and start transcription.

17.7 Organization of Protein-Encoding Regions and Their Regulatory Sequences

Bacterial DNA is organized into *operons*. Each operon contains DNA sequences that encode proteins and the upstream sequences for attachment of the RNA holoenzyme and regulatory proteins (Figure 17.2). Regulatory information is conveyed in several ways: It is imparted through the promoter recognition sequences that provide attachment sites for specific sigma factors: it is conveyed by degree of agreement between the promoter sequences and the consensus sequence, and by the length of the intervening spacer sequences, both influencing the strength of the transcription; and it is supplied by the presence of binding sites for transcription activators and repressors.

In contrast to transcription in eukaryotes, several genes can be transcribed from each promoter into an mRNA molecule. The protein-coding sequences follow one another and the protein-coding region terminates with a characteristic stop sequence that mediates the release of the polymerase from the DNA. Multiple gene transcription makes it possible for a promoter to regulate its own activity by implementing a simple negative feedback loop. Two kinds of proteins—*structural* and *regulatory*—can be

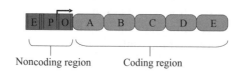

FIGURE 17.2. Bacterial operon: Shown are five genes, labeled A through E, controlled by a single gene regulatory region partitioned into repressor (O), core promoter (P), and enhancer (E) sections.

expressed at the same time. In negative feedback, regulatory proteins shut down transcription from their own promoters, thus providing a mechanism for automatically terminating transcription an appropriate time after the onset of transcription as determined by the threshold concentration. In many instances the regulatory proteins control the assembly of the cotranscribed structural proteins into membrane-bound structures.

Cassettes have the following properties. They are sets of operons that encode proteins performing a set of related functions. They are usually found on plasmids and enable the bacteria to respond to hostile conditions. The operons are positioned in the vicinity of one another, and because of their location on plasmids can be transferred from one bacterium to another. Some cassettes encode proteins involved in either antibiotic production or antibiotic resistance. Others, called *virulence cassettes* or *pathogenicity islands*, encode virulence factors such as toxins and proteins required for erection of morphological structures involved in the toxins' secretion and delivery to targets.

Regulons are sets of operons located at well-separated loci along the chromosome; they encode proteins involved in a common physiological response. Many regulons are organized by sigma factors. By binding to multiple sites along the chromosome they can either upregulate or downregulate 100 or more proteins. Examples of sigma factor associated regulons include the *E. coli* σ^S stationary phase regulon, the *E. coli* σ^{32} heat shock regulon, and the *B. subtilis* σ^B stress response regulon.

17.8 The Lac Operon Helps Control Metabolism in *E. coli*

Bacteria optimize their metabolism to take maximal advantage of their environments. Bacteria such as *E. coli* possess a broad range of metabolic strategies. They are able to metabolize a variety of sugars including glucose, lactose, and maltose, and make best use of the nutrients available in their environment. *E. coli* is the most studied bacteria and serves as a model system for how these organisms rapidly adjust their metabolism to environmental conditions. The genes involved in regulating whether glucose or lactose is metabolized form the *lac operon*. They are listed in Table 17.5, and their arrangement in the lac operon is illustrated in Figure 17.3.

TABLE 17.5. Genes belonging to the *E. coli* lac operon.

Gene	Function	Description
LacI	Repressor	Binds to transcription start site and part of the promoter, inhibiting RNAP
LacZ	β-galactosidase	Hydrolyzes lactose to glucose plus galactose
LacY	β-galactoside permease	Transports lactose into the cell
LacA	β-galactoside transacetylase	Nonessential enzyme

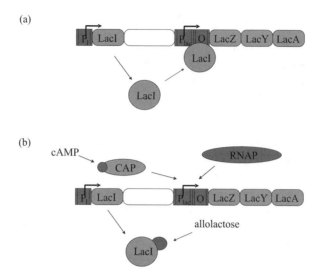

FIGURE 17.3. Organization of the lac operon: (a) Constitutive expression of the lac repressor protein (LacI) blocks transcription from the lac promoter when it binds to the operator. (b) Binding of LacI is blocked by allolactose allowing cAMP-bound CAP proteins to bind and stimulate transcription by the NRA polymerase (RNAP).

Glucose is the favored sugar and if it is plentiful the expression of proteins needed to metabolize alternative nutrients is inhibited. The LacI repressor is constitutively expressed and under plentiful glucose conditions binds downstream of the −10 and −35 sites and blocks RNAP from initiating transcription. When lactose is present in the environment, some of it will enter the cell and, in the form allolactose, will bind to the LacI repressor. The lactose binding induces structural changes in LacI and the repressor is no longer able to block RNAP.

The lac promoter is a weak one. The sequences at the −10 and −35 sites are not optimal for RNAP binding and the polymerase needs the binding activities of an activator, the *catabolite activator protein* (CAP), to efficiently transcribe the structural genes in the operon. CAP forms a complex with cAMP. When glucose levels are high the cAMP levels are low and the complex does not form. When glucose is absent, the cAMP levels are high enough for the CAP/cAMP complexes to form. The complexes bind at a site upstream of the RNAP binding site (CAP by itself does not), where they help RNAP to bind and start transcribing the genes. In sum, under the combined actions of the LacI (negative regulation) and CAP (positive regulation), the lac genes are transcribed when lactose is present and glucose is absent (Figure 17.3).

The structure of LacI is well suited for controlled interactions with DNA. As shown in Figure 17.4, two of its domains, the so-called Headpiece and

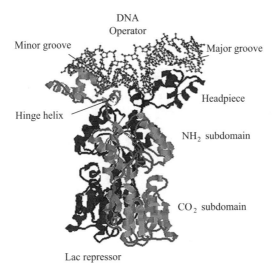

FIGURE 17.4. Crystal structure of the Lac repressor bound to a DNA operator determined using X-ray crystallography: Several domains of the Lac repressor are involved in the interactions. The headpiece domain binds to the DNA in the major groove while the Hinge helix binds DNA in the minor groove. The core domain is shown broken down into NH_2 and CO_2 subdomains that mediate interactions with inducers and repressors. One other domain, the tetramerization domain that mediates interactions with other LacI proteins, is not included in the figure, but can be seen in other crystal structures. The figure was generated using Protein Explorer with atomic coordinates deposited in the Protein Data Bank under accession code 1EFA.

Hinge helix domains, bind to the DNA in its major and minor grooves, respectively. The core domain enables the repressor to be regulated by allolactose, which induces allosteric transitions that prevent the protein from binding DNA.

Regulatory proteins such as CAP are able to bend DNA. This ability can be seen in Figure 17.5 where a CAP dimer is shown bound to its regulatory DNA regions. The bending of the DNA is quite pronounced, and by this means transcription factors and coactivators can make transcription either easier or harder.

17.9 Flagellar Motors Are Erected in Several Stages

Bacteria can alter their morphology, either building up or disassembling internal structures, such as gas vesicles, and membrane wall-attached structures, such as flagellar motors, pili, and virulence cassettes. It takes more than 50 genes to assemble a flagellar motor system. In species such as

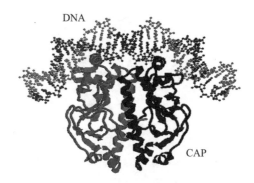

FIGURE 17.5. Crystal structure of DNA bending by a CAP dimer determined using X-ray crystallography: The figure was generated using Protein Explorer with atomic coordinates deposited in the Protein Data Bank under accession code 1RUN.

Escherichia coli and *Salmonella enterica,* these genes belong to a large flagellar motor regulon. The genes in this regulon are organized temporally into three classes—early, middle, and late (Figure 17.6).

A variety of cellular and environmental conditions are factored into the decision to begin biosysnthesis of the flagellar motor. The corresponding signals impinge on a master control point, namely the promoter for the flhCD master operon, from which the early genes flhC and flhD are transcribed. The transcriptional activation complex then regulates the transcription of middle genes. These genes encode structural proteins and assembly regulators for the basal body and hook of the flagella. The timing of events is carefully controlled during the assembly process. Two regulatory proteins are expressed during the middle phase. One of these, FlgM, binds to and inhibits the second, FliA (σ^{28}). The inhibition on σ^{28} by FlgM is relieved after the middle stage components of the flagellar motor have been assembled. The σ^{28} regulator then initiates transcription of the late genes including those for the chemotactic receptors Tar, Tsr and Aer, and the suite of Che proteins discussed in Chapter 6.

17.10 Under Starvation Conditions, *B. subtilis* Undergoes Sporulation

Under starvation conditions, *B. subtilis* undergoes a program of asymmetric cell division resulting in two daughters, a large "mother cell" and a small prespore that further develops into a spore that, in turn, aids in long term survival. When nutrients are plentiful *B. subtilis* enters a vegetative state where it multiplies exponentially. The vegetative growth sigma factor σ^A governs this state. Under starvation conditions, *B. subtilis* alters its meta-

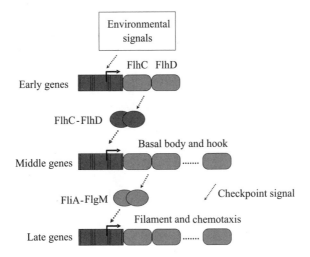

FIGURE 17.6. Assembly of the flagella motor: Environmental signals arriving at the master controller stimulate transcription of early (Class I) genes. FlhC and FlhD jointly stimulate transcription of middle genes from middle (Class II) promoters. FliA (σ^{28}) and FlgM (anti-σ^{28}) regulate transcription from late (Class III) promoters. Transcription of late genes is inhibited until completion of the basal body and hook assembly signified in the figure by the sending of a checkpoint signal.

bolic and reproductive strategy. It leaves the vegetative state and converts itself into an inactive spore that can survive for long periods of time. As indicated in Table 17.3 five sigma factors govern the transition from vegetative growth to spore. The first to come into play is σ^H. This sigma factor is followed in sequential fashion by σ^F, σ^E, σ^G, and finally σ^K. As the genes selected by these sigma factors are expressed the bacterium ceases exponential growth. Its plasma membrane invaginates, partitioning the cell into large and small compartments, each containing a chromosome generated during the last round of DNA replication. Once the septum is completed the large compartment, the mother cell, engulfs the small cell, the forespore, producing a cell within a cell. A series of coating stages follow where the forespore and its chromosome are protected, and lastly the spore is released through lysis of the mother cell; that is, the rupture of the plasma membrane/cell wall and the dissolution of the outer cell.

A number of signals, integrated together, govern the decision whether to undergo sporulation (Figure 17.7). These signals converge on the transcription factor Spo0A. This protein acts at the end of a phosphorelay system involving kinases such as KinA/KinB, a phosphorelay protein Spo0B and an upstream transmitter Spo0F. Signals sent through the phosphorelay, and associated protein phosphatases such as RapA, RapB, RapE, and Spo0E include metabolic information, DNA status, and cell density. These

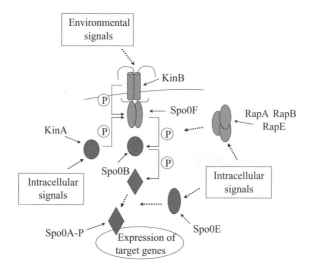

FIGURE 17.7. Integration of sporulation signals: Environmental and intracellular signals are integrated together into a phosphorelay consisting of at least two sensor (input) kinases (KinA and KinB) a Spo0F transmitter, Spo0B intermediate, and Spo0A response regulator, which functions as a transcription factor when phosphorylated. Negative regulatory signals are conveyed by several aspartyl phosphatases—Rap family members dephosphorylate Spo0F and Spo0E dephosphorylates Spo0A.

signals influence the phosphorylation status of Spo0A. The Spo0A transcription factor, along with KinA and Spo0F, is upregulated by σ^H. The Spo0A operon contains the genes for σ^F and σ^E, and starts the expression of spore and mother-specific genes.

17.11 Cell-Cycle Progression and Differentiation in *C. crescentus*

C. crescentus is a gram-negative bacterium adept at survival in a variety of aquatic environments. During every cell cycle it differentiates, producing as one offspring a motile swarmer cell equipped with a single flagellum at one pole, and as the second offspring an immobile (sessile) stalked cell possessing a stalk rather than a flagellum. While stalked cells are immediately competent to undergo cell division, the swarmer is not. It must first convert itself into a stalked cell. The swarmer possesses a single flagellum at one pole, and this flagellum is coupled to a chemotactic sensor system. During G1 phase, the swarmer can move about seeking nutrients using its chemotactic system. At the beginning of S phase (Figure 17.8), the swarmer degrades the flagellum and chemotactic sensor system. A stalk develops at the pole where the flagellum was sited, and the cell elongates. The DNA is

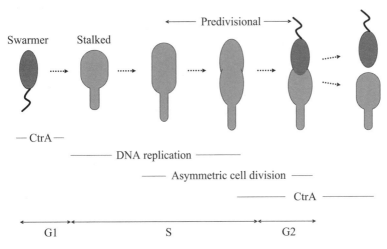

FIGURE 17.8. Asymmetric cell division in *Caulobacter crescentus:* A swarmer cell differentiates into a stalked cell, which then undergoes asymmetric cell division to produce a swarmer cell and a stalked cell. CtrA gene expression suppresses DNA replication. This master controller is shut down through regulated proteolysis at the end of G1 phase and is started up again in the new swarmer cell near the end of S phase.

replicated, the chromosomes move to opposite ends of the cell, and a new flagellum is formed at the other pole. In the next phase, G2, the cell divides into a swarmer cell and a stalked cell.

The key element in the regulatory network that coordinates the morphological changes with the cell cycle progression is CtrA. This regulatory protein is the counterpart to the Spo0A transcription factor discussed in the last section. The CtrA master cell cycle controller controls a large number of genes, in this case, about 25% of all the genes whose expression varies in a cell cycle-dependent way in *C. cresentus*. Cellular localization, proteolysis, and phosphorylation all have role in regulating the behavior of CtrA. The controller is expressed in the swarmer, but not in the stalker. The CtrA proteins that are present in the swarmer at the end of G1 phase undergo proteolytic destruction and thus are absent in the S phase starker. Pole proteins belonging to the flagellum and chemotactic sensor are also cleared by proteolysis. These and the subsequent regulatory processes are facilitated by a number of two-component signal transduction systems that localize to one or the other of the poles. One of these systems is DivJ/DivK. This two-component system facilitates the localization of CtrA to the pole that will produce a new flagellum and upon cell division will belong to the new swarmer. Another protein, the histidine kinase CckA phosphorylates CtrA after DNA replication takes place, thereby activating CtrA's transcriptional activities. In response, CtrA regulates the transcription of genes involved in flagellar biogenesis, cell division, DNA methylation, and DNA replication. In this way, localization, proteolysis, and phosphorylation pattern the spatial and temporal actions of CtrA, enabling it to coordinate morphogenic and cell cycle activities.

17.12 Antigenic Variation Counters Adaptive Immune Responses

Bacteria can up- and downregulate expression of their genes fairly rapidly and can move DNA about. One of the ways bacteria (and a number of single-celled eukaryotes) exploit such capabilities is through *antigenic variation*, the systematic alterations in antigens expressed on the cell surface. Antigenic variation is used by a number of vector-borne bacterial and protozoal pathogens to defeat the acquired immune response of vertebrates. The vectors are typically arthropods—tics, lice, flies, and mosquitoes. Well-studied pathogens include several species of *Borrelia* responsible for relapsing fever, African *Trypanosomas*, the protozoan causative agent of sleeping sickness, and *Plasmodium falciparum*, the protozoan responsible for malaria.

As discussed in Chapter 9, the acquired (adaptive) immune response takes several days to develop. It requires several steps: First there is antigen recognition, then communication and convergence of leukocytes to the site of infection, and finally clearance. In antigenic variation, the pathogen utilizes the differential expression and switching between members of a family of antigens among a population of clones. Multiple copies of the gene encoding the antigen, each differing somewhat in its recognition epitope, are maintained by the pathogen. By either differentially expressing different copies among different clones, or by switching from one copy to another more rapidly than the response time of the immune system, the pathogens evade recognition and simply wear out the immune system.

A variety of switching mechanisms are employed. Most are implemented at the transcription level. One method is to silence a promoter at one site and turn on expression at another. Alternatively, point mutations may be used, or a gene from a silent site may replace another at an active site of transcription through gene rearrangements. Still another method, one that does not operate at the transcription level, is to differentially transport the antigens to the surface. No matter what the mechanism, the result is the same. The population of pathogens survives.

17.13 Bacteria Organize into Communities When Nutrient Conditions Are Favorable

Bacteria talk to each other and share genetic material. Bacteria continually secrete communication molecules that are sensed by other bacteria. The signaling informs the bacteria that others are present in their vicinity and

TABLE 17.6. Bacteria and their collective behavior.

Bacterial species	Collective behavior
Gram-negative bacteria	
Photobacterium fischeri	Luminescence
Vibrio harveyi	Luminescence
Proteus mirabilis	Swarming, migratory rafts
Serratia liquefaciens	Swarming, migratory rafts
Myxococcus xanthus	Fruiting body
Chondromyces crocatus	Fruiting body
Pseudomonas aeruginosa	Antibiotic resistance, virulence factors
Escherichia coli	Virulence factors
Gram-positive bacteria	
Staphylococcus aureus	Virulence factors
Streptomyces coelicolor	Antibiotics, differentiation
Bacillus subtilis	Antibiotics, genetic competence
Streptococcus pneumoniae	Genetic competence

enables them to sense their numbers. If the density of bacteria exceeds a threshold for cooperative behavior then the bacteria express genes that support that form of behavior. Any of a number of collective behaviors may be exhibited, each depending on the species, its needs, and its environmental conditions. A sapling of these collective behaviors is presented in Table 17.6.

When nutrient conditions are favorable so that migration is not necessary bacteria organize into highly structured communities. In forming these associations the bacteria express sets of genes that differ from those utilized when free-living. The members of the bacterial communities differ in morphology and physiology from their planktonic counterparts. In communities that are anchored to solid surfaces, the bacteria can develop elaborate three-dimensional structures. These colonies take on characteristics usually associated with multicellular organisms. They exhibit simple forms of differentiation and cellular specialization. The bacteria excrete molecules that form a matrix, form channels that support the circulation of fluids, and share metabolic tasks. Some bacterial colonies are conical-shaped while others are mushroom-shaped and grow to sizes that allow the naked eye to observe them.

The first bacteria discovered to form communities were the luminescent marine organisms *Photobacterium fischeri* and *Vibrio harveyi*. These bacteria were found to form symbiotic associations with several species of marine fish and squids. The luminescent bacteria collect in the light organs of their marine hosts. The concentrations of the bacteria are generally low in seawater, typically on the order of 100 cells per milliliter, well below any threshold for luminescent behavior. However, the bacteria can reach high concentrations, 10^{10} to 10^{11} cells/ml under favorable nutrient conditions in the light organ. Once the threshold density is exceeded the bacteria collectively express genes for luminencence. The relationship between bacteria and animal host is a symbiotic one. The production of light by the bacteria masks the host from prey residing below them by reducing the host's shadow against the sky, while the hosts provide nutrients for the bacteria.

Table 17.6 lists several prominent forms of communal behavior. One of these is the production of antibiotics. These are natural byproducts of cellular metabolism. They are agents produced by microorganisms (fungi and bacteria) that kill or inhibit other microorganisms. Fungi produce penicillin and a closely related family of antibiotics, the cephalosporins. *Streptomyces* produce tetracycline, streptomycin, and erythromycin. These natural products interfere with protein synthesis and damage cellular membranes. *Bacillus subtilis* produces bacitracin, an antibiotic often added to animal feed.

Other widely utilized forms of coordinated behavior are the production of virulence factors and genetic competence. Virulence factors (e.g., toxins) are cellular products that enhance bacterial survival in hostile host environments. *Natural genetic competence* is the ability of bacteria to acquire large pieces of exogenous (extracellular) DNA from their local microenvironment.

17.14 Quorum Sensing Plays a Key Role in Establishing a Colony

Bacteria use cell-to-cell signaling to take a census of how many fellow bacteria are present in their local environment. The cell density-determining process is known as *quorum sensing*, and it operates in the following manner. The bacteria secrete signaling molecules called *autoinducers*. When the local environment is restricted in some manner so that the molecules do not simply diffuse away, the concentration of autoinducers will increase in a manner that reflects the number of bacteria secreting the molecules. The quorum sensing system responsible for secreting autoinducers and sensing their concentration is an integral part of the bacterial cell control system. Once the concentration exceeds a threshold level, a program of gene expression is initiated that leads to alterations in morphology and physiology, and supports communal behavior.

Many different kinds of signaling molecules are utilized in cell-to-cell communication. Some of these are species-specific while others enable the bacteria to communicate between members of different species. Gram-negative bacteria utilize short chain acetyl homoserine lactases (AHLs) to signal one another (Figure 17.9). These molecules are able to diffuse across the cell membranes and bind to cytoplasmic receptors of the recipient bacterium. Gram-positive bacteria utilize oligopeptide autoinducers (AIPs). Signaling by oligopeptides in gram-positive bacteria resembles signaling by peptide hormones in eukaryotes. The autoinducers do not diffuse across the

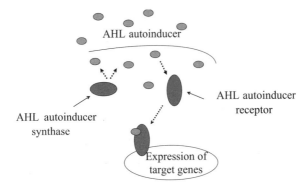

FIGURE 17.9. Quorum sensing in Gram-negative bacteria: One component of the system, the autoinducer synthase generates AHL signaling molecules that freely leave and enter bacterial cells. These are bound by autoinducer receptors, which can then stimulate transcription.

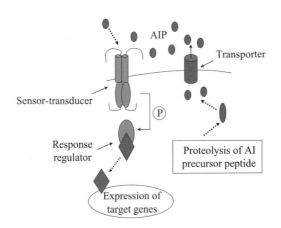

FIGURE 17.10. Quorum sensing in gram-positive bacteria: Signaling peptides are shipped out of the cell by means of a transporter. The signals (autoinducer peptides or AIPs) bind to bacterial two-component signals transduction systems, which transduce signals into the cell and stimulate gene transcription.

membranes. Instead, they are shipped out of the sender by ABC transporters, and diffuse over to and subsequently bind transmembrane receptors of the recipient. The latter receptors operate as the sensor components of two-component signal transduction systems (Figure 17.10).

In some species (e.g., *B. subtilis* and *S. pneumoniae*) quorum sensing triggers a bacterial state called *genetic competence* that is geared for the uptake of exogenous DNA. In others a threshold-exceeding density of cells induces a virulence response (*S. aureus*) and the expression of virulence genes (*P. aeruginosa*), or the formation of a biofilm, a colony of bacteria collectively exhibiting an increased resistance to antibacterial agents.

17.15 Bacteria Form Associations with Other Bacteria on Exposed Surfaces

Whenever possible, bacteria establish surface-attached communities known as *biofilms*. These structures can be formed by members of one species or by members of multiple species. Bacteria can freely switch between free-living and communal forms of organization, and communities can form, grow, collectively migrate, and disperse. When bacteria organize themselves into biofilms they become difficult to eradicate. The biofilms contain not only the bacterial cells but also a protective matrix of secreted molecules, secreted by the bacteria, that protects the members of the colony from harmful agents in their environment. *Actinomyces*, *Lactobacillus*, and *Streptococcus* films can appear as plaque on teeth; *Staphylococcus* and *Escherichia coli* films can emerge on medical instruments and surgical implants, and *Pseudomonas aeruginosa* and *Escherichia coli* colonies can form in the body.

Pseudomonas aeruginosa forms elaborate structures on surfaces. This highly adaptive species is an opportunistic pathogen, and is often encountered in the treatment of cystic fibrosis patients and burn victims. The formation of a biofilm protects the *P. aeruginosa* from antibiotic applications, increasing their resistance to the antibiotics by more than a hundredfold. These bacteria are highly resistant to antibiotic treatment due in part to the presence of their self-generated polymeric matrix that encases the colony and protects it from the actions of antibiotics.

Biofilms are also encountered in urinary tract infections. Most urinary tract infections are due to uropathogenic strains of *Escherichia coli* (UPEC). These bacteria are able to form podlike bulges on the surface of the bladder. Like the *P. aeruginosa* biofilms formed in lungs, the UPEC cells are encased in a protective shell that insulates them from antibiotics. They are able to persist as pods for several months undetected, and are utterly resistant to standard courses of antibiotics.

17.16 Horizontal Gene Transfer (HGT)

Bacterin share genetic information through horizontal gene transfer (HGT). Horizontal gene transfer differs from vertical gene transfer that takes place between parents and their offspring. In HGT genetic information is transferred between distantly related species. There are several different mechanisms for affecting the lateral transfer of genetic material. One of these is *conjugation*, in which two cells establish direct contact with one another. Sex pili are formed and genetic material in the form of a plasmid is sent from the donor bacterium to the recipient. This is perhaps the most common way of carrying out horizontal gene transfer. A second way of transferring genetic material is through *transduction*, where a bacteriophage picks up genetic material from one bacterium and delivers it to another. A third way of acquitting genetic material is through *transformation*, in which a bacterium becomes competent to capture exogenous DNA from its local environment as discussed in the previous section.

Resistance to antibiotics can be spread rapidly through horizontal gene transfer. One consequence of horizontal genetic transfer is that groups of genes that endow one bacterial species with antibiotic resistance can be transferred in minutes to hours to another species. The transfer of genetic material, that is, the sharing of blueprints for making useful proteins among bacteria, is a particularly potent adaptive response. It is far more efficient to simply transfer a complete set of genes than to have each species "invent" for itself the genes one by one by means of random mutations. In examining the complete genome sequences of these pathogens it becomes clear that entire regions of DNA have been acquired from other species in the past.

Antibiotics are agents synthesized by fungi and bacteria that kill competing microbes. There are several different kinds of antibiotics. Some antibiotics act on bacteria by causing damage to their cell walls. Other antibiotics impede the ability of the bacteria to carry out protein synthesis or to disrupt the replication of DNA or to interefere with metabolism. Penicillin-type antibiotics such as Ampicillin and Methicillin, cephalosporin antibiotics, and Vancomycin attack bacterial cells walls. Protein synthesis impeding antibiotics include tetracycline, aureomycin, streptomycin, and Erythromycin. Quinolones disrupt DNA replication, and sulfur drugs interfere with bacterial metabolism.

There are several ways of conferring resistance to antibiotics. One way is to synthesize biomolecules that destroys the antibiotics. Another approach is to pump the antibiotics rapidly out of the cell thereby keeping its intracellular concentration so low that it is ineffective. Still another way of dealing with the antibiotic is to synthesize cellular agents that bind antibiotics and neutralize them thereby acting as intracellular blockers.

17.17 Pathogenic Species Possess Virulence Cassettes

Most bacterial strains are harmless, but some can be deadly. In pathogenic strains additional proteins such as toxins and fimbriae components are typically encoded on plasmids and other mobile genetic elements that can be readily transferred and exchanged between species. It is these additional proteins that endow bacteria with their disease-causing capabilities. For example, the standard laboratory strain of *Escherichia coli K12* is harmless.

Its genome is 4.6 MBp long, while the pathogenic strain of *E. coli O157* is considerably larger—5.4 MBp in length. In *E. coli O157* there are "O" islands and "K" islands, clusters of genes that encode a total of 1387 proteins not found in the *K12* strain. The additional proteins encoded by the *E. coli O157* strain endow the bacterium with enterohaemorrhagic abilities.

One of the most striking findings from sequencing complete genomes of microbes is that bacterial genomes are riddled with phage genes. It appears that bacteriophages are a major source of mobile genetic material moving in and out of bacteria. In many species, pathogenicity and virulence of the bacterial strain are tied to the presence of a prophage, the inactive (lysogenic) bacteriophage whose genes have been integrated into the bacterial genome. An example of this coupling is provided by *group A Streptococcus* (GAS) strains. These bacteria promote sore throats, impetigo, acute rheumatic fever (the major heart disease of children), necrotizing fasciitis (the disease caused by flesh-eating bacteria), and toxic shock syndrome. The various strains (serotypes) of GAS are given "M" designations according to differences in a family of streptococcus cell surface proteins called *M proteins*, which impede the engulfment by phagocytes. The M18 strain, responsible for acute rheumatic fever contains phage genes that encode the disease-causing toxins. Another strain, M3, which produces toxic shock syndrome and necrotizing fasciitis, has a different set of phage genes from the M18 strain, and the strain most often responsible for sore throats, M1, has yet another set of phage genes.

Pathogenicity islands, or *virulence cassettes*, are clusters of functionally related virulence genes that are absent in nonpathogenic organisms. The term "virulence" denotes the ability to cause disease. A virulence factor is the designation given to an individual disease-causing or disease-promoting agent of an organism. Some examples of virulence factors are toxins, surface adhesion molecules, control molecules, and virulence regulatory proteins (transcription factors). Virulence cassettes, consisting of ensembles of these proteins, may be located on plasmids or on other mobile genetic elements that can be transferred from one bacterium to another. They may be found on the bacterial chromosome, as well. In all cases they encode a complete suite of proteins needed to promote virulence.

The virulence cassettes of enteropathogenic *E. coli* (EPEC) strains allow them to build their own signal transduction pathway in a host cell, which they use to remodel the host's cytoskeleton. The bacteria first send soluble proteins into the host cell membrane. The proteins are then modified to form physiologically functional receptors for its bacterial ligands, thereby establishing a firm contact between cells. Upon appropriating the communication machinery the bacterium uses the host's resources to erect a pedestal on the host cell's surface. This pedestal supports the subsequent assembly of a bacterial colony.

Virulence cassettes are used to transfer agents from bacterium to host. The most common forms, known as *Type III and IV secretion systems*, are encountered in many pathogenic bacterial species. They are similar enough to one another to be interchangeable and to have been acquired by horizontal transfer from species to species. The best characterized system of this sort is the *Yop virulon*, a Type III system encoded on a plasmid. It contains (i) a secretion apparatus, (ii) a delivery system that sends bacterial proteins into the eukaryotic host, (iii) a control unit, and (iv) an agent effector module that disarms the host signal transduction network. The objective of *Yersinia* is to survive and replicate in the lymphoid tissue of its host. The prime danger to survival is from phagocytes produced by the host organism's immune system. The virulons inactivate the phagocytes by activating the phagocytes' apoptosis (suicide) circuitry. *Yersinia* avoids attracting the attention of the immune system. The secretion apparatus and delivery system is activated by cell-to-cell contact and inserts agents into the target in a highly directional way so that there is little or no leakage into the intercellular spaces.

17.18 Bacterial Death Modules

Bacteria employ death modules to kill competing bacteria and in some instances to commit suicide. Bacteria are able to synthesize and release toxins without themselves being harmed. To accomplish this feat the bacteria synthesize the antidote to the toxin at the same time that they make the toxin. The genes for both toxin and antidote, or antitoxin, usually reside on a plasmid and are transcribed from the same promoter (Figure 17.11). The toxin is long-lived while the antidote is unstable and short-lived. As long as the plasmid is present and active both toxin and antidote get made and the bacterium is protected against the former.

Toxin-antidote pairings can serve as the core element in suicide modules. As is the case for self-protection, a long-lived toxin is paired with a short-

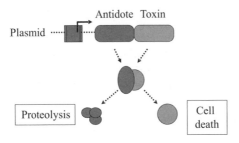

FIGURE 17.11. Bacterial death modules: The genes for toxin and antitoxin (antidote) reside side by side on a plasmid. When the genes are expressed the antidote binds the toxin and prevent it from binding to its substrate. In response to suicide signals, proteolytic enzymes degrade the antidote; the toxin then binds to its substrate leading to cell death.

lived antidote that antagonizes the activity of the toxin. If the plasmid is disabled the antidote will not be made, and because it has a short half-life the antidote will decay away rapidly. When the antidote is removed either by loss of the plasmid or by some other means such as proteolytic cleavage, the toxin is free to degrade its substrate, usually an enzyme critical for cellular survival, and the host cell dies. It forming colonies, bacteria sometimes use death modules of this type to selectively remove a potion of their population. This may be viewed as a bacterial form of programmed cell death.

17.19 Myxobacteria Exhibit Two Distinct Forms of Social Behavior

Myxobacteria are gram-negative, aerobic soil bacteria that feed on decomposing vegetation. When nutrient supplies are plentiful species such as *M. xanthus* form feeding swarms that glide together over solid surfaces and prey on other bacteria. The members of a swarm are free to come and go, explore other feeding sites, and establish colonies. This type of social behavior changes when nutrients are no longer available. Under starvation conditions the bacteria organize into a differentiated hemispherical structure called a *fruiting body*. Rod cells inside the fruiting body differentiate into spores that are subsequently dispersed. When these spores encounter a hospitable environment they reestablish a feeding colony of gliding bacteria.

The transition from feeding swarm to fruiting body is initiated by starvation signaling. When bacteria are deprived of essential nutrients they reduce or halt protein synthesis, DNA replication, and cell division and enter a stationary phase. Any such alteration in cellular physiology is referred to as a *stringent response*. One of the main elements of the stringent response is the accumulation of the highly phosphorylated, guanosine (penta)tetraphosphate in the cell. In bacteria, RelA (a synthase) and SpoT (a hydrolase) synthesize pppGpp and ppGpp from GTP and GDP using ATP as a donor. These nucleotides serve as intracellular second messengers of starvation conditions. Under adequate nutrient conditions these molecules are maintained in the bacterial cell at low, steady state concentrations. Starvation conditions, in the case of *M. xanthus*, a lack of amino acids or carbon compounds supplied by prey, stimulate an increased ppGpp production by RelA. The ppGpp molecules bind and alter the activity of RNA polymerase shifting the global pattern of gene expression from growth to a stationary phase. In *Streptomyces coelicolor*, the stringent response triggers an increase in antibiotic production and morphological differentiation. In *M. xanthus*, the buildup in ppGpp upregulates several asg genes, one result of which is to release proteases that generate a mixture of amino acids and peptides collectively referred to as *A-factors*. The A-factors are secreted from the myxobacteria and act as quorum-sensing signals. If the concentration of A-factors is adequate, fruiting body formation proceeds.

A-signaling in *M. xanthus* is followed by C-signaling that determines which cells in the nascent fruiting body are fated to become spores and which cells will retain their original rod shape. Transmission of C-signals occurs between cells that are in close, end-to-end contact. Transmission is ineffective when neighboring cells lie side by side. The circuitry involved in C-signaling closely resembles that used in Notch signaling in eukaryotes. C-signals transmitted between ends of cells activate FruA, a transcription factor that stimulates the expression of the csgA gene that encodes the C-signals. This positive feedback loop increases FruA activity above thresholds for its activation of frz genes and then dev genes. The frz genes encode proteins that mediate the aggregation of the cells into mounds, and the dev genes encode proteins that promote the differentiation of rod cells into spores. Because of the geometric constraint on C-signaling, cells in the periphery never experience threshold-exceeding concentrations of FruA while those in the nascent fruiting body reach critical signal densities.

17.20 Structure Formation by Heterocystous Cyanobacteria

Heterocystous cyanobacteria form differentiated structures with properties usually associated with multicellular eukaryotes. *Cyanobacteria* are phototrophs and, like plants, are capable of oxygenic photosynthesis in which CO_2 is removed and O_2 is produced. These organisms have minimal requirements for survival—water, light, CO_2, and some salts that supply N_2, iron, phosphorus, and sulfur. For many years the cyanobacteria were referred to as "blue-green algae," but are now recognized as being Gram-negative bacteria. Ancestral cyanobacteria first appeared some 3.5 billion years ago and are believed to be responsible for the oxygenation of the Earth's atmosphere. Photosynthesis in cyanobacteria is carried out in thylakoid membranes in the same way as it is done in plant cells. Similarities in size and structure, the presence of chlorophyll A, photosystems I and II, and similarities in control elements provide evidence that present day chloroplasts and cyanobacteria share a common ancestry.

Cyanobacteria are an essential part of the world's ecology. Not only do they release oxygen but they also fix nitrogen (and carbon). Most organisms cannot make use of atmospheric nitrogen. Cyanobacteria are able to fix nitrogen, that is, they can convert atmospheric nitrogen into inorganic nitrogen that can be used for biological purposes by other organisms. They are a main component of marine plankton, the dense collection of (mostly) small drifting organisms that cover the world's oceans and produce 90% of the planet's oxygen. They are also found in large numbers in the surface regions of lakes, are a major component of communities known as *micro-*

bial mats found in the hot springs of Yellowstone National Park, and populate nearly all ecosystems.

Members of genera such as *Calothrix* and *Anabaena* can form differentiated structures. The structures take the form of long filaments in which nitrogen-fixing heterocysts are interspersed at regular intervals among photosynthetic vegetative cells. The heterocysts reduce N_2 to ammonia but do not carry out photosynthesis or carbon fixation. The vegetative cells, on the other hand, perform photosynthesis and fix carbon. The technical problem being solved through multicellularity is the need to separate photosynthesis from the nitrogen-fixing due to the inactivation by oxygen of the nitrogenase complex responsible for nitrogen-fixing. The heterocysts possess a protective envelope and turn off their photosystem II. They export inorganic nitrogen to their neighbors and import nutrients. About 1 in 10 to 20 cells in a filament is a heterocyst. Cell-to-cell signaling is used to establish and maintain the pattern of cell differentiation just as it does in multicellular eukaryotes. Most strikingly, the heterocysts are terminally differentiated cells, like those in metazoan tissues.

17.21 Rhizobia Communicate and Form Symbiotic Associations with Legumes

Rhizobia is the name given collectively to several genera of rod-shaped, gram-negative bacteria that fix nitrogen and make it available for use by legumes. In poor soils they provide a major source of fixed nitrogen to grain legumes such as soybeans, peas, and beans, as well as to forage legumes such as alfalfa and clover. Rhizobia are free-living soil bacteria that enter and colonize plant root cells. They stimulate the formation of nodules, specialized root organs, and in the nodules they convert atmospheric nitrogen into ammonia and export it to their hosts, and receive carbon compounds in exchange.

The process whereby the bacteria and host plant establish their symbiotic association involves several stages of signaling between bacteria and plant root, rounds of gene expression, and differentiation and cellular/tissue remodeling. To initiate the process, the plant roots exude signaling compounds called *flavinoids* that attract the bacteria to the root surface. The rhizobia attach to young root hairs and signal the root cells by secreting glycolipids called nod factors. In response to nod factors, the plant cells starts to remodel their actin and microtubule cytoskeletons and form nodules. First, the bacteria become entrapped between cell walls of stimulated (deformed) root hairs, and the plant cell walls undergo hydrolysis resulting in lesions. New cell wall material is then deposited resulting in the formation of a tubule called an *infection thread* that is filled with rhizobia. The infection threads are the doorways to the root cells. The infection thread network expands into the root cortex while the bacteria multiply. Eventually, the bacteria exit the network of threads and enter the cytoplasm of the root cells. A nodule is formed containing a plant cell-derived perib-acteroid membrane. This membrane envelops the bacteria, which multiply and differentiate into bacteroids. In some symbiotic associations the bac-teroids differ greatly in morphology compared to their free-living forms, becoming several times larger and Y-shaped in form. Within the nodules the bacteroids and legumes exchange metabolites, which pass through pores and channels in the peribacteroid and bacteroid membranes.

Initial plant bacteria-signaling events not only launch the nodule-forming process but also confer specificity and filter out potentially harmful associa-tions from helpful symbiotic ones. Flavinoids secreted by the plants bind to NodD proteins situated in the cytoplasmic membrane of the bacteria. The NodD proteins are members of the LysR family of transcription factors. They serve as sensors of plant signals and as activators of genes whose promoters contain a 49 Bp sequence called a *nod box*. Binding by favinoid-NodD com-plexes results in the activation of the NodA, NodB, and NodC genes common to all rhizobia. They encode enzymes that catalyze the synthesis of the lipooligosaccharide core, or backbone, of the nod factors. Structural nod genes that are species-specific are expressed at the same time. These genes encode enzymes that modify or decorate the nod factor backbones in a manner that depends upon the structure of the flavinoids.

References and Further Reading

Gilmore MS, and Ferretti JJ [2003]. The thin line between gut commensal and pathogen. *Science*, 299: 1999–2002.

Whitman WB [1998]. Prokaryotes: The unseen majority. *Proc. Natl. Acad. Sci. USA*, 95: 6578–6583.

Xu J, and Gordon JI [2003]. Honor thy symbionts. *Proc. Natl. Acad. Sci. USA*, 100: 10452–10459.

Gene Organization

Busby S, and Ebright RH [1994]. Promoter structure, promoter recognition and transcription activation in prokaryotes. *Cell*, 79: 743–746.

Ptashne M, and Gann A [1997]. Transcription activation by recruitment. *Nature*, 386: 569–577.

Sigma Factors

Cannon WV, Gallegos MT, and Buck M [2000]. Isomerization of a binary sigma-pro-moter DNA complex by transcription activators. *Nature Struct. Biol.*, 7: 594–601.

Gralla JD [2000]. Signaling through sigma. *Nature Struct. Biol.*, 7: 530–532.

Sharp MM [1999]. The interface of σ with core RNA polymerase is extensive, con-served, and functionally specialized. *Genes Dev.*, 13: 3015–3026.

Flagellar Assembly

Aldridge P, and Hughes KT [2002]. Regulation of flagellar assembly. *Curr. Opin. Microbiol.*, 5: 160–165.

Chilcott GS, and Hughes KT [2000]. Coupling of flagellar gene expression to flagellar assembly in *Salmonella enterica, Servovar typhimurium.* and *Escherichia
coli. Micro. Mol. Biol. Rev.*, 64: 694–708.
Kalir S, et al. [2001]. Ordering genes in a flagella pathway by analysis of expression
kinetics from living bacteria. *Science*, 292: 2080–2083.

Sporulation

Burbulys D, Trach KA, and Hoch JA [1991]. Initiation of sporulation in *B. subtilis*
is controlled by a multicomponent phosphorelay. *Cell*, 64: 545–552.
Fabret C, Feher VA, and Hoch JA [1999]. Two-component signal transduction in
Bacillus subtilis: How one organism sees its world. *J. Bacterial.*, 181: 1975–1983.
Jiang M, et al. [2000]. Multiple histidine kinases regulate entry into stationary phase
and sporulation in *Bacillus subtilis*. *Mol. Microbiol.*, 38: 535–542.
Perego, M, Glaser P, and Hoch JA [1996]. Aspartyl-phosphate phosphatases deactivate the response regulator components of the sporulation signal transduction
system in *Bacillus subtilis*. *Mol. Microbiol.*, 19: 1151–1157.

Cell Differentiation in Caulobacter

Domain IJ, Quon KC, and Shapiro L [1997]. Cell type-specific phosphorylation and
proteolysis of a transcriptional regulator controls the G1-to-S transition in a bacterial cell cycle. *Cell*, 90: 415–424.
Jenal U [2000]. Signal transduction mechanisms in *Caulobacter crescentus* development and cell cycle control. *FEMS Microbiol. Revs.*, 24: 177–191.
Laub MT, et al. [2000]. Global analysis of the genetic network controlling a bacterial cell cycle. *Science*, 290: 2144–2148.
Quon KC, et al. [1998]. Negative control of bacterial DNA replication by a cell cycle
regulatory protein that binds at the chromosome origin. *Proc. Natl. Acad. Sci.
USA*, 95: 120–125.
Shapiro L, McAdams HH, and Losick R [2002]. Generating and exploiting polarity
in bacteria. *Science*, 298: 1942–1946.

Antigenic Variation

Barbour AG, and Restrepo BL [2000]. Antigenic variation in vector-borne
pathogens. *Emer. Infect. Dis.*, 6: 449–457.

Quorum Sensing

Davies DG, et al. [1998]. The involvement of cell-to-cell signals in the development
of a bacterial biofilm. *Science*, 280: 295–298.
Fuqua WC, Winans SC, and Greenberg EP [1994]. Quorum sensing in bacteria:
The LuxR-LuxI family of cell density-responsive transcriptional regulators.
J. Bacteriol., 176: 269–275.
Kleerbezem M, et al. [1997]. Quorum sensing by peptide pheromones and two-com-
ponent signal transduction systems in Gram-positive bacteria. *Mol. Miocrbiol.*, 24:
895–904.
Schauder S, and Bassler BL [2001]. The language of bacteria. *Genes Dev.*, 15:
1468–1480.
Van Delden C, and Iglewski BH [1998]. Cell-to-cell signaling and *Pseudomonas
aeruginosa* infections. *Emer. Infect. Dis.*, 4: 551–560.

Biofilms and Disease

Anderson GG, et al. [2003]. Intracellular bacterial biofilm-like pods in urinary tract infections. *Science*, 301: 105–107.

Costerton JW, Stewart PS, and Greenberg EP [1999]. Bacterial biofilms: A common cause of persistent infections. *Science*, 284: 1318–1322.

Singh PK, et al. [2000]. Quorum sensing signals indicate that cystic fibrosis lungs are infected with bacterial biofilms. *Nature*, 407: 762–764.

Horizontal Gene Transfer, Antibiotic Resistance, and Bacteriophages

Casjens S [2003]. Prophages and bacterial genomics: what have we learned so far? *Mol. Microbiol.*, 49: 277–300.

Frankel G, et al. [1998]. Enteropathogenic and enterohaemorrhagic *Escherichia coli*: More subversive elements. *Mol. Microbiol.*, 30: 911–921.

Hacker J, Hentschel U, and Dobrindt U [2003]. Prokaryotic chromosomes and disease. *Science*, 301: 790–793.

Karaolis DKR, et al. [1999]. A bacteriophage encoding a pathogenicity island, a type-IV pilus and a phage receptor in cholera bacteria. *Nature*, 399: 375–379.

Koch AL [2003]. Bacterial walls as target for attack: Past, present and future research. *Clin. Microbiol. Rev.*, 16: 673–687.

Lowy FD [2003]. Antimicrobial resistance: The example of *Staphylococcus aureus*. *J. Clin. Invest.*, 111: 1265–1273.

Ochman H, Lawrence JG, and Groisman EA [2000]. Lateral gene transfer and the nature of bacterial innovation. *Nature*, 405: 299–304.

Wagner PL, and Waldor MK [2002]. Bacteriophage control of bacterial virulence. *Infect. Immun.*, 70: 3985–3993.

Virulence Cassettes and Factors

Aizenman E, Engelberg-Kulka H, and Glaser G [1996]. An *Escherichia coli* chromosomal "addiction module" regulated by 3′,5′-bispyrophosphate: A model for programmed bacterial cell death. *Proc. Natl. Acad. Sci. USA*, 93: 6059–6063.

Cornelis GR, and Wolf-Watz H [1997]. The *Yersinia* Yop virulon: A bacterial system for subverting eukaryotic cells. *Mol. Microbiol.*, 23: 861–867.

Groisman EA, and Ochman H [1996]. Pathogenicity islands: Bacterial evolution in quantum leaps. *Cell*, 87: 791–794.

Death Modules and Programmed Cell Death

González-Pastor JE, Hobbs EC, and Losick R [2003]. Cannibalism by sporulating bacteria. *Science*, 301: 510–513.

Webb JS, et al. [2003]. Cell death in *Pseudomonas aeruginosa* biofilm development. *J. Bacteriol.*, 185: 4585–4592.

Problems

17.1 Nonfilamentous cyanobacteria solve the problem of separating photosynthesis from nitrogen-fixing in a way different from heterocystous cyanobacteria. The nonfilamentous forms carry out photosynthesis during the day and fix nitrogen at night. The separation of metabolic tasks is made possible by the presence of an internal clock that synchronizes the organism's metabolic processes with Earth's 24-hour rotation (circadian) period.

Circadian clocks are found in plants where they regulate the opening and closing of plant leaves with respect to the daily cycle of light and dark. They are found in protists, fungi, and animals, as well. All share a common set of properties such as the use of feedback loops and the introduction of time delays. How might a circadian clock be set up in a bacterium? How might the mechanism(s) used to fix the period and phase differ from those employed by eukaryotes?

Suggested Reading

Dunlap JC [1999]. Molecular bases for circadian clocks. *Cell*, 96: 271–290.

Ishiura M, et al. [1998]. Expression of a gene cluster KaiABC as a circadian feedback process in cyanobacteria. *Science*, 281: 1519–1523.

Xu Y, Mori T, and Johnson CH [2003]. Cyanobacterial circadian clockwork: Roles of KaiA, KaiB and the KaiBC promoter in regulating KaiC. *EMBO J.*, 22: 2117–2126.

Young MW, and Kay SA [2001]. Time zones: A comparative genetics of circadian clocks. *Nature Rev. Genet.*, 2: 702–715.

17.2 Some bacteria swim from place to place using a rotating flagellum at the back end to provide a driving force in the fluid medium. Swimming is not the only method exploited by bacteria for motility. Another means of locomotion is gliding motility. Social bacteria use this type of motion, sometimes called "twitching motility," quite often. Describe the components of a locomotion system that might allow a bacterium to glide along a surface.

Suggested Reading

Wall D, and Kaiser D [1999]. Type IV pili and cell motility. *Mol. Microbiol.*, 32: 1–10.

Wolgemuth C, et al. [2002]. How myxobacteria glide. *Curr. Biol.*, 12: 369–377.

18
Regulation by Viruses

Viruses consist of nucleic acid cores encased in protein sheaths. The protein sheaths surrounding and protecting the nucleic acid cores are referred to as *capsids*. Viral capsids enclose either DNA or RNA. As might be expected, the viruses belonging to the former class are known as *DNA viruses*, while members of the latter group are called *RNA viruses*. Some RNA viruses covert their RNA to DNA, and from there pass through both a transcription and a translation stage to make new viruses. These viruses are called *retroviruses*. Other RNA viruses do not convert their RNA to DNA. Instead, their RNA functions as messenger RNA molecules that are used directly in a translation stage to synthesize proteins.

Many viruses use lipids and carbohydrates acquired from their hosts to make an envelope. They encase their caspsids in lipids and carbohydrates picked up from plasma membrane or from organelles such as the nucleus, ER, or Golgi apparatus. These viruses are referred to as *enveloped viruses*. Glycoproteins are an important component of the viral envelopes. These proteins mediate contact with and entry into the host cell through their ability to bind to receptors on the host's plasma membrane.

Viral genomes vary greatly in size. Some, like those of the picornaviruses, are small, less than 10 kBp, and encode on the order of 10 proteins. Others, such as poxvirus genomes, are large. They range in size from 130 to 300 kBp and encode ~200 proteins, approaching those of the mycoplasmas in size of their genomes. In all cases, the viruses require assistance and resources from the host cell to replicate and propagate. For that reason their genomes encode not only genes of a structural character that encode proteins for the capsid and envelope, but also genes involved in replication and genes of a signaling and regulatory character. The viral signaling and regulatory genes encode proteins that regulate viral replication, disrupt and disable host immune/protective responses, and interact with cellular systems in order to use them for transport and replication.

In the first part of the chapter the viral cycle of entry, replication, and exit will be examined. This discussion will be followed by a general sketch of some of the ways viruses deal with the immune system. The remainder of the chapter is devoted to detailed discussion of how signaling and regulatory proteins make possible the survival of specific viruses. The hepatitis C virus responsible for chronic hepatitis, cancer-causing viruses, human immunodeficiency virus (HIV)—the causative agent of acquired immunodeficiency syndrome (AIDS), and bacteriophages will be examined.

18.1 How Viruses Enter Their Host Cells

Viruses enter their host cells using receptors and the cellular machinery for membrane fusion and endocytosis. Viruses cannot pass directly through the plasma membranes of cells. Instead they possess proteins in their exposed surfaces that bind to receptors on host cells triggering their ingestion into the cell. Enveloped viruses, for example, contain proteins in their envelopes that distinguish the correct host cells from non-host cells and initiate the process whereby the virus gains entry to the cell interior. There are two main ways of entering the host cell. In endocytic entry, the virus enters the cell in vesicles associated with clathrin-coated pits and caveolae. In this mode of entry, the viral nucleic acid and the capsid are enclosed in a transport vesicle that is pinched off from the plasma membrane. In the nonendocytic mode of entry, the viral envelope fuses with the plasma membrane leading to release of the nucleic acid encased in its capsid into the cytoplasm.

In order to replicate, viruses must first gain entry into the cell, and then disassemble, replicate DNA, synthesize components, reassemble, and exit. Viruses that use eukaryotic hosts for their replication have special challenges. They must navigate in an environment containing a dense cytoskeleton and numerous organelles partitioned into many subcompartments. The viruses, even small ones, are too big to traverse intracellular distances by means of passive diffusion, but instead must rely on the cell's machinery for directed active transport involving the use of the cytoskeleton and transport vesicles. Viral capsids and viruses packaged in transport vesicles attach to *dynein* (and kinesin) *motors* and move along microtubules to the nucleus. The viruses first move along actin fibers and then along microtubules to reach their sites of replication, and once replicated and assembled move again along the cell's active transport highway to reach the cell periphery.

18.2 Viruses Enter and Exit the Nucleus in Several Ways

Viruses that replicate in the nucleus must pass through the nuclear pore complex, first to get into the nucleus and then again to leave. In some cases, the viruses travel back and forth, in and out of the nucleus several times. Viruses exploit several mechanisms for passage through the nuclear pore complex, their choice depending on the virus' type and size. Virus particles of diameter less than 30 to 40 nm are small enough to pass through the NPC intact, without dissolution of the capsid, but larger viruses cannot pass through without some rearrangement and capsid disassembly. These larger viruses must either wait for dissolution of the nuclear envelope during mitosis, or they must at least partially disassemble to pass through the NPC during interphase.

Proteins that are actively transported into the nucleus are referred to as being *karyophilic*. These proteins rely on importins and Ran GTPases for passage through the NPC in the manner discussed in Chapter 11. These proteins have nuclear localization signals that interact with importins α and β. Many viruses enter and leave the nucleus in the same manner as cellular proteins. In the case of small viruses that can pass through the NPC intact, the capsids contain proteins bearing nuclear localization signals (NLSs).

Herpes simplex virus Type 1 (HSV-1) is an example of a virus that is too large to pass into the nucleus without disassembly. HSV-1 is an enveloped virus that fuses with the plasma membrane of the host cell. The intact capsid is released into the cytoplasm and is transported to the nucleus where it docks. In order to gain entry to the nucleoplasm of its nondividing host it expels its DNA directly into the nucleoplasm in a spring-loaded-like fashion, leaving its icosahedral 125-nm wide capsid outside the nucleus.

This is not the only mechanism utilized by large viruses to get their DNA into the nucleus. In the case of adenoviruses such as Ad2, the capsids are first disassembled and then the DNA is transported through the NPC into the nucleus. This is accomplished with the use of NLSs resident on proteins that chaperone the DNA through the NPC. Viruses such as HIV-1 utilize yet another mechanism. These viruses are able to bind directly to the nuclear pore, and their DNA can enter the nucleus without requiring the assistance of importins.

18.3 Ways that Viruses Exit a Cell

Viruses forms buds in endosomal vesicles that are transported to the plasma membrane. Viruses are able to exit a cell in at least two different ways. They may exit one at a time through membrane fusion, or they may egress in mass through cell lysis. They also may leave by some other less common means. Enveloped viruses acquire their envelopes from the host cells by *budding*. They may bud off of an internal membrane or generate a bud from the plasma membrane. These viruses typically exit the cell through membrane fusion where one virus at a time fuses with the plasma membrane and releases the virus into the extracellular spaces. The second common means for release, *cell lysis*, is typically associated with nonenveloped viruses. These viruses are released in large numbers all at once in a process that causes the rupture of the plasma membrane and death of the host cell.

In order to exit the cell, virus particles interact with the cellular machinery responsible for *multivesicular body* (MVB) biogeneisis. MVBs are a type of endosome having a multivesicular appearance; they play a role in receptor internalization, protein sorting operations, and viral budding. In their role as devices for internalization and protein sorting, the MVBs transport receptors from the plasma membrane to lysosomes for either degradation or recycling back to the plasma membrane. These transport and sorting operations are referred to as the *MVB sorting pathway*.

The proteins that participate in the vascular protein-sorting in MVBs are called *class E Vps's*. The reason for this name is that, in the absence of vascular protein-sorting (Vps) proteins, abnormal endosomes are formed called *class E compartments*. The class E Vps proteins form complexes called *endosomal-sorting complexes required for transport* (ESCRTs). In humans, there are four distinct ESCRT complexes. Members of these complexes interact with the exiting viruses through *late* (L) *domains*. Late domains are found in virus gag proteins in some viruses and in matrix proteins in others (virus gag and matrix proteins will be discussed shortly). These L domains mediate interactions with the ESCRTs. The motifs found in L domains are protein-protein interaction domains. The three most common binding motifs are PXXP, PPXY, and YXXL. The first of these binds to proteins with SH3 domains; the second binds proteins possessing WW domains, and the third provides attachment for clathrin-associated adapter proteins. The last named motif is often found in cytoplasmic tails of proteins involved in sorting and shipping.

18.4 Viruses Produce a Variety of Disorders in Humans

Viruses are responsible for colds and infections, immunodeficiency disorders, and cancers. Viruses that infect humans include both DNA and RNA viruses. One of the main families of DNA viruses infecting humans is the *herpesvirus* (Herpes simplex virus, or HHV) family. This family of viruses is a large one and includes the Epstein-Barr virus (HHV-4) and Kaposi's sarcoma associated herpes (HHV-8). Another common DNA family is the *adenovirus* family responsible for producing respiratory infections, and still another is the *poxvirus* family (vaccinia) responsible for smallpox and cowpox. Several families of RNA viruses cause diseases in humans, among which are colds, hepatitis, influenza, measles, mumps, polio, rabies, and AIDS. Members of the most prominent families of viruses and the diseases they cause are listed in Table 18.1. All of these viruses have been studied extensively in the laboratory along with viruses that infect bacteria (bacteriophages), plants, and arthropods (e.g., baculoviruses).

Viruses differ from bacteria in the way they damage hosts: Bacteria release toxins while viruses kill their host cells when they escape from them. In spite of differences in damage-causing mechanisms, both kinds of

TABLE 18.1. Viruses and some of the diseases they cause: Abbreviations—Reverse transcribing (RT); single-strand negative sense (–ss); single-strand positive sense (+ss); double strand (ds).

Virus family	Type of virus	Diseases
Adenovirus	DNA	Respiratory infections
Herpesvirus	DNA	Chickenpox, cold sores, encephalitis, genital sores, Kaposi's sarcoma, mononucleosis, roseola, shingles
Papovavirus	DNA	Warts
Poxvirus	DNA	Cowpox, smallpox
Calicivirus	+ssRNA	Norwalk
Filovirus	–ssRNA	Ebola, Marburg
Flavivirus	+ssRNA	Yellow fever, hepatitis C
Orthomyxovirus	–ssRNA	Influenza A and B
Paramyxovirus	–ssRNA	Measles, mumps, respiratory, colds (children), pneumonia (adults)
Picornovirus	+ssRNA	Polio, hepatitis A, common cold
Rhabdovirus	–ssRNA	Rabies
Rheovirus	dsRNA	Intestinal infections, respiratory infections
Togavirus	+ssRNA	Rubella
Lentivirus	RT RNA	AIDS
Oncornavirus	RT RNA	Human T cell leukemia

pathogens utilize many of the same strategies to evade detection and suborn host defenses. Like bacteria, viral genomes contain both structural and regulatory genes. The genomes of viruses encode proteins comprising their caspsids, envelopes, and other components. The genes encoding these structural proteins are referred to as *structural genes*. In addition, viral genomes contain a number of nonstructural genes. These genes encode proteins involved in replication and proteins of a regulatory nature.

18.5 Virus–Host Interactions Underlie Virus Survival and Proliferation

Viruses interact with their hosts in order to survive. They manipulate elements of the host control layer in order to create environments that support replication and proliferation. They turn off antiviral activities, prevent apoptosis, and alter the cell cycle progression. Shown in Table 18.2 are some examples of eukaryotic control proteins targeted by viruses. One of the groups of signaling proteins targeted by viruses is the cytokine group that coordinates the activities of leukocytes. The cytokine signaling events are disrupted by the viruses, thereby reducing the effectiveness of the immune system in responding to the viral invasion.

Another group of signaling elements targeted by the viruses are the antigen-presenting elements that present peptides on their surface derived

TABLE 18.2. Eukaryotic cell components targeted by some well-known viruses.

Control layer elements	Viruses (and signaling elements interfered with)
Antigen-presenting elements	Adenovirus (MHC class I, II), Epstein-Barr virus (MHC, class II)
Apoptosis system	Adenovirus (Fas, Bcl-2, caspases), baculovirus (caspases), cowpox (Fas, caspases), Epstein-Barr virus (Bcl-2), vaccinia (caspases)
Cytokine signaling	Cowpox (TNF), human herpesvirus-8 (CC and CXC chemokines and their receptors), vaccinia (IL-1β, IFNα/β)
Interferon signaling	Adenovirus (Jak/STAT), baculovirus (PKR), influenza virus (PKR), poliovirus (PKR), reovirus (dsRNA)

from the viruses of the invaded cells. A third set of targets is the proteins that regulate apoptosis. Cells infected by viruses are killed and induced to undergo apoptosis. The viruses manipulate this protective response; in some cases they turn off apoptosis and in others they use the process to hitch a ride out of the cell. Finally, receptors bound in the plasma membrane are targeted by viruses in order to gain entry into the cell.

18.6 Multilayered Defenses Are Balanced by Multilayered Attacks

Pathogens such as viruses and bacteria enter a body, or host, establish themselves in a specific locale, or niche, multiply, cause damage, and exit. Host defenses are mounted at multiple levels. The outermost layer is the erection of mechanical barriers such as the skin, and chemical barriers such as antibacterial defensins. Next, the innate immune response is triggered and then the adaptive immune response is.

The approach used by viruses to interfere with the host's control system is a multilayered one. Viruses, for example, overcome the acquired immunity response (antigen-presenting elements) by interdicting expression of the MHC Class I and II products at the transcription level at the translation level, by forcing the retention of the surface molecules in the cytoplasm, by accelerating their removal from the surface, and by stepping up their deactivation.

The survival of a pathogen in a host is critically dependent on its ability to deal with an arsenal of host defenses. Viruses and interferon systems have coevolved. One striking consequence of this coevolution is that practically all viruses have developed some level of resistance to the interferon system. The counterattacks take several forms and are multileveled, like the response to acquired immunity. Depending on the specific virus and target, the virus may

- block the production of interferons,
- supply interferon decoys,
- block interferon signaling, and
- block the ability of the interferon signaling targets to properly act.

At the most fundamental level, what is happening is that both host and pathogen (viral and bacterial) have coevolved. Both have developed control layers tuned to the other, promoting adaptation and evolution of each. The overall result is a balance, an equilibrium situation of coexistence of host and bacterial and viral pathogens.

18.7 Viruses Target TNF Family of Cytokines

The TNF family of cytokines and their receptors, and the apoptosis machinery, are targeted by viruses. The TNF family of cytokines and their receptors are central signaling elements of the adaptive immune response. Members of this family regulate secondary lymphoid organ development, lymph node formation, and apoptosis signaling. A variety of viruses including hepatitis C virus, Epstein-Barr virus, adenoviruses, and poxviruses elude destruction by the adaptive immune response by expressing proteins that alter the operation of the TNF machinery.

The hepatitis C virus core protein is a structural gene product that is able to bind to and modulate TNF family receptors. In response to these binding events cytokine signaling is impaired and cytotoxic T lymphocytes do not adequately respond to the viral invasion. Adenoviruses target the cellular apoptosis machinery triggered by the TNF family member Fas (Table 18.2). The viruses are able to induce the internalization of this receptor, thereby turning off the ability of the cells to kill the virus. Adenoviruses use other mechanisms as well to elude the adaptive immune response. For instance, they interfere with antigen presentation and T cell recognition. Each of the adenoviral evasion tactics utilizes a different subset of viral gene products.

Cells infected with a virus are usually induced to undergo apoptosis in order to prevent viral replication. Interestingly, some viruses have evolved ways to use this mechanism to their own advantage. They induce cell death at the end of their replication cycle with the result that they are encapsulated in a protective package that is taken up by neighboring cells. By this means they spread and elude detection at the same time.

18.8 Hepatitis C Virus Disables Host Cell's Interferon System

The hepatitis C virus is believed to infect more than 2% of the world's population. The virus causes chronic hepatitis and can lead to cirrhosis of the liver and/or liver cancer. The virus is a member of the *Flaviviridae*

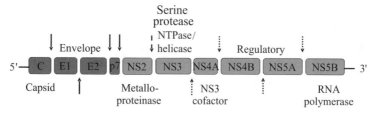

FIGURE 18.1. Organization of the genome of the hepatitis C virus: Proteins shown as dark-shaded boxes are structural (S) in character while those in light-shaded boxes are nonstructural (NS). From left to right the proteins are: C—nucleocapsid protein, sometimes called the "core protein"; E1 and E2—glycoproteins belonging to the viral envelope; p7—linker protein of unknown function; NS2 through NS5B— NS enzymes. Arrows denote cleavage sites. Solid arrows indicate sites cleaved by host cellular signalases; the dashed arrow indicates the site cleaved by NS2, and the dotted arrows denote sites cleaved by NS3.

family of viruses, which includes the yellow fever virus and several forms of encephalitis-inducing virus.

The hepatitis C virus (HCV) contains a single strand positive sense RNA (+ssRNA) molecule approximately 9.6 kBp in length that functions as a messenger RNA. The genome encodes a single open reading frame that is preceded at the 5′ end by a noncoding region and terminated at the 3′ end by another such region. The resulting polypeptide chain, or polyprotein, is proteolytically processed during and after translation into 10 proteins (Figure 18.1). The four amino-terminal-most are structural in character. The first protein, C, encodes the capsid protein, and the next two, E1 and E2, are glycoproteins that are inserted in the viral envelope. It is customary to refer to proteins forming the virus capsid and envelope as "structural" proteins and proteins involved in replication and signaling/regulation as "nonstructural." The fourth structural protein, p7, is a linker that separates the structural from nonstructural proteins.

The nonstructural genes encode several different kinds of enzymes. The NS5B gene encodes an RNA-dependent RNA polymerase that is responsible for viral replication. The NS2 and NS3 genes encode proteases that, along with cellular proteases, chop up the large polyprotein into the structural and nonstructural proteins at the sites indicated in Figure 18.1. The NS4B and NS5A gene products are regulatory in nature. The regulatory gene products along with the E2 glycoprotein target the interferon system of antiviral defense, allowing the viruses to survive and proliferate even when that system is activated.

Recall from Chapter 9 that the cell's first line of defense to a viral attack is the interferon system. During viral replication dsRNA is produced and the detection of these molecules alerts the cell that viruses are present. A sensor of dsRNA called *protein kinase R* (PKR) activates the interferon response. PKR shuts off translation and stimulates transcription of a host of antiviral factors. Interferons are produced that signal to other cells through interferon receptors and Jak/Stat proteins to generate a global response to the viral invasion. The favored treatment for patients with hepatitis C is IFNα, or α-interferon. However, this form of treatment, the best available at present, is ineffective in more than half of patients. In these many cases, the virus is able to overcome the antiviral activities of the interferons.

The interferon line of defense is breached by hepatitis C virus through its NS5A, NS4B, and E2 proteins, which inhibit the activity of PKR. The NS5A protein inhibits not only the PKR protein but also the IRF1 transcription factor, which acts downstream of PKR to trigger transcription of antiviral genes from the ISRE promoter. The E2, NS5A, and NS4B proteins enhance translation from what is termed an *HCV internal ribosomal entry site*, or IRES. The net effect of E2, NS4B, and NS5A is to throttle back expression of antiviral genes and maintain the translation capabilities of the host cell so that the viruses can replicate.

18.9 Human T Lymphotropic Virus Type 1 Can Cause Cancer

Certain retroviruses can induce the transformation of normal cells into cancerous ones. One group of retroviruses that can trigger cancers is the *oncogenic* (oncornaviruses) viruses, typified by the Rous sarcoma virus. This virus has picked up an oncogene from a host some time in its past, then deliveres it to its new hosts when it invades a cell. In the case of Rous sarcoma virus, the oncoprotein is Src, a tyrosine kinase involved in the growth pathway. This protein was discussed in several places in the text. To distinguish cellular from viral forms, one is designated as "c-Src" and the other as "v-Src." The v-Src oncoprotein is of considerable historical importance, as it was the first oncoprotein to be discovered.

A different mechanism is used to transform cells by viruses such as bovine leukemia virus (BLV) and human T lymphotropic virus type 1 (HTLV-1). The latter is associated with T cell leukemia. These retroviruses do not carry oncogenes. Instead regulatory proteins are encoded in their genomes, and these proteins along with promiscuous, virally supplied promoters that accompany them fatally alter the operation of the cellular control layer. Viruses belonging to this group cause a variety of cancers in livestock, while HTLV-1 and -2 promote cancer (adult T cell leukemia) in humans.

The organization and placement of control sites is an essential feature of a genome. Retorviruses bring their own control sites with them. These control sites are known as *long terminal repeats*, or LTRs. There is an LTR at the 5′ end and the other at the 3′ end. The one at the 5′ end is used for transcription control. Each LTR is composed of three regions: a U5 region, an R region, and a U5 region. The U3 region controls viral transcription

(a)

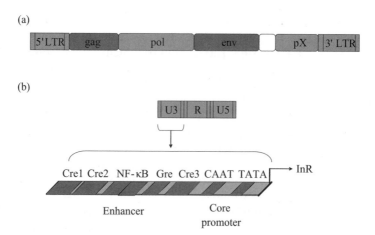

(b)

FIGURE 18.2. Retroviral control region: (a) A minimal retrovirus, containing gag, pol, and env genes, is shown bounded at both ends by long terminal repeats (LTRs) and a pX region near the 3′ end. (b) An expanded view of the 5′ LTR showing its U3, R, and U5 regions—A further expanded view of the U3 region depicts a canonical set of DNA binding sites (responsive elements) for transcription factors. Abbreviations: cyclic AMP responsive element (Cre); glucocorticoid responsive element (Gre); initiation, or start, site for RNA transcription (InR), located at the boundary between U3 and R.

while the R and U5 regions are essential for viral replication. As illustrated schematically in Figure 18.2, the U3 region is endowed with binding sites for many of the key families of eukaryotic transcription factors. These sites are capable of binding to both viral and cellular regulatory proteins.

A unique feature of HTLV-1 is the presence near the 3′ end of its genome of a region called *pX* that codes for several regulatory proteins. One of these proteins, called *Tax*, binds to CREB proteins at the various Cre sites not only in the viral genome but also in the cellular genome. Tax interacts with a variety of cellular regulators of transcription including not only CREB and cofactors but also NF-κB. Through its interactions with the cellular regulators, Tax changes the regulatory decisions of the T cell's control circuitry, immortalizing and transforming the cells.

18.10 DNA and RNA Viruses that Can Cause Cancer

Listed in Table 18.3 are a number of viruses that are known to transform cells in humans. The list includes both DNA and RNA viruses. The HTLV-1 virus appears in the table as a member of the retrovirus group that promotes T cell leukemias in adults. One of the recent additions to the list of viruses known to promote cancer is Kaposi's sarcoma-associated her-

TABLE 18.3. Tumor viruses.

Virus	Type of tumor
DNA tumor viruses	
Hepatitis B virus	Liver cancer
Herpesviruses	
Epstein-Barr virus	Burkitt's lymphoma, nasopharylgeal carcinoma
Kaposi's sarcoma virus	Kaposi's carcinoma
Papillomaviruses	Cervical cancer
RNA tumor viruses	
Hepatitis C virus	Hepatocellular carcinoma
Retroviruses	Adult T cell leukemias

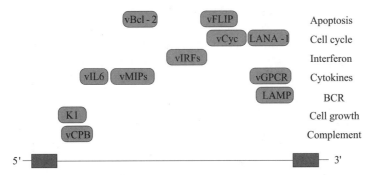

FIGURE 18.3. Approximate locations of regulatory genes encoded in the Kaposi's sarcoma-associated herpesvirus (KSHV) genome: The 141-kB genome is flanked by terminal repeat sequences (gray). The coding region (clear) contains two classes of regulatory genes. The first of these are the genes pirated from host genomes. The members of this group can be recognized by the presence of a letter "v" in front of familiar gene names. The second group consists of regulatory genes unique to the virus and its close relatives, such as the Epstein-Barr virus. Three examples are included in the figure: K1, latency-associated nuclear antigen type 1 (LANA-1), and latency-associated membrane protein (LAMP). Other abbreviations—Complement binding protein (CBP); macrophage inflammatory protein (MIP); interferon regulatory factor (IRF).

pesvirus (KSHV). Viruses such KSHV have pirated a large number of regulatory genes from their hosts, and through expression of these genes alter the cell cycle and growth responses of the host so that they serve the virus. They use the pirated genes, as well as genes that are unique to the viruses to degrade both the innate immune response and the adaptive immune response, and can transform cells, producing a number of cancers in humans.

Some of the most prominent regulatory genes in the KSHV genome are shown in Figure 18.3. The first group of genes depicted in the figure encodes two antiapoptotic proteins. The viral Bcl-2 protein acts at the PTPC to inhibit apoptosis, and the viral FLIP acts at the DISC to throttle back death receptor signaling by preventing activation of caspase 8. The next set of gene products regulates the cell cycle so that it becomes more conducive to viral replication. The viral cyclin protein (vCyc) is similar to the cellular D cyclins and primarily binds to Cdk6. The cell cycle genes are followed in the figure by a considerable number of genes that jointly regulate immune responses and cell growth. The LAMP protein is a 12-pass transmembrane protein that inhibits signaling from the B cell receptor. The other proteins further shift the life-death signaling balance towards growth and away from apoptosis, and suppress Th1 immune responses.

18.11 HIV Is a Retrovirus

The HIV-1 virus is a member of the retrovirus family. Like others in that family the packaging of the virus is fairly elaborate. The virus capsid is encapsulated in a matrix that, in turn, is enclosed inside an envelope. Two copies of an ssRNA molecule are contained within the capsid. Each ssRNA molecule encodes 9 polygenes and 15 genes overall. The structure of the virus is depicted in Figure 18.4.

The organization of the HIV-1 genome is depicted in Figure 18.5. As was the case for the other viruses, a combination of structural and nonstructural genes is encoded. The structural genes are located on group-specific antigen (gag) and envelope (env) genes. These are proteolytically processed to yield seven protein-coding genes. The core, or CA, gene encodes the capsid protein, and the MA gene encodes the matrix protein. The third gag gene is the nucleocapsid gene, NC. It encodes a protein that associates with the ssRNA molecule. The last gag segment is the p6 gene that contains a late domain that mediates docking, budding, and release of the HIV virus at the cell surface.

The envelope polygene encodes two glycoproteins—gp120 surface (SU) and gp41 transmembrane (TM). These proteins form complexes in the

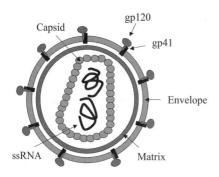

FIGURE 18.4. Structure of the human immunodeficiency virus (HIV): The virus is shown with its surface-bound gp120 and transmembrane gp41 proteins protruding out from the envelope of the virus.

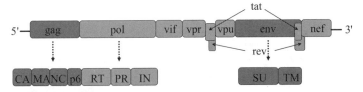

FIGURE 18.5. Organization of the HIV-1 genome: Abbreviations of regulatory genes—Negative effector (nef); regulator of viral expression (rev); transactivator of viral transcription (tat); virulence infectivity factor (vif); virulence protein regulatory (vpr); virulence protein unknown (vpu).

plasma membrane. They facilitate binding and entry of the virus into the host cell. All retroviruses contain gag, env, and pol genes. The pol gene encodes enzymes required for replication of the virus. One of the enzymes is a reverse transcriptase (RT). This enzyme catalyzes the polymerization of a DNA molecule from the RNA molecule. This is a reverse, or retro, information transfer operation, hence the name "retrovirus." The pol gene also encodes a protease (PR) that cleaves viral proteins following their translation, and an integrase (IN) that helps integrate the viral DNA into the host genome. There are six regulatory genes. These genes—vif, vpr, vpu, nef, tat, and rev—are unique to HIV and along with gp120 are responsible for HIV's lethal behavior. These virulence-promoting genes are discussed next in greater detail.

18.12 Role of gp120 Envelope Protein in HIV

The gp120 envelope protein mediates contact with and entry into the host cell. The gp120 protein is extensively modified in the Golgi where it acquires a set of complex sugars. The glycoproteins are then transported to the cell surface where they fuse with the budding viruses and are incorporated into their envelopes. The gp120 envelope protein interacts with receptors on the cell surface, most notably with CD4 receptor on helper T cells, and uses the chemokine CCR5 receptors as coreceptors, sometimes along with the CXCR4. Binding is sequential. Binding to CD4 triggers conformational changes leading to binding to the chemokine receptors. Membrane fusion and entry into the cell follows these binding events.

Neutralizing antibodies bind to epitopes on viral surface molecules such as gp120 and block the ability of those proteins to bind receptors on host cells. The viruses are prevented from receptor-binding and entry into the cell, and are thus neutralized. In the case of HIV, most neutralizing antibodies are unable to recognize the gp120 glycoprotein. This glycoprotein avoids detection by occluding its antibody-binding sites. Antibodies bind gp120 in two places. One of these overlaps the CD4 binding site and is

(a)   (b)

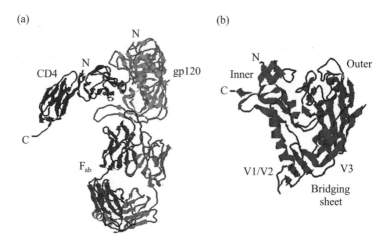

FIGURE 18.6. Crystal structure of the HIV-1 gp120 envelope glycoprotein in a complex with CD4 and a neutralizing antibody determined by X-ray crystallography: (a) The gp120 glycoprotein is shown in a bound to the CD4 receptor expressed on helper T cells and to a Fab antibody fragment called 17b. (b) Expanded view of the gp120 core—The structure is shown rotated 90 degrees about a vertical axis from the view shown in part (a). The figure was generated using Protein Explorer with atomic coordinates deposited in the Brookhaven Protein Data Bark under accession code 1GC1.

occluded by the V1/V2 loop. The other overlaps the chemokine binding site and is occluded by the V2 and V3 loops (Figure 18.6). This is not the only technique used. A second way HIV avoids neutralizing antibodies is by *conformational masking*. This is a thermodynamic effect, tied to conformational changes, that allows gp120 to bind to CD4 while simultaneously avoiding neutralization by the antibodies. Yet another way used by gp120 to avoid discovery is *glycosylation*. A large number of changing sugar groups are added, about 20 glycans for each gp120 molecule. These act as a constantly shifting shield against neutralizing antibodies.

18.13 Early-Acting tat, rev, and nef Regulatory Genes

Prior to entry into the nucleus the HIV-1 RNA is transcribed into double stranded DNA by the reverse transcriptase (RT). A preinitiation complex (PIC) is formed consisting of the viral DNA, integrase (IN), multiple matrix (MA) protein copies, nucleocapsid protein (NC), p6, and Vpr. The PIC binds to the pore and is then sent into the nucleus using NLSs embedded in viral and associated cellular proteins.

The 9-kB HIV genome encodes 15 proteins, but nascent transcripts are spliced in about 30 different ways to produce a variety of messenger RNA

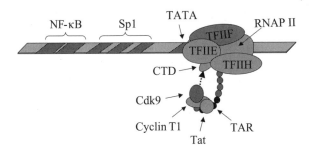

FIGURE 18.7. Regulation of transcription elongation by Tat: A portion of the HIV promoter is shown highlighting the upstream NF-κB and Sp1 binding sites, and TATA box. The RNAP II is depicted with several of its cofactors transcribing the DNA into an RNA chain. A portion of the RNA chain (shown in black), the TAR, is bound by Tat which then recruits the Cdk9/Cyclin T1 complex. The Cdk9 kinase hyperphosphorylates the C-terminal domain (CTD) of the RNAP II molecule.

molecules. In the early stages of HIV infection, short 2-kB mRNAs are produced and exported from the nucleus. These short transcripts encode Tat, Rev, and Nef proteins, and the genes for these are referred to as *early acting genes*. Later in the infection cycle longer 4-kB variants and full length mRNAs are made, and the genes transcribed at that time are termed *late acting genes*.

The HIV virus not only supplies its own promoter, as do other retroviruses, but also supplies its own transcription activator, Tat. The Tat protein is an RNA binding protein. It binds to a segment of nascent HIV RNA chain called the *transactivating response* (TAR) region located at the 5' end. A number of cellular proteins recognize the TAR sequence and are recruited to the site. As depicted in Figure 18.7, these include Cyclin T1 and its cyclin-dependent kinase Cdk9, which hyperphosphorylates the C-terminal domain of RNAP II. In the absence of hyperphosphorylation, the transcripts generated by RNAP II tend to be short. Hyperphosphorylation prevents premature elongation termination and enables the enzyme to catalyze full length transcripts.

Rev is a sequence-specific RNA binding protein that binds to a region called the Rev responsive region (RRE) located in env gene transcripts. The Rev protein contains a nuclear export sequence (NES) and mediates the export of all transcripts except those of tat, rev, and nef from the nucleus into the cytoplasm, where they are translated into the virus components. In the early stages of an HIV infection, before Rev levels are sufficient to support its export function, transcripts longer than 2 kB that get made collect in the nucleus. In the later stages of HIV infection, Rev facilitates the export of longer and full length transcripts from the nucleus.

The Nef protein helps the virus to control signaling by its host and creates an environment conducive to replication of the virus. One of the key targets of Nef is the MHC-I complex on the cell surface. Recall that MHC-I complexes are used to present antigens derived from, for example, viruses, on the surface where they can be recognized by killer cells of the immune system. Nef stimulates the internalization of these receptors and directs them to the Golgi. Another target of the Nef protein is the CD4 receptor. These are internalized and sent to digestive compartments. The MHC-I and CD4 receptors are not the only immune system signaling elements targeted by Nef. It also downmodulates the CD28 coreceptor. All of these actions help disable signaling between cells of the immune system, enabling infected cells to evade detection and destruction. In macrophages, which along with CD4 T cells are the main targets of HIV, Nef induces the release of chemokines that attract resting T cells and help to transmit the virus to uninfected cells.

18.14 Late-Acting vpr, vif, vpu, and vpx Regulatory Genes

Viral protein U (Vpu) is found exclusively in HIV-1, while viral protein X (Vpx) is restricted to HIV-2 and simian immunodeficiency virus (SIV). Vpu assists in the degradation of CD4 in the endoplasmic reticulum and in the exit of the virus from the cell. It binds CD4 in the ER and targets it for degradation in the ubiquitin-proteosome pathway thereby preventing its translocation to the plasma membrane.

Viral protein R (Vpr) encoded by the vpr gene, is a small protein, containing just 96 amino acids. It is soluble and is able to pass through membranes. Vpr acts in several ways to increase HIV survival in cells that it infects. It assists in the nuclear localization of the PIC, and triggers arrest of the cell cycle at G2/M transition by blocking Cdk1/Cyclin B complex formation that drives the cell into mitosis. The Vpr protein acts as a modulator of apoptosis. It migrates to the mitochondria where it works together with proteins resident in the PTPC to shift the balance either towards or away from apoptosis. It can work with the antiapoptotic Bcl-2 protein to inhibit apoptosis and work at the ANT and VDAC to increase mitochondrial membrane permeability and promote apoptosis.

Viral infectivity factor (Vif) counters the antiviral activities of a cellular agent with a long name—apolipoprotein B mRNA-editing enzyme, catalytic polypeptide-like 3G, or APOBEC3G. While the retroviral RNA is being reverse-transcribed into DNA this enzyme catalyzes the conversion of "C" nucleotides to "U" nucleotides. These alterations prevent replication of retroviruses in cells, referred to as being *nonpermissive*, that express the APOBEC3G enzyme. The Vif protein renders this antiviral enzyme ineffective by binding to it and speeding up its degradation in the ubiquitin-proteosome pathway.

18.15 Bacteriophages' Two Lifestyles: Lytic and Lysogenic

Bacteriophages, or phages, are bacteria-infecting viruses. Bacteriophage lambda is perhaps the best studied of this group of viruses. Like many others it has two alternative lifecycles—lytic and lysogenic. In its *lytic cycle*, the virus attaches to its host, and its nucleic acids—dsDNA—enter the cell while the head and tail sections of the capsid (Figure 18.8) remain outside. The DNA is rapidly replicated and capsid proteins are synthesized. The virus components are reassembled and the resulting virions are released from the host cell through lysis, in which the cell membrane ruptures, and the viruses escape leaving behind a dying cell.

In its *lysogenic cycle*, the phage DNA is integrated into the host cell DNA rather than being immediately replicated. It becomes a prophage, replicating as part of the bacterial host's chromosome. It continues to do so until environmental conditions are such that a lytic cycle is preferable. When those conditions arise it switches from the lysogenic cycle to the lytic cycle. It rapidly replicates itself and lyses the cell.

The organization of the bacteriophage lambda genome is depicted in Figure 18.9. Several sets of nonstructural genes are shown. These genes encode enzymes required for recombination (integration of the viral genome into the host's chromosome), replication, and lysis. Six regulatory genes are interspersed among the aforementioned nonstructural genes.

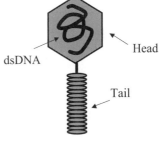

FIGURE 18.8. Bacteriophage lambda: The capsid of the bacteriophage is organized into a head section, which contains the DNA, and a tail, which assists in attachment of virus to its host and injection of the viral DNA into the bacterial host.

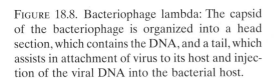

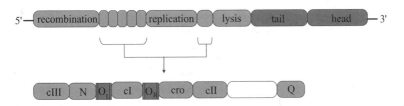

FIGURE 18.9. Organization of the bacteriophage lambda genome and its regulatory genes: The organization of the six regulatory genes and two noncoding operator regions are depicted in the lower portion of the figure.

Their gene products control the decision between the lytic and lysogenic life.

18.16 Deciding Between Lytic and Lysogenic Lifestyles

The regulatory network used by the bacteriophage lambda to switch between lysis and lysogeny is one of the best characterized genetic circuits. This genetic network consists of six regulatory genes (N, Q, cI, cro, cII, and cIII), three promoters (P_L, P_R, and P_{cI}), and two noncoding operator regions (O_L and O_R). The cI gene is called the *lambda repressor* because when it is expressed its gene product, the CI protein, represses transcription of the genes needed for viral replication thereby favoring the lysogenic lifestyle. For the same reason the Cro protein is called the *lambda counterrepressor* because when it is expressed it facilitates transcription of the genes needed for the lytic lifestyle. The decision process is represented schematically in Figure 18.10.

P_L and P_R are left and right promoters. P_L is responsible for transcribing the left hand portion of the network that includes cIII and N. P_R is responsible for transcribing the right hand portion that includes cro, cII and Q. O_L and O_R are noncoding operator regions each consisting of three cooperative binding sites. cI is the lambda repressor gene, which, when bound to the operator sites, shuts down transcription of all lambda genes except that of the repressor itself. It has its own promoters, P_{RE} and P_{RM}. These have been represented in Figure 18.10 as P_{cI}. When cI is expressed it binds to

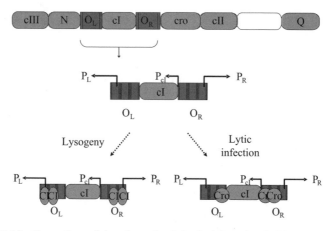

FIGURE 18.10. Operation of the phage lambda decision circuit: The central portion of the regulatory circuit in shown in an expanded view in the lower portion of the diagram. The lysogenic state is selected when CI binds to the operator sites and represses transcription from the left and right promoters. The lytic state is chosen when the Cro protein binds to the operator sites.

sites nearest the left and right promoters, thereby blocking transcription of the other genes, and the phage remains a prophage. The cro protein product is an antirepressor that shuts off cI when it is bound to the operator sites. As shown in Figure 18.10, it favors the innermost sites in the left and right promoters. When it binds to these sites it shuts off cI transcription while allowing transcription of the other genes from the left and right promoters. This circuit exploits both positive and negative feedback.

One of the most interesting aspects of the phage decision circuit is the occurrence of cooperativity in the two operator regions. Binding of a protein to a site in one of the operator regions increases the probability that a second protein will bind to a nearby site that would otherwise be unoccupied due to weak binding. The initial binding events thus serve to recruit additional proteins to the regulatory regions where they jointly control transcription. This type of control is not unique to bacteria or to phages; it was encountered earlier in the chapter on eukaryotic gene regulation.

The decision circuit operates under environmental control. The decision circuit is sensitive to good versus poor growth conditions and to exposure to ultraviolet radiation. The outcome of the race between lysogeny and lysis is determined by the concentration of CII in the cell. These proteins act as enhancers of cI transcription. The concentration of CII in a healthy cell is usually maintained at a low level by the activity of host proteases that are sensitive to glucose levels. When nutrients are plentiful, CII levels are low resulting in a stable lytic state of replication and release of new phages. The CII concentration under poor bacterial growth conditions is elevated leading to CI levels that are high enough to send the phage into a lysogenic state.

The phage can transition from lysogeny to lysis in response to DNA damage arising from UV or ionizing radiation. RecA proteins are synthesized by the host in response to ultraviolet damage to the cellular DNA. These proteins stimulate the repair of damaged DNA by inactivating the LexA repressor. The LexA repressor is similar to the lambda repressor CI. When phage lambda is present, RecA targets CI as well. There is an increased expression of Cro proteins leading to a lytic stage where the virus escapes from the UV damaged cell.

18.17 Encoding of Shiga Toxin in *E. coli*

Shiga toxin genes are encoded by a lambda-like phage in enterohaemor-rhegic *Escherichia coli*. In the last chapter, it was noted that harmless strains of bacteria such as the harmless K12 variety of *E. coli* differ from their pathogenic relatives in that the latter contain additional genes acquired mostly through horizontal gene transfer. Bacteriophages are a common means of transfer of DNA from one bacterial species to another. The Shiga

FIGURE 18.11. Phage genes incorporated into the genomes of several strains of enterohaemorrhagic *E. coli*: The toxin genes StxA and StxB are situated in between the replication and late antitermination (Q) genes on the left and lysis genes on the right.

toxin genes that render strains of *E. coli* such as O157:H7 so harmful provide an excellent example of this mechanism. As shown in Figure 18.11 the Shiga toxin genes StxA and StxB genes are encoded in *E. coli O157: H7* in a stretch of DNA almost identical to that of bacteriophage lambda. The gene sequence depicted in the figure is found not only in the O157:H7 strain but also in the several other *E. coli* and *Shigella dysenteriae* strains.

Shiga toxin genes are composed of catalytic A subunits and glycolipid-binding B subunits. They are regulated by factors intrinsic to the bacterial genome and by the bacteriophage life cycle. There is a binding site in the operator region upstream of the A gene subunit for the iron binding Fur protein. When iron is plentiful the protein is able to bind to the operator and block transcription. This repression is relived in low-iron environments. Production of toxins is also increased when the bacteriophage replicates and is released from the host cell through lysis. This places the genes for the toxins under the control of environmental factors such as increased H_2O_2 levels, which can damage DNA, and the presence of antibiotics.

References

General Reading

Schaechter M, Engelberg NC, Eisenstein BI, and Medoff G [1999]. *Mechanisms of Microbial Diseases* (3rd edition). Baltimore: Lippincott, Williams and Wilkins.
Walker TS [1998]. *Microbiology*. Philadelphia: W.B. Saunders and Company.

Viral Entry into the Cell

Greber UF [2002]. Signalling in viral entry. *Cell. Mol. Life Sci.*, 59: 608–626.
Overbaugh J, Miller AD, and Eiden MV [2001]. Receptors and entry cofactors for retroviruses include single and multiple transmembrane-spanning proteins as well as newly described glycophosphatidylinositol-anchored and secreted proteins. *Micro. Mol. Biol. Rev.*, 65: 371–389.
Sieczkarski SB, and Whittaker GR [2002]. Dissecting virus entry via endocytosis. *J. Gen. Virol.*, 83: 1535–1545.

Virus Cytoplasmic Transport

McDonald D, et al. [2002]. Visualization of the intracellular behavior of HIV in living cells. *J. Cell Biol.*, 159: 441–452.
Seisenberger G, et al. [2001]. Real-time single-molecule imaging of the infection pathway of an adeno-associated virus. *Science*, 294: 1929–1932.

Sodeik B [2000]. Mechanisms of viral transport in the cytoplasm. *Trends Microbiol.*, 8: 465–472.

Sodeik B, Ebersold MW, and Helenius A [1997]. Microtubule-mediated transport of incoming herpes simplex 1 capsids to the nucleus. *J. Cell Biol.*, 136: 1007–1021.

Viral Movement In and Out of the Nucleus

Salman H, et al. [2001]. Kinetics and mechanism of DNA uptake into the cell nucleus. *Proc. Natl. Acad. Sci. USA*, 98: 7247–7252.

Trotman LC, et al. [2001]. Import of adenovirus DNA involves the nuclear pore complex receptor CAN/Nup214 and histone H1. *Nature Cell Biol.*, 3: 1092–1100.

Viral Release and Budding

Freed EO [2002]. Viral late domains. *J. Virol.*, 76: 4679–4687.

Katzmann DJ, Odorizzi G, and Emr SD [2002]. Receptor downregulation and multivesicular-body sorting. *Nature Rev. Mol. Cell Biol.*, 3: 893–905.

McDonald D, et al. [2003]. Recruitment of HIV and its receptors to dendritic cell-T cell junctions. *Science*, 300: 1295–1297.

Pelchen-Matthews A, Kramer B, and Marsh M [2003]. Infectious HIV-1 assembles in late endosomes in primary macrophages. *J. Cell Biol.*, 162: 443–455.

Von Schwedler UK, et al. [2003]. The protein network of HIV budding. *Cell*, 114: 701–713.

Hepatitis C Virus

Bartenschlager R, and Lohmann V [2000]. Replication of hepatitis C virus. *J. Gen. Virol.*, 81: 1631–1648.

Goodbourn S, Didcock L, and Randall, RE [2000]. Interferons: Cell signalling, immune modulation, antiviral responses, and viral countermeasures. *J. Gen. Virol.*, 81: 2341–2364.

Katze MG, He YP, and Gale M, Jr. [2002]. Viruses and interferons: A fight for supremacy. *Nature Rev. Immunol.*, 2: 675–687.

Large MK, Kittlesen DJ, and Hahn YS [1999]. Suppression of host immune response by the core protein of Hepatitis C virus: Possible implications for Hepatitis C virus persistence. *J. Immunol.*, 162: 931–938.

HIV-1

Boshoff C, and Weiss R [2002]. Aids-related malignancies. *Nature Rev. Cancer*, 2: 373–382.

Emerman M, and Malim MH [1998]. HIV-1 regulatory/accessory genes: Keys to unraveling viral and host biology. *Science*, 280: 1880–1884.

Gu YP, and Sundquist WI [2003]. Good to CU, *Nature*, 424: 21–22.

Jacotet E, et al. [2000]. Control of mitochondrial membrane permeabilization by adenine nucleotide translocator interacting with HIV-1 viral protein R and Bcl-2. *J. Exp. Med.*, 193: 509–519.

Karn J [1999]. Tackling Tat. *J. Mol. Biol.*, 293: 235–254.

Kwong, PD, et al. [2002]. HIV-1 evades antibody-mediated neutralization through conformational masking of receptor binding sites. *Nature*, 420: 678–682.

Lum JJ [2003]. Vpr R77Q is associated with long-term nonprogressive HIV infection and impaired induction of apoptosis. *J. Clin. Invest.*, 111: 1547–1554.

Marin M, et al. [2003]. HIV-1 Vif protein binds the editing enzyme APOBEC3G and induces its degradation. *Nature Med.*, 9: 1398–1403.

Sheehy AM, Gaddis NC, and Malim MH [2003]. The antiretroviral enzyme APOBEC3G is degraded by the proteosome in response to HIV-1 Vif. *Nature Med.*, 9: 1404–1407.

Stevenson M [2003]. HIV-1 pathogenesis. *Nature Med.*, 9: 853–860.

Swigut T, Shohdy N, and Skowronski J [2001]. Mechanism for downregulation of CD28 by Nef. *EMBO J.*, 20: 1593–1604.

Swingler S, et al. [2003]. Nef intersects the macrophage CD40L signalling pathway to promote resting-cell infection. *Nature*, 424: 213–219.

Swingler S, et al. [1999]. HIV-Nef mediates lymphocyte chemotaxis and activation by infected macrophages. *Nature Med.*, 5: 997–1003.

Wei XP, et al. [2003]. Antibody neutralization and escapes by HIV-1, *Nature*, 422: 307–312.

Wyatt R, and Sodroski J [1998]. The HIV-1 envelope glycoproteins: Fusogens, antigens and immunogens. *Science*, 280: 1884–1888.

KSHV

Damania B, Choi JK, and Jung JU [2000]. Signaling activities of gammaherpesvirus membrane proteins. *J. Virol.* 74: 1593–1601.

Moore PS, and Chang Y [2001]. Molecular virology of Kaposi's sarcoma-associated herpesvirus. *Phil. Trans. R. Soc. Lond. B*, 356: 499–516.

Bacteriophage Lambda

Bell CE, et al. [2000]. Crystal structure of the λ repressor C-terminal domain provides a model for cooperative operator binding. *Cell*, 101: 801–811.

Campbell A [2003]. The future of bacteriophage biology. *Nature Rev. Genet.*, 4: 471–477.

McAdams HH, and Shapiro L [1995]. Circuit simulation of genetic networks. *Science*, 269: 650–656.

Ptashne M, and Gann A [1997]. Transcriptional activation by recruitment. *Nature*, 386: 569–577.

Shiga Toxin

O'Loughlin EV, and Robins-Brown RM [2001]. Effect of Shiga toxin and Shiga-like toxins on eukaryotic cells. *Microbes Infect.*, 3: 493–507.

Wagner PL, and Waldor MK [2002]. Bacteriophage control of bacterial virulence. *Infect. Immun.*, 70: 3985–3993.

Problems

18.1 Engineered viruses are being devised for use in cancer therapy. The goal of researchers is to find natural viruses, or create genetically modified ones, that can kill cancer cells while leaving normal cells unharmed. This main idea behind creation of these oncolytic viruses is to take advantage of the altered signaling pathways in the tumors by devising viruses that selectively target those cells.

The first viruses to be used in this way are the adenoviruses. These viruses and other like them stimulate host cells to pass through the cell cycle. If the host cell does not pass into S phase the viruses cannot replicate their DNA. Adenoviruses encode 30 to 40 proteins that are expressed sequentially. First immediate early genes, then early genes and finally late genes are expressed. The first protein to be made is E1A. This control protein binds to the pRB protein of the host cell and inactivates it (Figure 14.11). The E1B protein is made soon thereafter. It binds to p53 and inactivates it, too. These binding and inactivation events ensure entry of the host cells into S phase.

Recall from Chapter 14 that p53 and pRb central cell cycle controller circuitry and are mutated in most cancers. The inactivation of these genes is an important step in transforming the host cell into one that supports replication of the adenoviruses. How would you engineer an adenovirus to kill cancerous cells possessing mutated p53 proteins while leaving normal cells unharmed?

18.2 Some cells are permissive with respect to a virus while others are not. In a permissive cell, the virus can gain entry by binding a receptor and once inside the cell the viruses can replicate themselves. In nonpermissive cells, there aren't any receptors that can be exploited for entry, and if they do the signaling pathways do not support completion of the viral replication cycle.

One of the signaling pathways often altered in cancer cells is the Ras signaling pathway. One of the downstream targets in this growth-promoting pathway is the antiviral agent PKR. As mentioned earlier, when activated PKR shuts down proteins synthesis thereby preventing viral replication. Phosphorylation by one or more kinases acting downstream from Ras represses this activity. How would you engineer an oncolytic virus to kill cancerous cells possessing mutated Ras proteins while leaving normal cells unharmed? Note: dsRNA viruses often encode an anti-PKR protein.

19
Ion Channels

The plasma membranes of nerve and muscle cells differ from those of other cells in the body. Whereas the potential across the plasma membrane of any cell will change when the ion permeability across the membrane is altered, only in nerve and muscle cells will the converse occur. That is, in nerve and muscle cells, the ion permeability will change as a result of a change in membrane potential. This property is a consequence of the presence in the plasma membrane of large numbers of ion channels that open and close in response to changes in voltage. When an ion channel opens, up to 10 million ions can flow though its conducting pore per second, generating a few picoamperes (10^{-12} amps) of current. The presence of a few thousand of these channels endows the plasma membrane with the ability to actively conduct action potentials. Ion channels are responsible for all electrical signaling, and regulate a variety of processes including osmotic balance, hormone release, heartbeat, motor movement, learning, and memory.

Ion channels render the plasma membrane permeable to the passage of certain inorganic cations (Na^+, K^+, Ca^{2+}) and anions (Cl^-). The channels are composed of several subunits arranged in a way that creates a hollow pore that permits the one-way passive diffusion of ions from either outside the cell in, or inside the cell out. Some channels permit the passage of sodium, while others allow potassium ions or calcium ions or chloride ions to pass through. Ion channels are not open all the time, but rather open and close in response to extracellular ligand (neurotransmitter) binding, intracellular ligand binding, or to changes in membrane voltage. They possess one or more "gates" that mechanically open and close their pores to the passage of ions in response to the regulatory signals.

The first part of the chapter is devoted to an examination of how membrane potentials arise and how action potential can be generated from voltage-dependent changes in the permeability of the plasma membrane. A number of key biophysical concepts such as channel gating, and voltage-dependent activation and deactivation will be introduced along with the Nernst and Hodgkin–Huxley equations. The second part of the chapter is dedicated to the mechanisms that make properties such as gating possible.

These mechanisms are revealed by electron microscopy and X-ray crystallography studies of the atomic structure of the ion channel proteins.

19.1 How Membrane Potentials Arise

Membrane potentials arise from ion concentration differences and selective permeability. Nerve and muscle cells, unlike other cell types in the body, possess appreciable membrane potentials or membrane potential differences. Their membrane potentials arise from differences in ionic concentrations between outside and inside environments and from the selective permeability of the membrane to specific ion species due to the presence of ion channels. To see how ion concentrations and selective permeability give rise to membrane potentials four ion species will be considered—sodium (Na^+), potassium (K^+), chloride (Cl^-), and A^-. The A^- ion species represents the net negative charge of the large biomolecules residing within the cell. The four ion species are distributed outside and inside the cell subject to two conditions:

- Osmotic balance—The net concentration of ions outside the cell must equal the net concentration of ions inside the cell.
- Charge balance—The net charge both outside the cell and inside the cell must be zero.

The first condition ensured that the cell will not be subject to osmotic stresses and the second condition is a charge equalization requirement in each compartment. The concentrations of these four ion species for a representative nerve cell are depicted in Figure 19.1(a). As can be seen by

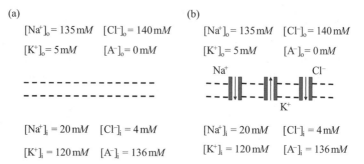

(a)

$[Na^+]_o = 135\,mM$ $[Cl^-]_o = 140\,mM$

$[K^+]_o = 5\,mM$ $[A^-]_o = 0\,mM$

$[Na^+]_i = 20\,mM$ $[Cl^-]_i = 4\,mM$

$[K^+]_i = 120\,mM$ $[A^-]_i = 136\,mM$

(b)

$[Na^+]_o = 135\,mM$ $[Cl^-]_o = 140\,mM$

$[K^+]_o = 5\,mM$ $[A^-]_o = 0\,mM$

Na^+ Cl^-

K^+

$[Na^+]_i = 20\,mM$ $[Cl^-]_i = 4\,mM$

$[K^+]_i = 120\,mM$ $[A^-]_i = 136\,mM$

FIGURE 19.1. Ion concentrations inside and outside the cell and membrane permeability: (a) Inside and outside are separated by an impermeable membrane represented by the double dashed lines. (b) The membrane separating the two compartments contains ion channels and is permeable to passage of potassium and chloride ions into the cell and sodium ions out of the cell. Ion concentrations are shown in units of millimoles per liter (mM/liter or mM).

adding up the concentrations the osmotic and charge conditions are satisfied.

The membrane separating inside and outside in part (a) of the figure is impermeable to the passage of ions. Since there is no membrane permeability there is no membrane potential. Consider now what happens when ion channels are added as in part (b) of Figure 19.1. Channels for K^+, Na^+, and Cl^- ions are now present, but not for A^- ions, which are too large to pass through an ion channel. The sodium and chloride channels will be neglected for the moment, with the main attention on the K^+ and the A^- ion concentration.

Potassium channels allow potassium ions to passively diffuse out of the cell. They will be driven to do so by the potassium concentration gradient set up because the potassium concentrations outside and inside the cell differ. As potassium ions diffuse out of the cell a net negative charge will build up within the cell since the negative charge associated with the A^- ions exceeds the positive charge of the remaining potassium ions. The buildup of negative charge will oppose the outward flow of positively charged potassium ions, eventually reaching the point where it is strong enough to halt the flow of ions. At this point an equilibrium situation is reached. There is a potential difference across the membrane and this potential just balances the concentration gradient. The potential difference required to halt the flow of potassium ions through the potassium channels is the potassium membrane potential, similar to the situation for other ion species. These potentials are given by the Nernst equation:

$$E_X = \frac{RT}{zF} \ln \frac{[X]_{out}}{[X]_{in}}. \qquad (19.1)$$

In the Nernst equation, R is the universal gas constant, T is the absolute temperature, F is Faraday's constant, z is the valence of the ion (+1 for sodium and potassium; +2 for calcium, and -1 for chloride), and X stands for the ion species under consideration. At a temperature of 37° Celsius, $RT/F = 26.73$ mV, and switching from natural logarithms to base 10 logarithms:

$$E_X = \frac{61.54}{z} \log \frac{[X]_{out}}{[X]_{in}}. \qquad (19.2)$$

The potentials given by the Nernst equation are commonly referred to as *reversal potentials*, or alternatively, as *equilibrium potentials*. They can be either positive or negative, depending on the sign of z and on which concentration is greater—inside or outside. As noted in Table 19.1, the reversal potentials for sodium and calcium are positive while those for potassium and chloride are negative.

The plasma membrane of cells is more permeable to some ions than to others. It is far more permeable to potassium ions than to sodium ions, and

TABLE 19.1. Values for typical concentrations of four main kinds of ions inside and outside mammalian skeletal muscle cells are presented along with the values for the equilibrium potential calculated from the Nernst equation.

Ion species	$[X]_{out}(mM)$	$[X]_{in}(mM)$	$E_x(mV)$
Na^+	145	12	+67
K^+	4	155	−98
Cl^-	123	4	−92
Ca^{2+}	1.5	10^{-4}	+129

somewhat more permeable to potassium ions than to chloride ions. The *resting potential* of a membrane (V_m) is given by combining the currents from the various ion species, taking into account not only the concentrations outside and inside but also the relative permeabilities. A simple expression, known as the *Goldman–Hodgkin–Katz equation*, incorporates both of these factors:

$$V_m = 61.54 \log \frac{[K]_{out} + (p_{Na}/p_k)[Na]_{out} + (p_{Cl}/p_k)[Cl]_{in}}{[K]_{in} + (p_{Na}/p_k)[Na]_{in} + (p_{Cl}/p_k)[Cl]_{out}}. \quad (19.3)$$

As is customary, the calcium contribution, which is small, has been neglected. In Eq. 19.3, p_K, p_{Na} and p_{Cl} are permeability constants. The appearance of their ratios in Eq. 19.3 takes into account the relative permeabilities of the membrane to the different ion species. Typical resting potentials are in the range −60 to −80 mV. These values are far closer to the potassium equilibrium potential than to the sodium equilibrium potential. The closeness of the membrane resting potential to the potassium reversal potential is a reflection of a far greater permeability for potassium than for either sodium or chloride. As their permeability ratios in the GHK equation become negligible, the expression for the resting potential approaches that of the Nernst equation for potassium. The reversal potential may be thought of as supplying a driving force for changes in resting potential.

19.2 Membrane and Action Potentials Have Regenerative Properties

Action potentials are generated by the joint action of fast-acting sodium channels and delayed-acting potassium channels. The generation of an action potential begins with a threshold-exceeding depolarization that opens sodium channels. When this happens the permeability of the membrane to sodium increases; according to the GHK equation the membrane depolarizes as its potential moves towards the equilibrium value for sodium. The open sodium channels permit the entry of sodium cations into the cell, which depolarize the membrane even further allowing for an opening of still more sodium channels. This continues for only a short while. The positive feedback of sodium channel openings stimulating the further opening of sodium channels is terminated when the channels inactivate.

At this point, the potassium channels, which also open in response to depolarization, start to dominate. The buildup of positive charge inside the cell through the entry of sodium cations is now countered by a decrease in positive charge through the exit of potassium cations. This slower activity moves the membrane potential towards the reversal potential for potassium ions. The opening of potassium channels continues for some time resulting in the repolarization of the membrane, and contributing to the shut down of the sodium channels. At the start of the cycle the resting potential is about −60 mV. The membrane depolarizes to the extent that its potential reaches positive values due to the sodium influx. It repolarizes to hyperpolarized values of more than −70 mV driven by the potassium efflux and then relaxes back to its resting potential. The hyperpolarization stage is important. It is needed to relieve the inactivation of the sodium channels, and prepare the membrane for the next round of activation and action potential generation.

As shown in Figure 19.2, the time course of the flow of sodium ions through their channels differs from that of potassium ions through their channels. Sodium and potassium channels are both activated by depolarization, but the sodium channels open for a short time and then shut down

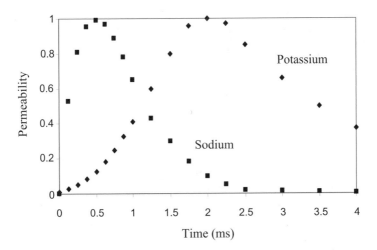

FIGURE 19.2. Time courses of sodium and potassium currents through a plasma membrane: Plotted are relative permeability values for the two ion species, that is, the number of open channels per unit surface area as a fraction of the peak number open channels/unit area for each ion species. Time is measured from initiation of an action potential by a threshold exceeding depolarization.

even if the membrane potential has not changed from its depolarized value. That is, the sodium channels inactivate after being activated for a short time. Potassium channels activate more slowly than sodium channels, reaching their peak conductivity values after the sodium channels shut down. However, once they reach their peak conductivity values, the potassium channels remain open for some time. That is, potassium channels are noninactivating.

To summarize, sodium channels launch and generate the rising part of the action potential; positive feedback sustains it, and potassium channels generate the slowly declining following portion of the pulse. The positive feedback loop of the sodium channels is the key to generation of action potentials that propagate down axons without diminishing. Once a small depolarization is started in a limited region of space, the spread of ions laterally to neighboring spaces triggers a growing depolarization in that space and from there to the next neighboring space and so on. The pulse does not die out since a small depolarization triggers a large one, that is, the signal is amplified, as a result of the positive feedback. Thus, once started, a pulse propagates along without losing strength, that is, the pulses are regenerative. The existence of a threshold all-or-nothing response is a consequence of the positive feedback from the sodium channels and negative feedback from the potassium channels. Negative feedback is present because the depolarization-induced exit of potassium hyperpolarizes the membrane, thereby acting negatively to shut down the potassium ion flow. As a result of both forms of feedback there is an all-or-nothing response. It takes a certain amount of depolarization to trigger the positive feedback dominated sodium influx, and once started the process runs to completion.

19.3 Hodgkin–Huxley Equations Describe How Action Potentials Arise

In 1952, Alan Hodgkin and Andrew Huxley constructed a mathematical model of the time course of sodium and potassium permeability changes in the giant axon of the squid. There are two ways to preserve and speed up the conduction of electrical pulses over long distances in an axon: either by increasing the diameter or by adding an insulating fatty sheath (myelination). The squid giant axon is unmyelinated and makes use of the first of the two solutions to the conduction problem. It is enormous, with a diameter that can reach 1mm. It can be easily manipulated in the laboratory, and is thus is an ideal system for laboratory studies of how pulses are transmitted down nerves. The Hodgkin–Huxley model represents the culmination of decades of study by Hodgkin, Huxley, Katz, and many others on how electrical pulses are produced, and how they propagate, in nerve tissue. The development of the model was preceded by their detailed experimental studies of the flow of electrical current through the surface membrane of the squid giant axon. In these studies, a method, the "voltage clamp," was developed that allowed researchers to directly measure current flow across the membrane of the axon.

The basic elements of the Hodgkin–Huxley model are a set of ionic currents separated into distinct contributions from sodium and potassium ions, and from the leakage current, representing the fairly constant and voltage-independent, background leakage of ions through the plasma membrane. These three currents are related to membrane permeabilities, which are expressed in terms of ionic conductances g_{Na}, g_K and g_L, and the displacements of the membrane potential from the reversal potentials for the ion species. The relationship between current, conductance, membrane potential, and reversal potential is from Ohm's law,

$$I_x = g_x (V_m - E_x). \tag{19.4}$$

In the above, X denotes the ion species under consideration, V_m the membrane potential, I_x is its current, E_x is its reversal potential, and g_x is its conductance, the inverse of its resistance R_x, that is,

$$g_x = 1/R_x. \tag{19.5}$$

In elementary circuits, R is a constant, and Ohm's law says that current and voltage are linearly related. This is not the case for a biological membrane. Although the linear form of Ohm's law is retained in the Hodgkin–Huxley equations, the sodium and potassium conductance are no longer constants, but instead are treated as functions of the membrane voltage.

Because of the nonconducting nature of the lipid bilayer the plasma membrane can be thought of as a capacitor that permits a buildup of charge on its surfaces. If a current is applied part will contribute to charging up the plasma membrane and part will flow through the membrane into the cell. The net ionic current is the sum of the sodium, potassium, and leakage currents plus any external applied or synaptic currents. The capacitive current I_C is, from Faraday's law,

$$I_C = C \frac{dV_m}{dt}. \tag{19.6}$$

In this expression, C is the capacitance, and dV_m/dt is the time rate of change in the voltage. The membrane voltage does not instantaneously adapt to the flow of charge in and out of the cell. Instead, it takes a certain amount of time for charge to build up on the membrane and the voltage to adjust to the current flow. The complete circuit equation, expressing the conservation of charge, is

$$C \frac{dV_m}{dt} = -g_K (V_m - E_K) - g_{Na} (V_m - E_{Na}) - g_L (V_m - E_L) + I, \tag{19.7}$$

where I is any applied or synaptic current and E_L is leakage current.

An equivalent circuit for the membrane and its ion channels is presented in Figure 19.3. The Hodgkin–Huxley equation describes an electrical circuit

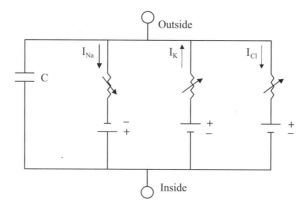

FIGURE 19.3. Electrical circuit described by the Hodgkin–Huxley equation: As is customary, the capacitor residing in the capacitive current branch is represented by a pair of short parallel lines, and the batteries in the ionic current branches by pairs of short-long lines denoting cathode (−) and anode (+). Arrows crossing through the resistors symbolize their non-ohmic character. Arrows show the direction of the current flow, sodium and chloride ions inward and potassium ions outward.

with branches for sodium current, potassium current, leakage current, capacitive current, and any external or synaptic currents. If desired, any other ionic currents such as chloride current can be added to the equation in the same way. Ionic currents are small compared to the ionic concentrations outside and inside a neuron and thus do not change the concentrations appreciably. Therefore, the concentration gradients are represented in the circuit diagram by simple batteries that supply currents, one type of battery for each ion species. The voltage across the terminals in these batteries is the reversal potential for that ion species represented by the Nernst equation: approximately +60 mV for sodium, −90 mV for potassium, and −75 mV for chloride. Time and voltage-varying resistors are present in each branch of the circuit.

19.4 Ion Channels Have Gates that Open and Close

The non-ohmic character of membrane resistance is of considerable significance. In a *non-ohmic*, or *rectifying*, resistor, the flow of ions through a particular kind of ion channel is directional. That is, ions will pass through the ion channels in one direction more easily than in the other direction. If the ions pass more easily into the cell, the ion channel is *inward rectifying*, and if they pass more readily out of the cell, the ion channel is *outward rectifying*. The rectifying character of the membrane permeability naturally leads to the introduction of gates that open and close to permit the flow of ions through the channels.

To model the opening and closing of sodium and potassium channels in response to depolarization of the nerve membrane, Hodgkin and Huxley introduced three gates. Recall that the sodium channels rapidly open as the membrane is depolarized and then shut after a short time interval. Hodgkin and Huxley describe this behavior in terms of two gates: an activation gate and an inactivation gate. Initially, at the resting membrane potential, the activation gate is shut and the inactivation gate is open. Once the threshold depolarization, for activation is exceeded, the activation gate opens allowing sodium ions to enter the cell. Further increases in depolarization allow the activation gates to open fully, but the inactivation gate, which operates more slowly, closes in response to the increasing depolarization, thereby shutting down the flow of sodium ions.

The third gate introduced by Hodgkin and Huxley is the activation gate for potassium. This gate is also opened by depolarization, but much more slowly than the activation gate for sodium. As the gate for the potassium channel opens in response to depolarization, potassium ions leave the cell, thus opposing the effect of the sodium entry, and eventually overcoming it to repolarize the cell. The repolarization accelerates the further closing of sodium channels and leads to the hyperpolarization of the membrane. When this occurs the inactivation gate is opened; the activation gate is shut, and the membrane relaxes back to the resting potential, ready for another round of excitation.

Activation and inactivation gates have sigmoidal (S-shaped) dependencies on voltage. The differences between activation and inactivation gates are illustrated in Figure 19.4. As shown in the figure, inactivation gates open when the membrane is hyperpolarized and shut when the membrane is depolarized. Activation gates work the opposite way. They open when the membrane is depolarized and shut when the membrane is hyperpolarized. Hodgkin and Huxley called the sodium and potassium gates m-, h-, and n-gates. That is, a fast m-gate describes the activation process and a slow h-gate describes the inactivation process for the sodium entry, while the potassium channels are controlled by a single n-gate. The sigmoidal dependence of their opening and closing was approximated by algebraic terms— cubic for the m-gate, linear for the h-gate, and quadratic for the n-gate. In mathematical terms, dimensionless variables m, h, and n were introduced, each a function of V_m. These variables assume values between 0 and 1 corresponding to completely closed and completely open. The sodium conductance was expressed as

$$g_{Na} = \bar{g}_{Na} m^3 h. \tag{19.8}$$

Similarly, the potassium conductance was written as

$$g_K = \bar{g}_K n^4. \tag{19.9}$$

The quantities $m^3 h$ and n^4 represent the fractions of open sodium and potassium channels, respectively, and the g-bars are constants representing the

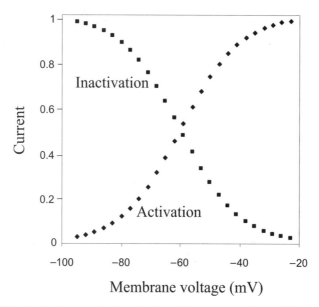

FIGURE 19.4. Activation and inactivation functions and resulting membrane currents: The activation (diamonds) and inactivation (squares) functions represent the amount of current flowing through the ion channels relative to the maximum possible currents. These quantities are plotted as functions of the membrane voltage.

maximum conductances occurring when the channels are completely open. These core equations are supplemented by a set of rate constants and rate equations for the m-, h-, and n-gate variables.

19.5 Families of Ion Channels Expressed in Plasma Membrane of Neurons

The human genome contains hundreds of genes that encode ion channel subunits. The repertoire of ion channels is further increased through extensive use of alternative splicing. Still more ion channel forms are generated by mixing together different combinations of subunits. The resulting ion channels differ from one another in their biophysical properties in several ways. The differ in their

- ion selectivity and direction of ion movement,
- gating mechanism,
- topology and assembly plan,
- time course of activation and inactivation, and
- reversal potentials for activation and inactivation.

TABLE 19.2. The four families of ion channels found in the plasma membrane of neurons: Except for the outward directed potassium channels, ion channels allow for the inward selective diffusion of the ion species listed in the second column. Abbreviations—Hyperpolarization-activated cyclic nucleotide gated (HCN); chloride channel of the CLC family (ClC); nicotinic acetylcholine receptor (nAChR); γ-aminobutyric acid Type A, Type C (GABA$_{A,C}$); 5-hydroxytryptamine Type 3 (5-HT$_3$); α-amino-3-hydroxyl-5-methyl-4-isoxazole propionate acid (AMPA); N-methyl-D-aspartate (NMDA).

Ion channel	Selectivity	Subunit topology	Channel assembly
6TM cation		6TM, loop	Tetrameric
		24TM, 4 loops	Monomeric
Calcium	Ca^{2+}		
HCN	Na$^+$, K$^+$		
Potassium	K$^+$		
Sodium	Na$^+$		
Voltage-gated anion		18IM	Dimeric
ClC	Cl$^-$		
Cys-loop receptor		4TM	Pentameric
nAChR	Cations		
GABA$_{A,C}$	Anions (Cl$^-$)		
Glycine	Anions (Cl$^-$)		
5-HT$_3$	Cations		
Glutamate receptor		3TM, loop	Tetrameric
AMPA	Na$^+$, K$^+$		
kainate	Na$^+$, K$^+$		
NMDA	Ca^{2+}		

Ion channels have several central properties. One of these is selectivity—the ability to pass some ion species through the membrane rapidly while preventing passage of other ion species. A closely related property is direction of ion movement. Ion channels let ions pass through in one way only. Some let ions into the cell while others let ions leave the cell, but none allow two-way passage. Another essential property of an ion channel is gating—the ability to open and close the pore through which the selected ions pass in response to voltage across the membrane and to ligands, intracellular and extracellular.

The different kinds of ion channels can be grouped into four large families. As can be seen in Table 19.2, two of the families open and close in response to changes in membrane voltage, and are thus said to *voltage-gated*. One of these families permits the selective passage of cations and the other family allows the passage of anions. The other two families are gated by ligands. For that reason they are designated as *receptor ion channels*. The voltage-gated ion channels, of which there are many different kinds, can be further distinguished by their time courses for activation and deactivation, and by their reversal potential. That is, by how fast they open and close, and at which membrane voltages, as was discussed earlier in the chapter. Finally, the different families of ion channels are characterized by their subunit composition (assembly plan) and by the manner in which the polypeptide chain for each subunit threads back and forth through the membrane (topology).

19.6 Assembly of Ion Channels

Ion channels are assembled from multiple subunits arranged in regular ways to form pores through which the ions diffuse. The threading patterns characteristic of the subunits belonging to each of the four families of ion channels are presented in Figure 19.5. In the case of the voltage-gated cation channels, the core unit, or basic module, consists of a polypeptide segment that passes back and forth through the plasma membrane six times so that the N- and C-terminals are both on the intracellular (cytoplasmic) side. As illustrated in Figure 19.5(a), there is a loop (helix) between the fifth the sixth transmembrane segments (TM5, TM6). The loop regions from each subunit jointly form the conducting pore and selectivity filter (the nar-

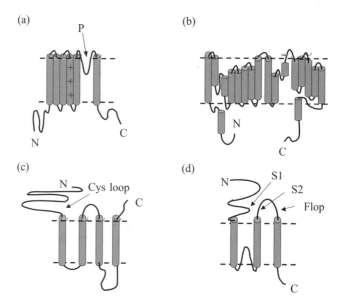

FIGURE 19.5. Ion channel subunit threading patterns: (a) Voltage-gated potassium channel—Transmembrane helices 5 and 6 along with the indicated loop form a conducting pore. The positively charged residues in transmembrane segment 4 plus portions of transmembrane helices 1–3 form the voltage sensor. (b) Bacterial ClC showing the threading of 18 alpha helices of varying lengths. (c) Acetylcholine receptor ion channel indicating the presence in the amino terminal portion of a conserved Cys loop near the membrane surface. (d) AMPA receptor ion channel showing the locations of S1 and S2 segments, which, along with the Flop segment, form the ligand-binding domain.

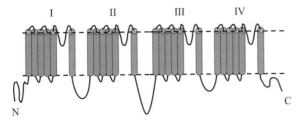

FIGURE 19.6. Single chain voltage-gated cation channel topology: Shown in the figure is a single 6 × 4-pass polypeptide chain topology characteristic of bacterial KscA potassium channel.

rowest part of the pore) that allows certain ion species to flow through. In channels that are voltage-gated, transmembrane segment 4 (TM4) contains a number of positive changes (basic arginine or lysine amino acid residues), and along with elements of TM1–3 functions as the voltage sensor. In channels whose gates are controlled by intracellular ligands, binding sites for the regulators are location on the C-terminal segment that follows TM6.

Voltage-gated cation channels are assembled in one of two ways. In the Shaker family of voltage-gated potassium channels, four of six TM core units are arranged into the ring. In many voltage-gated sodium channels and voltage-gated calcium channels, the amino acid residues replicated among the four subunits form domains belonging to a single, long (2000 amino acid residues), 6 × 4 = 24-pass, polypeptide chain. In either assembly mode, the result is a fourfold symmetric ring structure in the plasma membrane with a pore in the center. A typical threading pattern for a 24-pass transmembrane protein, the bacterial KscA potassium channel, is presented in Figure 19.6.

The threading pattern and assembly plan for voltage-gated anion channels are radically different from those of the cation channels. As shown in Figure 19.5(b), the polypeptide chain of a ClC chloride channel subunit consists of 18 distinct alpha helices of varying lengths. Some of the helices pass through the membrane, but others do not and one helix resides entirely in the cytoplasm. As will be discussed shortly, the anion channel is formed by two of these subunits, and in place of a single pore in the center there are two off-center pores.

The two families of ligand-gated ion channels can be further divided into several smaller families according to the types of ligands, or neurotransmitters, being bound. Some ligand-gated ion channels permit the passage into the cell of cations, while others allow anions such as chloride to enter the cell. When an anion enters the cell the net charge inside the cell becomes more negative and the membrane voltage shifts to more hyperpolarized values. Since action potentials are generated by threshold-exceeding depolarizations, the entry of negative ions such as chlorine suppresses firing. Neurotransmitters such as $GABA_{A,C}$ and glycine that act through anionic receptors are referred to as *inhibitory neurotransmitters*. Conversely, neurotransmitters such as glutamate and acetylcholine that act though cationic receptors are referred to as *excitatory neurotranmitters*.

19.7 Design and Function of Ion Channels

Ion channels are modular in design and function as small molecular machines. Voltage-gated ion channels are modular in design. Each ion channel is a small molecular machine with core (alpha) subunits that form its pore, voltage sensor, and selectivity filter. Voltage-gated sodium, potassium, and calcium channels possess one or two regulatory β subunits. Calcium channels contain not only a 6TM α1 core subunit and a cytoplasmic β subunit, but also an extracellular α2 subunit linked to a single-pass transmembrane δ subunit by disulfide bridges (the α2-δ subunit), and, in skeletal muscle and some neurons, a 4TM γ subunit. These subunits all have regulatory roles in the operation of the channels.

The gates of the ion channels are electro- and chemo-mechanical devices that open and close by means of conformational changes. The voltage sensor (or ligand-binding domain) is connected to the helices forming the pore. In response to voltage changes (or ligand binding), the sensor module does mechanical work on the helices lining the pore to alter its conformation from a closed to open one. A variety of regulatory mechanisms are encountered in the various ion channels. These include direction-sensitive blocking elements and regulation by second messengers, G protein βγ subunits, and protein kinases and phosphatases.

Receptor ion channels alter conformation in response to ligand binding. As was the case for other signaling receptors, the underlying mechanism is usually an allosteric one. The receptors contain multiple ligand-binding sites. Binding of a ligand at a regulatory site in the protein changes the electrostatic environment resulting in the stabilization of the protein in a different subset of conformations. As a result the binding properties at active sites are changed in a reversible way. The mechanism is an indirect one in that binding at the regulatory site influences binding at a physically removed primary, or active, site through conformational changes rather than by direct physical contact.

19.8 Gates and Filters in Potassium Channels

Potassium channels have spatially separated voltage gates and selectivity filters. A crucial feature of a voltage sensitive ion channel is its ability to sense changes in voltage and use these changes in voltage as an energy source to drive the opening and closing of the channel gate. The three-dimensional atomic structures of several potassium channels have been determined in a series of landmark studies. The resulting data reveal how

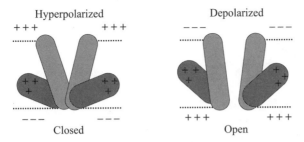

FIGURE 19.7. Potassium channel voltage sensor: In the absence of depolarizing currents, the membrane is hyperpolarized. The positive charges in helix S4, along with the other components of the S1 to S4 voltage sensor, maintain the channel in a closed configuration. When depolarizing currents are present the S1 to S4 helices move towards the outer leaflet, and as a result the pore helices S5 and S6 shift their orientation, thereby opening the pore.

the channels open and close in response to changes in membrane voltage and how these machines selectively permit potassium ions to flow through.

The representation of the subunit as being composed of equal-sized helices each passing through the membrane is a simplification that emphasizes the sequential connections between the helices. In the three-dimensional structure of the subunit some of the helices are not only not perpendicular to the plane of the membrane but actually lie within the membrane parallel to its surfaces. The *voltage-operate gate* works in the following way. Charged groups in S4 together with a portion of S3 and the loop connecting it to S4 form a charged paddle. This paddle lies entirely within the lipid bilayer, and changes its orientation in response to changes in membrane voltage. As illustrated in Figure 19.7, when the membrane is depolarized the paddle tugs on other helices pulling open the pore to let the ions through.

The selectivity filter is separated in position from the gate. The subunit 6-pass chain contains an intramembrane loop between segments 5 and 6. These last two segments and the loop from each of the four subunits of the protein form the ion selectivity filter and pore. As can be seen in Figure 19.8, the overall shape of the channel is that of an inverted tepee with the broad portion including the intramembrane loops near the top and the narrowest portion at the bottom.

19.9 Voltage-Gated Chloride Channels Form a Double-Barreled Pore

The second set of voltage-gated ion channels is the *anion, or chloride, channel family*. These ion channels are often referred to as the *ClC family* to distinguish them from the cystic fibrosis transmembrane conductance

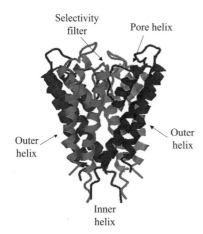

Selectivity filter   Pore helix

Outer helix   Outer helix

Inner helix

FIGURE 19.8. Potassium channel pore: Shown is the tetramer formed by membrane-spanning S5 and S6 segments and connecting intracellular loop from each domain from a bacterial (KscA) potassium channel. The data determined by means of X-ray crystallography were rendered as a ribbon diagram showing the α-helices and connecting loops. The top of the tetramer lies in the extracellular space and the bottom is in the intracellular region. The figure was generated using Protein Explorer with atomic coordinates deposited in the Brookhaven Protein Data Bank under accession code 1BL8.

regulator (CFTR) chloride channels implicated in cystic fibrosis, and the GABA and glycine receptors, which are ligand-gated and not voltage-gated. The ClC ion channels are found in the plasma membrane and in the membranes of internal organelles. These anion channels are not responsible for generating action potentials, but instead regulate the excitability of muscle and nerve cells, the acidification of internal organelles, and cell volume.

The X-ray crystal structure of two bacterial ClC family members reveal a radically different architecture and pore structure from the voltage-gated cation channels. In place of the four (or five) subunits used to form a centrally positioned conducting pore, only two subunits are utilized. Each subunit is quite large, and consists of 18 α-helices, some short and others long enough to pass through the membrane. Each subunit is large enough to form its own pore, and two of these subunits assemble into the chloride ion channel.

A side view of the ion channel formed by a pair of identical subunits is presented in Figure 19.9. As can be seen, the channel possesses a pair of off-center pores. As was the case for the cation channels the helices are oriented in a variety of ways. These varied orientations are not arbitrary but instead bring together the charged residues in the right way, from different positions in the polypeptide chain, to form the gate and selectivity filter.

19.10 Nicotinic Acetylcholine Receptors Are Ligand-Gated Ion Channels

Ligand binding controls the gates in the other two superfamilies of ion channels. One of these families is the *cys-loop, or acetylcholine, family*. In this family, the polypeptide chains comprising the subunits pass through the plasma membrane four times. The name for this family is derived from one

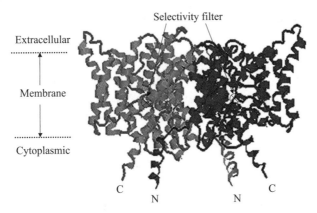

FIGURE 19.9. The ClC chloride ion channel determined by X-ray crystallography: A side view of the dimeric chloride-selective ion channel is presented. The approximate locations of the selectivity filter in the two subunits are denoted by the circled Xs. The figure was generated using Protein Explorer with atomic coordinates deposited in the Brookhaven Protein Data Bank under accession code 1KPK.

of the structural properties of the subunits. Their extracellular amino terminal contains a cys-loop—a pair of disulfide-bridged cysteines separated by a stretch of 13 amino acid residues.

Nicotinic acetylcholine receptor ion channels are responsible for excitatory signaling at the connections between nerve and muscle cells, or neuromusclular junctions, located throughout the body. There are two types of acetylcholine receptor ion channels—*nicotinic* and *muscarinic*. The former bind to the nicotine whereas the latter binds to the muscarine, like nicotine an alkaloid. Those nicotinic receptors, located at neuromuscular junctions, are termed *muscle-type nAChRs*, while those found at neuron-to-neuron contacts in the spinal cord ganglia and in the brain are referred to as *neuronal nAChRs*. As their name indicates, the neuronal nicotinic receptors mediate nicotine addiction in smokers, whereas muscle-relaxing, paralyzing, and spasm-promoting toxins target neuromuscular nAChRs. Examples of toxins that target nicotinic acetylcholine receptors are atropine, curare, and scopolamine, all from plants, and cone snail and cobra toxins.

The nAChRs are pentamers, assembled in a symmetric or nearly symmetric ring with a pore in the center. There are 17 known vertebrate subunits. These subunits belong to one of five groups—α, β, γ, δ, and ε. Neuronal receptors are typically homopentamers of a particular α subunit or heteropentamers of α and β subunits. The muscle-type of nAChR is less varied. These receptors consist of two α_1 subunits, plus one each of a β_1, δ, and γ or ε subunit.

Perhaps more has been learned about the acetylcholine receptor than any other neurotransmitter-gated ion channel. It has been studied with increasing resolution by electron microscopy for 30 years. Receptor ion channel subunits consist of several segments connected by flexible disordered loops. As a result of this architecture it is difficult to impossible to make crystals from entire subunit chains. However, crystals can be fabricated from portions of the subunits. Receptor ion channels, like so many other proteins, are highly modular, and this modularity can be exploited to arrive at a consistent picture of how the ion channels work by studying the channels part-by-part.

Each acetylcholine receptor subunit possesses an N-terminal ligand-binding domain, a transmembrane segment, and an intracellular region. As depicted in Figure 19.5(c), the extracellular, ligand-binding domain is quite large. There are four transmembrane segments and there is a substantial cytoplasmic region. The ligand binding domain is capable of binding not only to agonists, but also to antagonists. When an agonist binds the ligand binding domain the channel formed by the transmembrane segments rapidly opens to permit the passive diffusion of cations leading to a depolarization of the muscle cell's postsynaptic membrane and contraction of the muscle.

The acetylcholine-binding protein (AChBP) from the snail *L. stagnalis* is almost identical in size and in residues crucial for ligand binding to the N-terminal ligand binding domains of nAchRs. It lacks the transmembrane and cytoplasmic segments present in mammalian AChRs. It can be crystallized and serves as an archetypal example of a ligand binding domain. Shown in Figure 19.10 are two views of the pentamer formed by five iden-

(a)   (b)

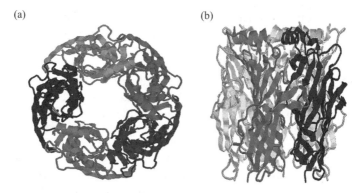

FIGURE 19.10. Structure of the ligand-binding domain of the acetylcholine-binding protein: (a) Top view showing the ringlike arrangement of the five subunits to form a conducting pore. Each subunit is shown in a different shade of gray. (b) Side view of the ring highlighting the two units located in the foreground. The N-terminus is at the top and the C-terminus is at the bottom of each subunit. The figure was generated using Protein Explorer with atomic coordinates deposited in the Brookhaven Protein Data Bank under accession code 1I9B.

tical AChBPs. The pore formed by the five subunits is clearly visible in this figure. As might be expected the residues lining the inner wall of the pore are highly hydrophilic, and many are charged. A ligand-binding site is present in a cavity formed at the interface between each adjacent subunit. The cavity, positioned towards the outside of the ring, is formed by a set of loops from one subunit and a series of beta strands from the other unit. Since there are five interfaces, the ligand binding domain can bind up to five ligands.

19.11 Operation of Glutamate Receptor Ion Channels

The second family of ligand gated ion channels is the *glutamate family*. Glutamate receptor ion channels are prominent components of sensory processing systems and contribute greatly to learning and memory processes. They are found on the postsynaptic side of sensory/information-processing neurons. Referring back to the Hodgkin–Huxley model, the cys-loop and glutamate receptor ion channels give rise to synaptic currents in the postsynaptic cells. The flow of ions through these channels leads to changes in membrane voltage. By this means, the ligand-gated ion channels communicate and work together with the voltage-gated channels in these cells to control membrane excitability and convey signals. In addition to mediating changes in membrane voltage, calcium is itself a second messenger and thus functions as a key signaling intermediary.

The core chains of glutamate receptors pass through the plasma membrane three times. The N-terminal is positioned in the extracellular spaces and the C-terminal is situated in the cytoplasm as illustrated in Figure 19.5(d). A pore-forming reentrant (M2) loop is situated between the first two transmembrane portions of the chain. Two extracellular sequences (S1 and S2) form the ligand-binding domain. One of these is located just after the amino terminus; the other occurs between M3 and M4. The conducting channel is assembled from four subunits. The subunits may be identical or they may be mixtures of different subunit types. Because of their modular structure, one kind of subunit may be swapped for another in response to regulatory signals thereby altering the biophysical properties of the channel.

Several kinds of receptors bind the neurotransmitter glutamate. Some ionotropic glutamate receptors (iGluRs) bind AMPA with high affinity while others bind either kainite or NMDA. The subunits belonging to each of the three groups of iGluRs are listed in the table. Like the acetylcholine receptors, the glutamate receptors are highly modular. They contain an N-terminal extracellular domain followed by an extracellular ligand binding domain, three transmembrane segments plus a reentrant loop, and a C-terminal cytoplasmic region.

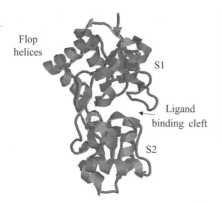

Flop
helices

S1

Ligand
binding cleft

S2

FIGURE 19.11. The S1S2 ligand-binding domain of the AMPA receptor: The S1, S2, and Flop regions of the polypeptide chain used in the S1S2 construct are identified in Figure 19.5(d). The figure was generated using Protein Explorer with atomic coordinates deposited in the Brookhaven Protein Data Bank under accession code 1FW0.

Glutamate receptor ion channels are responsible for most of the excitatory signaling in the brain. These receptors are most likely homo- and heterotetrameric assemblages of subunits. They assemble in several stages. Individual subunits first associate to form dimers, and these dimers then associate to form dimers of dimers. Like the voltage-gated ion channels, the receptor ion channels have open and closed conformations. Ligand binding results in stabilization of the open configuration through an allosteric mechanism. The receptors do not remain open indefinitely, but rather desensitize and go back to a closed conformation soon (within milliseconds) after ligand binding.

The S1S2 ligand-binding domain of an AMPA receptor is depicted in Figure 19.11. The S1S2 domain depicted in this figure is a laboratory construct made by cutting the polypeptide chain and then joining together the S1 and S2 + Flop segments using a 13-residue-long linker. The result is a ligand-binding domain that replicates the actions of the natural extracellular unit without the encumbrances of the other loops and transmembrane segments. As can be seen in Figure 19.11 the S1 and S2 semidomains form a pair of jaws that grip the ligand. In the absence of ligand binding, the jaws are open, but they close upon binding the ligand. The amount of jaw closing is variable, and depends on which of several different kinds of ligands are being bound. Binding by antagonists stabilizes the S1S2 domain in an open configuration; the jaws partially close in response to binding by partial agonists such as kainite, and close more fully in the presence of glutamate.

The jaws are mechanically coupled of the gate. A linker that connects the jaws to the membrane-spanning helices provides the mechanical coupling. In the dimeric form, there are two linkers, one for each subunit. The linkers move apart from one another as the jaws close and move nearer one another when the jaws open. This movement opens and closes the gate. When the jaws are open the gate is closed and when the jaws close through agonist binding the gate opens.

Ions flow into the cell when the gate opens. The gate does not remain open for very long, just a few milliseconds. The gate then closes even though the ligand remains bound. The closing of the gate is brought on by rearrangements of the dimer interface. The interface between subunits in the dimer is placed under strain by the gate opening. The rearrangements of the dimer interface relieve this strain by disengaging the conformation changes brought on by agonist binding and shutting the gate, resulting in receptor desensitization (to the ligand binding).

References and Further Reading

General References

Hille B [1992]. *Ionic Channels of Excitable Membranes* (2nd edition). Sunderland MA: Sinauer Associates, Inc.
Kaczmarek LK, and Levitan IB [1987]. *Neuromodulation.* New York: Oxford University Press, Inc.
Nicholls JG, Wallace BG, Fuchs PA, and Martin AR [2001]. *From Neuron to Brain: A Cellular Approach to the Function of the Nervous System* (4th edition). Sunderland MA: Sinauer Associates, Inc.

Voltage-Gated Potassium Channel

Doyle DA, et al. [1998]. The structure of the potassium channel: Molecular basis of K^+ conduction and selectivity. *Science*, 280: 69–77.
Jiang YX, et al. [2003]. X-ray structure of a voltage-dependent K^+ channel. *Nature*, 423: 33–41.
Jiang YX, et al. [2003]. The principle of gating charge movement in a voltage-dependent K^+ channel. *Nature*, 423: 42–48.
Perozo E, Cortes DM, and Cuello LG [1999]. Structural rearrangements underlying K^+-channel activation gating. *Science*, 285: 73–78.
Roux B, and MacKinnon R [1999]. The cavity and pore helices in the KcsA K^+ channel: Electrostatic stabilization of monovalent cations. *Science*, 285: 100–102.
Schrempf H, et al. [1995]. A prokaryotic potassium ion channel with two predicted transmembrane segments from *Streptomyces lividans*. *EMBO J.*, 14: 5170–5178.
Yellen G [2002]. The voltage-gated potassium channels and their relatives. *Nature*, 419: 35–42.

Voltage-Gated Sodium Channels

Caterall WA [2000]. From ionic currents to molecular mechanisms: The structure and function of voltage-gated sodium channels. *Neuron*, 26: 13–25.

Voltage-Gated Anion (Chloride) Channels

Dutzler R, Campbell EB, and MacKinnon R [2003]. Gating the selectivity filter in ClC chloride channels. *Science*, 300: 108–112.
Dutzler R, et al. [2002]. X-ray structure of a ClC chloride channel at 3.0 Å reveals the molecular basis of anion selectivity. *Nature*, 415: 287–294.

Acetylcholine Receptor Ion Channels

Brejc K, et al. [2001]. Crystal structure of an Ach-binding protein reveals the ligand binding domain of nicotinic receptors. *Nature*, 411: 269–276.

Cordero-Erausquin M, et al. [2000]. Nicotinic receptor function: New perspectives from knockout mice. *Trends Pharmacol. Sci.*, 21: 211–217.

Karlin A [2002]. Emerging structure of the nicotinic acetylcholine receptors. *Nature Rev. Neurosci.*, 3: 102–114.

Miyazawa A, Fujiyoshi Y, and Unwin N [2003]. Structure and gating mechanism of the acetylcholine receptor pore. *Nature*, 423: 949–955.

Miyazawa A, et al. [1999]. Nicotinic acetylcholine receptor at 4.6 Å resolution: transverse tunnels in the channel wall. *J. Mol. Biol.* 288: 765–786.

Glutamate Receptor Ion Channels

Armstrong N, and Gouaux E [2000]. Mechanism for activation and antagonism of an AMPA-sensitive glutamate receptor: Crystal structure of the GluR2 ligand binding core. *Neuron*, 28: 165–181.

Ayalon G, and Stern-Bach Y [2001]. Functional assembly of AMPA and kainite receptors is mediated by several discrete protein-protein interactions. *Neuron*, 13: 103–113.

Madden DR [2002]. The structure and function of glutamate receptor ion channels. *Nature Rev. Neurosci.*, 3: 91–101.

Sun Y, et al. [2002]. Mechanism of glutamate receptor desensitization. *Nature*, 417: 245–2253.

Problems

19.1 Using the Nernst equation with $R = 8.31$ Joule/deg (K) mol and $F = 9.65 \times 10^4$ C/mol, calculate the equilibrium potential for potassium, sodium, and chloride ions at 20 degree Celsius for the inside and outside concentrations presented in Figure 19.1.

19.2 Calculate the reversal potential using the Goldman–Hodgkin–Katz equation for the concentrations presented in Table 19.1 with the following ratios of permeability constants: $p_{Na}/p_K = 0.05$ and $p_{Cl}/p_K = 0.2$.

19.3 Sketch the time course of the sodium and potassium permeabilities using Figure 19.2 as the template. Sketch the behavior of the action potential produced by the sodium and potassium currents over that time period. Assume a maximum positive value of +20 mV and a maximum negative value of −70 mV, and indicate in the plot where the action potential will reach these maximum and minimum values.

20
Neural Rhythms

The focus in the last chapter was on the potassium and sodium ion channels responsible for generating action potentials in neurons. This subject is developed further in the present chapter where the emphasis shifts towards how ion channels generate *rhythmic discharges in neurons*. Voltage-gated ion channels, currents flowing through ligand-gated ion channels located in chemical synapses (synaptic currents), and electrical synapses (gap junctions)—all these contribute to the generation of neural rhythms. A variety of patterns can be produced, some in small populations of cells and others in large ones. Rhythmic patterns among large populations of neurons are associated with different sleep states and with states of arousal and attention. Small circuits of neurons known as *central pattern generators* generate rhythmic behavior in motor systems. These circuits control heartbeat, regulate respiration, and control locomotion, chewing, and digestion.

This subject of the nervous system concludes in the last chapter, where the focus will be on the chemical synapse—on how it is organized and how it promotes learning and memory formation. The organization of the sensory cortices will be looked at along with how they process sensory information that is relayed in from sensory organs. Special attention will be given to the glutamate receptor ion channels and their central role in sensory information processing, learning, and memory formation.

This chapter is divided into three parts. In the first part, pacemaker cells will be introduced in a discussion of heartbeat. In the middle part of the chapter, the low frequency synchronous firing of large populations of cells will be examined. The main focus will be on sleep oscillations and how these forms of rhythmicity may change into abnormal epileptic seizures. The third part of the chapter will deal with how central pattern generators drive muscle contractions, drawing on several invertebrate systems for examples.

20.1 Heartbeat Is Generated by Pacemaker Cells

Heartbeat is generated by rhythmic discharges from nerve cells situated in a small region of the heart called the *sinoatrial node* (SAN) located in the wall of the right atrium. Impulses from the SAN are distributed rapidly throughout the heart by a specialized conduction system and cause the rhythmic contractions of the heart muscle. Impulses from the SAN are sent to the *atrioventricular node* (AVN) and left and right bundles of Purkinje fibers, and then to locations throughout the heart. The cells in the SAN that generate the rhythmic impulses do not require rhythmic input, but instead generate the rhythmic patterns by themselves. Other neurons in the conduction system respond to the pacemaker cells by firing in synchrony with them. In the absence of input from the SAN, nerve cells in the AVN and the Purkinje fibers can act as pacemakers, but they fire at a lower frequency. These cells, along with other nonpacemaking cells, can be driven by the SAN neurons and made to fire in synchrony with them.

The ability to generate rhythmic discharges is due to the mix of ion channels expressed by the neurons in the SAN and other regions of the heart. Since the function of the neurons in each of the regions is slightly different, the neurons in these regions express slightly different mixes of ion channels. The main ion channels contributing to the action potentials in the heart are listed in Table 20.1. The ion channels each have a specific function, as listed in column 3. The meanings of the terms appearing in that column are illustrated in Figures 20.1 and 20.2.

The first entry in Table 20.1, the *sodium channel*, generates fast upstrokes that are pronounced in non-SAN neurons. *Funny (f) channels* in the heart are responsible for diastolic depolarization in SAN neurons. These same ion channels are called I_h *channels* in the brain and are referred to as *HCN channels* in the next section. The term *diastolic depolarization* refers to the rising (depolarization) portion of the action potential occurring at the very end of the SAN action potential. This depolarization allows the SAN neurons to immediately fire another action potential after completion of a previous one, and thus enables the SAN neurons to function as the pacemakers for the entire heart. The firing appears as the rising part of the action potential for the SAN neuron that occurs after the late repolarization stage, as shown in Figure 20.2.

Neurons express several different kinds of calcium channels. Some calcium channels are called "L-type" while others are referred to as being "N-" or "P/Q-" or "R-" or "T-type." Most of the calcium channels are acti-

TABLE 20.1. Ion currents contributing to the firing properties of neurons in the heart.

Cardiac current	Description	Function
I_{Na}	TTX sensitive, inward Na^+ current	Fast upstroke
I_f	Hyperpolarization-activated inward nonselective cation (Na^+, K^+) current	Diastolic depolarization
$I_{Ca,L}$; $i_{Ca,T}$	L- and T-type inward Ca^{2+} currents	Plateau
$I_{Kr,Ks}$	Rapid (r) and slow (s) delayed rectifying outward K^+ currents	Late repolarization
$I_{t0,sus}$	4-AP sensitive, transient (t0) and sustained (sus) outward K^+ currents	Early repolarization

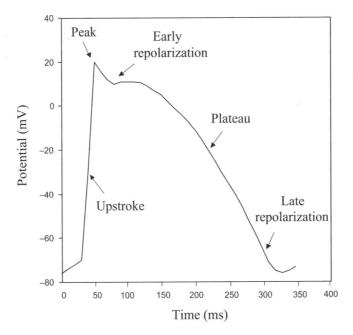

FIGURE 20.1. Stereotypic cardiac action potential: A stereotypic action potential of the form found in the heart is depicted. The rapid depolarization of the membrane, or upstroke, leads to a peak membrane potential of +10 to +20 mV. Once the peak depolarization is achieved there is an early phase of repolarization followed by a slow and sometimes flat regime called a "plateau," and then a late, or final, repolarization phase a few hundred milliseconds after the start of the upstroke.

vated by large depolarization while a few, most notably the T-type channels, require only mild depolarization of the membrane. The calcium channels requiring large depolarization are referred to as *high voltage-activated* (*HVA*) *calcium channels* whereas the T-type channels are termed *low voltage-activated* (*LVA*). The calcium currents found in the heart help maintain action potential plateaus, the fairly flat and slowly declining portions of the action potentials. Lastly, the potassium currents related in the last two sets of entries in Table 20.1 are responsible for the early and late phases of action potential repolarization.

20.2 HCN Channels' Role in Pacemaker Activities

Hyperopolarization-activated cyclic nucleotide-gated channels contribute to pacemaker activities in the sinoatrial node. The HCN (*hyperpolarization-activated cyclic nucleotide-gated*) channel is regulated both by cyclic nucleotides such as cAMP and by voltage. It is a member of the 6 TM family of ion channels and possesses a cAMP-binding domain in its cytoplasmic

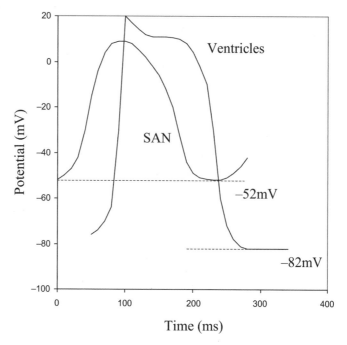

FIGURE 20.2. Action potentials in different parts of the heart: Shown are action potentials in the sinoatrial node (SAN) and in the ventricles. The latter are delayed by the amount of time required for the pulse from the SAN to reach the ventricular fibers. The minimum, hyperpolarized values of the membrane potentials are listed just below the dashed lines.

TABLE 20.2. Collective firing patterns observed in electroencephalograph recordings in the brain.

Type of activity	Frequency	Association
Delta waves	0.5 to 4 Hz	Deep sleep, abnormal states
Theta waves	4 to 8 Hz	Sleep onset, abnormal states
Alpha waves	8 to 12 Hz	Drowsiness and relaxed activity
Spindling oscillations	7 to 14 Hz	Quiescent sleep
Beta waves	12 to 30 Hz	Arousal, alert states, and information processing
Gamma band	30 to 60 Hz	Arousal, alert states, and information processing

C-terminal region. Its voltage regulation is unusual. Most ion channels are activated by depolarization. As discussed in the last chapter, their activation gates are closed at resting membrane potentials but open when the membrane is depolarized beyond some threshold level. The HCN channels work the opposite way. They are activated when the membrane potential is hyperpolarized beyond values in the range −50 to −70 mV. Whenever a cell fires an action potential the membrane hyperpolarizes and the HCN

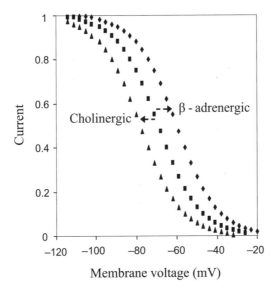

FIGURE 20.3. Activation functions for the hyperpolarization-activated cation current: The HCN channels are activated by hyperpolarization. β-adrenergic (adrenaline) stimulation shifts the activation function towards more depolarized values, while cholinergic (acetylcholine) stimulation does the opposite, shifting the activation function to more hyperpolarized values.

channel opens. Whereas the potassium channels responsible for the descending phase of the action potential has a reversal potential that is more negative than the resting membrane potential, the HCN channels exhibits the opposite voltage relationship. Its reversal potential is in the range from −20 to −30 mV, values that are less negative than that of the resting potential. When the HCN channel opens both Na⁺ and K⁺ ions flow into the cell to depolarize the membrane. The presence of HCN channels provides a means of moving the membrane potential in the direction needed to trigger action potentials and thus renders the cell more excitable.

The biophysical properties of ion channels can be altered by neuro-modulators acting in a variety of ways. Two examples of neurmodulation are included in Figure 20.3. Cardiac sinoatrial node neurons express β-adrenergic and cholinergic G protein-coupled receptors (GPCR). When adrenaline binds to a GPCR, G_s alpha subunits are activated. As discussed in Chapter 12, these subunits stimulate adenylyl cyclase to increase its the production of cAMP. The cAMP molecules diffuse to and bind the cytoplasmic portion of the HCN ion channels. This binding shifts the activation curve towards less negative membrane potentials. As a result the neurons are able to fire action potentials more rapidly and the heart rate goes up. Acetylcholine has the opposite effect. When it binds its GPCR, G_i alpha subunits are activated. These molecules bind to adenylyl cyclase and throt-

tle back its production of cAMP. As a result the activation function shifts
to more negative membrane potentials and the heart rate slows down.

20.3 Synchronous Activity in the Central Nervous System

Electroencephalographic (EEG) recordings of wave patterns in the brain
are a well-established tool for identifying abnormal conditions and various
stages of sleep and arousal. Several different kinds of wave patterns can be
produced in the brain, each characterized by a specific brain wave frequency
range (Table 20.2). These patterns fall into two groups—low frequency and
high frequency. The low frequency group includes delta waves, theta waves,
spindling oscillations, and alpha waves. These patterns are of large amplitude when observed in the EEG, and as such are produced by large populations of neurons undergoing synchronized firing. The high frequency
group includes the beta and gamma bands. The amplitudes of these rhythms
are far lower that those of the low frequency waves, indicative of participation by smaller populations of neurons. These patterns are not as sharply
delineated in the EEG recordings.

As indicated in Table 20.2, delta waves are associated with deep sleep and
with abnormal brain states such as coma and anesthesia. Theta oscillations
are most often associated with drowsiness and the onset of sleep. The amplitude of theta waves is elevated in a number of attention-related disorders.
Alpha waves are observed during drowsiness and relaxed activity, and
spindle activity associated with early stages of quiescent sleep. In contrast
to the low frequency regime, the high frequency activity is associated with
arousal and alert, attentive states, with complex motor and sensory processing tasks, and with rapid eye movement (REM) sleep.

20.4 Role of Low Voltage-Activated Calcium Channels

Neurons can fire action potentials in a number of ways. Neurons can be
silent and not fire at all. Alternatively, they can act as *spiking neurons* and
generate a small number, just one or two, of action potentials at isolated
time intervals. *Tonic firing* is yet another mode. In tonic firing, a train of
action potentials is generated. The action potentials may be generated at
regular time intervals or they may be irregularly spaced in time. This type
of firing pattern is illustrated in Figure 20.4(a) for the case of regular
spacing. Yet another kind of firing pattern is *rhythmic bursting*. In this kind
of firing, illustrated in Figure 20.4(b), multiple sequences of action potentials are produced. Each sequence, or burst, consists of a train of closely
spaced action potentials. The intervals between each spike in the burst are
small compared to the time intervals between successive bursts, which are

FIGURE 20.4. Rhythmic firing patterns, plots of membrane potential versus time: Stereotypic plots of current versus time are shown for (a) tonic firing and (b) rhythmic bursting.

(a)

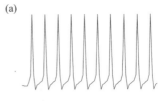

(b)

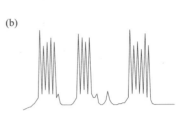

somewhat variable. The inverse of the average burst-to-burst time interval is the *burst frequency*. These frequencies can vary from less than one per second to 150 or 200 per second.

The creation of trains and bursts of action potentials rather than single spikes is an important component of the rhythmicity. One of the most important ion channels contributing to the rhythmicity is the low voltage-activated (T-type) calcium channel. Whereas high voltage-activated calcium channels require depolarization of about 40 mV away from resting membrane potentials in order to open, low voltage-activated calcium channels only need about 10 mV of depolarization to open. When they do open and let in calcium, the membrane is further depolarized, resulting in activation of sodium channels and the generation of bursts of action potentials.

The T-type calcium channels operate in a way closely resembling the sodium channels responsible for the upstroke of the action potential. That is, they possess activating and deactivating functions as do the sodium channels and like the sodium channels the T-type calcium channels are activated by depolarization. There are two main differences between the two kinds of channels. The first difference is that the calcium channel activation and deactivation functions are centered at larger hyperpolarizations than those for sodium. At the resting potential, the calcium channels are activating whereas the sodium channels are deactivating. This means that the calcium channels can open at and near the resting potential. The entry of calcium through these channels depolarizes the membrane and triggers the opening of the sodium channels leading to the firing of action potentials.

The second difference between the T-type calcium channel and the sodium channel is in how fast they respond to changes in membrane potential. The sodium channels respond quickly to changes in potential. The calcium channels activate more slowly, and deactivate considerably more slowly. As the membrane voltage moves up and down rapidly due to the sodium/potassium depolarization and repolarizations the calcium current increases and then decreases back to zero, terminating the burst of action potentials. The calcium channels that open to start the burst then require a large hyperpolarization in order to de-inactivate and start the next burst.

Tonic firing and rhythmic bursting are widely encountered in the brain, and neurons can switch from one form to the other. Large populations of neurons in the brain undergo slow rhythmic bursting during natural, slow wave sleep, but switch their firing patterns from bursting to tonic firing in waking states and rapid eye movement (REM) sleep. In order to transition from a slow wave sleep state to either awake or REM sleep states, the delta, alpha, and spindling patterns have to be abolished and replaced by the high frequency tonic firing in the beta band (Table 20.1).

The waking up and REM sleep tasks are carried out by collections of neurons that the wake-up system comprises, the *reticular formation*, located in the brain stem. Neurons in the reticular formation release a number of neuromodulators, the most prominent of which are acetylcholine (ACh), norepinephrine (NE), and serotonin (5-HT). These neuromodulators influence the activities of the neurons generating brain rhythms and also influence muscle states. In the brain, high frequency activity replaces the low frequency waves. The membrane potential is depolarized in response to the neuromodulators, and the T-type calcium channels responsible for bursting are shut down.

20.5 Neuromodulators Modify the Activities of Voltage-Gated Ion Channels

Up to this point, the discussion has been limited to voltage-gated ion channels and to how individual neurons fire. Except for two brief discussions of the effect of neuromodulators such as adrenaline and acetylcholine, no mention has been made of the role networks play in generating rhythms. Network properties such as the manner in which neurons are connected to one another and what signals are sent and exchanged, together with the voltage-gated ion channels intrinsic to the individual neurons, jointly determine the nature of the neural rhythms. Three forms of cell-to-cell communication contribute to rhythmicity. These are communication by neuromodulators, by neurotransmitters at chemical synapses, and by small signaling molecules at electrical synapses (gap junctions).

Neuromodulators act on ion channels to produce slow, long-lasting changes in the excitability of the neurons. Neuromodulators are secreted by many different neurons and bind to G protein-coupled receptors on target cells, in contrast to neurotransmitters, which bind to ion channels. Diffusible signaling molecules such as acetylcholine can act either as a neurotransmitter or as a neuromodulator. Acetylcholine acts as a neurotransmitter when binding to nicotinic (ion channel) receptors and as a neuromodulator when binding to muscarinic (G protein-coupled) receptors. Similarly, glutamate and GABA each can function either as a neuro-transmitter or as a neuromodulator, depending on whether it acts through ionotropic or metabotropic (G protein-coupled) receptors. The functional characterization of a molecule as a hormone, as a neurotransmitter, or as a neuromodulator is determined by the nature of its receptor and what kind of cellular response is elicited upon binding.

Recall from Chapter 12 that G protein-coupled receptors activate het-erotrimeric G proteins. Neuromodulators can exert their influences on ion channels in two ways. In direct regulation, G protein subunits, particularly the $G_{\beta\gamma}$ subunits, diffuse along the inner face of the plasma membrane and bind to the cytoplasmic portions of ion channel subunits, thereby modify-ing channel conductances. Alternatively, the neuromodulators can exert their influences indirectly by activating second messenger systems. The second messengers either bind the ion channels, thereby modifying their conductances, or they activate protein kinases and protein phosphatases, which, in turn, phosphorylate or dephosphorylate ion channel subunits, thereby modifying their conductances. The modified states of the ion channel subunits have lifetimes that are long compared to the time that it takes to generate and propagate action potentials, and thus the modulatory influences are long-lasting ones.

20.6 Gap Junctions Formed by Connexins Mediate Rapid Signaling Between Cells

Gap junctions are specialized communication channels that permit the rapid exchange of metabolites, ions, and second messengers such as Ca^{2+} and cAMP between pairs of neurons. The channels are nonselective and when open allow ions of mass less than about 1 kDa to pass directly from the cyto-plasm of one cell to another. These communication channels are often found in opposing dendritic processes and are widely distributed among neurons. Because of their direct connectivity, communication through gap junctions, or *electrical synapses*, as these pores are often called, is more rapid than is possible through chemical synapses. This rapid means of exchanging cyto-plasmic signaling elements enables the communication partners to coordi-nate their firing activities. This synchronizing capability is utilized heavily during the embryonic development of sensory circuits such as those found in the retina. Although there is no sensory input at that time, the neurons are spontaneously active. They continually communicate with one another in order to refine the circuits formed during the initial connecting stages. Com-munication through gap junctions plays a similar role during development of motor systems, and contributes to adult motor behavior as well.

Gap junctions are formed by *connexons*, hemipores the span the plasma membrane of a cell. A gap junction between two adjacent cells is created when a connexon hemipore embedded in the plasma membrane of a cell aligns and docks with a connexon situated in the plasma membrane of an opposing cell. Once joined the two connexons form a communication channel permitting one- and two-way flow of small charged and uncharged molecules. The size restriction prevents the passage of larger molecules such as nucleic acids and proteins. Gap junctions tend to cluster together into local regions of the plasma membranes of adjacent cells. The two cells remain separated by a gap of from 2 to 4 nm, hence the name "gap junction."

Each connexon is erected from six connexins arranged in a symmetric fashion to form a hollow pore through which small molecules can diffuse. Some connexons contain a single type of connexin and the connexon mates with an identical connexon. In these cases two homomeric connexons form a homotypic gap junction. Alternatively, a connexon can be assembled from more than one kind of connexin to form a heteromeric connexon. Such a connexon can dock with either the same kind of connexon or with a different kind of connexon on the opposing cell membrane. Thus, there are a variety of homomeric and heteromeric connexons, and these can combine in different homotypic and heterotypic ways to form gap junctions.

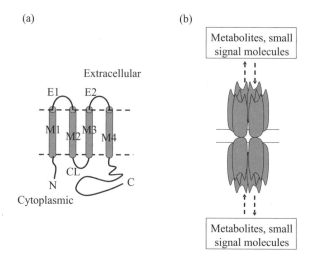

FIGURE 20.5. Connexin topology and assembly into gap junctions: (a) Connexins pass back and forth through the plasma membrane four times. Starting from the N-terminal end the transmembrane segments are designated as M1 through M4. The N- and C-terminal portions of the polypeptide chains are situated in the cytoplasm of the host cell. The C-terminal end is longer than the N-terminal end and may form a loop. The two extracellular loops, E1 and E2, mediate connexin to connexin contacts while the three cytoplasmic portions—the N- and C-terminal ends and the cytoplasmic loop (CL) connecting M2 to M3—are sites for regulation of the pore. (b) Six connexins arrange themselves in a symmetric fashion to form a hemipore called a "connexon." Two connexons, one from each cell, form the conduction pore, or gap junction. These remain separated from one another by a 2- to 4-mm gap, denoted in the figure by the pair of parallel lines.

Like ion channels, gap junctions are voltage-gated. Gap junctions will open and close, and change their permeabilities, in response to alterations in potential difference. The state of the pore is responsive to differences in membrane potential of the two cells (transjunctional voltage differences) and to differences in potential across plasma membranes, i.e., between the cytoplasm and extracellular spaces (inside-outside voltage differences). Gap junctions are also responsive to biochemical changes in the intracellular environment—for instance, the pores will close when exposed to low pH levels. In addition, gap junctions are subject to regulation by phosphorylation.

20.7 Synchronization of Neural Firing

Gamma-aminobutyric acid (GABA) is the preeminent inhibitory neurotransmitter in the brain. As indicated in Table 19.2, GABA binds to receptors belonging to the cys-loop family of ion channels. When these ion channels open, Cl^- anions enter the cell and drive the membrane potential towards more negative, hyperpolarizing values. There are three kinds of GABA receptors: $GABA_A$ and $GABA_C$ receptors are chloride ion channels, while $GABA_B$ receptors are 7TM, G protein-coupled receptors. GABA-releasing neurons are found throughout the CNS interspersed among a larger population of excitatory, glutamate-releasing neurons. Glutamate acts through members of the glutamate family of receptor ion channels to allow Na^+, K^+, and Ca^{2+} cations to enter the cell. Most cells in the central nervous system (CNS), inhibitory and excitatory, express GABA receptors. These along with glutamate-binding NMDA and AMPA receptor ion channels are the predominant receptor ion channels.

GABA-releasing inhibitory neurons, referred to as *interneurons*, make contacts with other interneurons to form interneuron circuits. They also make contacts with excitatory cells, especially those expressing *N*-methyl-D-aspartate (NMDA) receptors and capable of repetitive bursting behavior. They excitatory cells make contacts with each other, as well, and together the excitatory and inhibitory cells form large networks of interconnected cells that exhibit the population oscillations seen in the EEGs.

Inhibition mediated by GABAergic neurons and by gap junctions synchronizes the firing of neurons. The GABAergic interneurons synchronize their firing through release of GABA and the subsequent activation of the chloride channels. Their synchrony is greatly improved, and in some situations predominantly mediated, by gap junctions. There are a number of different kinds of interneurons. Each population of interneurons expresses a particular kind of gap junction that enables it to communicate with other interneurons of the same kind. For example, one kind of interneuron, the *fast spiking* (FS) interneuron, will communicate through gap junctions with other FS interneurons. The networks of interneurons collectively act as a pacemaker circuit for the larger circuits involving the excitatory cells. In these situations, the gap junctions operate in combination with NMDA receptors that endow the neuron with intrinsic rhythmic bursting proper-ties. In the absence of the gap junctions and the inhibitory connections, each excitatory neuron would carry out its program of rhythmic bursting in a dis-tinct and uncoordinated manner. When the inhibitory circuitry is present, the cells synchronize their firing patterns. They can undergo rhythmic dis-charges at low frequencies and also at the higher frequencies characteris-tic of the beta and gamma bands of the EEG.

20.8 How Spindling Patterns Are Generated

Spindle waves (Figure 20.6) are generated during the early stages of quies-cent sleep. Large populations of neurons in the thalamus and cortex undergo rhythmic discharges in the 7 to 14 Hz range. These oscillations wax and wane with a 1- to 3-second period. That is, the oscillations steadily increase in mag-nitude up to a maximum and then decrease until they vanish some 1 to 3 seconds after they started. After a silent period of 3 to 20 seconds they start up again, and the waxing and waning pattern repeats itself.

Spindle rhythms are generated in thalamic neurons. Two populations of cells are involved—*inhibitory neurons* (GABAergic) in the thalamic reticu-lar nucleus that make reciprocal connections with *excitatory neurons* in the thalamic relay nucleus. When pulses of GABA are released at the presynap-tic terminal of thalamic reticular neurons, inhibitory postsynaptic potentials (IPSPs) are produced in the relay nuclei. These hyperpolarizing events acti-vate the I_h channels, which permit the generation of action potentials in the relay neurons, which through their reciprocal synapses with the reticular neurons trigger membrane depolarization and re-excite the reticular cells.

20.9 Epileptic Seizures and Abnormal Brain Rhythms

Excitation mediated through glutamergic cationic channels must be in balance with inhibition mediated by GABAergic anionic channels. One of the consequences of an imbalance between excitation and inhibition is the

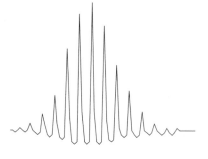

FIGURE 20.6. Plot of membrane potential versus time for 7- to 14-Hz spindle oscilla-tions: The train of action potentials first waxes and then wanes in amplitude.

occurrence of epileptic seizures. These abnormal events are generated when there is too much excitation (hyperexcitabilty) leading to aberrant rhythms and epileptic seizures.

Because of their ability to generate rhythmic discharges, neurons in the neocortex, thalamus, and hippocampus are ideal sites for aberrant rhythms leading to epileptic events and seizures. These regions of the brain contain large numbers of excitatory, intrinsically bursting pyramidal cells that are massively interconnected with one another, and smaller populations of several different kinds of inhibitory interneurons that, as just discussed, not only make connections with one another but also make contact with large numbers of excitatory pyramidal cells.

One of the best-studied forms of epilepsy is the absence seizure, or *petit mal epilepsy*. This type of seizure appears in EEGs as an aberrant 3-Hz spike-and-wave pattern. Absence seizures occur most often in children and adolescents, and are characterized behaviorally as brief (~10s), nonconvulsive interruptions in consciousness. The spike-and-wave patterns are generated by the general thalamic/thalamocortical circuitry that produces the spindling patterns. One of the likely causes of this type of aberrant brain wave pattern is malfunctioning $GABA_A$ receptors. These receptors are erected from mixtures of different GABA subunits. The different kinds of $GABA_A$ receptor subunits are designated as $\alpha, \beta, \gamma, \delta, \epsilon, \pi$, or ρ. Several different isoforms exist for many of the subunit types. Ion channels built from ρ subunits are called *$GABA_C$ ion channels*. The GABA channels are constructed from five subunits and, like the other ion channels, its subunits are arranged to form a hollow conducting pore. A stoichiometric combination that is widely distributed throughout the brain is one α subunit, two β subunits and two γ subunits. Mutations in the γ_2 subunit have been implicated as a possible cause of absence seizures in several experimental studies, while theoretical modeling efforts have demonstrated how malfunctions in $GABA_A$ and $GABA_B$ receptors can generate aberrant 3-Hz spike-and-wave rhythms.

That mutations in an ion channel subunit can cause aberrant rhythms and have behavioral consequences is not unique to the GABA and absence seizures. Rather, mutations in other ion channel subunit exits have similar behavioral consequences. These include mutations in the channel-forming α subunit of high voltage-activated, P/Q-type calcium channels, in other GABA subunits, and in potassium and sodium channel subunits. Thus, there is an entire ensemble of ion channel disorders, or "channelpathies."

20.10 Swimming and Digestive Rhythms in Lower Vertebrates

Central pattern generators located in the spinal cord of lower vertebrates generate swimming and digestive rhythms. The rhythmic movements of muscles are controlled by neurons located in the brain stem and spinal cord.

These neurons are organized into small neural circuits called *central pattern generators* (CPGs). The CPGs may be thought of as elementary circuits, autonomous modules built from small numbers of neurons that are used to drive locomotion activities such as walking and swimming, breathing, and chewing and digestion. Many of the principles governing their operation are shared by other rhythm-producing systems such as those responsible for sleep and alert behavior. Processes that generate rhythms, processes such as reciprocal connections involving inhibition, and burst-promoting voltage-gated ion channels, are common to all rhythm-producing circuits.

Motor neurons supply input to muscles that drives their contractions. They are the last stage of neural connectivity before the muscles. Inputs from different drivers such as the CPGs converge upon these cells. Some of the regulatory inputs come directly to the motor neurons while others converge upon the CPGs to regulate their firing patterns.

Invertebrates are widely used as model systems for studying central pattern generators. The invertebrate CPGs exhibit all of the main properties of vertebrate CPGs, yet are small enough to be well characterized at the network and ion channel levels. The invertebrates selected as model systems include the nudibranch *Tritonia diomedea*, the sea snail *Aplysia californica*, the decapod crustaceans *Panulirus interruptus* (spiny lobster) and *Cancer borealis* (rock crab), and the medicinal leech *Hirudo medicinalis*.

The *stomatogastric ganglion* (STG) of the lobster and crab is the site of a pair of central pattern generators that supply digestive rhythms. These crustaceans swallow their food whole. The material ingested is broken down in the stomach through rhythmic contractions of gastric teeth and the stomach. Several different kinds of rhythms are produced during digestion. One of these is the gastric rhythm that controls the gastric teeth, and another is the pyloric rhythm that regulates the food sorting and sifting machinery. The rhythms are supplied by two CPGs. The CPG responsible for the pyloric rhythm contains 14 neurons. Of these 13 are motor neurons whose output drives the muscle contractions, and one is a pacemaker neuron. The CPG responsible for gastric rhythms contains just 11 neurons. However, its wiring and operation is more complex than that of the pyloric circuit, and the discussion will be limited to the latter.

The pyloric circuit is presented in Figure 20.7. There are six different kinds of neurons in the circuit. One each of these neuron types and its connectivity is shown in the figure. Connection leading to cells outside the CPG are omitted to maintain clarity. As can be seen in the diagram the neurons in the pyloric central pattern generator are connected to one another by electrical synapses and by chemical synapses. The AB neuron functions as the pacemaker neuron for the circuit. It produces bursts of action potentials at regular intervals. Electrical synapses link the AB and PD neurons to one another. As a result the PD neuron fires in synchrony with the AB neuron, and these two neurons may be regarded jointly as the network pacemaker. As is shown in Figure 20.7, these two neurons establish contact with all the downstream motor neurons.

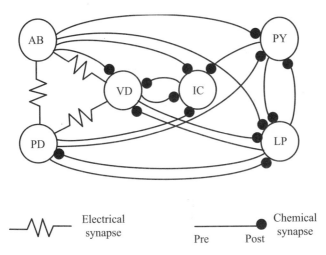

FIGURE 20.7. Crustacean pyloric circuit: The central pattern generator responsible for pyloric rhythms consists of anterior burster (AB), pyloric dilator (PD), ventricular dilator (VD), inferior cardiac (IC), lateral pyloric (LP), and pyloric (PY) neurons. Resistor symbols denote electrical synapses, and lines terminating in filled circles represent inhibitory chemical synaptic connections.

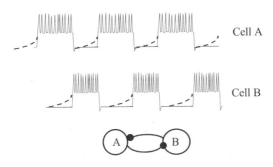

FIGURE 20.8. Mutual inhibition: Two cells, A and B, are coupled to one another by means of inhibitory synaptic connections. Because of this coupling the two cells alternatively fire synchronized sequences of action potentials. The dashed lines denote the gradual depolarization of the membrane potential towards the plateau potential and threshold for firing action potentials.

The second main feature of the network is wide usage of mutual inhibition in which pairs of cells are reciprocally connected to each other and inhibit one another. Mutual inhibition produces alternative firing pattern that can drive the phased contractions of opposing muscles. A stereotypic pattern of alternating firing by a pair of cells mutually inhibiting one another is depicted in Figure 20.8. In the pyloric circuit, the AB and PD neurons fire first, inhibiting the firing of the downstream neurons until the bursting is completed. The downstream neurons then fire in a sequence determined by the synaptic connections and by the mix of ion channels expressed by each neuron.

Two kinds of neurotransmitters are released at the presynaptic terminals—glutamate and acetylcholine. These neurotransmitters diffuse across the synaptic cleft and bind to receptors expressed on the postsynaptic cell. In crustaceans, glutamate and acetylcholine both act as inhibitory, not excitatory, neurotransmitters. The glutamate receptors are members of an inhibitory glutamate receptor ion channel (IGluR) family. They resemble the GABA and glycine receptors, and open to permit the entry of Cl⁻ ions into the cell, as do the crustacean acetylcholine receptor ion channels.

20.11 CPGs Have a Number of Common Features

Many of the key design elements found in the stomatogastric ganglion (STG) pyloric circuit are encountered in other central pattern generators. Some of these properties, such as the occurrence of certain sets of ion channels, operate at the cellular level. Other properties, such as mutual inhibition, are related to how the circuits are built, and operate at the network level. A summary of design properties common to many CPGs is presented in Table 20.3.

One of the most common properties is *spike frequency adaptation*. Small conductance, calcium activated potassium channels are quite sensitive to small changes in calcium concentration. Although the effect of sodium and potassium is small compared to their concentrations within the cell, the same is not true for calcium. Its concentration in the cell in quite low, and small changes in calcium concentration can be sensed and responded to. The small conductance, calcium-activated potassium channels open in response to changes in calcium concentration. The effect of these openings is to hyperpolarize the membrane and decrease its excitability in response to sustained firing. The sequence of pulses exhibits an increasing time lag between successive pulses, and eventually reaches a steady state firing frequency.

TABLE 20.3. Amine and peptide neuromodulators expressed in humans: Abbreviations for specific neuromodulators (excepting the general opioid and tachykinin categories) are shown in parentheses.

Amines	Peptides
Acetylcholine (ACh)	Corticotropin-releasing factor (CRF)
Dopamine (DA)	Cortistatin (CST)
Epinephrine (E)	Opioids (dynorphins, endorphins)
Histamine (HA)	Oxytocin (OT)
Norepinephrine (NE)	Somatostatin (SST)
Octopamine (OCT)	Tachykinins (substance P, neurokinins)
Serotonin (5-HT)	Vasopressin (AVP)

TABLE 20.4. Properties of central pattern generators.

Property	Description
Cellular	
Rhythmic bursting	Pacemaker and other rhythm-producing neurons express HCN ion channels that support rhythmic bursting
Plateau potentials	Increase excitability resulting from a stable state lying close to thresholds for action potential generation
Spike frequency adaptation	Small conductance, calcium-activated potassium channels allow for adaptation of the firing rate
Post-inhibitory rebound	Low voltage-activated (T-type) calcium channels permit the temporary increase in excitability
Network	
Input signals that initiate rhythms	Input from sensory neurons, and from neurons in brain stem control areas
Feedback signals that regulate rhythms	Mechanical and chemical signals that are used to fine tune the rhythms
Highly interconnected circuits that generate rhythms	Mutual inhibition and gap junctions ensure the proper firing sequences and produce synchrony
Pacemaker cells that generate rhythms	Cells that entrain other cells in the CPG
Regulatory	
Extrinsic	Neuromodulatory signals sent from neurons outside the pattern generation circuit
Intrinsic	Neuromodulatory signals sent and received within the rhythm-producing circuit

Reciprocal (mutual) inhibition is another widely encountered network property. The underlying phenomenon is a simple one. Suppose that two cells, A and B, expressing the ion channels mentioned in Table 20.4, mutually inhibit one another. Further suppose that A fires first. It will inhibit B from firing while it is bursting, but then will exhibit spike frequency adaptation or simply cease to fire. This cessation will release B from its inhibition. B will fire and inhibit A from firing until it too has completed its firing sequence. A will then be released from inhibition and the cycle will repeat. Alternatively, if the B cell being inhibited exhibits a gradual depolarization of its membrane, as illustrated in Figure 20.8, it will escape from the inhibition when its membrane potential depolarizes sufficiently. When it starts firing it will inhibit the *A* cell from firing and shut down its burst. This will continue until the membrane of the *A* cell depolarizes and so on.

Another property commonly encountered in rhythmically firing neurons is the presence of *plateau potentials*. Many motor neurons exhibit bistability. Recall that stable states are long-lived states. If a system is in a stable state it will return to that state whenever it is subjected to small perturbations and disturbances. Plateau potentials are membrane potentials that, like the resting membrane potentials, are stable states. Plateau states occur at depolarizations greater than that of the resting membrane potential. When a membrane is at a plateau potential, it is far more able to repeatedly fire action potentials, even in the absence of sustained excitatory input from synapses. As a result neurons possessing plateau potentials exhibit two types of behavior. When at their resting membrane potential, the neurons do not fire action potentials in the presence of modest sustained depolarizations, and thus are silent. In contrast, when they at their more depolarized plateau potential, even modest sustained depolarizations will trigger the repeated firing of action potentials. Compare the firing pattern shown in Figure 20.4(b) to the one presented in Figure 20.8. The cells producing the bursts in Figure 20.8 possess plateau potentials; their membrane potentials remain elevated after each action potential. This situation differs from the one depicted in Figure 20.4 in which the potential returns to values close to the resting potential after each action potential.

Last, referring again to Table 20.4 *post inhibitory rebound* refers to the paradoxical situation that inhibition can actually promote excitability. Some ion channels require hyperpolarization to de-inactivate and others need hyperpolarization in order to activate. If a neuron expressing these ion channels is subjected to a strong hyperpolarization that is quickly terminated, its membrane can rapidly rebound, depolarizing and firing an action potential.

20.12 Neural Circuits Are Connected to Other Circuits and Form Systems

Central pattern generators do not work in isolation. Rather, they receive inputs from sensory and higher information-processing areas and send outputs to motor neurons controlling muscles. In situations where multiple circuits must work together to generate motor rhythms, the CPGs communicate laterally to other CPGs. Properties of a typical mammalian network such as the convergence of multiple input signals and feedback control are listed in Table 20.4. The human respiratory system provides an excellent example of both of these properties.

In humans, neurons organized into a CPG in a section of the medulla called the *Prebötzinger complex* generate respiratory rhythms. This CPG receives input from a variety of sources denoting emotional, sleep-related, environmental, and motor activity states. The output from the CPGs drives pools of motoneurons controlling the respiratory muscles—the diaphragm, and the intercostal and abdominal muscles—that pump air in and out of the lungs. The degree of resistance to airflow is regulated by smooth muscle in the bronchial walls, and by muscles controlling the diameter of the upper airways, namely, the laryngeal, pharyngeal, and hyperglossal muscles.

The respiratory system is a feedback control system. Respiratory neurons send output to the diaphram, intercostal muscles, abdominal muscles, and accessory muscles of respiration. Mechanoreceptors supply feedback information on the effect of the firing patterns on the muscles. Information is supplied by stretch receptors, located in the bronchial smooth muscle, by a variety of irritant receptors located in the airway epithelium, and by C-fiber receptors located within the walls of the alveoli. Information on dissolved gases is provided by peripheral (arterial) oxygen and carbon dioxide receptors, and by central carbon dioxide (acidification) medullary receptors that monitor the cerebrospinal fluid (CSF).

Several different patterns can be generated by the respiratory CPG. Under normal breathing conditions the CPG generates a fairly regular, eupneic firing pattern. However, under conditions of oxygen starvation/ hypoxia, the patterns generated within the CPG will change to either a sighing rhythm or to a gasping rhythm. One of the features that characterize neurons lying within the CPG is the presence of neurokinin-1 receptors, the receptors for the peptide neuromodulator, substance P. The neurons expressing these receptors not only generate respiratory rhythms, but also alter their firing frequency in response to the presence or absence of substance P and other neuromodulators that, in turn, reflect oxygen status and other types of regulatory information.

20.13 A Variety of Neuromodulators Regulate Operation of the Crustacean STG

The neurons that make up the crustacean stomatogastric ganglion express receptors for a number of amine and crustacean-specific peptide (CSP) neuromodulators. Receptors are present for (amines) serotonin, dopamine, and octopamine, and for (CSPs) crustacean cardioactive peptide (CCAP), proctolin, and cholecystokinin (CCK). These neuromodulators not only alter the firing patterns of the central pattern generators, but also can trigger aggressive behavior. Most of the CSPs are released by neurons that send their axons to the STG. The usual effect of the peptides is to initiate one or more of the rhythms—gastric, pyloric, and cardiac sac. In some instances, a neuropeptide such as red pigment-concentrating hormone (RPCH) triggers the functional reconfiguration of the circuits. The reconfiguration may take the form of circuit fusion or appear as alterations in circuit membership. In the former case, two or more circuits may fuse together so that they work together rather than independently. In the latter instance, a neuron, previously firing as part of one circuit may switch its firing properties and become part of another circuit.

The amine neuromodulators act on the ion channels, enhancing some currents and inhibiting others. They exert their influences by either shifting the activation and or deactivation curves or by increasing the maximal conductances for the ion channels. These changes lead to alterations in phase of one neuron relative to others within the CPG and change the number of action potentials generated in a given time interval.

As indicated in Table 20.4, neuromodulatory signals can originate either within a CPG or externally to it. There are many sources of external inputs in the central nervous system. In vertebrates these sources are organized into centers that project out to, and make contact with, neurons in most areas of the brain. These centers are able to influence the response properties of many circuits simultaneously. Intrinsic modulation is far more specific in its actions. In the stomatogastric ganglion, modulatory inputs are supplied by higher brain centers and by feedback from sensory neurons. These modulatory inputs are sensitive to the outputs of the circuit being modified and are local to that circuit. These influence the rhythms being produced and the attendant behaviors.

20.14 Motor Systems Adapt to Their Environment and Learn

The nudibranch gastropod *Tritonia diomedea* is preyed upon by starfish. When *Tritonia* comes into contact with a foot of a starfish, it changes its behavior from a slow-moving feeding mode to a fast-moving escape swimming mode. The circuitry that makes this escape response possible includes, besides the sensory neurons, a command interneuron that receives sensory information, and a central pattern generator that receives information from the *command dorsal ramp interneuron* (DRI) to start the escape swim rhythm. The CPG contains three kinds of interneurons—an interneuron called C2, a pair of ventral swim interneurons (VSIs), and three dorsal swim interneurons (DSIs). The output from the CPG is sent to an array of left and right flexion motor neurons that drive the muscles responsible for escape swimming (Figure 20.9).

In *Tritonia*, the swim CPG produces either a swimming rhythm or a nonswimming withdrawal response. The transformation of nonswimming into swimming rhythms involves modulatory signaling by serotonin. The command interneuron sends swim signals to the DSIs. In response to these initiating signals, the DSIs release serotonin, which binds to receptors on the C2 neuron. In response to serotonin binding, the C2 interneuron increases the amount of neurotransmitter released from its presynaptic terminals and exhibits an increased excitability. The effect of the increased neurotransmitter release is to strengthen the synaptic connection from the C2 interneuron back to the DSIs thereby increasing the DSIs excitability through this positive feedback loop. (In the absence of the increased neurotransmitter release, the DSIs are repressed by inhibitory signaling between themselves). The increased excitability of the C2 neuron enables it to trigger and maintain the DSIs' swim rhythm. If the firing rate of the C2 neuron falls below the swim threshold, escape swimming will not occur. In the terminology of

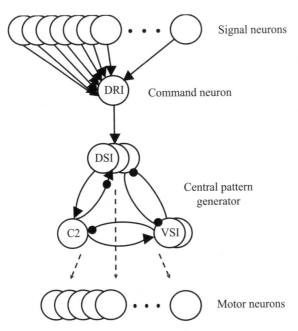

FIGURE 20.9. *Tritonia* escape circuit: Upstream signals conveyed by a set of about 80 S cells are relayed to the command neuron, the dorsal ramp interneuron (DRI), which sends its synaptic output to a set of three dorsal swim interneurons (DSIs) belonging to the central pattern generator. The CPG contains, besides the DSIs, a pair of ventral swim interneurons (VSIs) and a single cerebral cell 2 (C2) neuron. The three kinds of CPG neurons make synaptic contacts with a set of about 55 dorsal and ventral flexion motor neurons. In the diagram, sharp arrows denote excitatory connections; filled circles represent inhibitory connections, and the hybrid arrow/circle symbol denotes a multiple component (excitatory and inhibitory) monosynaptic contact.

Table 20.4, the alteration in the firing pattern of C2 by serotonin released by the DSIs is an example of *intrinsic neuromodulation*.

The serotonin-mediated changes in rhythmicity are long-lasting in that they outlast the stimuli that produced them. When the serotonin molecules bind to GPCRs on the postsynaptic cell, they activate G_i alpha subunits of the heterotrimeric G proteins docked nearby. These subunits stimulate adenylyl cyclase to increase its synthesis of cAMP second messenger molecules leading to alterations in the way the ion channels operate and the cell responds to further stimuli. The organism will exhibit an increased responsiveness to further noxious stimuli and a reduced responsiveness to harmless ones. The process of more rapidly responding to the dangerous stimuli is called *sensitization*, and the process of responding less strongly to benign stimuli is known as *habituation*. Both types of adaptive change are examples of learning and have been extensively studied in the snails, crustaceans, and leech.

References and Further Reading

General References

Hille B [1992]. *Ionic Channels of Excitable Membranes* (2nd edition). Sunderland MA: Sinauer Associates, Inc.

Kaczmarek LK, and Levitan IB [1987]. *Neuromodulation.* New York: Oxford University Press, Inc.

Llinás RR [1988]. The intrinsic electrophysiological properties of mammalian neurons: Insights into central nervous system function. *Science,* 242: 1654–1664.

Nicholls JG, Wallace BG, Fuchs PA, and Martin AR [2001]. *From Neuron to Brain: A Cellular Approach to the Function of the Nervous System* (4th edition). Sunderland MA: Sinauer Associates, Inc.

Heartbeat

Boyett MR, Honjo H, and Kodama I [2000]. The sinoatrial node, a heterogeneous pacemaker structure. *Cardiovasc. Res.,* 47: 658–687.

Schram G, et al. [2002]. Differential distribution of cardiac ion channel expression as a basis for regional specialization in electrical function. *Circ. Res.,* 90: 939–950.

HCN Channels

DiFrancesco D [1999]. Dual allosteric modulation of pacemaker (f) channels by cAMP and voltage in rabbit SA node. *J. Physiol.,* 515: 367–376.

Ludwig A, et al. [1998]. A family of hyperpolarization-activated mammalian cation channels. *Nature,* 393: 587–591.

Zagotta WN, et al. [2003]. Structural basis for modulation and agonist specificity of HCN pacemaker channels. *Nature,* 425: 200–205.

Thalamocortical Rhythms

Huguenard JR, and Prince DA [1992]. A novel T-type current underlies prolonged Ca^{2+} dependent burst firing in GABAergic neurons of rat thalamic reticular nucleus. *J. Neurosci.,* 12: 3804–3817.

McCormick DA, and Huguenard JR [1992]. A model of the electrophysiological properties of thalamocortical relay neurons. *J. Neurophysiol.,* 68: 1384–1400.

Munk MHJ, et al. [1996]. Role of reticular activation in the modulation of intracortical synchronization. *Science,* 272: 271–274.

Steriade M, Amzica F, and Contreras D [1996]. Sunchronization of fast (30–40 Hz) spontaneous cortical rhythms during brain activation. *J. Neurosci.,* 16: 392–417.

Steriade M, McCormick DA, and Sejnowski TJ [1993]. Thalamocortical oscillations in the sleeping and awake brain. *Science,* 262: 679–685.

von Krosigk M, Bal T, and McCormick DA [1993]. Cellular mechanisms of a synchronized oscillation in the thalamus. *Science,* 261: 361–364.

Epilepsy

Crunelli V, and Leresche N [2002]. Childhood absence epilepsy: genes, channels, neurons and networks. *Nature Rev. Neurosci.,* 3: 371–382.

Velazquez JPL, and Carlen PL [2000]. Gap junctions, synchrony and seizures. *Trends Neurosci.,* 23: 68–74.

Gap Junctions

Bruzzone R, White TW, and Paul DL [1996]. Connections with connexins: The molecular basis of direct intercellular signaling. *Eur. J. Biochem.*, 238: 1–27.

Evans WH, and Martin PEM [2002]. Gap junctions: Structure and function. *Mol. Mem. Biol.*, 19: 121–136.

Kumar NM, and Gilula NB [1996]. The gap junction communication channel. *Cell*, 84: 381–388.

Simon AM, and Goodenough DA [1998]. Diverse functions of vertebrate gap junctions. *Trends Cell Biol.*, 8: 477–483.

Unger VM, et al. [1999]. Three-dimensional structure of a recombinant gap junction membrane channel. *Science*, 283: 1176–1180.

GABAergic Neurons and Gap Junction-Mediated Synchrony

Galarreta M, and Hestrin S [2001]. Electrical synapses between GABA-releasing interneurons. *Nature Rev. Neurosci.*, 2: 425–433.

Galarreta M, and Hestrin S [1999]. A network of fast-spiking cells in the neocortex connected by electrical synapses. *Nature*, 402: 72–75.

Gibson JR, Belerlein M, and Connors BW [1999]. Two networks of electrically coupled inhibitory neurons in neocortex. *Nature*, 402: 75–79.

Tamás G, et al. [2000]. Proximally targeted GABAergic synapses and gap junctions synchronize cortical interneurons. *Nature Neurosci.*, 3: 366–371.

Central Pattern Generators

Grillner S, et al. [1998]. Intrinsic function of a neuronal network—A vertebrate central pattern generator. *Brain Res. Revs.*, 26: 184–197.

Grillner S, et al. [1995]. Neural networks that co-ordinate locomotion and body orientation in lamphrey. *Trends. Neurosci.*, 18: 270–279.

Marder E, and Bucher D [2001]. Central pattern generators and the control of rhythmic movement. *Curr. Biol.*, 11: R986–R996.

Marder E, and Calabrese RL [1996]. Principles of rhythmic motor pattern generation. *Physiol. Rev.*, 76: 687–717.

Nusbaum MP, and Beenhakker MP [2002]. A small-systems approach to motor pattern generation. *Nature*, 417: 343–350.

Respiration

Gray PA, et al. [1999]. Modulation of the respiratory frequency by peptidergic input to the rhythmogenic neurons in the Prebötzinger complex. *Science*, 286: 1566–1568.

Lieske SP, et al. [2000]. Reconfiguration of the neural network controlling multiple breathing patterns: Eupnea, sighs and gasps. *Nature Neurosci.*, 3: 600–607.

Ramirez JM, and Richter DW [1996]. The neuronal mechanisms of respiratory rhythm generation. *Curr. Opin. Neurobiol.*, 6: 817–825.

Motor Learning and Memory

Kandel ER [2000]. The molecular biology of memory storage: A dialogue between genes and synapses. *Science*, 294: 1030–1038.

Problems

20.1 A small number of ion channels have particularly prominent roles in generating rhythmic patterns of action potentials. One of these, the HCN channel, was discussed in Section 20.2 while another, the T-type calcium channel, was discussed in Section 20.4. Make a sketch of the activation and deactivation functions for the T-type calcium channel and the sodium channels responsible for action potential upstrokes. Recall from the discussion in Section 20.4 that at the resting potential the T-type calcium channel is activating while the sodium channel is deactivating. In the plots of current versus membrane potential, indicate the location of a resting potential and position the four curves so they satisfy the relationship just mentioned.

20.2 The utility of a gradually depolarizing membrane potential was discussed in Section 20.11 and illustrated in Figure 20.1. As noted in that discussion, this behavior greatly enhances excitability in cells possessing plateau potentials. When two cells of this type are coupled through mutual inhibition they can synchronize their firing acting as oscillatory units that produce motor rhythms. List some types of currents that promote gradual declining membrane potentials of the sort depicted in the figure.

21
Learning and Memory

The learning and memory formation processes carried out by the snails and other invertebrates is certainly more modest than that of humans, yet the two are similar. Each involve alterations in behavioral responses to stimuli brought on by experience. Learning is the adaptive process whereby changes in behavior are induced in response to experience. Memory is the record, or trace, underlying the changes in behavior. The records in all organisms, snails and mice, flies and humans, are stored in the patterns of connection between neurons. During learning and memory formation some connections are strengthened while others are weakened. For instance, in the case of sensitization discussed at the end of the last chapter, connections between neurons were strengthened. This led to behavioral changes. The snails were able to respond with increased rapidity to noxious stimuli.

The term *efficiency of synaptic transmission* refers to the magnitude of the response generated in a postsynaptic neuron when an action potential is generated in the presynaptic neuron. In many situations, the efficiency of the transmission is not fixed but instead varies in ways that depend on use. The changes in efficiency can be transient, lasting for a few milliseconds, or can be permanent, lasting for a lifetime. Connections may be strengthened or be weakened or disappear altogether. *Synaptic plasticity* is the term used to describe the use-dependent, or adaptive, changes in the efficiency of synaptic transmission between pre- and postsynaptic cells.

The ability to vary connection strengths is an important one. It allows the nervous system to form its precise connections and neural circuits during development; it gives rise to a unique personality for each individual; and it makes possible learning throughout life. Cortical circuits develop over time in several stages. The first stage of circuit development involves neurite outgrowth, growth cone pathfinding and then synaptogenesis. As the circuits become functional and can respond to synaptic input, ineffective and improper connections are pruned, and neurons that are no longer needed die off through programmed cell death. The process of varying the strengths of synaptic connections continues throughout life. It underlies the development of simple motor skills as well as highly coordinated eye and hand movements. It allows associations to be made that are needed for survival (for example, of sights, sounds, touches and odors indicative of danger), and makes possible learning and remembering and forgetting.

The pre- and postsynaptic terminals of nerve cells are highly enriched in signaling molecules. Most major classes of signaling molecules discussed in the preceding chapters are present in the synapses. Thee include ion channels, enzymes such as protein kinases, protein phosphatases and GTPases, G protein-coupled receptors, cell adhesion molecules, and nonenzymatic scaffold, adapter, and anchor proteins. The molecular composition of the synapses will be examined following a brief overview of how neurons are organized in the brain. The remainder of the chapter is devoted to explorations of learning and memory formation. The preeminent experimental model of how this is accomplished is called *long-term potentiation*. It is studied most often in vitro using slice preparations from the hippocampus of rodents. Findings from these experiments as well as from experiments of fear conditioning and drug dependence, which can be thought of as an aberrant form of learning and memory, will be explored, as well.

21.1 Architecture of Brain Neurons by Function

Neurons in the brain are organized into architecturally distinct functional areas. The large-scale division of the brain into brain stem, limbic system, cerebellum, and cerebral cortex is depicted in Figure 21.1. The first of the aforementioned large-scale regions, the brain stem, contains a number of structures whose activities were discussed in the previous chapter. Neurons in the medulla oblongata and pons regulate rhythmic movements associated with breathing and heartbeat, while cells in the reticular activating complex control alertness and the sleep/wake cycle. The limbic system is situated just above the brain stem. It includes the hippocampus, and the amygdala, which mediates emotional responses such as fear and aggression.

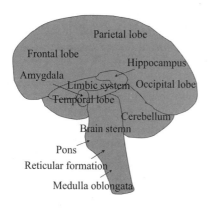

FIGURE 21.1. Architecture of the brain showing its major divisions: Shown is a side view with the front of the head on the left and the back of the head on the right. The partitioning of the cerebral cortex is depicted along with the locations of the brain stem, cerebellum, and limbic system.

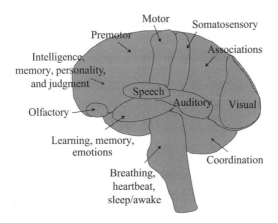

FIGURE 21.2. Architecture of the brain showing the sequestering of brain functions within specific areas of the brain: The locations of the major sensory areas are shown along with motor areas and sites where forms of learning and memory, and higher processing functions, as most strongly associated.

The cerebellum is a large region that sits behind the brain stem and below the occipital lobe. Neurons in this region of the brain integrate and coordinate motor actions and mediate motor learning. Cerebellar neurons control balance and posture, and ensure that fine motor actions are accurate. The large section of cortex sitting above and about these structures is partitioned into frontal, parietal, occipital, and temporal lobes. The temporal lobe sits above and to the outside of the brain stem at about the level of the ears. As indicated in Figure 21.2, neurons in this region are involved in the senses of sound, taste, and smell, and in memories thereof. Information of a tactile character, that is, the sense of touch, is processed by neurons located in the somatosensory cortex that lies above in the parietal lobe. This region is situated next to regions where associations and interpretations are made from information sent from the primary visual cortex and other primary sensor cortices. Behind this region, at the back of the head in the occipital lobe, (primary) visual information is received and processed. Lastly, the nonmotor portions of the frontal cortex are used for making judgments and intelligent decisions. The premotor cortex controls head and eye movements, while speech is centered in Broca's area and smell in the olfactory center.

The presence of a large region devoted to motor and sensory information processing is unique to mammals. This region is highly differentiated radially into layers and laterally into regions. As first noted by Brodmann in 1909, regions of cortex involved in processing sensory information are organized in the radial direction into six layers and in the lateral direction into at least 50 distinct areas. The numbers of sublayers, and their thicknesses and arrangements, vary from area to area in the cortex. Within each area there are cells of a particular size and arborization, with a distinct set of inputs, outputs, and intercortical connections. The overall organization of the neocortex is that of a patchwork of areas each with its own cellular organization, or *cytoarchitecture*. Brodmann based his partitioning of the cortex upon differences in appearance that he was able to observe using a light microscope. In more modern approaches, functional MRI and other biophysical methods are used to assign functions to anatomical regions. The anatomical regions, called *Brodmann's areas*, correspond approximately to functional areas, so that each anatomically distinct area carries out some unique function.

21.2 Protein Complexes' Structural and Signaling Bridges Across Synaptic Cleft

Communication occurs when neurotransmitters released from the presynaptic terminal diffuse across the synaptic cleft and bind to receptors on the postsynaptic side. Before this communication can occur, a minimal set of molecular components must be assembled into a functional unit, the synapse. Active zones on the presynaptic side that contain the machinery for neurotransmitter release must align with the postsynaptic density (PSD) on the postsynaptic side containing the receptors for the neurotransmitters. These activities must take place as the two parts of the signaling unit are formed.

These pre- and postsynaptic portions of the synapse are aligned and stabilized by proteins that form bridges from one side of the cleft to the other. Several of the most prominent bridging pairs—Neurexins-to-Neuroligins, SynCAMs-to-SynCAMs and EphB2s-to-EphrinBs—are depicted in Figure 21.3. Neurexins are brain-specific cell surface proteins containing one or more LNS (laminin, neurexin, sex hormone-binding globulin) domains. The LNS domains, approximately 190 amino acid residues in length, act as cell surface recognition modules. They are found in a variety of extracellular matrix proteins and in cell surface receptors, typically in combination with multiple EGF repeats. Neuroligins act as ligands for Neurexins. They are brain-specific, single-pass, transmembrane proteins found on the opposing surface to the neurexins. Neurexin-Neuroligin cell-cell adhesion complexes form a bridge between the pre- and postsynaptic terminals. Neurexin interacts with CASK on the presynaptic side, and neuroligin binds to the PSD-95 proteins on the postsynaptic side. The interactions with each another and with the cytoplasmic CASK and PSD-95 proteins stabilize and maintain the alignment between active zones on the presynaptic side and the postsynaptic density at the postsynaptic terminal.

More than one kind of signaling complex ties together the active zone on the presynaptic side and the PSD at the postsynaptic terminal. Members of the immunoglobulin superfamily of cell adhesion molecules called "synCAMs" establish homophilic contacts between proteins expressed on the two opposing surfaces. As is the case for other proteins that perform a structural role mediating surface adhesion, the synCAMs also have a

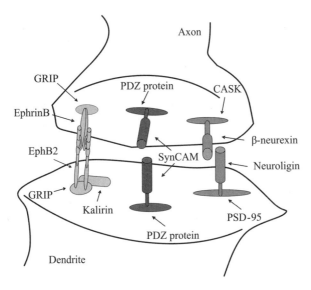

FIGURE 21.3. Transsynaptic protein–protein interactions: Protein pairs form bridges that physically link pre- and postsynaptic terminals. The protein pairs are tied to scaffolding proteins on the two sides of the synapse. Many, if not most, of the scaffolding proteins have PDZ domains. Kalirin, which binds EphB2, is a Rho GEF.

signaling role. They appear to supply signals telling nascent presynaptic terminals to differentiate into a mature signaling form. The neurexin-neuroligin complexes serve a similar signaling role. These too instruct the cells to form differentiated presynaptic terminals. Several families of cell adhesion molecules participate in bridging actions. Neuron-specific cadherins form bridging complexes, too. These proteins are anchored at both sides of the synaptic cleft to β-catenins.

Another family of proteins having an important role in synapse formation is the EphB receptor tyrosine kinases and their Ephrin ligands. The EphB2 receptor signals through the Rho GEF kirilin to the actin cytoskeleton, thus helping in dendritic spine morphogenesis. In synapses that signal using glutamate, *N*-methyl-D-aspartate (NMDA) receptors are an early occupant of the postsynaptic density. Kirilin helps establish NMDA receptor aggregates at newly created synapses.

21.3 The Presynaptic Terminal and the Secretion of Signaling Molecules

When action potentials arrive at the presynaptic terminal, voltage-gated calcium channels open and calcium ions flow into the terminal. The calcium ions bind to calcium sensors attached to the neurotransmitter-filled vesicles.

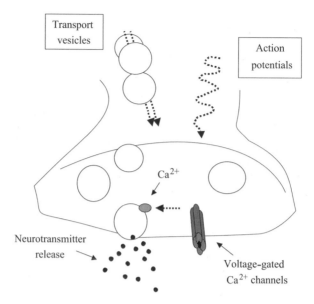

FIGURE 21.4. Neurotransmitter release at the presynaptic terminal: In response to the arrival of an action potential, calcium enters the cell through voltage-gated Ca^{2+} channels (VGCCs) and triggers fusion and release of neurotransmitters from neurotransmitter-filled synaptic vesicles. The neurotransmitter molecules diffuse across the synaptic cleft and bind to receptors embedded in the terminal membrane of the postsynaptic cell.

Binding of the calcium to the sensor triggers conformational changes leading to vesicle fusion and release of the contents. Once released, the neurotransmitters diffuse across the synaptic cleft and bind to receptors embedded in the postsynaptic membrane. This set of steps is depicted schematically in Figure 21.4.

The amount of neurotransmitter released at a presynaptic terminal depends on several factors. Synapses in the central nervous systems are, in general, small. Their amount of neurotransmitter released at one time is not only small but also can be quite variable. The amount of neurotransmitter released can, for example, depend on timing. If a large number of pulses arrive at the presynaptic terminal in a short period of time, the pool of available neurotransmitter vesicles can be depleted. The release of neurotransmitter will therefore monotonically decline with each later-arriving action potential. Synapses with large pools will release more neurotransmitter than those with small pools. The term "active zone" refers to the exact site along the presynaptic terminal where the vesicles fuse and release neurotransmitter. The amount of neurotransmitter release will depend on the number of active sites per terminal and on the number and size of the vesicles.

Several protein complexes regulate vesicle movements and neurotransmitter release in the presynaptic terminal. These have been organized under three headings in Table 21.1. The first set of signaling elements in the table is the voltage-gated calcium channel grouping that converts membrane depolarization into a calcium signal. Included under this heading are a number of signaling proteins that allow neuromodulators to influence the strength of the calcium signals and thus the amount of neurotransmitter released into the cleft.

The second grouping contains vesicle regulators such as the soluble NSF attachment protein receptor (SNARE) complex. SNARE proteins regulate the final stages of neurotransmitter release—docking and fusion of the vesicle with the plasma membrane resulting in release of the vesicle contents into the synaptic cleft. There are two classes of SNARE proteins— those that bind to the vesicle surface (v-SNAREs) and those that attach to the plasma membrane (t-SNAREs). Interactions between the two kinds of SNARES regulate the release of neurotransmitter. One of the v-SNAREs, synaptotagmin, contains a calcium-binding C2 domain and functions as the calcium sensor. Conformational changes brought on by calcium binding are conveyed to the t-SNAREs with which it interacts promoting fusion and vesicle content release.

The other signaling complexes in the vesicle regulator grouping help shepherd vesicles into the presynaptic terminal and to the plasma membrane, and then regulate membrane fusion. Neurotransmitter-filled vesicles

TABLE 21.1. Main components of the presynaptic terminal at CNS synapses: Abbreviations—*N*-ethylmaleimide-sensitive fusion protein (NSF); soluble NSF-attachment protein (SNAP); soluble NSF attachment protein receptor (SNARE); vesicle-associated membrane protein (VAMP); CaMK/SH3/guanylate kinase domain protein (CASK).

Presynaptic terminal component	Function
Channels and their modulators	
Voltage-gated calcium channels	N- and P/Q-type calcium channels; enable calcium to enter the cell in response to membrane depolarization
GPCRs	Gβγ subunits negatively regulate N-type channels
Protein kinase C	Positively regulates N-type channels
Vesicle regulators	
v-SNAREs	Synaptobrevin (VAMP) and synaptotagmin mediate vesicle docking and fusion. Synaptotagmin functions as a calcium sensor
t-SNAREs	Syntaxin and SNAP-25 mediate vesicle docking and fusion
Sec6/8 complex	Targeting of vesicles to the plasma membrane
Sec1/Munc18	Regulators of Syntaxin
NSF, α-SNAP	Dissociation of the SNARE complex
Scaffolds and adapters	
CASK	Attaches to N- and P/Q-type channels and Mint1; aligns vesicles
Mint1	Attaches to CASK and N- and P/Q-type channels
Veli	Member of the CASK, Mint1, Veli adapter complex

are transported along the cytoskeleton into the presynaptic terminal. Upon arrival at the terminal the vesicles are targeted to the plasma membrane by a multiple subunit protein complex called Sec6/8. Another set of proteins, Sec1/Munc18, act both as positive and as negative regulators of SNARE-mediated fusion through interactions with Syntaxin. Yet another set of molecules, NSF and α-SNAP, direct the dissociation of the SNAREs permitting their reuse.

The third grouping presented in Table 21.1 encompasses the scaffold and anchor proteins. These proteins help organize the complexes in the presynaptic terminal. As is the case in any cell, these proteins provide attachment sites where proteins that must work together can bind in close proximity to one another.

21.4 PSD Region Is Highly Enriched in Signaling Molecules

The region just below the postsynaptic membrane is the *postsynaptic density* (PSD). The PSD is a disclike structure that sits opposite the active zone in the presynaptic terminal (Figure 21.5). It is highly enriched in struc-

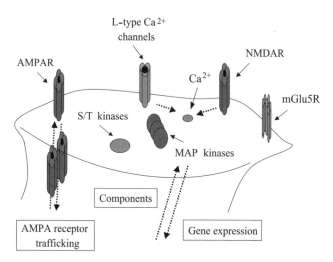

FIGURE 21.5. The postsynaptic density: Shown as some of the types of signaling proteins that are found in the postsynaptic density at excitatory synapses in the central nervous system. Glutamate released at the presynaptic terminal diffuses across the synaptic cleft and binds to NMDA and AMPA receptors leading to membrane depolarization, calcium influx, alterations in AMPA receptor population, and gene expression. Short-term modulatory effects are mediated by GPCRs such as mGlu5R, leading to G protein subunit- and second messenger-binding to ion channels, and phosphorylation of ion channels by protein kinases.

TABLE 21.2. Main components of the PSD at excitatory glutamergic synapses: Abbreviations—α-amino-3-hydroxy-5-methyl-4-isoxazolepropionate (AMPA); N-methyl-D-aspartate (NMDA); calcium/calmodulin-dependent protein kinases II and IV (CaMKII and CaMKIV); glutamate receptor interacting protein (GRIP); guanylate kinase-associated protein (GKAP).

PSD component	Function
Receptors	
AMPA receptor	GluRs: Key component of active synapses
NMDA receptor	NRs: Dual voltage and ligand-gated allowing entry of calcium into the cell
Metabotropic receptors	mGluRs: Signal to IP$_3$ regulated intracellular calcium stores
Kinases and phosphatases	
αCaMKII and CaMKIV	Central signaling element, involved in both short and long term forms of synaptic plasticity
Protein kinase A	Central signaling element, involved in both short and long term forms of synaptic plasticity
Protein kinase C β, γ isoforms	Signaling element, involved in short term forms of synaptic plasticity
ERK2 type MAPKs	Signaling element, involved in both short and long term forms of synaptic plasticity
Protein phosphatase 1	Regulator of homeostasis and learning
Anchors, scaffolds, and adapters	
AKAP79	Sequesters protein kinase A, protein kinase C, and calcineurin near receptors and ion channels
GRIP	Attaches to GluR2 and GluR3
PSD-95	Attaches to NR2s
SynGAP	Activates MAP kinase signaling
GKAP	Link between PSD-95 and GKAP
Shank	Link between Homer and GKAP
Homer	Attaches to mGluRs and to Shank

tural (cytoskeleton) and regulatory (receptor, ion channel, scaffolding, signaling) proteins. The several different kinds of regulatory proteins have been placed into three groupings in Table 21.2. The first group contains the receptors for neurotransmitters. Glutamate is the predominant neurotransmitter at excitatory synapses in the central nervous system. Several different kinds of glutamergic receptors are found in the PSD of excitatory synapses in the CNS. Some glutamergic receptors are ion channels (ionotropic) while others are G protein coupled receptors (metabotropic). Two kinds of ionotropic glutamate receptors—the α-amino-3-hydroxy-5-methyl-4-isoxazolepropionate (AMPA) receptors and the N-methyl-D-aspartate (NMDA) receptors—are intimately connected with synaptic plasticity. Rather than listing a large number of receptors in the table, only the three main classes of glutamergic receptors have been included in the table. These are the principal receptors involved in learning and memory formation.

The second group in the table consists of several different kinds of serine/threonine kinases and phosphatases that are central signaling elements in the postsynaptic density. The third grouping contains connecting and supporting agents such as GTPases functioning as molecular adapters, and anchoring and scaffolding proteins that help organize the signaling complexes. These signaling elements form a matrix within the PSD that allows ion channels to cluster together in close proximity to second messenger and downstream signaling elements. The postsynaptic density is enriched in cytoskeleton proteins, especially in actin and actin-related cytoskeleton components.

21.5 The Several Different Forms of Learning and Memory

There are several different kinds of learning. The forms of learning described earlier with regard to *Tritonia* and *Aplysia*—sensitization and habituation—are nonassociative in character. Repeated stimulation of noxious stimuli produce sensitization while repeated exposures to a harmless stimulus lead to habituation, a reduced responsiveness. These forms of learning do not require a second pairing stimulus to elicit behavioral changes.

Associative learning, in contrast, involves a pairing of stimuli. The earliest and still most famous example of associative learning is Pavlov's dog experiments. In these and similar experiments, an initial nondangerous event, called the conditioned stimulus (CS), is paired repeatedly with a biologically significant (potentially dangerous) event called the unconditioned stimulus (US). Over time, the relationship between the CS and US is learned; that is, the behavioral responses adapt in response to the experiences. In a typical laboratory experiment, a rat is given a tone, a neutral nondangerous CS, which is paired with an electrical shock, the dangerous US. After as few as one or two exposures, the tone will elicit a number of physiological and behavioral adaptations. The rat will freeze in place and exhibit reflex actions. Heart rate and blood pressure will change, and stress hormones will be released. In *Aplysia*, the associative learning counterpart is called *long-term facilitation*. In this process, an electric shock to the tail serves as the US, and a weak touch to the siphon functions as the CS. The snail learns to associate the two stimuli and, because of the association with the electrical shock, alters its behavioral response to the weak touch over time.

The circuitry responsible for the *Aplysia* gill withdrawal response is depicted in Figure 21.6. The axon of the sensory neuron that conveys touch signals contacts a dendrite on a motor neuron. The sensory neuron that conveys the electric shock signals also makes contact with the motor neuron. As shown in the figure the neuron has a synaptic connection with an interneuron that releases serotonin from its axon terminal. The serotonin

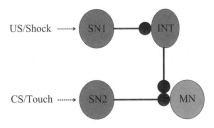

FIGURE 21.6. Circuitry underlying the associative *Aplysia* gill withdrawal response: Three different kinds of neurons form the circuit—sensory neurons that detect and respond to electrical shock (SN1) and touch (SN2), a serotonin-releasing interneuron (INT), and a motor neuron (MN).

diffuses across the synaptic cleft to the axon terminal of the touch sensory neuron where the two sets of signals converge.

Just as there are several types of learning, there are several kinds of memory. *Short-term memory* refers to memories that are *labile* (responsive to changes in conditions under which they are created) and are only transiently formed. They are lost if not converted into stable memory traces. Short-term labile memories involve celluar changes that utilize cellular resources that are already available and are local to the synapse or synapses undergoing modification. In contrast, *long-term stable memory traces* typically involve gene expression and protein synthesis leading to architectural modifications that produce the lasting changes in synaptic transmission.

This events stimulated by the release of serotonin are depicted in Figure 21.7. These include stimulation of adenylyl cyclase leading to the production of cAMP and its subsequent activation of the catalytic subunits of protein kinase A. Protein kinase A phosphorylates the nearby potassium channels and, as a result, currents through potassium channels are reduced. The depolarization accompanying arrival of the action potential in the axon terminal lasts longer. More calcium is able to enter the terminal resulting in a greater release of neurotransmitter. Two routes are illustrated in the figure. One of the routes leads to ion channel modification, a short-term memory trace. The other route leads to the nucleus and results in changes in the pattern of gene expression, producing a longer lasting memory trace.

21.6 Signal Integration in Learning and Memory Formation

Synapses were first described by Ramón y Cajal in the 1890s. It took several years for the idea to be accepted that nerve cells were not hard-wired to one another, but instead communicated using chemical messengers that diffuse across a narrow gap between pre- and postsynaptic terminals. The

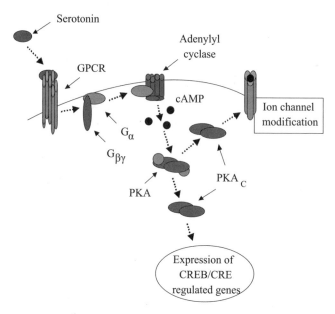

FIGURE 21.7. Signaling in the learning pathway in *Aplysia* leading to short- and long-term facilitation: Serotonin stimulates an increased production of cAMP by adenylyl cyclase through actions of GPCRs and G_α subunits. A transient facilitation of the signal response is produced by phosphorylation of ion channels by the catalytic subunit (PKA_C) of protein kinase A. This modification alters the biophysical properties of the ion channel. Long-term responses are elicited by sustained serotonin signaling resulting in gene transcription and morphological changes at the synaptic terminal.

efficiency of the transmission process depends on both presynaptic and postsynaptic factors. First, it depends on the amount of neurotransmitter release from the presynaptic terminal for a given depolarization or action potential. Second, it depends on how strong a response that release elicits at the postsynaptic side. The strength of the presynaptic signal and the post-synaptic response depends on the mix of ion channels and signal receptors and on their biophysical states.

In the case of short-term facilitation in *Aplysia*, the increased efficiency in synaptic transmission is implemented at the presynaptic side of the cleft. Calcium influx resulting from membrane depolarization and serotonin induced G protein signals converge on adenylyl cyclase, which responds by increasing cAMP production leading to changes in the amount of neuro-transmitter being released from the terminal. Adenylyl cyclase acts as a signal integrator, serving as a moleculer device for detecting the temporal coincidence of electric shock and weak touch stimuli.

NMDA receptors perform a similar function, but, instead of acting on the presynaptic side of the cleft, the NMDA receptors act at the postsynaptic terminal. The key biophysical property of these receptors is that they are both voltage-dependent and glutamate-dependent thereby allowing them to integrate signals and to evaluate associativity. These receptors work in the following way: At the membrane's resting potential, the NMDA receptor ion channel is closed—it is physically blocked by an Mg^{2+} ion that sits inside the opening and blocks the pore. When the membrane is depolarized sufficiently the Mg^{2+} block is relieved, and once glutamate is bound, calcium ions can pass through the channel and enter the cell. The voltage dependence is the source of the receptor's integrative property. No single synapse by itself can depolarize the postsynaptic membrane sufficiently to relive the magnesium block. Rather, many synaptic inputs acting in close spatial and temporal proximity to one another are needed to adequately depolarize the membrane. When depolarization and ligand binding occurs within an appropriate time window, the NMDA channel will open. Thus, the NMDA receptor provides synapses with a way of determining whether the associativity conditions for synaptic modification have been satisfied or not.

One of the ways that both the efficiency of synaptic transmission and the ability to vary this property can be controlled is through the balance between AMPA and NMDA receptors. At many postsynaptic terminals only NMDA receptors are active during development. These synapses remain silent at resting potentials because of the Mg^{2+} block, but are capable of transmitting signals when the membrane is depolarized by one means or another. In response to entry of calcium into the terminal through the NMDA receptors, AMPA receptors become active. The presence of active AMPA receptors allows the synapses to respond rapidly to glutamate and generate stronger responses at the postsynaptic terminal to the release of glutamate. AMPA receptors and AMPA receptor trafficking have a prominent role in learning and memory formation and for that reason are highlighted in Figure 21.5.

21.7 Hippocampal LTP Is an Experimental Model of Learning and Memory

The hippocampus is a horseshoe-shaped structure belonging to the limbic system. The hippocampus receives input from a number of sensory information processing areas. It, along with its surrounding structures, the entorhinal cortex, perirhinal cortex, and parahippocampal region, are believed to be the place where short term memories are stored and then shipped out to long term, more permanent memory storage locations through a process called *memory consolidation*. The hippocampal system is thought to be responsible for recalling spatial relationships between objects in the environment and for spatial navigation.

Hippocampal long-term potentiation (LTP) is a widely studied mammalian form of synaptic plasticity. This process is elicited in the laboratory

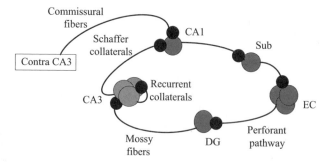

FIGURE 21.8. Hippocampal circuit in the rodent used in studies of LTP: The core circuit includes a set of one-way connections from the entorhinal cortex (EC) to the hippocampal dentate gyrus (DG), cornu ammonis 3 (CA3) region, CA1 region, subculum (Sub), and back to the EC and to other cortical regions. LTP is studied in the CA3 region, where there is a dense network of recurrent connections, and in the connections between the CA3 and CA1 neurons. Two sets of connections converge on the CA1 neurons—the Schaffer collaterals from the (ipsilateral) CA3 and the commissural fibers from the contralateral (contra) CA3. The names commonly given to the various axons are shown.

in brain slices prepared from portions of the rodent hippocampus called the CA1 and CA3 regions. The general architecture of the hippocampus showing the locations of the CA1 and CA3 regions is presented in Figure 21.8. Two sets of connections are noted in the figure—recurrent connections between cells in CA3, and connections from CA3 to CA1. In either circuit, when a tetanic stimulus (i.e., a train of high frequency pulses) is delivered to the presynaptic cell, or alternatively a series of pulses at low frequency is delivered and the event paired with a laboratory-supplied depolarization of the postsynaptic membrane, the efficiency of synaptic transmission is increased. The increase, or potentiation, of the synaptic efficiency can be generated in milliseconds, yet it lasts for minutes or even hours.

21.8 Initiation and Consolidation Phases of LTP

The initiation phase of LTP is followed by a consolidation phase involving gene expression. The transition from a short lived form of memory formation to a more permanent one lasting for days, weeks, and even years is through a process of memory consolidation. Gene expression is required during this phase of memory formation. The cyclic AMP (cAMP) signaling pathway is activated in which protein kinase A and downstream MAP kinases interact with cAMP response element-binding protein (CREB) transcription factors that bind to cAMP response elements (CREs) in promoters. This pathway can be activated in several ways. It can be activated through calcium regulation of adenylyl cyclases, and by neuromodulators

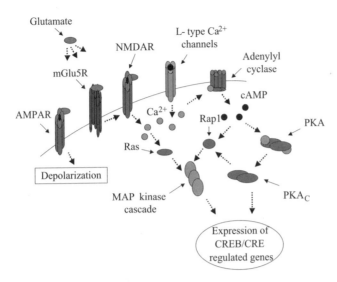

FIGURE 21.9. Signaling in the learning pathway in hippocampal neurons leading to consolidation of LTP: Calcium entry into the cell through NMDA receptors and L-type calcium channels leads to activation of downstream acting serine/threonine kinases. Key signaling intermediates include the GTPases Ras, and Rap1. These are activated in a Ca^{2+} and cAMP-dependent manner, respectively, thereby integrating and routing signals from these second messengers to the nucleus via a Raf-MEK-ERK MAP kinase complex.

such as serotonin and dopamine that bind to GPCRs, which act through heterotrimeric G proteins to stimulate ACs to produce cAMP.

The MAP kinase and PKA routes are highlighted in Figure 21.9. As shown in this figure, calcium entry either through NMDA receptors or through L-type calcium channels leads to the activation of a number of serine/threonine kinases—protein kinase A, protein kinase C, MAP kinases, and calcium/calmodulin-dependent protein kinases II and IV (CaMKII and CaMKIV), not all of which are shown the figure. Several different actions can occur depending on the duration of the synaptic signaling. Short-term actions can take place in the neighborhood of the receptors using resources already available. If the synaptic signaling is sustained over time, long-term changes can occur involving changes in gene expression.

21.9 CREB Is the Control Point at the Terminus of the Learning Pathway

The learning pathway terminates when the serine/threonine kinases arrive in the nucleus and phosphorylate the cAMP response element-binding protein (CREB) on Ser133. In more detail, a number of kinases translocate

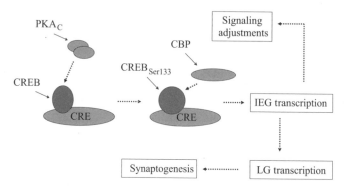

FIGURE 21.10. Downstream signaling events leading to consolidation of memories: Abbreviations—Immediate early gene (IEG); late gene (LG); cAMP response element-binding protein (CREB); p300/CREB-binding protein (CBP).

to promoter sites in the nucleus containing cAMP responsive elements (CREs). The kinases so involved include protein kinase A, protein kinase C, MAP kinases, and CaM kinase IV. Members of the CREB family of transcription factors bind to these sites. The CREB proteins remain inactive if not phosphorylated on Ser133. Phosphorylation on Ser133 by kinases such as PKA results in binding by p300/CREB-binding protein (CBP). As noted earlier, these are histone acetylases (HATs) that work together with the basal transcription machinery to initiate a program of transcription leading to formation of new synapses and morphological changes to old ones (Figure 21.10). As is the case for other transcriptional control points, a variety of factors acting positively and negatively regulate CREB-mediated transcription.

Transcription takes place in several stages, starting with immediate early genes and ending with late genes. For example, among the first genes to be transcribed in serotonin-mediated facilitation in *Aplysia* are those whose products extend to duration of the PKA signaling. This stage is followed by the transcription and synthesis of transcription factors such as C/EBP, which leads to the creation of new synaptic connections. This same overall set of signaling activities is observed in other forms of learning and memory, and in other test animals, so that there is a common family of mechanisms that extends across phyla from *Aplysia* to flies to mice.

21.10 Synapses Respond to Use by Strengthening and Weakening

In 1947, Donald Hebb formulated an associative rule governing use-dependent changes in synaptic transmission. Hebb's rule is couched in terms of cellular firing by a pair of cells (Figure 21.11). It says: When a when an axon of cell A is near enough to excite cell B and repeatedly and persistently takes

(a)

(b)

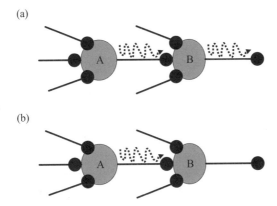

FIGURE 21.11. Hebb's rule: (a) In response to the firing of an action potential by cell A, cell B fires an action potential. In this situation, the depolarization of the postsynaptic membrane contributes to the firing of an action potential by cell B. The strength of that synapse is increased. (b) Cell B does not fire an action potential when cell A does. The depolarization at the postsynaptic membrane is not effective in eliciting action potentials. In situations of this sort the strength of the synapse is reduced.

part in firing it, some growth process or metabolic change takes place in one or both cells such that A's efficiency, as one of the cells firing B, is increased. The rule in this form is incomplete. It does not state what happens when the postsynaptic cell fails to fire. To complete the rule one adds a subrule that says: When a presynaptic axon of cell A repeatedly and persistently fails to excite the postsynaptic cell B, while cell B is firing under the influence of other presynaptic axons, metabolic changes takes place in one or both cells such that A's efficiency, as one of the cells firing B, is decreased. In other words, synapses whose activation is strongly correlated with the firing of the postsynaptic cell are strengthened, while those synapses whose activation is poorly correlated with the firing of the postsynaptic cell, for example, by being silent while the cell fires, are reduced in efficiency.

According to the expanded form of Hebb's rule two types of changes in the efficiency of synaptic transmission can occur: the synapse can be strengthened or it can be weakened. The efficiency of synaptic transmission increases when a synaptic connection is strengthened and decreases when the connection is weakened. Modifications in synaptic efficiency depend upon the timing of the action potentials in the pre- and postsynaptic cells. Repetitive pre- and postsynaptic firings occurring within 10 to 50 msec of one another are able to influence the efficiency of synaptic transmission. If cell A fires its action potential before cell B fires its action potential, then the synaptic connections between the two cells will be increased. If, on the other hand, cell B repeatedly fires before cell A, then the strength of the synapse will decrease. The situation is intrinsically asymmetric so that there is a preferred direction of information flow.

21.11 Neurons Must Maintain Synaptic Homeostasis

If the efficiency of transmission increases in too many synapses of a neuron, excitation-induced toxic effects, such as excessive calcium levels, may set in. As a result the neuron may be harmed or even killed. In a typical Hebbian or associative process, each synapse acts in a fairly independent manner. There has to be some additional mechanism or constraint operating cell-wide that can throttle back runaway positive or negative changes in synaptic efficiency. There are two, not necessarily distinct, ideas of how this may occur, both supported by a body of experimental data. One of these is the notion of a *sliding modification threshold*; the other is the idea of *synaptic scaling*.

In the sliding threshold model (Figure 21.12), a cell B possesses a threshold for synaptic modification. If the postsynaptic cell's firing rate exceeds that threshold, paired pre- and postsynaptic firing will increase the efficiency of the synapse. If the postsynaptic cell's firing rate falls below the threshold value, the same correlated activity will diminish the efficiency of that synapse. The effect is a dynamic one. Cell B will continually adjust its threshold according to the changes in the cell's firing rate suitably averaged over time. The threshold will slide up or down as the firing rate increases or decreases. By moving up it becomes more difficult to produce further increases, and becomes easier to elicit decreases. Similarly, if the threshold goes down because of weak firing activity, it becomes easier to strengthen synapses and harder to weaken them.

In synaptic scaling, like the sliding threshold model, the total activity in the postsynaptic cell determines how that cell responds to neurotransmitter release. The firing rate reflects the sum of all synaptic activities in the postsynaptic cell. If the firing rate in the postsynaptic cell is already high all synaptic strengths will be scaled back, and conversely if the firing rate is too low the strength of each synapse will be scaled up. In mathematical terms, the effect is a multiplicative one. The strength at a particular synapse is proportional to the product of the firing rate and the synaptic strength at that

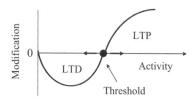

FIGURE 21.12. Sliding threshold model: Shown is a plot of the modification predicted by the sliding threshold model as a function of the total activity in the postsynaptic cell. Activity levels below that of the threshold produce long-term depression, or LTD (negative modification), while those that exceed the threshold lead to long-term potentiation, or LTP (positive modification). The threshold adapts to the overall activity in the postsynaptic cell averaged over a suitable time interval.

synapse. Since the scaling affects all synapses in the same way it may be best thought of as a form of cellular homeostasis operating in addition to Hebbian mechanisms.

21.12 Fear Circuits Detect and Respond to Danger

The forms of synaptic plasticity such as LTP that are observed in the hippocampus are not limited to that region. They are encountered in regions of the brain ranging from the visual cortex to the amygdala to regions of the brain involved in drug addiction. The amygdala is the central player in fear conditioning. In fear conditioning experiments using rats, tone signals are relayed through the thalamus to the auditory cortex and from the thalamus and auditory cortex to the amygdala. Foot shock signals are similarly relayed into the somatosensory cortex and from there to the amygdala.

The amygdala is partitioned into several regions. Some signals first enter the lateral nucleus (LN) of the amygdala and then pass to the basal nucleus (BN), while other signals enter the basal nucleus directly. These areas along with the accessory basal nucleus function as the central input unit of the amygdala that receive auditory, visual, somatosensory, and other forms of sensory information from the thalamic relay nuclei and higher sensory cortical areas. These nuclei also receive input information from the hippocampus of a contextual character, that is, environmental cues about, for example, place and space, associated with the experience. As shown in Figure 21.13 connections from the input unit are made to several regions. These include connections between the input unit and central nucleus, which sends projections to a number of brain stem regions and by that means controls the behavioral and physiological fear responses.

Fear conditioning produces synaptic modifications quite similar to hippocampal LTP. As is the case for LTP in hippocampal slices, LTP in the lateral and basal nuclei of the amygdala is dependent on postsynaptic depolarization. Calcium entry through NMDA receptors plays an important role, as does activation of and signaling by protein kinase A and MAP kinases. As was the case for LTP, there are several different forms ranging from short-term transitory changes to long lasting ones requiring CREB transcriptional activity and protein synthesis.

21.13 Areas of the Brain Relating to Drug Addiction

Drug addiction involves areas of the brain that shape mood and generate emotional responses. There are many different kinds of addictive drugs. Prominent examples of addictive drugs are alcohol, amphetamines, cocaine, morphine, and nicotine. Although the chemical makeup of each of these substances is different, all drugs target populations of neurons situated in

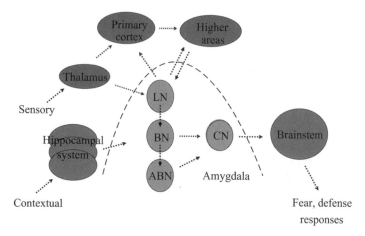

FIGURE 21.13. Circuitry involved in responses to fearful and dangerous sensory and contextual stimuli: Abbreviations: LN—Lateral nucleus of the amygdala; BN—Basal nucleus; ABN—Accessory basal nucleus; CN—Central nucleus. The hippocampal system responsible for relaying contextual information into the amygdala from sensory regions consists of the perirhinal cortex, entorhinal cortex, and hippocampus.

areas involved in mood and emotional responses and exhibit the following set of characteristics:

- the compulsion to take drugs,
- loss of control in limiting intake,
- entry into a negative emotional state when access to drugs is prevented, and
- relapses into addiction after periods of abstinence.

The changes induced by drug use are long-lived and involve changes at both the cellular and molecular levels. The cells and circuits that underlie drug addiction are located in several regions of the brain. They include cells that express acteylcholine receptors and communicate using dopamine (dopaminergic) and serotonin (serotonergic), and also include cells that release glutamate (glutamergic) or GABA (GABAergic). The corresponding brain regions are associated with reward, mood, arousal, and cognition. Long-lasting changes in synaptic transmission similar to those associated with learning and memory, but leading to addiction, take place in these brain regions.

The sites of action of addictive drugs such as cocaine, amphetamines, and nicotine lie deep in the brain, in regions associated with mood, emotions, and learning. As shown in Figure 21.14, drugs act on neurons in the upper part of the brain stem (ventral tegmental area and substantia nigra), the limbic system (amygdala and hippocampus), and the mesolimbic system

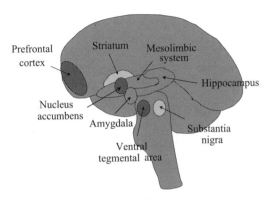

FIGURE 21.14. Side view of the brain showing the main sites of drug action: The regions of drug action—the ventral tegmental area, substantia nigra, nucleus accumbens, striatum, and prefrontal cortex—are shown in lighter and darker gray shades. Areas that form the nigrostriatal system are shown in light gray and those that the mesocorticolimbic system comprises are presented in dark gray.

(nucleus accumbens). The amygdala and hippocampus are ancient structures. The amygdala, which means "almond-shaped," connects to hippocampus and prefrontal areas. Activity in this area of the brain is associated with strong emotions such as fear, aggression, and the fight or flight response. The hippocampus, meaning "seahorse-shaped," is located just behind the amygdala. As discussed earlier in this chapter, this structure is associated with learning and memory, and especially with the transfer of information into memory.

21.14 Drug-Reward Circuits Mediate Addictive Responses

Under the influence of drugs, neurons involved in mood, arousal, and cognition organize into drug-reward circuits. The core circuitry involved in drug addiction contains two systems—the nigrostriatal system and the mesocorticolimbic system. The substantia nigra (SN) is located in the upper part of the brain stem. It contains two groups of cell. One group, the dorsally located cells (the SN pars compacta, or SNc), uses dopamine to communicate with the corpus striatum. The other group, the ventrally positioned neurons (the SN pars reticulata, or SNr), utilizes GABA to communicate with the thalamus. The nigrostriatal system consists of neurons in the substantia nigra pars compacta and the corpus striatum, or striatum. The mesocorticolimbic system consists of cells in the ventral tegmental area (VTA) located near the SN, the nucleus accumbens (NAc) situated just above the amygdala, and the prefrontal cortex that encompasses the nonmotor portions of the frontal lobe.

The most striking feature of cells in the two systems is the prominence of dopamine-releasing neurons. Neurons in the ventral tegmental area and in the substantia nigra contain large numbers of dopamine-releasing neurons. When stimulated by drugs these neurons elevate the extracellular dopamine levels in the nucleus accumbens and striatum. Dopamine promotes reinforcement, in which behavioral responses linked to rewards increase in frequency over time. In the drug-reward circuits, dopamine functions as the reward signal in response to the taking of drugs such as cocaine, and also as a stimulant of reward-seeking behavior.

21.15 Drug Addiction May Be an Aberrant Form of Synaptic Plasticity

Drug addiction appears to be an aberrant form of synaptic plasticity, in which circuits involved in LTP and LTD are altered. Glutamate and dopamine work together in the VTA to promote addiction (Figure 21.15). Addictive drugs such as nicotine act in at least three different ways in the VTA. They bind to nicotinic acetylcholine receptors (nAChRs) on dopamine-releasing cells triggering the release of dopamine. They also bind to nAChRs on presynaptic terminals of glutamate-releasing cells resulting in an increased release of glutamate. The glutamate, in turn, triggers an increased production of dopamine by dopamine-releasing cells receiving

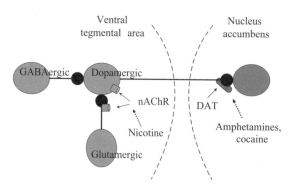

FIGURE 21.15. Schematic representation of a portion of the VTA-NAc drug-reward circuitry: Shown is a representative VTA dopamergic neuron whose axon projects into the nucleus accumbens and releases dopamine there. As explained in the text, the dopamine release is regulated and, under the influence of nicotine, potentiated by a GABAergic neuron and by a glutamergic neuron, here depicted as located in the VTA. Also included in the figure is the action of amphetamines and cocaine on the dopamine transporter (DAT).

the glutamate signals. The effect of the glutamate is to prolong the release of dopamine results in an LTP-like potentiation of synaptic transmission and increase in duration of the reward signal. (When nAChRs found on dopamergic neurons in the VTA bind nicotine they depolarize the membrane. The nAChRs located on presynaptic terminals enhance glutamate release, thus stimulating further dopamine release. When NMDA receptors are present, long-term potentiation of synaptic transmission is produced.) Thirdly, nicotine binds to GABA-releasing cells, which for a while throttle back the actions of dopamine-releasing cells, but desensitization turns this off over time, further promoting addictive effects of the drug.

A definition of synaptic plasticity useful in discussions of drug addiction is one describing it as the ability of circuits and systems to modify their responses to a stimulus brought on by prior stimulation, either of the same type or of an associated form. The connections between drug addiction and synaptic plasticity have been studied most intensively in the ventral tegmental area and the nucleus accumbens. Opiates and nicotine exert their influences through interactions with neurons of the VTA, while cocaine and amphetamines act on neurons of the NAc. As indicated in Figure 21.15, cocaine and amphetamines impede the ability of the dopamine transporter (DAT) to take up dopamine from the extracellular spaces. Homeostatic regulation is lost and the dopamine levels become excessive.

Neurons in the prefrontal cortex send and receive signals from VTA and NAc neurons, and these regions are the locus of feelings of pleasure associated with drug addiction. The networks in these areas serve as the core of the brain's reward circuitry. Again, a key element in this reward circuitry, that is, a feature that is common to all forms of drug abuse, is the elevation of dopamine levels leading to pleasurable sensations.

21.16 In Reward-Seeking Behavior, the Organism Predicts Future Events

To survive an organism must be able to decide between alternative responses whenever it is searching for food or water, or avoiding danger, or seeking a mate. In reward-driven behavior a positive reinforcement is supplied to behavioral acts in order to strengthen future response to the associated stimuli. Reward-driven behavior is a type of learning, quite like the Hebbian form, in which a stimulus is paired with a reward or a punishment. This kind of learning is called *predictive learning*, the classical example of which is Pavlov's experiments with dogs.

Pairing a reward/punishment with a stimulus does not always result in learning. There is a second requirement: Some element of uncertainty or novelty must be present. More formally, there must be some discrepancy, or error, between the reinforcement predicted by the stimulus and the actual reinforcement. If this reinforcement error is zero there is no learning since everything turns out as expected. The dopamine release can be regarded as a teaching signal that triggers an increased alertness whenever something interesting or unexpected has occurred. In doing so, it strengthens connections between neurons leading to LTP and drug addiction.

References and Further Reading

General References

Bear MF, Connors BW, and Paradiso MA [2001]. *Neuroscience: Exploring the Brain* (2nd ed.) Baltimore, MD: Lippincott, Williams and Wilkins.
Kandel ER [2001]. The molecular biology of memory storage: A dialogue between genes and synapses. *Science* 294: 1030–1038.
Kandel ER, Schwartz JH, and Jessel TM [2000]. *Principles of Neural Science* (4th edition). Norwalk CT: Appleton and Lange.
Lamprecht R, and LeDoux J [2004]. Structural plasticity and memory. *Nature Rev. Neurosci.*, 5: 45–54.

Synaptic Assembly

Biederer T, et al. [2002]. SynCAM, a synaptic adhesive molecule that drives synapse assembly. *Science*, 297: 1525–1531.
Dalva MB, et al. [2000]. EphB receptors interact with NMDA receptors and regulate excitatory synapse formation. *Cell*, 103: 945–956.
Dean C, et al. [2003]. Neurexin mediates the assembly of presynaptic terminals. *Nature Neurosci.*, 7: 708–716.
Penzes P, et al. [2003]. Rapid induction of dendritic spine morphogenesis by *trans*-synaptic EphrinB-EphB receptor activation of the Rho GEF Kalirin. *Neuron*, 37: 263–274.
Takasu MA, et al. [2002]. Modulation of NMDA receptor-dependent calcium influx and gene expression through EphB receptors. *Science*, 295: 491–495.

The Presynaptic Active Zone

Atwood HL, and Karunanithi S [2002]. Diversification of synaptic strength: Presynaptic elements. *Nature Rev. Neurosci.*, 3: 497–516.
Chen YA, and Scheller RH [2001]. SNARE-mediated membrane fusion. *Nature Rev. Mol. Cell Biol.*, 2: 98–106.
Dobrunz LE, and Stevens CF [1997]. Heterogeneity of release probability, facilitation and depletion at central synapses. *Neuron*, 18: 995–1008.
Jahn R, Lang T, and Südhof TC [2003]. Membrane fusion. *Cell*, 112: 519–533.
Rizo J, and Südhof TC [2002]. SNAREs and Munc18 in synaptic vesicle fusion. *Nature Rev. Neurosci.*, 3: 641–653.
Yoshihara M, and Littleton JT [2002]. Synaptotagmin I functions as a calcium sensor to synchronize neurotransmitter release. *Neuron*, 36: 897–908.

The Postsynaptic Density

Bredt DS, and Nicoll RA [2003]. AMPA receptor trafficking at excitatory synapses. *Neuron*, 40: 361–379.

Impey S, Obrietan K, and Storm DR [1999]. Making new connections: Role of ERK MAP kinase signaling in neuronal plasticity. *Neuron*, 23: 11–14.

Kennedy MB [2000]. Signal-processing machines at the postsynaptic density. *Science*, 290: 750–754.

Kim JH [2003]. Presynaptic activation of silent synapses and growth of new synapses contribute to intermediate and long-term facilitation in *Aplysia. Neuron*, 40: 151–165.

Kim JH, and Huganir RL [1999]. Organization and regulation of proteins at synapses. *Curr. Opin. Cell Biol.*, 11: 248–254.

Malinow R [2003]. AMPA receptor trafficking and long-term potentiation. *Phil. Trans. R. Soc. Lond. B*, 358: 7–7–714.

Sheng M, and Kim MJ [2002]. Postsynaptic signaling and plasticity mechanisms. *Science*, 298: 776–780.

Sweatt JD [2001]. The neuronal MAP kinase cascade: A biochemical signal integration system subserving synaptic plasticity and memory. *J. Neurochem.*, 76: 1–10.

Thomas GM, and Huganir RL [2004]. MAPK cascade signalling and synaptic plasticity. *Nature Rev. Neurosci.*, 5: 173–182.

Ziff EB [1997]. Enlightening the postsynaptic density. *Neuron*, 19: 1163–1174.

Long-Term Potentiation

Bliss TVP, and Collingridge GL [1993]. A synaptic model of memory: Long-term potentiation in the hippocampus. *Nature*, 361: 31–39.

Lisman J, Schulman H, and Cline H [2002]. The molecular basis of CaMKII function in synaptic and behavioral memory. *Nature Rev. Neurosci.*, 3: 175–190.

Malenka RC, and Nicoll RA [1999]. Long-term potentiation—A decade of progress? *Science*, 285: 1870–1874.

McGaugh JL [2000]. Memory—A century of consolidation. *Science*, 287: 248–251.

Bidirectional Synaptic Plasticity

Abraham WC, and Bear MF [1996]. Metaplasticity: The plasticity of synaptic plasticity. *Trends Neurosci.*, 19: 126–130.

Shouval HZ, Bear MF, and Cooper LN [2002]. A unified model of NMDA receptor-dependent bidirectional synaptic plasticity. *Proc. Natl. Acad. Sci. USA*, 99: 10831–10836.

Fear Conditioning

Blair HT, et al. [2001]. Synaptic plasticity in the lateral amygdala: A cellular hypothesis for fear conditioning. *Learn. Mem.*, 8: 229–242.

McKernan MG, and Shinnick-Gallagher P [1997]. Fear conditioning incuces a lasting potentiation of synaptic currents in vitro. *Nature*, 390: 607–611.

Rogan MT, and LeDoux JE [1996]. Emotion: Systems, cells and synaptic plasticity. *Cell*, 85: 469–475.

Rogan MT, Stäubli UV, and LeDoux JE [1997]. Fear conditioning induces associative long-term potentiation in amygdala. *Nature*, 390: 604–607.

Schafe GE, et al. [2001]. Memory consolidation of Pavlovian fear conditioning: A cellular and molecular perspective. *Trends Neurosci.*, 24: 540–546.

Shumyatsky GP, et al. [2002]. Identification of a signaling network in lateral nucleus of the amygdala important for inhibiting memory specifically related to learned fear. *Cell*, 111: 905–918.

Drug Addiction

Berke JD, and Hyman SE [2000]. Addiction, dopamine, and ther molecular mechanism of memory. *Neuron*, 25: 515–532.

Fiorillo CD, Tobler PN, and Schultz W [2003]. Discrete coding of reward probability and uncertainty by dopamine neurons. *Science*, 299: 1898–1902.

Hyman SE, and Malenka RC [2001]. Addiction and the brain: The neurobiology of compulsion and its persistence. *Nature Rev. Neurosci.*, 2: 6695–703.

Koob GF, Sanna PP, and Bloom FE [1998]. Neuroscience of addiction. *Neuron*, 21: 467–476.

Mansvelder HD, and McGehee [2000]. Long-term potentiation of excitatory inputs to brain reward areas by nicotine. *Neuron*, 27: 349–357.

Nestler EJ [2001]. Molecular basis of long-term plasticity underlying addiction. *Nature Rev. Neurosci.*, 2: 119–128.

Saal D, et al. [2003]. Drugs of abuse and stress trigger a common synaptic adaptation in dopamergic neurons. *Neuron*, 37: 577–582.

Schultz W, Dayan P, and Montague PR [1997]. A neural substrate of prediction and reward. *Science*, 275: 1593–1599.

Waelti P, Dickinson A, and Schultz W [2001]. Dopamine responses comply with basic assumptions of formal learning theory, *Nature*, 412: 43–48.

Problems

21.1 How does the neuron know which synapses to strengthen and which synapses to weaken when making long-term changes involving gene expression and protein synthesis? In a Hebbian type of rule for changing the strength of synapses, only those synapses exhibiting correlated firing are strengthened. Referring back to Figure 21.11(a), only the middle synapse onto cell B is strengthened. The middle synapse must be "tagged" in some fashion or other, so that the results of gene expression can be applied to that synapse alone. How might that be done?

21.2 In a Hebbian type of rule the postsynaptic cell fires an action potential. This is a cell-wide signal that a set of associations has been made. This may be viewed as a strong form of Hebb's rule. The weak form of the rule would only require postsynaptic membrane depolarization by some amount. In a weaker form of Hebbian rule, an action potential is not required, but rather the convergent signals produce a depolarization. How might action potential information get conveyed to the dendrites to elicit an immediate strengthening of the synapse?

21.3 Several different experimental protocols are utilized in the hippocampal slice preparations. All of these involve pairing of spikes in the pre- and postsynaptic cells. One of the most effective protocols is that of tetanic stimulations. In the last chapter, at least four different kinds of firing modes were discussed. Which of the different modes of firing would be most effective in eliciting LTP in vivo and in an adult where strong signaling is needed?

Suggested Reading

Martin KC, and Kosik KS [2002]. Synaptic tagging: Who's it? *Nature Rev. Neurosci.*, 3: 813–820.

Paulsen O, and Sejnowski TJ [2000]. Neural patterns of activity and long-term synaptic plasticity. *Curr. Opin. Neurobiol.*, 10: 172–179.

Glossary

Acetylation   The addition, catalyzed by acetyltransferases, of acetyl groups to amino acid side chain nitrogens on target proteins.

Action potential   Self-regenerating pulse of electrical activity that propagates down axons and is generated by the opening and closing of sodium and potassium channels.

Active zone   Region of the presynaptic axon terminal containing the machinery for neurotransmitter release.

Adaptive immune response   Immune response unique to vertebrates involving antigen recognition and the production of antibodies.

Adapter proteins   Nonenzymatic proteins that contain protein-protein interaction domains and serve as intermediaries that allow signaling proteins that would otherwise not be able to communicate to do so.

Agonist   A molecule that binds a receptor and induces the same response in the receptor as the one triggered by the natural ligand.

Allosteric modification   Shifts in equilibria between two preexisting populations of conformational states, brought on either by ligand binding or by covalent attachment of groups. In an allosteric modification, binding at one location in the molecule is able to alter how other portions of the molecule respond to their binding partners because of the conformational changes that accompany the shifts in equilibrium.

Anchor proteins   Nonenzymatic proteins that attach to the plasma membrane and to membranes of organelles, and provide platforms for signaling proteins to dock in close proximity to receptors and ion channels.

Angiogenesis   The late stages of vasculogenesis in which an initial set of tubules is refined through further differentiation, sprouting, and branching to form a mature vascular system containing arterial and venous structures.

Antagonist   A molecule that binds a receptor, but the receptor does not transmit a signal in response to the binding event. Drugs that bind in an antagonistic fashion are known as **blockers**.

Antibiotic   Biomolecules synthesized by fungi and bacteria that kill competing microbes.

Antibodies   Receptors synthesized by B cells that recognize and bind antigens.

Antigen (antibody generator)   Foreign substance, derived from a pathogen and expressed either on the outer surface of the pathogen or on the surface of an antigen-presenting cell, that triggers the production of antibodies.

Antigenic variation   Systematic alterations in the antigens expressed on the outer surface of a pathogen.

Apoptosis   Programmed cell death, in which there is an orderly disassembly of the cell that avoids harming neighboring cells. Also called **cell suicide**.

Apoptosome   The major control point for converting internal stress signals into apoptotic responses. It is located just outside the mitochondria and is activated by the release of cytochrome c.

Associative learning   Changes in the behavioral response to the weak stimulus that has been paired with a strong (positive or negative) stimulus.

Autocrine   A signaling mode in which hormones secreted from a cell act back on the cell releasing them.

Auxiliary spice sites   Sites where splicing regulators bind. Auxiliary sites located within exons are called **exonic splice enhancer** (ESE) and **exonic splice inhibitory** (ESI) sites, depending on which regulatory outcome is supported. Similarly, intronic sites are termed **intronic splice enhancer** (ISE) and **intronic splice inhibitory** (ISI) sites.

Bacteriophage   Virus that infects bacteria; also called a phage.

Biofilm   A bacterial colony formed on exposed surfaces and exhibiting cooperative behavior between members.

Branching morphogenesis   Growth, invasion, and proliferation of cells that form branched tubular structures that carry fluids in the vasculature, lungs, kidneys, and mammary glands.

Caspases   Proteolytic enzymes that catalyze the cleavage of specific molecules in response to apoptosis signals.

Catch bond   A bond that is strengthened by the external forces. The force-driven enhancements in the lifetime of these bonds allow leukoctytes to be captured by the walls of the blood vessels and begin rolling.

Caveolae (little caves)   Tiny flask-shaped invaginations in the outer leaflet of the plasma membrane that are detergent-insoluble, enriched in glycosphingolipids, cholesterol, and lipid-anchored proteins, and in **caveolins**, a coatlike material.

Cell adhesion   Cell-to-cell and cell-to-ECM attachment mediated by long modular and flexible glycoproteins expressed on opposing surfaces acting as receptors and counterreceptors or ligands.

Cell fate   The determination of which tissue or organ a particular cell becomes a member of during embryonic development.

Cell polarity   The asymmetric distributions of cellular components that arise during development. In the case of nerve cells it produces striking differences in morphology—axons at one end, dendrites at the other; in epithelial cells it gives rise to apical and basolateral plasma membrane domains.

Central pattern generators   Circuits built from small numbers of neurons that are used to drive the rhythmic firing of muscles responsible for activities such as walking and swimming, breathing, and chewing and digesting.

Chaperones   Proteins that help other proteins to fold into their native state, shuttle proteins to their correct locations in the cell, prevent unwanted aggregation, and assist in recovering and refolding proteins that have become misfolded due to cellular stresses.

Checkpoints   Signaling pathways that ensure that a cell cycle or assembly process does not begin before a prior necessary process is completed.

Chemotaxis   The process whereby a unicellular organism senses nutrients and noxious substances in its local environment, and, in response, moves towards the nutrients and away from the harmful chemicals.

Chromatin   In eukaryotes, material from which chromosomes are made, consisting of DNA wrapped around proteins called histories.

Chromophore   Groups of atoms or molecules that act as pigments, imparting color to the materials in which they reside by absorbing light at some wavelengths and scattering it at others.

Competence   The ability of a bacterium to take up exogenous DNA from its environment.

Conjugation   A form of horizontal gene transfer in which bacteria establish direct contact with one another; sex pili are formed, and genetic material in the form of plasmid are sent from donor to recipient.

Control point   Locations in the cell where environmental and regulatory signals converge, integrate, and convent to cellular responses.

Cytokines   Small signaling proteins synthesized and secreted by leukocytes, most commonly, macrophages and T cells. They convey a variety of instructions to leukocytes and to other cells such as neurons.

Death-inducing signaling complex (DISC)   The name given to the control point responsible for converting external death signals into apoptotic responses. It is organized by death receptors at and just below the plasma membrane.

Denatured state   The ensemble of states that a newly synthesized protein, or an unfolded protein, populates.

Dephosphorylation   The removal catalyzed by protein phosphatases of phosphoryl groups previously added to amino acid side chain hydroxyls on protein substrates by protein kinases.

Desensitization   The process whereby a G protein-coupled receptor, or any other receptor, loses its responsiveness to binding by its ligand.

Diffraction   Scattering of light by atoms, molecules, and larger objects resulting in departures from rectilinear motion other than reflection or refraction.

Diffusion   Thermally driven movement of particles in a fluid from one locale to another produced by random collisions of the particles with the molecules of the fluid.

Distal sites   DNA regulatory regions where long-range interactions between regulatory proteins and the basal transcription machinery take place. They may be located upstream of the core promoter, downstream of the core promoter, in between coding regions, and inside introns. Positively acting transcription factors that bind at these sites are called **enhancers**, while negatively acting transcription factors are referred to as **silencers**.

Domain fold   Stable arrangements of multiple secondary structure elements and of two or more structural motifs into independent folding units.

Efficiency of synaptic transmission   Magnitude of the response generated in a postsynaptic neuron when an action potential is generated in the presynaptic neuron.

Electrostatic complementarity   The matching of hydrophobic patches, the complementary pairing of hydrogen bond donors and acceptors, and the matching of positive and negative charges of basic and acidic polar residues from one surface to the other of the interface.

Endocrine   A signaling mode in which hormones are secreted into the bloodstream and other bodily fluids by specialized cells and travel large distances to reach multiple target cells.

Endocytosis   The process whereby plasma membrane proteins and materials captured at the cell surface are packaged into vesicles and shipped to digestive compartments for processing and recycling.

Energy landscape   A graphical depiction of how the number of states available to a protein at each value of the potential energy varies as a function of a few significant degrees of freedom.

Envelope   Lipid/carbohydrate membrane derived from the host cell that surrounds a viral capsid.

Euchromatin   Transcriptionally active chromatin with an open shape that permits transcription factors and the basal transcription machinery to access promoters.

Exocytosis   The packaging and shipping of newly synthesized proteins destined for export and use in the plasma membrane in vacuoles that move over the rail system and fuse with membranes at their destination.

Exons   Short coding sequences separated from one another by introns in pre-messenger RNAs.

Focal adhesions   Points of contact and adhesion between the cell and the supporting extracellular matrix. They serve as control points where growth and adhesion signals are integrated together to govern the overall growth and movement of the cell.

Folding funnel   The shape of the potential energy landscape that arises because there are many high energy states and few low energy ones.

Gate   The part of the ion channel that opens and closes the pore through which ions pass.

GDP dissociation inhibitors (GDIs)   Enzymes bind and maintain pools of inactive GTPases by inhibiting the dissociation of GDP from the GDPase.

Glycosylation   A common posttranslational modification to proteins destined for insertion in the plasma membrane in which covalently linked oligosaccharides that extend out from their extracellular side are added. The modified proteins are referred to as **glycoproteins**.

Glycosyl phosphatidylinositol (GPI) anchors   Posttranslational modifications to proteins that allow them to attach to the outer, or exoplasmic, leaflet of the plasma membrane. GPI anchors are made from a complex sugar plus a phosphatidylinositol grouping.

Growth cones   Sensory structures located at the tip of advancing axonal and dendritic processes sent out from neurites. They explore, interact with, interpret, and respond to signaling molecules in their local microenvironment.

Growth factors   Molecules that stimulate growth and development.

GTPase-activating proteins (GAPs)   Enzymes that catalyze the hydrolysis of the GTPase-bound GTP to GDP.

GTPases   Small enzymes that hydrolyze GTP and function as molecular switches and timers. They are activated when bound to GTP and deactivated when bound to GDP.

Guanine nucleotide exchange factors (GEFs)   Enzymes that catalyze the dissociation of GDP from the GTPase.

Habituation   Weakening of a behavioral response to a harmless stimulus through repeated exposures to that stimulus.

Heterochromatin   Transcriptionally inactive chromatin with a tightly compacted shape that prevents transcription factors and the basal transcription machinery from accessing promoters.

Holoenzyme   Inactive enzyme (apoenzyme) plus additional noncovalently bonded biomolecules, either loosely bound cofactors or tightly bound prosthetic groups, required for full catalytic activity.

Homeostasis   Physiological process by which a cell, tissue, or organism balances and stabilizes internal conditions such as temperature, pH, excitability, and cell type in the presence of external perturbations.

Horizontal gene transfer   Lateral transfer of genetic information between distantly related species through conjugation, transduction, and transformation.

Hydrophilic   A water-loving amino acid; a water molecule would rather bind to this amino acid than to another water molecule.

Hydrophobic   A water-hating amino acid; a water molecule would rather bind to another water molecule than to this amino acid.

Inflammatory response   Set of physiological responses including fever and pain, redness and swelling. A local environment is formed that promotes migration of leukocytes to the infection site, the destruction of the invasive agents, and the repair of damaged tissues.

Innate immune response   Immune response involving recognition by leukocytes of molecules situated on the outer surface of pathogens that are characteristic of the pathogen.

Insulators   DNA sequences that mark boundaries between independent sections of DNA.

Interface   Region of surface contact between two macromolecules through which binding and communication take place.

Internalization   The removal of a receptor from the plasma membrane through endocytic mechanisms.

Introns   Long noncoding, or intervening, sequences situated in between exons in pre-messenger RNAs.

Ion channels   Membrane-spanning proteins forming narrow pores that enable specific inorganic ions, typically Na^+, K^+, Ca^{2+} or Cl^-, to passively diffuse in a directional manner through cell membranes.

Ionotropic   Receptor ion channel that opens and closes in response to neurotransmitter binding.

Juxtacrine   A signaling model in which messages are conveyed by direct contact between a receptor on one cell and a cell surface-bound ligand or counterreceptor on an adjacent cell.

Kinetic proofreading   A cellular mechanism for improving the fidelity of a process by tying it to a series of intermediate time- and energy-consuming steps. In receptor-ligand binding, differences in affinity are converted to differences in signaling because of the intermediate time- and energy-consuming steps.

Kinetic trap   A set of states forming a local minimum in the energy landscape and enclosed by energy barriers large compared to the thermal energy.

Labile   Readily undergoes change or breakdown.

Lateral inhibition   Process whereby a cell adopting a particular cell fate inhibits its neighbors from adopting the same fate.

Learning   The adaptive process whereby changes in behavior are induced in response to experience.

Leukocytes   White blood cells; highly motile and short-lived cells that move through the cardiovascular and lymphatic systems into damaged tissues where they kill bacterial, protozoan, fungal and multicellular pathogens, destroy cells infected with viruses and bacteria, and eliminate tumor cells.

Lipid rafts   Plasma membrane microdomains that are detergent insoluble and enriched in cholesterol and sphingolipids. Unlike caveolae, they do not contain caveolins and are not cavelike in shape but instead are flat.

Long-term facilitation   Strengthening of the behavioral response to a mild touch to the tail that has been paired with an electric shock to the siphon, an experimental model of associative learning in the *Aplysia* siphon withdrawal circuit.

Long-term potentiation   Strengthening of the postsynaptic response to a presynaptic action potential brought on by pairing a series of presynaptic action potentials with a postsynaptic depolarization or action potential, an experimental model of associative learning in the rodent hippocampal CA1–CA3 regions.

Lysogenic Life cycle in which phage DNA is integrated into the host cell DNA and the bacteriophage becomes a prophage, replicating as part of the bacterial host's chromosome.

Lytic Life cycle in which phage DNA is replicated and multiple virus particles are formed and escape from the cell by rupturing the cell's plasma membrane.

Matrix protein Associates with the inner layer of the viral envelope and is situated in between the inner layer and the capsid.

Memory The record underlying the changes in behavior brought on by learning.

Metabotropic receptors G protein-coupled receptors (GPCRs) that activate their cognate heterotrimeric G proteins in response to neuromodulator binding.

Metastasis The process whereby cancer cells break away from their point of origin, the primary tumor, enter the circulatory system, and invade other organs, where they form secondary tumors.

Methylation The addition catalyzed by methyltransferases of methyl groups to amino acid side chain nitrogens on target proteins.

Morphogens Signaling proteins expressed either on cell surfaces or secreted into the extracellular spaces in the form of concentration gradients that are read by other cells to determine their developmental fate.

Motor neurons Supply input to muscles that drives their contractions, and receive input from upstream sensory and control neurons.

Mutual inhibition Pairs of neurons that are reciprocally connected and sequentially inhibit each other's firing activities.

Native state A stable state of the folded protein. It is a state of minimum Gibbs free energy at physiological temperatures and conditions.

Neuromodulators Signaling molecules secreted in a broader manner than neurotransmitters; they modify the excitability of large numbers of target cells by regulating the activities of their ion channels.

Neurotransmitters Signaling molecules released from the presynaptic terminal of a neuron and diffuse in a directional manner across the synaptic cleft to the postsynaptic terminal of a neighboring neuron, where they bind receptor ion channels.

Nucleocapsid The viral capsid plus its nucleic acid core.

Nucleosome   The fundamental repeating unit of chromatin. Nucleosomes, 146 base pairs of DNA wrapped about a histone octamer, are strung together like beads on the string by means of linker segments.

Oncoproteins   Proteins that operate in the signal transduction, integration, and regulatory pathways involved in cellular growth, multiplication, differentiation, and death, that when mutated stimulate unregulated cell growth and proliferation thus promoting the development of cancer.

Operon   DNA sequences that encode one or more proteins and the upstream sequences for attachment of the RNA holoenzyme and regulatory proteins in bacteria.

Organizing centers   Localized groupings of cells that secrete morphogens that impart patterns of cell fates to fields of progenitor cells.

Pacemaker neuron   Neurons that generate the rhythmic firing patterns without requiring any rhythmic input.

Paracrine   A signaling mode in which molecules are secreted into the extracellular spaces by an originating cell and travel no more than a few cell diameters to reach their target cell.

Pathogen (bacterial)   Organisms possessing genes that encode virulence factors and are situated on plasmids and other mobile genetic elements that can be readily transferred and exchanged between species.

Permeability   The propensity of a membrane to allow passage of certain ions and molecules.

Permeability transition pore complex (PTPC)   Also known as the **permeability transition pore (PTP)**, this control point is formed at points of contact between the inner and outer mitochondrial membranes. The PTPC is a conduit for the passage of agents such as cytochrome c and Smac/DIABLO that trigger apoptosome assembly and activation of caspase 9, and is the major site for regulation by Bcl-2 proteins.

Phosphorelay   A signal transduction system consisting of a hybrid sensor unit, a histidine phosphotransfer protein, and a response regulator. Compared to the two-component system, the hybrid sensor unit contains an extra module, an aspartate-bearing receiver, and a histidine phosphotransfer protein is situated in between the sensor unit and the response regulator.

Phosphorylation   The reversible addition, catalyzed by protein kinases, of phosphoryl groups to amino acid side chain hydroxyls on target proteins.

Plasmid   Extrachromosomal DNA found in bacteria and encoding sets of functionally related genes.

Plateau potentials   Stable membrane potentials occurring at depolarizations greater than that of the resting membrane potential. When a membrane is at a plateau potential it is far more excitable and can repeatedly fire action potentials even in the absence of sustained excitatory input from synapses.

Platelets   Cytoplasmic fragments of bone marrow cells called **megakaryocytes** that form clots that block blood flow at sites of injury.

Pleiotropic   Multifunctional.

Polypeptide hormones   Small compact polypeptide growth factors that bind to receptor tyrosine kinases.

Pores   Membrane-spanning proteins found in the outer membrane of Gram-negative bacteria, mitochondria, and chloroplasts forming channels that enable hydrophilic molecules smaller than about 600 Da to pass through.

Post-inhibitory rebound   A strong hyperpolarization that is quickly terminated and followed by a rapid depolarization leading to the firing of an action potential.

Postsynaptic density   Region of the postsynaptic dendrite terminal containing the machinery for neurotransmitter signal transduction.

Pre-initiation complex (PIC)   Also known as the basal transcription machinery the PIC consists of RNA polymerase II and a set of general transcription factors.

Pre-messenger RNA (pre-mRNA)   Eukaryotic RNA molecule produced by transcription from DNA. It contains exons, introns, and regulatory sequences that provide binding sites for the splicing machinery and regulatory proteins.

Primary splice sites   Consist of (i) the **5′ splice site** characterized by the presence of a binding sequence containing a guanine-uracil (GU) pair within a longer GURAGU-like sequence, where R is a purine; (ii) the **branch site** characterized by the nucleotide sequence YNYURAY, where Y is a pyrimidine; (iii) the **pyrimidine tract**, a string of pyrimidine nucleotides; and (iv) the **3′ splice site** characterized by either a CAG sequence or a UAG sequence.

Primary structure   The protein's covalent structure, the linear sequence of amino acids linked to one another by peptide bond plus all disulfide bonds formed during folding.

Promoter   Transcriptional regulatory region of DNA containing binding and start sites for RNA polymerase II and binding sites for the transcription control elements.

Protease   See proteolysis.

Protein backbone   The main chain, the set of repeating $NC_\alpha C$ units covalently linked to one another by peptide bonds.

Protein folding   The process whereby newly synthesized linear polypeptide chains spontaneously fold into functional three-dimensional forms.

Protein kinases   Enzymes that catalyze the transfer of phosphoryl groups to amino acid side chain hydroxyls on protein substrates using ATP as the donor.

Protein phosphatases   Enzymes that catalyze the removal of phosphoryl groups previously added to selected amino acid side chain hydroxyls on protein substrates by protein kinases.

Proteolyis   The process of chopping up proteins by proteolytic enzymes, or proteases, which cleave peptide bonds at specific residues.

Proteosomes   Multisubunit complexes situated in the cytosol that degrade ubiquitin-tagged proteins.

Proximal sites   DNA regulatory regions situated upstream of the core promoter. Proteins that stimulate transcription when they bind at these sites are called **activators** while those that impede transcription are called **repressors**. DNA sequences that provide sites for attachment of coactivators and corepressors and mediate long-range enhancer-promoter interactions are called **tethering elements**.

Pumps   Membrane-spanning proteins that actively transport ions and molecules across cellular and intracellular membranes.

Quaternary structure   In multisubunit (chain) proteins, the ensemble of subunits and how they are arranged.

Quorum sensing   Cell-to-cell signaling used by bacteria to determine the density of fellow bacteria in their local environment.

Reactive oxygen species (ROS)   Molecules possessing unpaired electrons (free radicals) involving oxygen. These molecules have a tendency to take electrons from other molecules, in many cases breaking bonds to acquire them.

Receptors   Transmembrane proteins that function as sensors of environmental stresses and as receivers of chemical messages. They transmit and, in the process, convert the signals from an external outside-the-cell form to an internal inside-the-cell one that can be understood and further processed.

Rectifying   Ion channels that allow the passage of ions in one direction only.

Regulon   Sets of operons located at well-separated loci along the chromosome-encoding proteins involved in a common physiological response.

Response regulator   In a two-component system, this unit functions as the receiver of the transferred phosphoryl group and as an output unit for the signal. Most output response regulators function as transcription factors.

Responsive elements   Transcriptional control points, consisting of short DNA sequences located in promoters, where transcription factors come together and bind in a sequence-specific manner to regulate transcription.

Reversal potential   The value of the membrane potential for a particular ion species that exactly cancels out the flow of those ions through the membrane arising from concentration differences in that ion.

Rhythmic bursting   Multiple sequences of action potentials in which each sequence, or burst, consists of a train of closely spaced action potentials separated by large interbust intervals.

Robustness   A property of a system with respect to one or more of its parameters in which feedback damps out the effect of variations in the value of the parameter(s) on the performance of the system.

Scaffold proteins   Nonenzymatic proteins that enable signaling proteins that must work together to attach in close proximity to one another.

Secondary structure   The ensemble of short segments of the polypeptide chain that fold into a geometrically regular, repeating structure, such as alpha helices and beta sheets, that are stabilized by networks of hydrogen bonds.

Second messengers   Signaling intermediaries that tie together events taking place at and just below the plasma membrane subsequent to ligand binding. Acting as coactivators and allosteric regulators they help to recruit and organize the proteins that function as receptors and intracellular signal transducers.

Selectivity   The ability of an ion channel to pass some ions through while preventing passage of other ions.

Selectivity filter   The part of the ion channel that selects which ions are able to pass through the pore.

Senescence   A nondividing stage of cellular life entered into when a cell's telomeres become critically shortened.

Sensitization   Strengthening of a behavioral response to a noxious stimulus through repeated exposures to that stimulus.

Sensor unit   In a two-component system, this unit functions as a receptor and plasma membrane signal transducer. It possesses a histidine kinase activity that promotes autophosphorylation on a histidine residue, followed immediately by a phosphotransfer operation.

Shape complementarity   The propensity of the surfaces of two molecules to geometrically fit together so that multiple contacts can be established at their interface.

Signal transduction   The process of relaying messages and, in the process, converting them from one form to another that can be understood by the downstream signaling targets.

Slip bond   A bond that is weakened by external applied forces. The force-driven reductions in lifetime of these bonds allow rolling leukocytes to detach from surface tethers at the right time while maintaining good adhesion to that place on the surface at earlier times.

Spike frequency adaptation   Hyperpolarization of the postsynaptic membrane brought on by calcium entry into the cell decreases its excitability. The sequence of pulses exhibits an increasing time lag between successive pulses and eventually reaches a steady-state firing frequency.

Spindle oscillations   7- to 14-Hz oscillations that wax and wane with a 1- to 3-second period and are associated with early stages of quiescent sleep.

Spliceosome   Machinery consisting of a family of small nuclear RNA molecules responsible for removing introns and selected exons from pre-mRNAs.

Stable (equilibrium) state   Long-lived states of a system. In these states, small perturbations and thermal fluctuations are rapidly damped out so that the behavior of the system is not appreciably altered.

Steroid hormones   Small lipophilic growth factors that are synthesized from cholesterol and bind to nuclear receptors.

Structural motif   Stable arrangement of secondary structure elements into small compact structures.

Synaptic   A signaling mode in which neurotransmitters and neuromodulators diffuse across the synaptic cleft between the membranes of a pair of pre- and postsynaptic cells.

Synaptic plasticity   The use-dependent, or adaptive, changes in the efficiency of synaptic transmission between pre- and postsynaptic cells.

Telomere   A capping structure consisting of a series of TTAGGG repeats and associated proteins that screen the ends of DNA strands from double-strand DNA repair machinery.

Tertiary structure   The ensemble of structural motifs and domains in the protein and how they are arranged.

Tonic firing   Trains of unitary action potentials generated either at regular time intervals or irregularly spaced in time.

Transcription factors   Proteins that bind to DNA in a sequence-specific manner and regulate the transcription of protein-coding and associated sequences into RNA molecules.

Transduction   A form of horizontal gene transfer in which a bacteriophage picks up genetic material from one bacterium and delivers it to another.

Transformation   A form of horizontal gene transfer in which bacteria become competent to capture exogenous DNA from their local environment.

Trophic factors   Molecules that stimulate growth and development.

Tumor suppressors   Similar to oncoproteins. They, too, are proteins that operate in the signal transduction, integration, and regulatory pathways involved in cellular growth, multiplication, differentiation, and death, except, unlike oncoproteins, they normally act as brakes on growth. Whey they suffer critical mutations these brakes on growth are removed.

Two-component system   A signal transduction system consisting of a **sensor unit** and a **response regulator**. The sensor unit possesses a histidine kinase activity and catalyzes the transfer of the phosphate group to the response regulator, which contains an aspartate residue that receives the transferred phosphate group.

Ubiquitination   The process of tagging and preparing proteins for proteolytic destruction by the proteosome. It involves sequential operations by E1 ubiquitin-activating enzyme, E2 ubiquitin-conjugating enzyme, and E3 ubiquitin-ligase enzyme.

Vasculogenesis   Creation of blood vessels to supply oxygen and nutrients to newly formed tissues and remove waste products from them.

Virulence factors   Bacterial products such as toxins and surface adhesion molecules that enhance bacterial survival in hostile host environments and cause disease.

Zymogen   Enzyme synthesized as an inactive precursor that is made into an active form by proteolytic cleavage and removal of a prodomain.

Index

Volumes Published in This Series: